PROCEEDINGS OF SYMPOSIA
IN PURE MATHEMATICS
Volume XXXII, Part 2

ALGEBRAIC AND GEOMETRIC TOPOLOGY

AMERICAN MATHEMATICAL SOCIETY
PROVIDENCE, RHODE ISLAND
1978

PROCEEDINGS OF THE SYMPOSIUM IN PURE MATHEMATICS
OF THE AMERICAN MATHEMATICAL SOCIETY

HELD AT STANFORD UNIVERSITY

STANFORD, CALIFORNIA
AUGUST 2–21, 1976

EDITED BY

R. JAMES MILGRAM

Prepared by the American Mathematical Society
with partial support from National Science Foundation grant MCS 76-01696

Library of Congress Cataloging in Publication Data

Symposium in Pure Mathematics, Stanford University, 1976.
Algebraic and geometric topology.

(Proceedings of symposia in pure mathematics; v. 32)

"Proceedings of the Symposium in Pure Mathematics of the American Mathematical Society, held at Stanford University, Stanford, California, August 2–21, 1976."

Bibliography: v. 1, p. v. 2, p.

1. Algebraic topology–Congresses. 2. Manifolds (Mathematics)–Congresses. 3. Global analysis (Mathematics)–Congresses. I. Milgram, R. James. II. American Mathematical Society. III. Title. IV. Series.

QA612.S93 1976 514′.2 78–14304

ISBN 0-8218-1432-X (v. 1)

ISBN 0-8218-1433-8 (v. 2)

AMS (MOS) subject classifications (1970). Primary 53C15, 53C30, 53C35, 53C40, 55-XX, 57Axx, 57Bxx, 57Cxx, 57Dxx, 58Bxx, 58C25, 58Dxx, 58Gxx.

Math Sci
Sep

PREFACE

The American Mathematical Society held its 24th Summer Research Institute at Stanford University from August 2–21, 1976. The topic of the meeting was Algebraic and Geometric Topology. Particular emphasis was placed on Algebraic *K*- and *L*-Theory, Surgery and Surgery Classifying Spaces, Group Actions on Manifolds, and 3 and 4 Manifold Theory.

The organizing committee consisted of Raoul Bott, William Browder (Co-chairman), Pierre Conner, Robion Kirby, Richard Lashof, R. James Milgram (Chairman), Daniel Quillen, and P. Emery Thomas (Co-chairman).

The main lecturers were Raoul Bott, Richard Lashof and Mel Rothenberg, Sylvain Cappell and Julius Shaneson, C. McA. Gordon, Robert Edwards, Allen Hatcher and Jack Wagoner, Wu. C. Hsiang, Max Karoubi, R. James Milgram, Daniel Quillen, Laurence Siebenmann, and C. T. C. Wall.

Special hour lectures were given by Gregory Brumfiel, William Jaco, Robion Kirby, Ronnie Lee, James Lin, Ib Madsen, John Morgan, Robert Oliver, Ted Petrie, P. Emery Thomas, F. Waldhausen, and James West.

There was also a series of problem sessions run by Raoul Bott (foliations), W. C. Hsiang (group actions and surgery), R. Kirby (3 and 4 manifolds), and C. T. C. Wall (algebraic *K*- and *L*-theory).

Among the more active seminars were 3 and 4 manifolds (Kirby), Kervaire invariant (Browder—E. Brown), Homotopy theory (E. Thomas—M. Mahowald), Algebraic *K*- and *L*- theory (J. Wagoner), Group actions (W. C. Hsiang), Foliations (Bott), Geometry of manifolds (L. Siebenmann), and Surgery and surgery classifying spaces (R. Lashof).

These proceedings include write ups of most of the main lectures, as well as selected seminar talks, and most of the problem sessions.

The institute was sponsored by the National Science Foundation under contract number MCS 76–01696.

The organizing committee wishes to thank Dorothy Smith of the American Mathematical Society for her help in running and organizing the conference.

R. James Milgram
Stanford University

TABLE OF CONTENTS

Structure of Topological Manifolds

Low Dimensional Manifolds

Geometry of Differential Manifolds and Algebraic Varieties

H-Spaces, Loop Spaces, and CW Complexes

Problems

STRUCTURE OF TOPOLOGICAL MANIFOLDS

Proceedings of Symposia in Pure Mathematics
Volume 32, 1978

TRIANGULATION OF MANIFOLDS

VIA A STUDY OF THE RELATION BETWEEN POLYHEDRAL HOMOLOGY MANIFOLDS AND TOPOLOGICAL MANIFOLDS

TAKAO MATUMOTO

Is any (metrizable) topological manifold homeomorphic to some locally finite simplicial complex? This triangulation problem is still unsolved, although Kirby and Siebenmann established a fine obstruction theory for the existence of a *combinatorial* triangulation of manifolds of dimension $\geqq 5$. Recall that a locally finite simplicial complex M is called a polyhedral (integral) homology n-manifold if the local homology group $H_*(M, M - x; Z)$ is isomorphic to $H_*(R^n, R^n - 0; Z)$ for any point x of M.

Our study starts from the converse question: Does every polyhedral homology manifold admit an underlying structure of topological manifold? Of course, there are many polyhedral homology manifolds which are not topological manifolds; for example, the (single) suspension of a non-simply-connected combinatorial homology sphere. The answer is, however, "yes" in the following sense. From every polyhedral homology n-manifold M, we can construct, in a standard way, a topological $(n + 1)$-manifold N unique up to a notion of blocked homeomorphism, and a simple homotopy equivalence $g\colon N \to M \times S^1$. This construction is accomplished by translating the s-decomposition of $M \times S^1$ given by the products of dual cells of M with S^1 into the topological manifold category. The argument is analogous to one used by Sullivan, Cohen and Sato for PL resolution of polyhedral homology manifolds, but it needs no obstructions in this context. Moreover, if $\dim M \neq 4$ and $\dim \partial M \neq 4$, we obtain a topological n-manifold M_{TOP} unique up to homeomorphism by a standard splitting of the projection $N \simeq_s M \times S^1 \to S^1$. We can regard M_{TOP} with a preferred simple homotopy equivalence

AMS (MOS) subject classifications (1970). Primary 57A99, 57B99, 57C15.
Key words and phrases. Triangulating, homology manifolds, topological manifolds.

$f: M_{\text{TOP}} \to M$ as an underlying topological manifold for M. As a consequence, one can introduce a standard map $\mu: BHML_r \to \boldsymbol{BTOP}_r$, $r \geqq 3$, from the classifying space of normal homology cobordism bundles into that of topological (regular) neighborhoods of codimension $r \geqq 3$. When stabilized, we have also $\mu: BHML \to \boldsymbol{BTOP} \simeq BTOP$, where $BTOP$ is the classifying space of stable topological microbundles.

We will use the surgery theory of Wall's type on compact polyhedral homology manifolds. Since this theory as given previously by the others, for example, by Matsui or Maunder, is not fully satisfactory for our purpose, we present a simple proof of its existence as a good application of M_{TOP}.

A surgery datum x consists of a Poincaré complex X of formal dimension n, a homology cobordism bundle ξ over X, a compact polyhedral homology n-manifold M and a degree one map $\phi: M \to X$ with a stable equivalence $F: \tau(M) \cong \phi^*\xi$ of the homology cobordism bundles. The obstruction $\theta(x)$ in the Wall group $L_n(\pi_1(X))$ is defined to be the obstruction $\theta(x_{\text{TOP}})$ of the underlying topological datum x_{TOP}.

THEOREM. *Let $n \geqq 5$. The datum x is normally cobordant to a simple homotopy equivalence if and only if $\theta(x) = 0$.*

With the help of the stability theorem proved by Matumoto-Matsumoto for $\mu_*: HML/HML_r \xrightarrow{\simeq} \boldsymbol{TOP}/\boldsymbol{TOP}_r$ $(r \geqq 3)$, the following embedding lemma reduces all the steps of the homology manifold surgery to that of topological manifold surgery which is established by Wall and Kirby-Siebenmann.

LEMMA. *Let $\gamma \in \pi_{r+1}(\phi)$ and $r \leqq n-2$. Assume that an embedding $j: S^r \times D^{n-r} \to M_{\text{TOP}}$ is in the regular homotopy class for doing the surgery on γ_{TOP}. Then, there exists a homology manifold s-cobordism $(W; M, \bar{M})$ with $\bar{M} \supset S^r \times D^{n-r}$ so that the natural inclusion $S^r \times D^{n-r} \to \bar{M}_{\text{TOP}} = M_{\text{TOP}}$ is regular homotopic to j.*

In fact, if $r \leqq 2$, then γ is representable by a PL embedding in the PL submanifold which is the neighborhood of the dual 3-skeleton. If $r \geqq 3$, we consider the homotopy inverse of f, $\psi: M \to M_{\text{TOP}}$. By the Williamson's simplicial transversality theorem, ψ may be supposed to be transverse regular to the axis of $S^r \times D^{n-r}$ and we put $K = \psi^{-1}(S^r \times 0)$. Then, the tangent homology cobordism bundle of K is stably trivial so that K is s-cobordant to a PL manifold. By modifying M by an s-cobordism we may assume that K is a PL submanifold. Moreover, we can observe that $\psi|K: K \to S^r \times 0$ completes to a PL surgery datum whose obstruction vanishes and $\psi|K$ is normally cobordant to the identity even when $r \leqq 4$. If the cobordism contains only handles of index $\leqq 2$, each of these handles can be realized in the zero-codimensional PL submanifold in M as before. If not, $r \geqq 3$ and we identify M_{TOP} with $(K \times D^{n-r}) \cup (M - K \times \text{int } D^{n-r})_{\text{TOP}}$ and embed each normally framed handle $D^s \times D^{r-s}$ in M_{TOP}; we get the PL surgery datum for $(K_1, \partial D^s) \to (D^s, \partial D^s)$ relative to the boundary with $s + 1 \leqq r$. The situation is the same as before when we wanted to do ambient surgery on K in M and hence by an induction on the dimension of the submanifold K in the ambient surgery data we can achieve the ambient surgery on K in M. This completes the proof of the lemma and consequently the surgery theorem.

In this setting, we can classify the polyhedral homology manifold structures on a given topological manifold: Let $\mathscr{H}^3$ be the smooth H_*-cobordism classes of oriented smooth homology 3-spheres and let $\alpha: \mathscr{H}^3 \to Z_2$ be the Milnor-Kervaire-Rohlin homomorphism.

THEOREM. *A topological manifold V of dimension* $\geqq 5$ *is homeomorphic to* M_{TOP} *for some polyhedral homology manifold M if and only if* $\beta k(V) \in H^5(V; \ker \alpha)$ *vanishes, where* $k(V)$ *is the Kirby-Siebenmann obstruction class to admit a combinatorial triangulation and* β *is the Bockstein operation associated to the exact sequence* $0 \to \ker \alpha \to \mathscr{H}^3 \to^{\alpha} Z_2 \to 0$.

For the proof, let X be the Poincaré complex with a simple homotopy equivalence $\nu: V \to X$. We consider the following homotopy-commutative diagram each sequence of which is a homotopy fibration.

$$\begin{array}{ccccc} BPL & \longrightarrow & BHML & \longrightarrow & B(HML/PL) \\ \| & & \downarrow \mu & & \downarrow \alpha_* \\ BPL & \longrightarrow & BTOP & \xrightarrow{k} & B(TOP/PL) \\ \downarrow & & \downarrow \nu & & \downarrow \beta_* \\ * & \longrightarrow & B(TOP/HML) & = & B(TOP/HML) \end{array}$$

Since HML/PL and TOP/PL are the Eilenberg-Mac Lane spaces $K(\mathscr{H}^3, 3)$ and $K(Z_2, 3)$ respectively and the canonical map: $HML/PL \to TOP/PL$ is identified with the map induced by α, β_* in the diagram corresponds to the Bockstein operation associated to α. So, we understand that the obstruction for the stable bundle reduction through μ is $\nu(\tau(V)) = \beta k(\tau(V))$.

Using the transversality theorem due to Williamson, we can prove that the normal cobordism classes of homology manifold surgery data correspond to the concordance classes $\mathscr{T}_{HML}(X)$ of the homology tangential structures on X. So, we have the following commutative diagram:

$$\begin{array}{ccccccc} L_{n+1}(\pi_1(X)) & \longrightarrow & \mathscr{S}_{HML}(X) & \xrightarrow{\eta} & \mathscr{T}_{HML}(X) & \xrightarrow{\theta} & L_n(\pi_1(X)) \\ \| & & \downarrow \mu_* & & \downarrow \mu_* & & \| \\ L_{n+1}(\pi_1(X)) & \longrightarrow & \mathscr{S}_{TOP}(X) & \xrightarrow{\eta} & \mathscr{T}_{TOP}(X) & \xrightarrow{\theta} & L_n(\pi_1(X)) \end{array}$$

The horizontal Sullivan-Wall sequences are exact in the sense that $L_{n+1}(\pi_1(X))$ operates on $\mathscr{S}_{CAT}(X)$ and $\ker \theta = \text{image } \eta \cong \mathscr{S}_{CAT}(X)/L_{n+1}(\pi_1(X))$ where $CAT = HML$ or TOP. Now, we do diagram chasing, assuming $\beta k(V) = 0$. Let $a \in \mathscr{T}_{TOP}(X)$ be the class of $\nu: V \to X$. Then, by the observation above, $\eta(a)$ has a lifting b such that $\mu_* b = \eta(a)$. There exist $c \in \mathscr{S}_{HML}(X)$ and $\lambda \in L_{n+1}(\pi_1(X))$ so that $\eta(c) = b$ and $\lambda \cdot \mu_* c = a$. Hence, $\mu_*(\lambda \cdot c) = a$. Thus, corresponding to $\lambda \cdot c$ there exists a polyhedral homology manifold M and a simple homotopy equivalence $m: M \to X$ such that the induced $m_{\mathrm{TOP}}: M_{\mathrm{TOP}} \to X$ is equivalent to $\nu: V \to X$; in particular, V is homeomorphic to M_{TOP}. End of proof!

Now we remark that the standard construction $M \mapsto M_{\mathrm{TOP}}$ gives $M = M_{\mathrm{TOP}}$ if the polyhedral homology manifold M is itself a topological manifold. And, if all the double suspensions of the polyhedral homology manifolds, that are homology spheres, are homeomorphic to the standard sphere, then every polyhedral homology manifold of dimension $\geqq 5$ is locally homeomorphic to euclidean space except at the vertexes with non-simply-connected links. Because R. D. Edwards gave an affirmative answer to the double suspension problem for combinatorial homology spheres of dimension $\geqq 4$, we can reduce the question of simplicial triangulation of topological manifolds to (1) the structure of ker α and (2) the multiple suspension problem for smooth homology 3-spheres.

In fact, we can state as follows (in a simplified form).

THEOREM. *Let $n \geqq 6$. Then, all the topological n-manifolds without boundary are simplicially triangulable if and only if there exists a smooth homology 3-sphere H^3 which satisfies the following three conditions*:

(i) *H^3 bounds a parallelizable smooth 4-manifold with signature 8,*

(ii) *the $(n-3)$-ple suspension of H^3 is homeomorphic to the n-sphere, and*

(iii) *the oriented connected sum $H^3 \# H^3$ bounds an acyclic smooth 4-manifold.*

We can remark that the existence of H^3 with (i) and (ii) is enough to triangulate V when $\beta_0 k(V) = 0$, where $\beta_0\colon H^4(V; Z_2) \to H^5(V; Z)$ is the integral Bockstein operation. In the case $n = 5$, we need possibly some modification of the condition (iii). And the case $n = 4$ is still mysterious. Recently D. Galevski and R. Stern obtained a similar result by different methods, which will be described in these PROCEEDINGS.

This note summerizes the results obtained in my thesis, which is supervised by Professor L. C. Siebenmann to whom I would like to express my deepest thanks.

REFERENCES

1. R. C. Kirby and L. C. Siebenmann, *Foundational essays on topological manifolds smoothings and triangulations*, Ann. of Math. Studies **88**, Princeton Univ. Press, 1977.

2. N. Martin and C. R. F. Maunder, *Homology cobordism bundles*, Topology **10** (1971), 93–110. MR **42** #8511.

3. T. Matumoto, *Variétés simpliciales d'homologie et variétés topologiques métrisables*, Thèse, Orsay, 1976.

4. T. Matumoto and Y. Matsumoto, *The unstable difference between homology cobordism and piecewise-linear bloc bundles*, Tôhoku Math. J. **27** (1975), 57–68. MR **53** # 4077.

5. A. Matsui, *Surgery on 1-connected homology manifolds*, Tôhoku Math. J. (2) **26** (1974), 169–172. MR **49** #3990.

6. C. R. F. Maunder, *Surgery on homology manifolds*. I, II, Proc. London Math. Soc. **32** (1976), 480–520; J. London Math. Soc. **12** (1976), 169–175.

7. H. Sato, *Constructing manifolds by homotopy equivalences*. I. *An obstruction to constructing* PL *manifolds from homology manifolds*, Ann. Inst. Fourier (Grenoble) **22** (1972), fasc. 1, 271–286. MR **49** #1522.

8. D. Sullivan, *Singularities in spaces*, Proc. Liverpool Singularities Sympos., II (1969/1970), Lecture Notes in Math., vol. 209, Springer, Berlin, 1971, pp. 196–206. MR **49** #4002.

9. C. T. C. Wall, *Surgery on compact manifolds*, Academic Press, London, 1971.

10. J. Williamson, *Cobordism of combinatorial manifolds*, Ann. of Math. (2) **83** (1966), 1–33.

KYOTO UNIVERSITY

Proceedings of Symposia in Pure Mathematics
Volume 32, 1978

SIMPLICIAL TRIANGULATIONS OF TOPOLOGICAL MANIFOLDS

DAVID E. GALEWSKI* AND RONALD J. STERN**

In this lecture we will motivate and outline our work concerning simplicial triangulations of topological manifolds. Details of these and related results will appear in [**8**], [**9**], and [**10**].

The primary question we are concerned with is when can a given topological manifold M be triangulated as a simplicial complex, and if so, in how many "different" ways can it be triangulated? The work of R. Kirby and L. Siebenmann ([**11**], [**12**]) shows that in each dimension greater than four there exist closed topological manifolds which admit no piecewise linear manifold structure and hence cannot be triangulated as a combinatorial manifold. However, R. D. Edwards [**5**] has recently demonstrated the existence of noncombinatorial triangulations of S^n, $n \geqq 5$. It is still unknown whether or not every topological manifold can be triangulated as a simplicial complex.

Let us first determine what restrictions are put on a triangulation of a topological manifold. Note that if X is a compact space, then the $(n-k)$-suspension of X, denoted $\Sigma^{n-k} X$, is homeomorphic ($\approx$) to the n-sphere S^n if and only if $c'X \times R^{n-k}$ is an open topological n-manifold, where $c'X$ denotes the open cone over X. Thus K is a triangulation of a topological n-manifold M without boundary if and only if the link L^k of an $(n-k-1)$-simplex in the first barycentric subdivision K' has the homology of S^k and $\Sigma^{n-k} L^k \approx S^n$. We improve this as follows.

Recall that a (polyhedral) *closed homology manifold* is a compact polyhedron with the property that the links of $(n-k)$-simplices have the homology of S^{k-1}.

THEOREM 1. *A closed homology n-manifold M is a topological n-manifold if and only if*

AMS (MOS) subject classifications (1970). Primary 57A15.

*This research was supported in part by National Science Foundation grant GP29585–A4.
**This research was supported in part by National Science Foundation grant MCS76–06393.

(1) *for every 3-dimensional link* L^3 *of* M, $\Sigma^{n-3} L^3 \approx S^n$, *and*
(2) *every* $(n - 1)$*-dimensional link of* M *is* 1*-connected.*

OUTLINE OF THE PROOF. By our observation above we need only check that $\Sigma^{n-k} L^k \approx S^n$ for every k-dimensional link L^k of M for $4 \leqq k \leqq n - 1$.

Case 1. $k = 4$. We show that $c'L^4 \times R^{n-5}$ is an open topological n-manifold. Now L^4 is a closed homology 4-manifold, so the only non PL sphere links are the links of a finite number of vertices. For simplicity suppose there is only one such bad link and call it $\bar{L}$. Then $L^4 = P^4 \cup c\bar{L}$, where the union is taken along $\partial P^4 = \partial(c\bar{L}) = \bar{L}$. The double Q^4 of P^4 is a PL homology 4-sphere so that a recent result of R. D. Edwards [6] implies that $\Sigma^{n-4}Q^4 \approx S^n$; hence $c'Q^4 \times R^{n-5}$ is an open topological n-manifold. By the codimension one approximation theorem of Bryant-Edwards-Seebeck or of Ancel-Cannon [1], we can re-embed $c'P^4 \times R^{n-5} \subset c'Q^4 \times R^{n-5}$ via an embedding h so that its complement in $c'Q^4 \times R^{n-5}$ is 1-ULC. Since $c'\bar{L} \times R^{n-5}$ is an open topological manifold by (1), the taming theorem of J. Cannon [4] implies that $h(c'\bar{L} \times R^{n-5})$ is collared in the closure of the complement of $h(c'P^4 \times R^{n-5})$ in $c'Q^4 \times R^{n-5}$. This then implies that $h(c'P^4 \times R^{n-5}) \cup (h(c'\bar{L} \times R^{n-5}) \times [0, 1]) \approx c'L^4 \times R^{n-5}$ is an open topological n-manifold as required.

Case 2. $k \geqq 5$. Suppose inductively that $\Sigma^{n-k+1}L^{k-1} \approx S^n$. This then implies that $L^k \times R^{n-k}$ and hence $L^k \times T^{n-k}$ is a topological manifold for every k-dimensional link L^k of M. By the results of [7] or [14], there exist a topological homology k-sphere H^k and a simple homotopy equivalence $f: H^k \times T^{n-k} \to L^k \times T^{n-k}$ which is homotopic to a homeomorphism h. As $k \geqq 5$, the Kirby-Siebenmann obstruction to putting a PL manifold structure on H^k is zero, so that we can assume that H^k is a PL homology k-sphere. Now lift h to a bounded homeomorphism $h': H^k \times R^{n-k} \to L^k \times R^{n-k}$ which therefore extends to a homeomorphism $h': H^k * S^{n-k-1} \to L^k * S^{n-k-1}$ (cf. [16]). By a recent result of R. D. Edwards [6] $H^k * S^{n-k-1} \approx S^n$, so that $\Sigma^{n-k}L^k \approx S^n$ as required. □

So we now know how to identify a simplicial triangulation of a topological manifold. What "nice" properties of a simplicially triangulated topological manifold would one like? Note that if K is a polyhedron, then $K \times R$ is a PL $(n + 1)$-manifold if and only if K is a PL n-manifold. This is a fundamental transversality property for PL manifolds. However, if $K \times R$ is a topological $(n + 1)$-manifold it is not necessarily the case that K is a topological n-manifold. But observe that the links of K have all the suspension properties of the n-skeleton of a simplicially triangulated topological $(n + 1)$-manifold. This then motivates the following definition.

A TRI$_n$ *m-manifold* is a homology m-manifold M such that if $k < n$ and L is a k-dimensional link of M, then $\Sigma^{n-k}L^k \approx S^n$ or $\Sigma^{n-k+1}L^k \approx D^{n+1}$, where D^{n+1} is the $(n + 1)$-disk. We now list some facts about TRI$_n$ manifolds. Let K be a polyhedron.

(1) $K \times R$ is a TRI$_n$ manifold if and only if K is a TRI$_n$ manifold.

(2) If K is a TRI$_n$ m-manifold without boundary and with $n \geqq m$, then $K \times R^{n-m}$ is a topological n-manifold without boundary.

(3) If K is a TRI$_n$ m-manifold with $m > n \geqq 6$, then there exists a TRI$_n$ m-manifold $\bar{K}$ which is also a topological manifold and a PL contractible map $f: \bar{K} \to K$. (By Theorem 1, K is a topological manifold except that the $(m - 1)$-dimensional

links of K need to be 1-connected and we blow up these links via a PL contractible map to be 1-connected.)

We now wish to construct a "normal" bundle theory for TRI_n manifolds similar to PL block bundles. A TRI_n q-sphere is a TRI_n q-manifold H^q having the homology of S^q and if $q < n$ we further require that $\Sigma^{n-q}H^q \approx S^n$. A TRI_n *cell complex* is then a cone complex whose cones are cones on TRI_n spheres. A TRI_n cone q-bundle ξ^q/K over a TRI_n cell complex K assigns to each p-cell α of K a block B_α which is the cone on a $\mathrm{TRI}_n(p + q - 1)$-sphere and these cones fit together like the eclls in a cell complex. Using the mock bundle recipe of Buoncristiano, Rourke and Sanderson [3] for representing homotopy functors, Theorem 1 and facts (1)—(3) above, there exists a classifying space $B\,\mathrm{TRI}_n(q)$ for TRI_n cone q-bundles, $q \geqq n \geqq 6$. Let $B\,\mathrm{TRI}_n = \lim_{q\to\infty} B\,\mathrm{TRI}_n(q)$. Using fact (3) above, one shows that every TRI_n cone q-bundle, $q \geqq n \geqq 6$, is concordant to a topological block bundle, so that there is a natural map $j\colon B\,\mathrm{TRI}_n \to B\,\mathrm{TOP}$, where $B\,\mathrm{TOP}$ classifies stable topological block bundles.

We now return to our primary question, when can a given topological m-manifold M be triangulated as a simplicial complex, and if so, in how many "different" ways? Let N be a codimension zero submanifold of ∂M and let Σ_0 be a TRI_n manifold structure on N which extends to a neighborhood of N in M. Let $\mathscr{S}_{\mathrm{TRI}_n}(M \text{ rel } N, \Sigma_0)$ denote the set of TRI_n manifold structures on M agreeing with Σ_0 near N modulo the equivalence relation (called TRI_n *concordance*) that two such structures Γ_0 and Γ_1 on M are TRI_n concordant if there exists a TRI_n manifold structure Γ on $M \times I$ agreeing with $\Sigma_0 \times I$ near $N \times I$ and $\Gamma| M \times \{i\} = \Gamma_i$ for $i = 0, 1$.

Similarily let Lift(τ rel N, F_0) denote the set of lifts of the map $\tau\colon M \to B\,\mathrm{TOP}$, which classifies the stable topological tangent bundle of M, to $B\,\mathrm{TRI}_n$ through $j\colon B\,\mathrm{TRI}_n \to B\,\mathrm{TOP}$ that agree near N with a fixed lift F_0 of τ near N induced by Σ_0, modulo the equivalence relation of vertical homotopy rel N.

THEOREM 2 (CLASSIFICATION THEOREM). *Let M, Σ_0, and F_0 be as above. If $m > n \geqq 6$ ($m \geqq n \geqq 6$ if $N = \partial M$), then M admits a* TRI_n *manifold structure agreeing with Σ_0 near N if and only if τ has lift $M \to B\,\mathrm{TRI}_n$ equaling F_0 near N. In fact there is a bijection $\mathscr{S}_{\mathrm{TRI}_n}(M$ rel N, $\Sigma_0) \to$ Lift(τ rel N, F_0).*

TOWARDS A PROOF OF THEOREM 2. Assume $\partial M = \emptyset$ and suppose $\tau\colon M \to B\,\mathrm{TOP}$ lifts to $B\,\mathrm{TRI}_n$. Then embed M in R^s for some large s and let Q be a PL manifold neighborhood of N equipped with a deformation retraction $r\colon Q \to M$. Then τr classifies a topological bundle over Q whose total space is homeomorphic to $M \times R^k$, for some k. As τr lifts to $B\,\mathrm{TRI}_n$, $M \times R^k$ is a TRI_n manifold. We now wish to show that this implies that M has a TRI_n manifold structure. It clearly suffices to show that if $M \times R$ is a TRI_n manifold, then so is M. This is accomplished via

THEOREM 3 (PRODUCT STRUCTURE THEOREM). *Let M^m be a connected topological m-manifold and let Θ be a* TRI_n *manifold structure on $M \times R$. Let N be a codimension zero submanifold of ∂M and Σ_0 a* TRI_n *manifold structure on N which extends to a neighborhood of N in M such that $\Sigma_0 \times R$ agrees with Θ near $N \times R$. If $m > n \geqq 6$ ($m \geqq n \geqq 6$ if $N = \partial M$), then there exists a* TRI_n *manifold structure Γ*

on M agreeing with Σ_0 near N, unique up to concordance rel Σ_0, *such that $\Gamma \times R$ is concordant* rel $\Sigma_0 \times R$ *to Θ.*

TOWARDS A PROOF OF THEOREM 3. Our proof is modeled on W. Browder's *Structures on $M \times R$* [2]. Assume M is closed. Triangulate $M \times R$ and R so that there is a simplicial map $\pi: M \times R \to R$ homotopic to the projection of $M \times R$ onto R. Let $*$ be a point interior to a simplex of R. Then $\pi^{-1}(*) \times R$ is a codimension zero TRI_n submanifold of $M \times R$, so that by fact (1) above $K = \pi^{-1}(*)$ is a TRI_n manifold. We can assume K is connected, so let W be the cobordism between K and M. By doing a series of handle exchanges we wish to make W into a topological manifold and the inclusion of K into W a simple homotopy equivalence. Then the topological s-cobordism theorem would yield a TRI_n manifold structure on M.

Step 1. We first do the handle exchanges in the homology manifold category. To do this we need surgery below the middle dimension for homology manifolds, a Whitney type trick, and some algebra. The first requirement is accomplished by Matsui [13]; the second is accomplished by using the topological Whitney trick in $M \times R$ and then making it polyhedral by using the homology transversality theorem of [7] and the established surgery below the middle dimension; and the last requirement is purely formal. Thus by doing a series of homology handle exchanges we arrive at a homology m-manifold K' and a cobordism W between K' and M with $K' \subset W$ a simple homotopy equivalence. Also $W = W' \cup W''$ where W' union a collar is a topological manifold and W'' is a homology manifold cobordism from K to K'.

Step 2. We observe that as K is a TRI_n manifold, by using Theorem 1 and fact (3) we can resolve the singularities of W'' via a simple homotopy equivalence so that W is in fact a TRI_n manifold which is a topological manifold. Thus W is our desired topological s-cobordism. $\square$

We now discuss the (homotopic) fiber $\mathrm{TOP}/\mathrm{TRI}_n$ of $j: B\,\mathrm{TRI}_n \to B\,\mathrm{TOP}$. Let $\theta_3^{\mathrm{TRI}_n}$ denote the group of oriented PL homology 3-spheres modulo those which bound acyclic TRI_n 4-manifolds; let $\Theta_3^{\mathrm{TRI}_n/\mathrm{PL}}$ denote the group of oriented PL homology 3-spheres which bound acyclic TRI_n 4-manifolds modulo those which bound acyclic PL 4-manifolds; and let θ_3^H denote the group of PL homology 3-spheres modulo those which bound acyclic PL 4-manifolds. The only concrete theorem known about θ_3^H is the existence of the Kervaire-Milnor-Rochlin surjection $\alpha: \theta_3^H \to Z_2$. From the definitions we have the short exact sequence $0 \to \theta_3^{\mathrm{TRI}_n/\mathrm{PL}} \to \theta_3^H \to \theta_3^{\mathrm{TRI}_n} \to 0$.

THEOREM 4. *If $n \geqq 6$, the homotopy groups of* $\mathrm{TOP}/\mathrm{TRI}_n$ *are zero except possibly for π_3 and π_4. Furthermore there are two exact sequences*

(1) $0 \to \pi_4 \to \mathit{kernel}(\alpha: \theta_3^H \to Z_2) \to \theta_3^{\mathrm{TRI}_n} \to \pi_3 \to 0$,

(2) $0 \to \pi_4 \to \theta_3^{\mathrm{TRI}_n/\mathrm{PL}} \to^{\alpha} Z_2 \to \pi_3 \to 0$,

where α is the Kervaire-Milnor-Rochlin map.

COROLLARY 5. (1) $\pi_3(\mathrm{TOP}/\mathrm{TRI}_n)$ *has at most* 2 *elements.*

(2) $\pi_3(\mathrm{TOP}/\mathrm{TRI}_n) = 0$ *if and only if there exists a* PL *homology* 3-*sphere with* $\alpha(H^3) = 1$ *and* $\Sigma^{n-3} H^3 \approx S^n$.

(3) $\pi_4(\mathrm{TOP}/\mathrm{TRI}_n) = 0$ *if and only if any* PL *homology* 3-*sphere with* $\alpha(H^3) = 0$ *and* $\Sigma^{n-3}H^3 \approx S^n$ *bounds an acyclic* PL 4-*manifold.*

We also have the following existence theorem.

THEOREM 6. *Every topological m-manifold has a* TRI_n *manifold structure for* $m > n \geqq 6$ ($m \geqq n \geqq 6$ *if* $\partial M = \emptyset$) *if and only if there exists a* PL *homology 3-sphere* H^3 *satisfying the following* 3 *properties.*

(1) $\alpha(H^3) = 1$,
(2) $\Sigma^{n-3}H^3 \approx S^n$,
(3) $H^3 \# H^3$ *bounds a* PL *acyclic 4-manifold.*

REMARK. When $m = 5$, L. Siebenmann demonstrated in **[16]** that the existence of a PL homology 3-sphere H^3 satisfying (1) and (2) above implied that all closed oriented 5-manifolds can be triangulated as simplicial complexes. If $H^3 \# H^3$ bounds a contractible PL 4-manifold, then he also shows that all 5-manifolds can be triangulated. For closed topological m-manifolds M with $6 \leqq n \leqq 8$ and with the integral Bockstein of the Kirby-Siebenmann obstruction to putting a PL manifold structure on M being zero, M. Scharlemann **[15]** has shown that (1) and (2) above imply that M is triangulable as a simplicial complex. T. Matumoto **[14]** proves a version of the sufficiency of Theorem 6 with (2) replaced by the condition that $\Sigma^{n-4}H^3 \approx S^{n-1}$.

REMARK. Our proof of Theorem 6 actually shows that if there exists a PL homology 3-sphere satisfying (1)—(3) above, then every topological m-manifold has a TRI_n manifold structure in which the 3-sphere links are PL homeomorphic to connected sums of H^3, $-H^3$, and S^3.

TOWARDS A PROOF OF THE SUFFICIENCY OF THEOREM 6. Let H^3 be a PL homology 3-sphere satisfying (1)—(3) of Theorem 6. One can consider TRI_n manifolds M whose 3-dimensional sphere links in M and ∂M are PL homeomorphic to connected sums of H^3, $-H^3$, and S^3. Call such manifolds H^3 manifolds. One can construct a classifying space BH^3 for stable TRI_n cone bundles based on H^3 manifolds. There are natural maps $i_0\colon BH^3 \to B\,\mathrm{TOP}$, $i_1\colon B\,\mathrm{PL} \to BH^3$ and $i_2\colon BH^3 \to B\,\mathrm{TRI}_n$. The fiber of i_1 is a $K(Z_2, 3)$ so that by considering the homotopy exact sequence of the triple ($B\,\mathrm{TOP}$, BH^3, $B\,\mathrm{PL}$) we have that i_0 is a homotopy equivalence. The result now follows from Theorem 2. □

More generally we have the following existence theorem. Let $Sq_k\colon H^4(\ ; Z_2) \to H^5(\ ; Z_k)$ denote the Bockstein associated with the short exact coefficient sequence $0 \to Z_k \to^{\times 2} Z_{2k} \to Z_2 \to 0$. Also let $\Delta(M) \in H^4(M; Z_2)$ denote the Kirby-Siebenmann obstruction to the existence of a PL manifold structure on M.

COROLLARY 7. *If there exists a closed topological m-manifold M with a* TRI_n *manifold structure,* $m \geqq n \geqq 6$, *and if* $Sq_{2k}\Delta(M) \neq 0$, *then there exists a* PL *homology 3-sphere* H^3 *such that*

(1) $\alpha(H^3) = 1$,
(2) $\Sigma^{n-3}H^3 \approx S^n$, *and*
(3) *the 2k-fold connected sum of* H^3 *bounds a* PL *acyclic 4-manifold.*

Also, if there exists a PL *homology 3-sphere* H^3 *satisfying* (1)—(3), *then every topological m-manifold M with* $Sq_k\Delta(M) = 0$ *has a* TRI_n *manifold structure if* $m > n \geqq 6$ ($m \geqq n \geqq 6$ *if* $\partial M = \emptyset$).

We also remark that there is a surgery theory for TRI_n manifolds completely analogous to topological surgery theory. This is given in **[9]**.

We also investigate the question of whether a given topological n-manifold, $n \geqq 5$, can be triangulated as a simplicial homotopy manifold. For example,

PROPOSITION 5. *Suppose that every* PL *homotopy* 3-*sphere bounds a contractible* PL 4-*manifold. Then there is a one-to-one correspondence between the set of concordance classes of simplicial homotopy manifold triangulations of a topological n-manifold M, $n \geqq 5$, and concordance classes of* PL *manifold structures on M.*

PROPOSITION 6. *Suppose there exists a bad counterexample to the* 3-*dimensional Poincaré conjecture; namely suppose there exists a* PL *homotopy* 3-*sphere H^3, with*

(i) $\alpha(H^3) = 1$, *and*

(ii) $H^3 \# H^3$ *bounds a contractible* PL 4-*manifold.*

Then every topological n-manifold, $n \geqq 5$, can be triangulated as a simplicial homotopy manifold.

BIBLIOGRAPHY

1. F. Ancel and J. Cannon, *Any embedding of S^{n-1} in $S^n (n \geqq 5)$ can be approximated by locally flat embeddings*, Notices Amer. Math. Soc. **23** (1976), A-308.
2. W. Browder, *Structures on $M \times R$*, Proc. Cambridge Philos. Soc. **61** (1965), 337–345. MR **30** #5321.
3. S. Buoncristiano, C. P. Rourke, and B. J. Sanderson, *A geometric approach to homology theory*, J. London Math. Soc., Lecture Notes Series 18, Cambridge Univ. Press, London and New York, 1976.
4. J. Cannon, *Taming codimension one generalized manifolds* (preprint).
5. R. D. Edwards, *The double suspension of a certain homology* 3-*sphere is S^5*, Notices Amer. Math. Soc. **22** (1975), A-334.
6. ———, *Stabilizing cell-like maps of manifolds onto ANR's to homeomorphisms.*
7. D. Galewski and R. Stern, *The relationship between homology and topological manifolds via homology transversality* (preprint).
8. ———, *Classification of simplicial triangulations of topological manifolds* (preprint).
9. ———, *Surgery on compact simplicially triangulated topological manifolds* (preprint).
10. ———, *Geometric transversality and bordism theories* (to appear).
11. R. Kirby and L. Siebenmann, *Essays on topological manifolds, smoothings and triangulations*, Ann. of Math. Studies No.88, Princeton Univ. Press (to appear).
12. ———, *On the triangulation of manifolds and the Hauptvermutung*, Bull. Amer. Math. Soc. **75** (1969), 742–749. MR **39** #3500.
13. A. Matsui, *Surgery on* 1-*connected homology manifolds*, Tôhoku Math. J. (2) **26** (1974), 169–172. MR **49** #3990.
14. T. Matumoto, *Variétés simpliciales d'homologie et variétés topologiques métrisables*, Thesis, University de Paris-Sud, 91405 Orsay, France, 1976.
15. M. Scharlemann, *Simplicial triangulations of non-combinatorial manifolds of dimension less than nine* (preprint).
16. L. Siebenmann, *Are nontriangulable manifolds triangulable*? Topology of Manifolds (Proc. Inst., Univ. of Georgia, Athens, Ga., 1969), pp. 77–84. Markham, Chicago, Ill., 1970. MR **42** #6837.

UNIVERSITY OF GEORGIA

UNIVERSITY OF UTAH

Proceedings of Symposia in Pure Mathematics
Volume 32, 1978

FIBERING TOPOLOGICAL MANIFOLDS

ALLAN L. EDMONDS[1]

1. Introduction. Suppose a compact topological manifold M can be expressed as the total space of a locally trivial fiber bundle $F \subsetneq M \to B$, with no additional assumptions on the spaces F and B. Is M necessarily expressible as such a bundle where F and B are manifolds? In particular, if $M \cong F \times B$ can M be factored as a product of manifolds?

In this paper we answer these questions affirmatively under the additional assumptions that F is known to be a manifold and that B has dimension at least 5. The general case remains unsettled.

The original motivation for the questions considered here was the following: Suppose that a compact Lie group G acts freely (or, more generally, with just one orbit type) on a compact manifold M. Call the action *tame* if M/G is a manifold and *wild* otherwise. It is known (cf. **[Li]**) that if M admits a tame G action and $\dim G \geqq 1$ and $\dim M - \dim G \geqq 3$, then M also admits uncountably many distinct wild actions, obtained, for example, by collapsing out noncellular arcs from M/G. The present techniques can be applied to show that if M admits a given wild action then it also admits a tame action which can be associated to the given action in a rather canonical way.

The usefulness of the solution of the problem on group actions is that various classification theorems available in the tame case (e.g. **[W**, pp. 191–194**]**) can be interpreted in the general case as well.

2. The fibering theorem. The following is the precise statement of the main result; it should be compared with Theorem A of N. Levitt **[Le]**.

THEOREM 2.1. *Let $F^k \subsetneq M^m \to^p B$ be a locally trivial fiber bundle where M and F are closed m- and k-manifolds, respectively. If $m - k \geqq 5$ there is a locally trivial fiber*

AMS (MOS) subject classifications (1970). Primary 57A99; Secondary 57E10, 57D65.

[1]Supported in part by NSF.

bundle $F \subsetneq M \to N$ *where* N *is a closed manifold with the property that there is a bundle map*

$$\begin{array}{ccc} M & \xrightarrow{f'} & M \\ \downarrow & & \downarrow \\ N & \xrightarrow{f} & B \end{array}$$

(f' *is a homeomorphism on each fiber) such that* f' *is homotopic to the identity* 1_M.

REMARK. The map f will actually be constructed to be a simple homotopy equivalence and to possess certain tangential properties described below. This will be important in the proof and in §3.

PROOF OF (2.1). The essential idea is to use surgery theory **[W]** in its topological formulation **[KS]** to find a manifold simple homotopy equivalent to B. The proof proceeds in five steps. For simplicity we shall assume M and B are connected. The general case follows easily from this.

Step 1. B is a finite simple Poincaré duality space. Since B is not a C.W. complex one must take a little care similar to that taken in **[KS**, Essay III, §5] for topological manifolds.

Observe that $B \times \boldsymbol{R}^k$ is a topological manifold (since $B \times \boldsymbol{R}^k$ is locally homeomorphic to M) and so is $B \times D^{k+1}$. According to **[KS]**, the compact manifold $B \times D^{k+1}$, and hence B, has the homotopy type of a finite C.W. complex and in fact has a well-defined simple homotopy type obtained, say, by finding a handle decomposition of $B \times D^{k+1}$. Compare the arguments of Chapman **[C]**.

Note that B satisfies Poincaré duality in the sense of Wall. This follows immediately from a result of Quinn [**Q**, 1.6]. An alternate proof in this case is to note that since $B \times \boldsymbol{R}^k$ is a manifold, B is necessarily a homology n-manifold, $n = m - k$, and hence satisfies duality. (An orientation homomorphism $w\colon \pi_1(B) \to \{\pm 1\}$ is determined by that for $B \times \boldsymbol{R}^k$ and will be suppressed from the notation in this discussion.) Better one can easily deduce duality for B from that for the manifold $B \times S^r$, $r > \max\{k, n\}$.

Now let X be any finite C.W. complex simple homotopy equivalent to B. Then X also satisfies Poincaré duality in dimension $n = m - k$. We claim that the duality isomorphism is induced by a *simple* chain homotopy equivalence of chain complexes.

Let $C_*(X)$ and $C^*(X)$ be the cellular chain and cochain complexes of X obtained from the universal cover $\tilde{X}$ as in **[W]**. Represent the fundamental class of X by an n-chain ξ and consider the cap product $\xi \cap\colon C^i(X) \to C_{n-i}(X)$. We assert that this chain homotopy equivalence is simple with respect to the standard bases determined by the cell structure of X. To see this consider $X \times S^r$ where $r > \max\{k, n\}$. Then $X \times S^r$ is simple homotopy equivalent to the compact manifold $B \times S^r$. Here S^r is given the C.W. structure with one 0-cell and one r-cell. By **[KS**, Essay III, 5.13] duality is simple for any C.W. complex simple homotopy equivalent to $B \times S^r$. Hence duality is simple for $X \times S^r$ (with the natural product cell structure). Consider the commutative diagram

$$\begin{array}{ccc} C^i(X) & \xrightarrow{\xi\cap} & C_{n-i}(X) \\ \big\downarrow & & \big\downarrow \times [S_r] \\ C^i(X \times S^r) & \xrightarrow{(\xi\times[S^r])\cap} & C_{n+r-i}(X \times S^r) \end{array}$$

The vertical homomorphisms are simple isomorphisms since $r > n$ and $X \times S^r$ has a product cell structure. $(\xi \times [S^r])\cap$ is simple by the previous remark. Therefore $\xi \cap$ is simple. □

Step 2. The Spivak normal fibration ν_B of B admits a canonical reduction to a topological euclidean space bundle η.

Now $B \times \boldsymbol{R}^k$ is a manifold and so has a stable normal bundle $\nu(B \times \boldsymbol{R}^k)$. Set $\eta = \nu(B \times \boldsymbol{R}^k)|B$. It follows from the fact that B has an actual trivial bundle neighborhood in $B \times \boldsymbol{R}^k$ that η determines a well-defined reduction of ν_B.

This reduction to η is canonical in the sense that if V^v is any v-manifold, $v \geqq k$, then $\nu(B \times V^v)|B = \eta$ stably, since B has a neighborhood of the form $B \times \boldsymbol{R}^v$ in $B \times V$. □

Recall that via transversality the reduction of ν_B to η corresponds to the normal cobordism class of a surgery problem (or degree 1 normal map) $f\colon N^n \to B$, covered by a bundle map $\nu(N) \to \eta$. This surgery problem is obtained in this case by homotoping the identity $B \times \boldsymbol{R}^k \to B \times \boldsymbol{R}^k$ to a map transverse to $B \times O$ with respect to the trivial normal microbundle for $B \times O$ in $B \times \boldsymbol{R}^k$.

Step 3. The surgery obstruction $\theta(f)$ in $L_n^s(\pi_1 B)$ is zero.

This will be seen to be a consequence of the periodicity of Wall surgery obstructions [**W**, (9.9)]. Let $V^v = (CP^2)^r$, the product of r copies of the complex projective plane, where $4r > k$. Notice that the surgery problem $f \times 1_V : N \times V \to B \times V$ is normally cobordant to the identity $1_{B\times V} : B \times V \to B \times V$ (recall $B \times V$ is a manifold). For $f \times 1_V$ is obtainable by homotoping the identity $1\colon B \times \boldsymbol{R}^k \times V \to B \times \boldsymbol{R}^k \times V$ to a map transverse to $B \times O \times V$ with respect to the trivial normal microbundle. But 1 is already transverse to $B \times O \times V$. Hence by relative transversality there is a homotopy between $f \times 1_V$ and $1_{B\times \boldsymbol{R}^k \times V}$ which is transverse to $B \times O \times V$, and which provides the needed normal cobordism.

Now by periodicity $\theta(f) = 0$ if and only if $\theta(f \times 1_V) = 0$. But $\theta(f \times 1_V) = \theta(1_{B\times V}) = 0$. □

Thus in what follows we may assume that $f\colon N \to B$ is a simple homotopy equivalence defining the canonical reduction η of ν_B.

Step 4. Completion of the proof when $\pi_1 M = 0$.

Form the pullback diagram

$$\begin{array}{ccc} \bar{N} & \xrightarrow{\bar{f}} & M \\ {\scriptstyle \bar{p}}\downarrow & & \downarrow{\scriptstyle p} \\ N & \longrightarrow & B \end{array}$$

Now $\bar{f}$ is a homotopy equivalence and determines the trivial reduction of the Spivak normal fibration ν_M, as one can easily see by crossing the entire diagram with $\boldsymbol{R}^k$. Thus by standard Browder-Novikov-Sullivan surgery theory $\bar{f}$ is homotopic to a homeomorphism $h\colon \bar{N} \to M$. (The surgery obstruction of any normal cobordism of $\bar{f}$ to 1_M can be killed by connected sum with an appropriate plumbing manifold [**BS**, II.3.8].)

Finally the homeomorphism h induces from the fibering $F \subsetneq \bar{N} \to N$ the desired fibering of M with $\bar{f}h^{-1}\colon M \to M$ the desired bundle map such that $\bar{f}h^{-1} \simeq 1_M$:

$$\begin{array}{ccc} M & \xrightarrow{\bar{f}h^{-1}} & M \\ {\scriptstyle \bar{p}h^{-1}}\downarrow & & \downarrow{\scriptstyle p} \\ N & \xrightarrow{f} & B \end{array} \qquad \square$$

Step 5. Completion of the proof in the general case.

When $\pi_1 M \neq 0$, $\bar{f}$ above need not be homotopic to a homeomorphism. We correct this by altering the choice of $f\colon N \to B$, using periodicity again.

As before let $V^v = (CP^2)^r$, $v = 4r > k$. According to Step 3 we may choose a normal cobordism

$$(H; f \times 1, 1)\colon (W; N \times V, B \times V) \to B \times V \times (I; 0,1).$$

This has some surgery obstruction $\theta(H)$ in $L^s_{n+v+1}(\pi_1 B)$. Let α in $L^s_{n+1}(\pi_1 B)$ correspond to $\theta(H)$ under periodicity.

Form another normal cobordism

$$(G; f, f')\colon (U; N, N') \to B \times (I; 0, 1)$$

from f to a new simple homotopy equivalence f' such that $\theta(G) = -\alpha$ in $L^s_{n+1}(\pi_1 B)$, by [**W**, (5.8) and (6.5)].

Then the normal cobordism

$$(G \times 1_V; f \times 1_V, f' \times 1_V)\colon (U \times V; N \times V, N' \times V) \to B \times V \times (I; 0, 1)$$

has $\theta(G \times 1_V) = -\theta(H)$, by periodicity.

Glue W and $U \times V$ together along $N \times V$, obtaining

$$H \cup (G \times 1_V)\colon W \cup (U \times V) \to B \times V \times I,$$

a normal cobordism from $1_{B \times V}$ to $f' \times 1_V$. By additivity of obstructions in unions $\theta(H \cup (G \times 1_V)) = 0$. Thus $f' \times 1_V$ is homotopic to a homeomorphism, since there is then an s-cobordism between $f' \times 1_V$ and $1_{B \times V}$.

As in Step 4 construct a pullback diagram

$$\begin{array}{ccc} \overline{N'} & \xrightarrow{\overline{f'}} & M \\ \downarrow & & \downarrow \\ N' & \xrightarrow{f'} & B \end{array}$$

We assert that $\overline{f'}$ is homotopic to a homeomorphism. To see this first of all notice that by the Covering Homotopy Theorem $\overline{f'} \times 1_V : \overline{N'} \times V \to M \times V$ is homotopic to a homeomorphism since we arranged that $f' \times 1_V\colon N' \times V \to B \times V$ is. This easily implies that $\overline{f'}$ is also homotopic to a homeomorphism: We have a normal cobordism of $\overline{f'} \times 1_V$ to $1_{M \times V}$ with 0 surgery obstruction in $L^s_{n+v+1}(\pi_1 M)$; periodicity says this normal cobordism is normally cobordant rel the ends to one which is a product with V and still has 0 surgery obstruction; thus $\overline{f'}$ is homotopic to a homeomorphism.

Now the argument is completed just as in Step 4. □

COROLLARY 2.2. *If $M \cong F \times B$ where M and F are m- and k-manifolds, respectively, and $m - k \geqq 5$, then $M \cong F \times N$ where N is a manifold.*

PROOF. Apply (2.1) to obtain a pullback diagram

$$\begin{array}{ccc} M & \longrightarrow & M \\ {\scriptstyle p'}\downarrow & & \downarrow{\scriptstyle p} \\ N & \longrightarrow & B \end{array}$$

Since p is the projection map of a trivial bundle, so is p'. □

3. Application to group actions. In the context of group actions the fibering theorem (2.1) can be interpreted as follows. Recall that a free action of a compact Lie group G on a manifold M is *tame* if the orbit space M/G is a manifold.

THEOREM 3.1. *If φ is a free action of a compact Lie group G on a closed manifold M such that* $\dim M - \dim G \geqq 5$, *then there is a free tame action ψ of G on M and an equivariant homotopy equivalence $(M, \psi) \to (M, \varphi)$ which is homotopic to 1_M.*

PROOF. The orbit map $p: M \to M/\varphi$ is a locally trivial fiber bundle with fiber G by a theorem of Gleason (or its generalization, the Slice Theorem [**B**, II.5.4]).

Applying (2.1) we obtain a pullback diagram

$$\begin{array}{ccc} M & \xrightarrow{f'} & M \\ {\scriptstyle p'}\downarrow & & \downarrow{\scriptstyle p} \\ N & \xrightarrow{f} & M/\varphi \end{array}$$

where N is a manifold and $f' \simeq 1_M$. Since this is a pullback diagram and p is an orbit map, there is a free (necessarily tame) action of G on M with orbit map p' such that f' is equivariant. □

Define free actions φ and ψ of G on M to be *tangentially equivariantly homotopy equivalent* if there is an equivariant homotopy equivalence $(M, \psi) \to (M, \varphi)$ such that the induced map of orbit spaces $M/\psi \to M/\varphi$ preserves the canonical reductions of Spivak fibrations to topological bundles, from (2.1), Step 2.

COROLLARY 3.2. *If G is a connected compact Lie group and M is a* 1-*connected closed manifold, then there is a one-to-one correspondence between the set of tangential equivariant homotopy equivalence classes of free actions of G on M and the set of equivariant homeomorphism classes of free tame actions of G on M.*

PROOF. *The proof of* (2.1) shows that the $f': (M, \psi) \to (M, \varphi)$ of (3.1) can be chosen to be an equivariant tangential homotopy equivalence. Applying (2.1) and (3.1), then, it suffices to show that any tangential equivariant homotopy equivalence $f: (M, \psi) \to (M, \varphi)$ between tame actions is G-homotopic to an equivariant homeomorphism.

Consider the induced map $f^*: M/\psi \to M/\varphi$. By hypothesis f^* preserves the canonical reductions of Spivak fibrations. Thus f^* is normally cobordant to $1_{M/\varphi}$. Now, since M is 1-connected and G is connected, M/φ is also 1-connected. By Browder-Novikov-Sullivan theory f^* is homotopic to a homeomorphism. An application of the covering homotopy property completes the proof. □

Presumably a more awkward and less easily applied version of (3.2) can be formulated to avoid the connectivity assumptions on M and G.

The main consequence of (3.2) is that tame classification theorems become applicable in more general cases. For example, Wall [**W**, p. 192] has written down the equivariant homeomorphism classification of tame free actions of the circle group S^1 on spheres S^n, $n \geqq 7$. He conjectured that the classification could be interpreted in the general case. By (3.2) the same classification applies to the tangential S^1-homotopy classification of arbitrary free S^1-actions on spheres. An alternative way of reducing the classification of free S^1-actions to the tame case was given in [E].

References

[B] G. E. Bredon, *Introduction to compact transformation groups*, Academic Press, New York, 1972.

[BS] W. Browder, *Surgery on simply-connected manifolds*, Springer-Verlag, Berlin and New York, 1972.

[C] T. A. Chapman, *Simple homotopy theory for compact Hilbert cube manifold factors*, Lecture Notes in Math., vol. 375, Springer-Verlag, Berlin and New York, 1974, pp. 53–63. MR **51** #4254.

[E] A. L. Edmonds, *Taming free circle actions*, Proc. Amer. Math. Soc. **62** (1977), 337–343.

[KS] R. C. Kirby and L. C. Siebenmann, *Foundational essays on topological manifolds, smoothings, and triangulations*, Ann. of Math. Studies, no. 88, Princeton Univ. Press, Princeton, N.J., 1977.

[Le] N. Levitt, *A necessary and sufficient condition for fibering a manifold*, Topology **14** (1975), 229–236.

[Li] L. L. Lininger, *On topological transformation groups*, Proc. Amer. Math. Soc. **20** (1969), 191–192.

[Q] F. Quinn, *Surgery on Poincaré and normal spaces*, Bull. Amer. Math. Soc. **78** (1972), 262–267. MR **45** #6014.

[W] C. T. C. Wall, *Surgery on compact manifolds*, Academic Press, New York, 1970.

Cornell University

LOW DIMENSIONAL MANIFOLDS

Proceedings of Symposia in Pure Mathematics
Volume 32, 1978

RECENT RESULTS ON SUFFICIENTLY LARGE 3-MANIFOLDS

FRIEDHELM WALDHAUSEN

This is an expository paper, an expanded version of the talk actually given (which went only to what is §2 of this paper). The topics discussed are:

Johannson's *classification of exotic homotopy equivalences*;

Hemion's *classification of homeomorphisms of a 2-manifold* (compact, with nonempty boundary).

It will be indicated that (and in what sense) some of the main problems on sufficiently large irreducible 3-manifolds can now be considered solved: Classification, classification up to homotopy type, classification of manifolds homotopy equivalent to a given one, classification of knots, classification of knot groups.

The plan of the paper is as follows.

§1 gives background material on exotic homotopy equivalences and in particular some examples.

§2 introduces the *characteristic submanifold*; this notion is needed in the statement of Johannson's result. The result is then discussed.

§3 introduces *manifolds with boundary pattern*, a relativization of 3-manifolds required for inductive proofs. A rough indication of proof of Johannson's result is included.

§4 discusses Haken's approach to classification. The language of the preceding sections is used (at least part of this was indeed implicitly used by Haken). It is indicated how Hemion's result provides the missing step in Haken's theory. Some related results are also discussed.

1. Prelude to homotopy equivalences. The question is: If $f : M \to N$ is a homotopy equivalence of 3-manifolds, what conditions guarantee that f is homotopic to a homeomorphism?

One sufficient set of conditions is the following **[24]** (everything PL, say).

AMS (MOS) subject classifications (1970). Primary 55D10, 57A05, 57A10; Secondary 57C25.

(1) M should be compact, orientable, and irreducible (that is, every 2-sphere in M bounds a 3-ball in M).

(2) If M is closed it should be sufficiently large (that is, there should exist an embedding of a closed orientable 2-manifold in M whose fundamental group is nontrivial and injects).

(3) If the boundary $\partial M \neq \emptyset$ the map f should actually be given as a homotopy equivalence of pairs $M, \partial M \to N, \partial N$.

The status of these conditions is, roughly, the following: 'compact' and 'orientable' are mainly asked for convenience. That is, each can be replaced by a considerably weaker but more technical condition, and everything goes through without essential change. For example in the nonorientable case one may simply define the projective planes away; and in the noncompact case one may insist on maps being proper and manifolds being 'sufficiently large at infinity' (one way to put the latter is to ask that the pro-object of fundamental groupoids at infinity be isomorphic to one in which all maps of vertex groups are injective).

'Irreducibility' is justified by the Kneser-Haken-Milnor unique decomposition theorem. It is of great importance, both technically and otherwise. Its purpose is twofold. It serves to get around the unresolved Poincaré conjecture, and it serves to avoid splitting problems at 2-spheres. The latter have actually been solved by Laudenbach, Swarup, and, ultimately, Hendriks.

The condition 'sufficiently large' is being discussed elsewhere **[27]**. Notice it is only asked for closed manifolds.

We are here interested in the problem of omitting the condition prescribing f on the boundary.

EXAMPLE 1.1. Suppose there is a component G of ∂M so that $\ker(\pi_1 G \to \pi_1 M) \neq 0$. By the loop theorem of Papakyriakopoulos, cf. **[20]**, this means that we can write $M = M' \cup$ 1-handle attached at $\partial M'$, and it is obvious that there are many homotopy equivalences from M to itself which map this 1-handle through the interior. The situation is messy but comparatively easy to analyse. For example the homotopy equivalences fixing M' are given by group theory.

To avoid this phenomenon we consider henceforth only 3-manifolds whose boundary is (nonempty and) incompressible, that is, for any component G of ∂M, $\pi_1 G \to \pi_1 M$ is injective. We also insist on condition (1) above. The class of manifolds still being considered includes some of the most interesting 3-manifolds, in particular it includes the knot spaces of nontrivial knots.

For this class of manifolds one knows **[24]**: If $f: M \to N$ is a homotopy equivalence, and if there exists f' homotopic to f with $f'(\partial M) \subset \partial N$, then f is homotopic to a homeomorphism.

So we have focussed our attention, for the class of manifolds being considered, on

Problem. Suppose $f: M \to N$ is a homotopy equivalence. Suppose there does not exist f' homotopic to f with $f'(\partial M) \subset \partial N$. What can one say?

We will call such homotopy equivalences *exotic*. Here are some examples of exotic homotopy equivalences.

EXAMPLE 1.2. Let F_1 and F_2 be, respectively, the 2-torus with one open disk removed and the 2-sphere with three open disks removed. There is a homotopy equivalence $F_1 \to F_2$. Then $F_1 \times S^1 \to F_2 \times S^1$ is exotic.

EXAMPLE 1.3. Assume given M_1, M_2 and embeddings of the annulus, f_i: $S^1 \times I \to \partial M_i$ with f_{i*} injective on π_1. Let h: $S^1 \times I \to S^1 \times I$ be given by the flip of the interval. Form M (resp., N) by gluing M_1 and M_2 at $S^1 \times I$ by means of f_1 and f_2 (resp., $f_1 \circ h$ and f_2). Then M and N are homotopy equivalent but in general not homeomorphic.

Note that in Example 1.2 the number of boundary components changes. In Example 1.3 their type may change as well.

EXAMPLE 1.4. Let M have a Seifert fibre space structure with decomposition surface B (and $\partial B \neq \emptyset$ since $\partial M \neq \emptyset$), or to use current language, M has a (stable) foliation by circles, and B is the space of leaves. Then M may admit exotic (self-) homotopy equivalences of the type of Example 1.2, i.e., exotic homotopy equivalences which are induced from 2-manifold phenomena. But there may also be other ones. Specifically, the foliation is characterized by B and the nontrivial monodromy; this occurs at isolated singular leaves, say r in number, and in each case may be specified by a coprime pair (α_i, β_i), $0 < \beta_i < \alpha_i$. Then, except for the cases (B = 2-disk and $r \leqq 1$) and (B = Möbius strip and $r = 0$), the type of B and the set of the (α_i, β_i) are an invariant of the oriented homeomorphism type of M **[22]**. But the homotopy type is only given by the homotopy type of B and the set of the α_i (no β_i) **[23]**.

The additional phenomena in Example 1.4 may be traced to the following

EXAMPLE 1.5. Let $M = S^1 \times D^2$ be the solid torus, and f: $S^1 \times I \to \partial M$ an embedding of the annulus, with winding number $\alpha > 1$. Letting $F = f(S^1 \times I)$, the oriented homeomorphism type of (M, F) is characterized (among such pairs) by an integer β, coprime with α, and $0 < \beta < \alpha$. But the homotopy type of the pair (M, F) is characterized by α alone.

This last example does not really fit into the framework we have been considering so far. It just illustrates that 'relative' phenomena may manifest themselves in a nonrelativized framework.

The reader may amuse himself in looking for more examples of exotic homotopy equivalences. Leaving aside modifications of the examples given, he will probably find the search rather difficult—with reason, as we shall see later.

All of the examples given have one thing in common. Underlying any of them is a very simple, and very special, geometric phenomenon. It will turn out there is a system. One may compare the situation to the ancient myth of the stable of one Augias. In that tale, after a considerable effort to dispose of the obvious, what remained was much cleaner.

2. Classification of homotopy equivalences. The results to be described are due to Johannson **[9]**, **[10]**, **[11]** and partly myself **[26]**; partial results have been rediscovered by Feustel, Jaco, Shalen, and others, in a large number of papers.

To begin with, one considers a special case. Suppose f: $M \to N$ is an exotic homotopy equivalence, and each component of ∂M is a torus. Then by definition of the terms involved there exists in N an *essential singular torus* which cannot be deformed into the boundary, in the following sense.

DEFINITION. g: $S^1 \times S^1 \to N$ is *essential* if g_* is injective on π_1.

This draws attention to essential singular tori. One would like to analyse them à la loop theorem and sphere theorem. However the cut and paste technique turns out to be inadequate, mainly because the naive conjecture has easy counterexamples.

EXAMPLE 2.1. In the knot space of a torus knot, any nonsingular essential torus is parallel to the boundary [**22**].

EXAMPLE 2.2. In any of the Seifert fibre spaces which furnish the known examples of nonsufficiently large 3-manifolds (closed, irreducible, with infinite fundamental group) there is an essential singular torus, in fact there are infinitely many such, but there is no incompressible surface whatsoever [**23**]. Also there is an infinite number of manifolds M which are Seifert fibre spaces over S^2 with precisely three exceptional fibres, and which satisfy $H^1(M) \neq 0$ and are hence sufficiently large. Any such M contains a unique incompressible surface, up to isotopy (it is a fibre in some fibration over S^1), but for only finitely many M can this surface be a torus [**22**].

Turning hindsight into foresight one decides to consider *all* essential tori, singular or not, all at once, hoping to force them into a pattern. Having decided this far, it is clearly unreasonable not to consider at the same time essential annuli, singular or not.

DEFINITION. $g: (S^1 \times I, S^1 \times \partial I) \to (N, \partial N)$ is *essential* if

$$g_*: \pi_1(S^1 \times I) \to \pi_1 N, \qquad g_*: \pi_1(S^1 \times I, \partial) \to \pi_1(N, \partial)$$

are both injective.

One considers now submanifolds of a given manifold M which in a sense can be manufactured out of essential tori and annuli. It is convenient to change here our conventions about M. In addition to the manifolds considered up to this point we also admit manifolds that are closed (orientable and irreducible) and sufficiently large.

DEFINITION 2.3. A compact codimension zero submanifold V of M is an *essential F-manifold* ('F' for 'fibering') if and only if for each component W of V at least one of the following holds: either

(a) (i) W admits a structure of Seifert fibre space, $p: W \to B$, such that $p^{-1}(p(W \cap \partial M)) = W \cap \partial M$, and

(ii) each component of $\mathrm{Cl}(\partial W - \partial M)$ is an essential annulus or torus, or

(b) (i) W admits a structure of line bundle, $p: W \to B$ such that $W \cap \partial M$ is the associated 0-sphere bundle, and

(ii) each component of $\mathrm{Cl}(\partial W - \partial M)$ is an essential annulus.

DEFINITION 2.4. A *characteristic submanifold* of M is an essential F-manifold V in M satisfying

(i) if X is any essential F-manifold in M then X can be properly isotoped into V,

(ii) if Y is any union of components of $\mathrm{Cl}(M - V)$ then $V \cup Y$ is not an essential F-manifold.

A condition equivalent to (ii) is that one cannot throw away a component of V and still have (i). Thus V is definable by a universal property.

THEOREM 1. *The characteristic submanifold of M exists and is unique up to ambient isotopy.*

The status of the theorem is this. Once conceived of, the definition of characteristic submanifold may easily be reformulated to involve some kind of 'complexity'. Existence is then provable by the method of the Kneser-Haken finiteness theorem [**5**]. Given the existence, uniqueness is not hard to show.

EXAMPLE 2.5. With the present (nonrelativized) definition of the characteristic submanifold one has $V = M$ if and only if M is either a Seifert fibre space or a line bundle over a closed 2-manifold.

EXAMPLE 2.6. Let M be a *graph manifold* in the sense of **[22]**. Then, in general, the following is true: There exists a system T of incompressible tori, unique up to isotopy, with the following properties: (i) if $U(T)$ denotes a regular neighborhood then each component of $\mathrm{Cl}(M - U(T))$ admits the structure of a Seifert fibre space, (ii) no subsystem of T has property (i). In this case $V = \mathrm{Cl}(M - U(T))$. It is disputable if one should not rather adjust the definition of characteristic submanifold so that $V = M$ in this situation. However the smaller V the better are the results one formulates using it. The thing to remember from this example is the following. If one removes V from M, i.e., forms $\mathrm{Cl}(M - V)$, then there may be some 'trivial components' left over, such as $U(T)$ in the example.

THEOREM 2. *Essential singular annuli and tori can be deformed into the characteristic submanifold.*

For example suppose there exists at least one essential singular torus in M. Then the characteristic submanifold V of M cannot be empty, by the theorem. In general one will have $V \neq M$, thus $\mathrm{Cl}(\partial V - \partial M) \neq \emptyset$. But by definition of V, any component of $\mathrm{Cl}(\partial V - \partial M)$ is an essential torus or annulus. Therefore such must exist in M. In the special case when M equals its characteristic submanifold, it is a Seifert fibre space or a line bundle. In Seifert fibre spaces, in general, essential tori do exist in large numbers. But in very special cases there may be none at all, cf. Example 2.2.

The slogan is that it takes many nonsingular annuli or tori to manufacture one singular one. Also one manufactures them in a very special way and still gets them all. That is, essential singular annuli and tori in Seifert fibre spaces and line bundles can be fairly explicitly classified, and in particular any such map can be deformed into the composition of a covering map and an immersion without triple points.

Working in a suitable relative framework, and using the notion of essential map, both of which will be discussed later, one may formulate a corollary giving a version of the theorem for maps of Seifert fibre spaces and line bundles. One way to put the corollary is to say that the universal property of the characteristic submanifold continues to hold for 'singular essential F-manifolds'. In a very special case this amounts to the following: If M has a finite covering which is a Seifert fibre space then M must be a Seifert fibre space itself.

To formulate the main theorem it is convenient to make the following definition.

DEFINITION. Let $f: M \to N$ be a homotopy equivalence, and $M' \subset M$. One says f has *singular support in* M' if and only if there exist $N' \subset N$ and $f': M \to N$ homotopic to f with the following properties

(i) $f'(M') \subset N'$,

(ii) $f'(M - M') \subset N - N'$,

(iii) $f'|M': M' \to N'$ is a homotopy equivalence,

(iv) $f'|\mathrm{Cl}(M - M'): \mathrm{Cl}(M - M') \to \mathrm{Cl}(N - N')$ is a homeomorphism.

Let M, N be as specified earlier.

THEOREM 3. *Every homotopy equivalence* $f: M \to N$ *has singular support in the characteristic submanifold of* M.

The theorem admits an immediate strengthening. Namely if W is a component of V, the characteristic submanifold, and W is in the interior of M, then for the f' above it is true that $f'|W$ is a boundary preserving homotopy equivalence, hence deformable into a homeomorphism. Hence

COROLLARY. *Every homotopy equivalence has singular support in V', the union of those components of V which meet the boundary of M.*

Thus all exotic homotopy equivalences are just modifications of the Examples 1.2, 1.3, 1.4 described earlier.

A special case is when V' is 'trivial', that is, contained in a neighborhood of ∂M (there could be boundary tori). Obviously this is the case if and only if there is no essential annulus in M.

COROLLARY. *No essential annulus, no exotic homotopy equivalence.*

EXAMPLE. In the knot space of a nontrivial knot there exists an essential annulus if and only if the knot is either a composite knot or a cable or torus knot, respectively.

Conversely, in these special cases, it is in general easy to exhibit exotic homotopy equivalences. One may expect to do better if one restricts attention to homotopy equivalences *between knot spaces.* This is clear in the case of torus knots. In the case of cable knots the matter depends on the unresolved status of the unique embedding conjecture, that any embedding of a nontrivial knot space into S^3, can be extended to an automorphism of S^3. In writing the announcement **[26]** I was under the impression that one could get around the unique embedding conjecture by a trick, but in fact one cannot as was shown to me years ago by an explicit construction of hypothetical counterexamples, by John Hempel (the examples were not published then, they have subsequently been rediscovered by J. Simon). At any rate it is not difficult to prove the following; notice the unique embedding conjecture is used twice.

COROLLARY. *If it is true that nontrivial knot spaces have the unique embedding property then noncomposite knots are characterized by their groups.*

Here are some comments on the status of Theorem 3. Given Theorem 2, the proof of Theorem 3 is fairly easy in the case of manifolds whose boundary consists of tori only. The point is that in this case ∂M is actually contained in the characteristic submanifold, so special arguments apply. Unfortunately nothing like this is true in the general case. Indeed the proof of Theorem 3 is quite complicated in the general case.

The main trick by which one makes Theorem 3 provable at all is to formulate a more general assertion. This involves the relative framework of 'manifolds with boundary pattern'. A bonus is that the proof of Theorem 2 becomes relatively easy, in that framework.

3. Manifolds with boundary, revisited. Let M be a compact n-manifold, $n \leqq 3$. A *boundary pattern* for M consists of a set of compact connected $(n-1)$-manifolds in the boundary ∂M which meet nicely, that is, the intersection of any two of them is an $(n-2)$-manifold, the intersection of any three is an $(n-3)$-manifold, and so on.

The boundary pattern $\{F_i\}$ of M is *complete* if $\bigcup F_i = \partial M$. In general one may define the *completed boundary pattern* to consist of $\{F_i\}$ plus the set of connected components of $\mathrm{Cl}(\partial M - \bigcup F_i)$.

A *map* from $(M, \{F_i\})$ to $(N, \{G_j\})$ is a pair of maps $f: M \to N$, $v: \{F_i\} \to \{G_j\}$ such that $f(F_i) \subset G_{v(i)}$ and such that

$$\{F_i\} = \bigcup_j \text{(set of connected components of } f^{-1}(G_j) \cap \partial M).$$

A loose way to phrase this is to say that $\{F_i\}$ must be induced, by means of f, from the boundary pattern $\{G_j\}$. In particular it is never possible that nondisjoint members of $\{F_i\}$ are mapped to the same G_j.

A *homotopy* is a continuous family of maps, in the sense just defined, satisfying that the map of index sets does not change. Having defined 'homotopy' one also has defined 'isotopy', 'homotopy equivalence', and so on.

The unit interval is canonically a 1-manifold with complete boundary pattern. A *singular arc* in $(M, \{F_i\})$ is, by definition, a map $f: (I, \{0, 1\}) \to (M, \{F_i\})$. It is called *inessential* if f can be deformed to a point map (note this may happen even if $f(0)$ and $f(1)$ are in distinct, but adjacent, elements of $\{F_i\}$), otherwise it is called essential. An essential singular curve is a map $f: S^1 \to M$ that cannot be deformed to a point map. Using these notions we may define a map $(M, \{F_i\}) \to (N, \{G_j\})$ to be *essential* if it preserves essential curves and arcs.

DEFINITION 3.1. A boundary pattern $\{F_i\}$ of M is *useful* if and only if for any j the embedding

$$\left(F_j, \bigcup_{i \neq j} \{\text{connected components of } F_i \cap F_j\}\right) \longrightarrow (M, \{F_i\})$$

is an essential map.

This is the proper notion to work with, whence the name.

EXAMPLE. Let *i-faced disk* denote a 2-disk with complete boundary pattern of i elements. This is a 2-manifold with useful boundary pattern only if $i \geqq 4$. A 4-faced disk will be referred to as a *square*; this is isomorphic to $I \times I$ as a manifold with boundary pattern.

REMARK. Call an embedding of an i-faced disk, $i \leqq 3$, in $(M, \{F_i\})$ 'uninteresting' if it is isotopic to an embedding into ∂M so that $D \cap \bigcup \partial F_i$ is isomorphic to the cone on $\partial D \cap \bigcup \partial F_i$. There is a version of the loop theorem for 3-manifolds with boundary pattern (it is more or less equivalent to the main technical result of **[25]**). It says that (except for a few degenerate cases) usefulness is equivalent to the nonexistence of interesting i-faced disks, $i \leqq 3$.

EXAMPLE. Let $M = S^1 \times D^2$, the solid torus. Let $F \subset \partial M$ be an annulus, with winding number w (that is, $\pi_1 F$ has index w in $\pi_1 M$). Then the completed boundary pattern of $\{F\}$ is useful if and only if $w \geqq 2$.

REMARK. Still following our earlier convention that the 3-manifolds under consideration are compact, orientable, irreducible, and sufficiently large, let f: $(M,\{F_i\}) \to (N, \{G_j\})$ be a homotopy equivalence of 3-manifolds with boundary patterns which are both complete and useful. Then f is homotopic to a homeomorphism. This can be seen by an adaptation of the argument of **[24]**. Indeed the adaptation clarifies the argument.

We will now adapt the notion of essential F-manifold to 3-manifolds with boundary pattern. To avoid introducing even more language this will be done only in the case of complete boundary patterns.

DEFINITION 3.2. Let M be a 3-manifold with boundary pattern $\{F_i\}$, both complete and useful. An *essential F-manifold* in $(M, \{F_i\})$ is an embedded 3-manifold with boundary pattern

$$\left(V, \bigcup_i \{\text{connected components of } V \cap F_i\}\right)$$

whose completed boundary pattern is useful. Furthermore for each component W of V, and its induced boundary pattern, at least one of the following must be true; either

(a) W admits a structure of Seifert fibre space, with fibre projection $p\colon W \to B$, such that the boundary pattern of W is induced, by means of p, from some boundary pattern of B; or

(b) W admits a structure of line bundle, $p\colon W \to B$, such that the boundary pattern of W consists of the components of the associated 0-sphere bundle, plus the boundary pattern induced, by means of p, from some boundary pattern of B.

One defines a *characteristic submanifold* of $(M, \{F_i\})$ as an essential F-manifold having a certain universal property, just as before.

THEOREM 1′. *The characteristic submanifold exists and is unique up to isotopy (in fact, ambient isotopy of manifolds with boundary pattern).*

EXAMPLE. If $\{\text{connected components of } \partial M\}$ is a useful boundary pattern for M (that is, if ∂M is incompressible) the characteristic submanifold in the present sense coincides with the one defined previously, except that now the set of connected components of its intersection with ∂M has been designated as boundary pattern. Indeed the later fact is crucial in translating the notion of 'essential' from one setting to the other.

DEFINITION 3.3. Let $(M, \{F_i\})$ be any manifold with boundary pattern. Let N be a codimension zero submanifold of M such that $N \cap \partial M$ is a codimension zero submanifold of ∂M, in general position with respect to $\{F_i\}$. Then N is naturally endowed with what we refer to as its *proper boundary pattern*, given by

$$\bigcup_i \{\text{connected components of } V \cap F_i\}$$
$$\cup \{\text{connected components of } \mathrm{Cl}(\partial N \cap \mathrm{Int}(M))\}.$$

Note that the inclusion of N is not a map of manifolds with boundary patterns, in general.

Notation 3.4. Let $(M, \{F_i\})$ be a 3-manifold with boundary pattern, not necessarily complete. Assume the completed boundary pattern is useful. Then the characteristic submanifold V may be constructed, with respect to this completed boundary pattern. V may be endowed with its proper boundary pattern, with respect to $\{F_i\}$. It is this, V together with its proper boundary pattern, which by an abuse of language we will refer to as the *characteristic submanifold* of $(M, \{F_i\})$.

3.5. Similarly, $\mathrm{Cl}(M - V)$ may be endowed with its proper boundary pattern. It will, in general, have certain 'trivial components' of the type considered in Example 2.6; any such trivial component (with its completed boundary pattern) is isomorphic to either $S^1 \times S^1 \times I$, or $S^1 \times I \times I$, or $I \times I \times I$, respectively.

Throwing away the trivial components one obtains a manifold with boundary pattern, $(M^*, \{F_j^*\})$, say, which is referred to as being obtained from $(M, \{F_i\})$ by *splitting at its characteristic submanifold.*

LEMMA. *$(M^*, \{F_j^*\})$ is a manifold with useful boundary pattern. Furthermore it is simple in the sense that any component of its characteristic submanifold is contained in a neighborhood of one of the F_j^*, or of a component of* $\mathrm{Cl}(\partial M^* - \bigcup F_j^*)$.

REMARK. If $(M, \{F_i\})$ is simple then splitting at its characteristic submanifold does not change it, up to isomorphism.

3.6. Let $(M, \{F_i\})$ be a 3-manifold with useful boundary pattern. We consider incompressible surfaces S in M. We insist on considering only surfaces S not separating M (resp., ∂S not separating ∂M) provided there is at least one such surface. The latter of these is the case if M has at least one component that is neither closed nor a ball. We also insist that S be in general position with respect to $\{F_i\}$. A numerical function $c(S)$, the *complexity*, is defined by

$$c(S) = (\text{number of points } S \cap \bigcup \partial F_i) + 5 \cdot (\text{first Betti number of } S).$$

Let $U(S)$, a regular neighborhood, and $\mathrm{Cl}(M - U(S))$ both be endowed with their proper boundary patterns. To the latter we refer as the manifold obtained by *splitting* $(M, \{F_i\})$ *at* S.

LEMMA. *Let S be such that $c(S)$ is minimal. Then the proper boundary patterns of $U(S)$ and* $\mathrm{Cl}(M - U(S))$, *respectively, are useful.*

From now on, the surfaces involved in a 3-manifold with boundary pattern will be dropped from the notation.

Let M_1 be a 3-manifold with boundary pattern, satisfying that the completed boundary pattern is useful. One forms M_2 by splitting M_1 at its characteristic submanifold, as in 3.5. In M_2 one picks some incompressible surface, nonseparating (etc.) if possible, of minimal complexity, and forms M_3 by splitting M_2 at S, as in 3.6. In general one forms M_{2i} by splitting M_{2i-1} at the characteristic submanifold, and M_{2i+1} by splitting M_{2i} at some S of minimal complexity. The process must stop after a finite number of steps (in the sense that any component of the manifold left over is some ball with boundary pattern); in fact, the argument of Haken [**4**] gives an explicit upper bound for this number. All the M_j in the sequence satisfy that the completed boundary pattern is useful; all the M_{2i} are simple.

DEFINITION 3.7. Any sequence $M_1, M_2, M_3, \cdots$ obtained in this way is called a *great hierarchy* for M_1.

It is by induction on a great hierarchy that one proves Theorems 2 and 3. In the inductive step one uses the following notion about 2-manifolds.

DEFINITION AND LEMMA. *Let F be a 2-manifold with complete boundary pattern, and let F_1 and F_2 be essential 2-submanifolds of F. Then there exists an essential 2-submanifold F_0 of F, unique up to isotopy, with the following properties*:

(i) *F_0 can be isotoped into both of F_1 and F_2;*

(ii) *any essential curve or arc in F, possibly singular, that can be deformed into both F_1 and F_2, respectively, can also be deformed into F_0;*

(iii) *no proper subcollection of components of F_0 has property* (ii).

F_0 is called the virtual intersection of F_1 and F_2.

One uses it in the following way.

3.8. Let $M_1, M_2, M_3, \cdots$ be a great hierarchy. Let M be the component of M_{2i} that contains the surface S, let $M' = \mathrm{Cl}(M - U(S))$ and V' the characteristic submanifold of M'. One desires to construct a submanifold P of M that consists of 'nice' pieces of components of V', fitting properly together across $U(S)$, and so that P is as large as possible. Here is a rough sketch of the construction. The components of V' that are Seifert fibre spaces can meet $\partial U(S)$ in a very special way only (in a neighborhood of a system of curves); for simplicity we assume there are none. For simplicity we assume further that any component of V' is a trivial line bundle rather than a nontrivial one (since S has minimal complexity this is in fact automatically true if S is nonseparating). Identify $U(S)$ with $S \times I$. Then by inductive application of the preceding lemma one finds that there is a largest sub-line-bundle V'' of V' satisfying that the virtual intersection, in S, of $V'' \cap (S \times 0)$ and $V'' \cap (S \times 1)$, is represented by these two surfaces themselves. Thus the components of V'' can now be fitted together, across $U(S)$, to form P. It is immediate from the construction that any component of P fibres over a 1-manifold, with fibre a 2-manifold. But M was assumed simple. So looking at ∂P one sees that in fact only two outcomes of the construction are possible: Either P is 'trivial' (that is, contained in a neighborhood of some surfaces of the completed boundary pattern), or P is essentially all of M, and M fibres over S^1. The preceding process will be referred to as *combing* of V'. In general, without the special assumptions we made, the process is more complicated. But one can still conclude that either P is trivial, in the above sense, or that P is essentially all of M; and in the latter case M either fibres over S^1, or is a union of two twisted line bundles glued at S (and in particular, some 2-sheeted covering of M fibres over S^1).

THEOREM 2′, *Let M_1 be a 3-manifold with boundary pattern, both complete and useful. Then any essential singular torus, annulus, or square in M_1 can be deformed into the characteristic submanifold.*

REMARK. This includes Theorem 2 as the special case where the boundary pattern equals the set of connected components of ∂M_1.

INDICATION OF PROOF. One uses induction on a great hierarchy $M_1, M_2, M_3, \cdots$. The induction beginning is with a ball with boundary pattern. But here any essential square (the only case that can occur!) can be deformed into a nonsingular essential square, and hence into the characteristic submanifold, in view of the universal property defining the latter. The inductive step from M_{2i} to M_{2i-1} is of similar calibre. Where one really has to prove something is the step from M_{2i+1} to M_{2i}. Recall that M_{2i+1} is obtained from M_{2i} by splitting at some incompressible surface S, of minimal complexity, and nonseparating if possible. One assumes, contrary to the assertion, that there is an essential singular torus, say, call it f, that can neither be deformed off S, nor into a nonsingular torus in M_{2i}, this being necessarily some surface of the boundary pattern since M_{2i} is simple. In view of the induction hypothesis one may assume that the image of f is contained in the union of the regular neighborhood $U(S)$ and the characteristic submanifold of M_{2i+1}. In fact, one can apply the process of 'combing' of 3.8, and finds $\mathrm{Im}(f)$ can be contained in the manifold P constructed by combing. Thus P is nontrivial. Thus the component of M_{2i} that contains $\mathrm{Im}(f)$ fibres over S^1 (or at least some 2-sheeted covering does).

In this special case one produces a special proof, mainly by reading Nielsen [**14**]. Thus M_{2i} is not simple, a contradiction.

THEOREM 3′. *Let* f_1: $M_1 \to N_1$ *be a homotopy equivalence of manifolds with boundary patterns. One assumes the completed boundary patterns are useful. Then* f_1 *has singular support in the characteristic submanifold.*

REMARK. This includes Theorem 3 as the special case of an empty boundary pattern.

INDICATION OF PROOF. One uses induction on a great hierarchy M_1, M_2, M_3, $\cdots$. Actually additional conditions of a technical nature have to be asked of the surfaces S involved in the great hierarchy; these will here be tacitly assumed. One first has to go down the hierarchy. This uses, inductively,

LEMMA. f_1 *can be deformed to be a homotopy equivalence both on the characteristic submanifold and its complement, either being endowed with the proper boundary pattern.*

LEMMA. f_2: $M_2 \to N_2$ *can be deformed to be a homotopy equivalence both on* $U(S)$ *and* $\mathrm{Cl}(M_2 - U(S))$, *either being endowed with the proper boundary pattern.*

Next one has to establish the induction beginning, i.e., prove the theorem in the case when the manifolds are balls with boundary patterns, and simple. This is not entirely trivial (the argument is a pleasant exercise on the Jordan curve theorem). And finally one has to work up the hierarchy again, i.e., assuming the theorem is true for f_{2i+2} (and hence also for f_{2i+1}), one must show that f_{2i}: $M_{2i} \to N_{2i}$ can be deformed into a homeomorphism. Again one invokes the notion of virtual intersection, on S, the surface such that $M_{2i+1} = \mathrm{Cl}(M_{2i} - U(S))$. This uses

LEMMA. *Let* f: $F \to G$ *be a homotopy equivalence of* 2*-manifolds with boundary patterns. Let* F' *denote* F *with its completed boundary pattern, and let* F_1, F_2 *be essential* 2*-submanifolds of* F'. *If* F_1 *and* F_2 *are singular supports for* f *then so is their virtual intersection.*

By the lemma, the homotopy equivalence $f_{2i}|S$ has a unique minimal singular support. The problem is to show this is empty.

One now applies the process of 'combing' of 3.8. The submanifold P produced must contain the minimal singular support of $f_{2i}|S$, by the lemma. As pointed out in 3.8, only two cases are possible for P since M_{2i} is simple. If P is trivial we are done. If not, P is essentially all of the component of M_{2i} that contains S, and this component is of a very special kind. So a special argument applies.

The burden of the proof is in establishing the above lemmas.

4. Classifications. We must insist here that 3-manifolds shall be given in some particular, and effective, way. Thus a 'compact 3-manifold' shall mean a finite simplicial complex of a particular kind [**19**]. These form a recursive set, i.e., they 'can be listed', M_1, M_2, $\cdots$; furthermore, given two, it is trivial to decide, by inspection, if these two are isomorphic.

The *homeomorphism problem* is to give a recipe which, given M, M', decides if M and M' are PL isomorphic, i.e., if M and M' have isomorphic subdivisions.

The *classification problem* is to give a list of representatives, one from each PL isomorphism class.

The two problems are equivalent, by what Haken calls the cheapological trick. Indeed, given a solution to the former, one constructs the list $M_1, M_2, \cdots$, above, but at each step one inquires if a PL isomorphic manifold has been listed before, and if so, drops the new one. Conversely, given a solution of the latter in terms of representatives $M_1, M_2, \cdots$, one from each PL isomorphism class, then, given M', one generates the list of manifolds PL isomorphic to M' (by subdivision and its converse) until one finds the place of M' in the list $M_1, M_2, \cdots$; similarly one finds the place of M and thus sees if it is PL isomorphic to M' or not.

To clarify the meaning of this kind of classification one best considers an example pointed out by H. Schubert. Namely a finite group may be specified by a particular kind of multiplication table. It can be decided by inspection if two such multiplication tables define isomorphic groups. Hence one can make a list of finite groups $G_1, G_2, \cdots$, one from each isomorphism class. Similarly one can make a sublist giving the simple groups, or sporadic simple groups, respectively. But neither is this procedure very practical for groups of order exceeding 10^{80}, say, nor is it surprising that, in this case, the procedure just fails to answer any interesting question whatsoever. For example, from the recipe how to make the list of sporadic simple groups one cannot infer, even theoretically, if this list is finite.

In contradistinction to this example there are, in the case of 3-manifolds, at least two reasons for attempting just this kind of classification. Firstly there is no classification whatsoever of finitely presented groups and hence (Markov [**13**], cf. also [**1**]) of compact manifolds of dimension exceeding 3. Secondly, any kind of classification of a sufficiently large class of 3-manifolds must invariably be tied up with some interesting structure theory.

The idea of Haken's approach to the homeomorphism problem [**4**] may be put as follows. Let M be given and suppose it is actually 'known' that M has certain desirable properties, in particular that it is irreducible and sufficiently large (as there is no algorithm yet to decide if these properties hold, they must be 'known' in advance so that one may use them in constructions involved in algorithms).

Step 1. Consider the hierarchies of M which are as simple as possible, show there are only finitely many, up to isomorphism, and produce a list. (Note that one does not classify here up to isotopy as the number of hierarchies would then be infinite, in general. To have this stronger finiteness one needs to work with something like great hierarchies, cf. below.)

Step 2. Given N similarly, consider any pair of hierarchies of M and N, respectively, and decide, by inspection, if they match.

Haken has shown [**4**], [**6**] that this idea can be made to work 'in general'. The exceptional phenomenon is very special indeed, but unfortunately one does not have much control on its occurrence. It concerns embedded submanifolds which fibre over S^1, have incompressible boundary, and do not contain any incompressible surfaces apart from those isotopic to a fibre or a component of the boundary.

To illustrate the phenomenon suppose that M itself fibres over S^1, that ∂M is incompressible (possibly empty), and that the only incompressible surfaces are those isotopic to the fibres or the boundary components, respectively. If M' is similar then any homeomorphism $M \to M'$ must be isotopic to a fibre preserving one. So, presenting M as the mapping torus of a homeomorphism $f\colon F \to F$

(where F is the fibre of M), and M' similarly, the problem to decide if M and M' are homeomorphic is equivalent to deciding the following problem.

Problem. Given (F, f) and (F', f'), does there exist a homeomorphism $h: F \to F'$ so that hf is isotopic to $f'h$?

That is, one wants a solution to the conjugacy problem in the group of isotopy classes of automorphisms of a compact 2-manifold. Actually, one needs the solution only in a special case, but the special assumption seems hard to use, in general.

This problem eluded solution for a long time until very recently G. Hemion solved it at least in the case of nonempty boundary [7]. The solution is as follows.

Let $\tilde{F} \to F$ be a universal covering for F, and $\Delta \subset \tilde{F}$ a *fundamental domain*, that is, an embedding of the 2-disk so that $\Delta \to F$ is surjective. It is convenient to assume that $\Delta \to F$ is particularly nice, but this is not really relevant. For any f : $F \to F$, and any lifting $\tilde{f}: \tilde{F} \to \tilde{F}$, there exist elements $g_1, \cdots, g_n$ of the covering translation group so that $\tilde{f}(\Delta) \subset \bigcup_j g_j(\Delta)$. The minimal number of such group elements is denoted $d_\Delta(f)$, the *diameter of f with respect to Δ*. It depends on $\Delta \to F$, but not on any other choices.

If f fixes a point x, and $\tilde{x} \subset \tilde{F}$ is the pre-image of x, then f is determined, up to isotopy, by the permutation of $\tilde{x}$ induced by some lifting $\tilde{f}$. This makes it clear that for any given n' there exist only finitely many f, up to isotopy, with $d_\Delta(f) \leqq n'$, and furthermore that there is an effective procedure (by trial and error, say) to construct them all (or at least, to construct a slightly larger set).

THEOREM 4 (HEMION). *Suppose $\partial F \neq \varnothing$. Let $f, f': F \to F$ be given. Suppose f has neither periodic arcs, up to homotopy, nor periodic curves not deformable into ∂F. Suppose $h: F \to F$ satisfies that hf is isotopic to $f'h$. Then for some integer m, and some h' isotopic to hf^m, it is true that*

$$d_\Delta(h') \leqq \$(d_\Delta(f), d_\Delta(f'))$$

where \$ is some explicitly given function of two variables.

In view of known results (implicitly used below) the general solution of the above problem follows from this theorem if $\partial F \neq \varnothing$.

The solution of the homeomorphism problem will now be described in a way that explicitly uses great hierarchies. As a bonus, additional results can be obtained on the classification of homeomorphisms. Let M be as before: It is (known to be) irreducible, and if it is closed it is also (known to be) sufficiently large. If $\partial M \neq \varnothing$ we assume M equipped with some boundary pattern $\{F_i\}$ which is both complete and useful (an algorithm of Haken can be used to check the latter). As before, S denotes an incompressible surface in M (nonseparating, etc., if such exist at all) with complexity $c(S)$ defined by

$$\left(\text{number of points } S \cap \bigcup \partial F_i\right) + 5 \cdot (\text{first Betti number of } S).$$

THEOREM 5 ([2], [3], [4]). (i) *There is an algorithm to construct the characteristic submanifold of M.*

(ii) *If M is simple, cf.* 3.5, *then there is only a finite number, up to ambient isotopy,* *of surfaces S of minimal complexity. There is an algorithm which produces one such surface from each isotopy class.*

REMARK. An algorithm of Haken can be used to decide if two incompressible surfaces are isotopic. Thus the final assertion of the theorem may be strengthened to say that precisely one S is produced from each isotopy class.

Let $(M_1, M_2, M_3, \cdots)$ and $(M'_1, M'_2, M'_3, \cdots)$ be great hierarchies, in the sense of 3.10. An *isomorphism* from one to the other is a sequence of PL isomorphisms $f_j\colon M_j \to M'_j$ such that

(i) for $j = 2i$, and the surfaces S and S' used in passage from $2i$ to $2i+1$, the surfaces $f_j(S)$ and S' in M'_j are ambient isotopic;

(ii) for any j, f_{j+1} is induced, up to isotopy, from f_j.

If $M_1 = M'_1$ and f_1 is the identity, the isomorphism will be called an *isotopy*. In view of the fact that the lengths of great hierarchies of M are uniformly bounded above [**4**], Theorem 5 thus gives

COROLLARY. *There is only a finite number, up to isotopy, of great hierarchies of M. There is an algorithm which produces precisely one from each class.*

Let $\mathscr{H}(M, N)$ denote the set of isotopy classes of homeomorphisms from M to N, and $\mathscr{H}(M)$ the group $\mathscr{H}(M, M)$. If $h\colon M \to M$ is a homeomorphism, we will say h has *support in* $M' \subset M$ if there exists h' isotopic to h so that $h'|M - M'$ is the identity; these form the subgroup $\mathscr{H}_{M'}(M)$ of $\mathscr{H}(M)$. For example, a Dehn twist is an automorphism of a 2-manifold with support in the neighborhood of an embedded circle. Similarly, an automorphism of a 3-manifold may be called a *Dehn twist* if it has support in the neighborhood of an essential annulus or torus.

We denote by $bp(M)$ the set of surfaces involved in the boundary pattern of M, and by $hier(M)$ the set of isotopy classes of 'oriented' great hierarchies (that is, M, and each of the surfaces S involved, is endowed with some orientation).

THEOREM 6. *Let M_2 be connected and simple. Suppose M_2 does not fibre over S^1, nor, if it is closed, that it is the union of two twisted line bundles. Then*

$$\mathscr{H}(M_2) \to \mathrm{Aut}(bp(M_2)) \times \mathrm{Aut}(hier(M_2))$$

is injective. If N_2 is similar then the image of $\mathscr{H}(M_2, N_2)$ in

$$\mathrm{Hom}(bp(M_2), bp(N_2)) \times \mathrm{Hom}(hier(M_2), hier(N_2))$$

is a computable set; in particular it can be decided whether or not $\mathscr{H}(M_2, N_2)$ is empty.

INDICATION OF PROOF. Let $f_2\colon M_2 \to M_2$ be such that $bp(f_2)$ is the identity, and $hier(f_2)$ has a fixed point. By induction on the length of the great hierarchy, f_2 must then be isotopic to the identity, as follows. Let $M_2, M_3, \cdots$ represent this fixed point, where $M_3 = \mathrm{Cl}(M_2 - U(S))$, etc. It may be assumed that $f_2(S) = S$, $f_2(M_3) = M_3$, $f_2(M_4) = M_4$, and, by the inductive hypothesis, that $f_2|M_4$ is the identity. It is clear, more or less, that it suffices to prove $f_2|S$ is isotopic to the identity. One uses

LEMMA. *Let $f\colon S \to S$ be an automorphism of a 2-manifold with complete bound-*

ary pattern. Let F_1, F_2 be essential 2-submanifolds of S. If F_1 and F_2 are supports for f then so is their virtual intersection.

One may thus apply the process of combing the characteristic submanifold of M_3, cf. 3.8, and the submanifold P produced will contain the minimal support of $f_2|S$, by the lemma. But P must be trivial, in view of the hypothesis that M_2 does not fibre, etc.

The second part of the theorem follows since the combing process is really a constructive method, and one can thus compare, by inspection, the data in M_2 and N_2, respectively.

THEOREM 7. *Let M_2 be connected and simple. Suppose that $\partial M_2 \neq \emptyset$ and that M_2 fibres over S^1. Let N_2 be similar. Then $\mathscr{H}(M_2, N_2)$ is a finite computable set.*

PROOF. By Haken, there are, up to isotopy, only finitely many fibrations of M_2 whose fibre is of minimal complexity, and all of these can be constructed. It thus suffices to consider only fibre preserving homeomorphisms. Let M_2 be the mapping torus of a homeomorphism $f\colon F \to F$. Then f does not have periodic arcs or curves, up to homotopy, which cannot be deformed into ∂F because otherwise M_2 would not be simple (by Theorem 2, or really Nielsen [**14**]). The assertion is thus immediate from Theorem 4.

In the following we let M be a connected 3-manifold as specified earlier, but we exclude these two cases:

(i) M is closed and simple and fibres over S^1;

(ii) M is closed and simple and is the union of two line bundles.

Note that it can be checked by Haken's algorithm if M is one of these. $\mathscr{H}_V(M)$ denotes the normal subgroup of $\mathscr{H}(M)$ of homeomorphism classes with support in V, the characteristic submanifold of M.

THEOREM 8. *If M is as just specified, and N similarly, then $\mathscr{H}_V(M)\backslash\mathscr{H}(M, N)$ is a finite computable set.*

INDICATION OF PROOF. Let M' be obtained from M by splitting at the characteristic submanifold, and N' similarly. The hypothesis about M implies that each component of M' satisfies the hypothesis of either Theorem 6 or 7. Thus $\mathscr{H}(M', N')$ is a finite computable set. By definition, M' was obtained from $\mathrm{Cl}(M - V)$ by discarding certain trivial components. Thus we must now add V to M', and those trivial components. For example, one component of $\mathrm{Cl}(M - M')$ could be a graph manifold in the sense of [**22**], cf. Example 2.6 above. To proceed one can, e.g., use explicit knowledge about such special manifolds and their homeomorphisms.

COROLLARY. *Let M denote a compact orientable irreducible 3-manifold with nonempty incompressible boundary.*

1. *These M can be classified* (*by Theorem* 8).
2. *The M' homotopy equivalent to M can be classified* (*by Theorem* 3).
3. *These $\pi_1 M$ can be classified* (*by* 1 *and* 2).
4. *Knot groups can be classified* (*by* 3).
5. *Knots can be classified* (*by* 1, *and inclusion of a meridian in the data*).

COROLLARY. *If M is as in Theorem* 8, $\mathscr{H}(M)/\mathscr{H}_V(M)$ *is a finite computable group. In particular if M is simple,* $\mathscr{H}(M)$ *is a finite computable group and hence so is the group of automorphism classes of* $\pi_1 M$, *by Theorem* 3.

The following remarks show that $\mathscr{H}_V(M)$ is also computable though in general not finite. Let N be a Seifert fibre space with decomposition surface B. Let G denote the group of fibre preserving homeomorphisms modulo fibre preserving isotopy. It was indicated in [**24**] that, in general, the map $G \to \mathscr{H}(N)$ is an isomorphism (the exceptions are (i) those of Example 1.4, (ii) the Seifert fibre spaces of 2.2, plus alternative Seifert fiberings of these, (iii) finitely many others, e.g., the 3-torus). Let the chain of subgroups $G_3 \subset G_2 \subset G_1 \subset G$ be defined by

(G_1) h is orientation preserving,

(G_2) h maps each exceptional fibre to itself, by an orientation preserving map,

(G_3) h maps each fibre to itself, by an orientation preserving map.

Then G_1/G_2 is always finite. In the case where B is orientable, it is a product of permutation groups (namely those exceptional fibres may be permuted that can be permuted) and possibly a Z_2. In the other case, the structure is slightly more complicated in that each exceptional fibre may be flipped individually.

$G_2/G_3 \xrightarrow{\approx} \mathscr{H}(B)$ where B is to be considered as a pointed 2-manifold, pointed by the exceptional fibres. Generators for this group have long been known, in particular $\mathscr{H}(B)$ has a subgroup of finite index which is generated by Dehn twists. A system of relators has recently been obtained by Hatcher and Thurston (not yet published), in particular it is now known that $\mathscr{H}(B)$ is finitely presented.

$G_3 \xrightarrow{\approx} H_1(B, \partial B)$ (with two exceptions; cf. below). A cocycle in the dual group $H_1(B; Z)$ is represented by a section in a certain S^1-bundle over B; the section can be interpreted to measure how the associated element of G_3 rotates the nonexceptional fibres. If k is a nonsingular arc in B not containing an exceptional point then the element of $H^1(B, \partial B)$ represented by k corresponds to a primitive Dehn twist along the annulus over k; similarly, if k is a closed curve, it corresponds to a primitive Dehn twist at a torus, in fibre direction.—The exceptions are given by the S^1-bundles over the torus and the Klein bottle, respectively. The exceptional phenomenon is that in these cases there exist nontrivial isotopies which slide the fibres around (such exceptional isotopies exist in three more cases, but here they do not do anything). The effect is that $H_1(B, \partial B)$ has to be replaced by the quotient group

$$H_1(B, \partial B) \otimes Z/nZ$$

where n is the Euler number in the case where B is the torus, and twice the Euler number in the case of the Klein bottle.

Similar but simpler considerations apply to line bundles. Putting together these considerations involving the work of Haken, Hemion, Johannson, Hatcher-Thurston, one has

COROLLARY. *Let M be as in Theorem* 8, *but in addition exclude the case where M is a Seifert fibre space over* S^2 *with precisely three exceptional fibres* (*that is, one must explicitly exclude now only those which are sufficiently large*). *Then* $\mathscr{H}(M)$ *is a finitely presented computable group.*

REMARK. In the case additionally excluded in the corollary, it is true that $\mathscr{H}(M)$ is isomorphic to the group of automorphism classes of $\pi_1 M$. Therefore results of Zieschang **[29]** show that the corollary extends to this case.

COROLLARY. *Let M be as in the preceding corollary. Then the normal subgroup of $\mathscr{H}(M)$ generated by Dehn twists at essential annuli and tori is of finite index.*

REFERENCES

1. W. W. Boone, W. Haken and V. Poénaru, *On recursively unsolvable problems in topology and their classification*, Contributions to Math. Logic, North-Holland, Amsterdam, 1968, pp. 37–74. MR **41** #7695.

2. W. Haken, *Theorie der Normalflächen*, Acta Math. **105** (1961), 245–375. MR **25** #4519A.

3. ———, *Ein Verfahren zur Aufspaltung einer* 3-*Mannigfaltigkeit in irreduzible* 3-*Mannigfaltigkeiten*, Math. Z. **76** (1961), 427–467. MR **25** #4519C.

4. ———, *Über das Homöomorphieproblem der* 3-*Mannigfaltigkeiten*. I, Math. Z. **80** (1962), 89–120. MR **28** #3410.

5. ———, *Some results on surfaces in* 3-*manifolds*, Studies in Modern Topology, MAA Studies in Math., vol. 5, Prentice-Hall, Englewood Cliffs, N. J., pp. 39–98. MR **36** #7118.

6. ———, *Connections between topological and group theoretical decision problems*, Word Problems, North-Holland, Amsterdam, 1973, pp. 427–441.

7. G. Hemion, *On the classification of homeomorphisms of* 2-*manifolds and the classification of* 3-*manifolds*, Univ. Bielefeld (In prep.).

8. K. Johannson, *On essential embeddings of annuli and tori in irreducible* 3-*manifolds which are sufficiently large*, Univ. Bielefeld, Aug. 1974 (preliminary version of **[10]**).

9. ———, *Equivalences d'homotopie des variétés de dimension* 3, C. R. Acad. Sci. Paris Sér. A **281** (1975), 1009–1010. MR **52** #11918.

10. ———, *Homotopy equivalences of knot spaces*, Univ. Bielefeld, Feb. 1976.

11. ———, *Homotopy equivalences of* 3-*manifolds with boundary*, Univ. Bielefeld, Aug. 1976.

12. H. Kneser, *Geschlossene Flächen in dreidimensionalen Mannigfaltigkeiten*, Jber. Deutsch. Math.-Verein. **38** (1929), 248–260.

13. A. A. Markov, *Insolubility of the problem of homeomorphy*, Proc. Internat. Congress Math. (1958). Cambridge Univ. Press, London and New York, 1960, pp. 300–306. (Russian) MR **22** #5962.

14. J. Nielsen, *Untersuchungen zur Topologie der geschlossenen zweiseitigen Flächen*. II, Acta Math. **53** (1929), 1–76.

15. ———, *Abbildungsklassen endlicher Ordnung*, Acta Math. **75** (1943), 23–115. MR **7**, 137.

16. C. D. Papakyriakopoulos, *On Dehn's lemma and the aspherity of knots*, Ann. of Math. (2) **66** (1957), 1–26. MR **19,** 761.

17. ———, *On solid tori*, Proc. London Math. Soc. (3) **7** (1957), 281–299. MR **19,** 441.

18. H. Seifert, *Topologie dreidimensionaler gefaserter Räume*, Acta Math. **60** (1933), 147–238.

19. H. Seifert and W. Threlfall, *Lehrbuch der Topologie*, Teubner, Leipzig, 1934.

20. J. R. Stallings, *On the loop theorem*, Ann. of Math. (2) **72** (1960), 12–19. MR **22** #12526.

21. ———, *On fibering certain* 3-*manifolds*, Topology of 3-manifolds and Related Topics, Prentice-Hall, Englewood Cliffs, N. J., 1962, pp. 95–100. MR **28** #1600.

22. F. Waldhausen, *Eine Klasse von* 3-*dimensionalen Mannigfaltigkeiten*. I, II, Invent. Math. **3** (1967) 308–333; ibid. **4** (1967), 87–117. MR **38** #3880.

23. ———, *Gruppen mit Zentrum und* 3-*dimensionale Mannigfaltigkeiten*, Topology **6** (1967), 505–517. MR **38** #5223.

24. ———, *On irreducible* 3-*manifolds which are sufficiently large*, Ann. of Math. (2) **87** (1968), 56–88. MR **36** #7146.

25. ———, *The word problem in fundamental groups of sufficiently large* 3-*manifolds*, Ann. of Math. (2) **88** (1968), 272–280. MR **39** #2167.

26. F. Waldhausen, *On the determination of some bounded* 3-*manifolds by their fundamental groups alone*, Proc. Sympos. on Topology and its Applications, Herceg-Novi, 1968 (Beograd, 1969), 331–332. MR **42** #2416.

27. ———, *Some problems on* 3-*manifolds*, these PROCEEDINGS, part 2, pp. 313–322.

28. J. H. C. Whitehead, *On* 2-*spheres in* 3-*manifolds*, Bull. Amer. Math. Soc. **64** (1958), 161–166. MR **21** #2241.

29. H. Zieschang, E. Vogt and H. D. Coldewey, *Flächen und ebene diskontinuierliche Gruppen*, Lecture Notes in Math., vol. 122, Springer-Verlag, Berlin and New York, 1970. MR **41** #6986.

UNIVERSITÄT BIELEFELD

Proceedings of Symposia in Pure Mathematics
Volume 32, 1978

ON SLICE KNOTS IN DIMENSION THREE

A. J. CASSON AND C. McA. GORDON*

1. Introduction. Under the equivalence relation of concordance (sometimes called cobordism), smooth knots in the 3-sphere S^3 form an abelian group with respect to connected sum **[4]**. The knots K representing the zero class are precisely those which are *slice*, that is, satisfy $(S^3, K) = \partial(B^4, D)$ for some smooth 2-disc D in the 4-ball B^4. Now associated with any knot K and a Seifert surface V spanning K, is a bilinear Seifert pairing $\theta_V: H_1(V) \times H_1(V) \to \boldsymbol{Z}$ **[12]**, **[6]**. We say that K is *algebraically slice* if θ_V vanishes on a subgroup of $H_1(V)$ whose rank is $\frac{1}{2}$ rank $H_1(V)$ (this condition is independent of the choice of V). It is known that a necessary condition for K to be slice is that it be algebraically slice. Moreover, in higher (odd) dimensions analogous definitions may be made, and there the conditions are equivalent **[6]**. We shall show that this is not the case in dimension 3.

The Seifert pairing (up to appropriate equivalence) and a fortiori the 'algebraic concordance' class of K, is determined by the Blanchfield linking pairing on $H_1(\tilde{X})$, where $\tilde{X}$ is the universal abelian cover of the complement X of K **[14]**. Our 'second order' obstructions may be regarded as arising from certain cyclic covers of $\tilde{X}$, or (as in the present paper), from certain metacyclic branched covers of (S^3, K). In particular, our method provides potentially nontrivial obstructions to null-concordance for any knot with Alexander polynomial $\Delta(t) \neq 1$. (Whether or not there exist knots with $\Delta(t) = 1$ which are not slice is an interesting open question.)

The present paper and the earlier account **[2]** are related as follows. First, a fairly simple method was found for showing that certain (algebraically slice) knots K were not ribbon knots, using signatures associated with certain cyclic covers of (say) the 2-fold branched cover of (S^3, K) (see **[2]**). This was extended to give a necessary condition for K to be slice, in terms of the behavior as $n \to \infty$ of the corresponding invariants associated with the 2^n-fold branched cyclic cover of (S^3, K) (see §4). Calculations for certain specific examples, however, disclosed a multiplicativity in

AMS (*MOS*) *subject classifications* (1970). Primary 55A25; Secondary 57D90.

*Supported in part by NSF grant MCS72–05055 A04.

the invariants which showed that this behaviour was determined by the 2-fold branched cover (see §5). The search for an explanation of this phenomenon led to the approach in [2]. Thus the purpose of the present paper is to fill the existing historical gap, to motivate [2], and to provide variety. It should also make clear the relationship to earlier work of Massey [8], Hsiang-Szczarba [5], and Rohlin [10].

Organization is as follows. In §2 we use the Atiyah-Singer G-signature theorem [1] to associate with a 3-manifold M and an epimorphism $\varphi\colon H_1(M) \to \boldsymbol{Z}_m$, certain rational numbers $\sigma_r(M, \varphi)$, $0 < r < m$. In §3 we show that if M is obtained by surgery on a link L in S^3 (and if φ is appropriately related to L), then $\sigma_r(M, \varphi)$ may be expressed in terms of standard invariants of L, in particular, signatures of the type introduced by Tristram [13]. In §4 we establish a necessary condition, in terms of certain $\sigma_r(M, \varphi)$, for a knot K to be slice. More precisely, we consider, for some fixed prime q, the q^n-fold branched cyclic cover M_n of (S^3, K), and show that if K is a slice knot, then, for suitable $\varphi_n\colon H_1(M_n) \to \boldsymbol{Z}_m$, $\sigma_r(M_n, \varphi_n)$ must remain bounded as $n \to \infty$. Finally, in §5, we study the class of knots consisting of the various doubles of the unknot, and use the result of §4 to show that although there are infinitely many algebraically slice knots in this class, only two are slice. The calculation of the relevant invariants $\sigma_r(M_n, \varphi_n)$ is based on §3.

We work throughout in the smooth category. In the absence of evidence to the contrary, manifolds are to be assumed compact and oriented, and homology to be with integer coefficients.

2. An invariant. Let $\tilde{N} \to N$ be an m-fold cyclic branched covering of closed 4-manifolds, branched over a surface $F \subset N$ with inverse image $\tilde{F} \subset \tilde{N}$. The (symmetric) intersection form on $H_2(\tilde{N})$ extends naturally to a nonsingular Hermitian form $\cdot$ on $H = H_2(\tilde{N}) \otimes \boldsymbol{C}$. Let $\tau\colon H \to H$ be the automorphism induced by the covering translation of $\tilde{N}$ which rotates each fibre of the normal bundle of $\tilde{F}$ through $2\pi/m$. Note that τ is an isometry of $(H, \cdot)$, and that $\tau^m = \text{id}$. Write $\omega = e^{2\pi i/m}$, and let E_r be the ω^r-eigenspace of τ, $0 \leqq r < m$. Then $(H, \cdot)$ decomposes as an orthogonal direct sum $E_0 \oplus E_1 \oplus \cdots \oplus E_{m-1}$. Let $\varepsilon_r(\tilde{N})$ be the signature of the restriction of $\cdot$ to E_r.

The following identity is proved by Rohlin in [10]. For the convenience of the reader we include a proof, which follows closely that of Rohlin.

LEMMA 2.1. $\varepsilon_r(\tilde{N}) = \text{sign}\, N - 2[F]^2 r(m-r)/m^2$.

PROOF. We can write $E_r = E_r^+ \oplus E_r^-$, where $\cdot$ is $\pm$ definite on $E_r^\pm$. Then $H = H^+ \oplus H^-$, where $H^\pm = E_0^\pm \oplus E_1^\pm \oplus \cdots \oplus E_{m-1}^\pm$, and for $0 \leqq s < m$ we have the τ^s-signatures

$$\begin{aligned}\text{sign}(\tau^s, \tilde{N}) &= \text{trace}(\tau^s | H^+) - \text{trace}(\tau^s | H^-)\\ &= \sum_{r=0}^{m-1} \omega^{rs}\, \varepsilon_r(\tilde{N}).\end{aligned}$$

A standard transfer argument gives $\varepsilon_0(\tilde{N}) = \text{sign}\, N$. Thus

$$\text{sign}(\tau^s, \tilde{N}) - \text{sign}\, N = \sum_{r=1}^{m-1} \omega^{rs}\, \varepsilon_r(\tilde{N}).$$

Inverting, we obtain, for $0 < r < m$,

$$\varepsilon_r(\tilde N) = \frac{1}{m}\sum_{s=1}^{m-1}(\omega^{-rs}-1)(\operatorname{sign}(\tau^s,\tilde N) - \operatorname{sign} N)$$

$$= \operatorname{sign} N + \frac{1}{m}\sum_{s=1}^{m-1}(\omega^{-rs}-1)\operatorname{sign}(\tau^s,\tilde N).$$

By the G-signature theorem [1, Proposition 6.18], $\operatorname{sign}(\tau^s,\tilde N) = [\tilde F]^2 \operatorname{cosec}^2(\pi s/m)$, $0 < s < m$. We see geometrically that the self-intersection number $[\tilde F^2]$ is equal to $[F]^2/m$. Therefore

$$\varepsilon_r(\tilde N) = \operatorname{sign} N + \frac{[F]^2}{m^2}\sum_{s=1}^{m-1}(\omega^{-rs}-1)\operatorname{cosec}^2\frac{\pi s}{m}.$$

Now

$$\sum_{s=1}^{m-1}(\omega^{-rs}-1)\operatorname{cosec}^2\frac{\pi s}{m} = -2\sum_{s=1}^{m-1}\sin^2\frac{\pi rs}{m}\operatorname{cosec}^2\frac{\pi s}{m}$$
$$- i\sum_{s=1}^{m-1}\sin\frac{2\pi rs}{m}\operatorname{cosec}^2\frac{\pi s}{m}.$$

The second sum must vanish, and one may easily verify that it does. To evaluate the first sum, let $\xi = e^{\pi i/m}$. Then

$$\sum_{s=1}^{m-1}\sin^2\frac{\pi rs}{m}\operatorname{cosec}^2\frac{\pi s}{m} = \sum_{s=1}^{m-1}\left(\frac{\xi^{rs}-\xi^{-rs}}{\xi^s-\xi^{-s}}\right)^2$$
$$= \sum_{s=1}^{m-1}(\xi^{s(r-1)}+\xi^{s(r-3)}+\cdots+\xi^{-s(r-1)})^2$$
$$= \sum_{s=1}^{m-1}P(\xi^s), \quad \text{say.}$$

Now $P(z) = P(z^{-1})$, and $\xi^{2m} = 1$. Therefore

$$\sum_{s=1}^{m-1}P(\xi^s) = \frac{1}{2}\sum_{s=0}^{2m-1}P(\xi^s) - \frac{1}{2}(P(1)+P(-1))$$
$$= \frac{1}{2}\left(2m\sum_t \text{coefficient of } z^{2mt} \text{ in } P(z)\right) - r^2$$
$$= r(m-r)$$

as the only contribution to the sum of coefficients comes from $t = 0$, and is r. Hence $\sum_{s=1}^{m-1}(\omega^{-rs}-1)\operatorname{cosec}^2(\pi s/m) = -2r(m-r)$, and the proof is complete.

Now let M be a closed 3-manifold, and $\varphi\colon H_1(M)\to \boldsymbol{Z}_m$ an epimorphism. φ induces an m-fold cyclic covering $\tilde M \to M$, with a canonical generator, corresponding to $1 \in \boldsymbol{Z}_m$, for the group of covering translations.

Suppose that for some positive integer n, there is an mn-fold cyclic branched covering of 4-manifolds $\tilde W \to W$, branched over a surface $F \subset \operatorname{int} W$, such that $\partial(\tilde W \to W) = n(\tilde M \to M)$, and such that the covering translation of $\tilde W$ which induces rotation through $2\pi/m$ on the fibres of the normal bundle of $\tilde F$ restricts on each component of $\partial\tilde W$ to the canonical covering translation of $\tilde M$ determined by φ. Let this covering translation induce τ on $H = H_2(\tilde W)\otimes \boldsymbol{C}$. As in the closed case, $(H, \cdot)$ is an orthogonal direct sum of eigenspaces of τ, and again we have the eigenspace signatures $\varepsilon_r(\tilde W)$, the only difference being that the form $\cdot$ will not now in general be nonsingular. Define, for $0 < r < m$, the rational number

$$\sigma_r(M, \varphi) = \frac{1}{n}\left(\text{sign } W - \varepsilon_r(\tilde{W}) - \frac{2[F]^2\, r(m-r)}{m^2} \right).$$

It follows readily from Lemma 2.1, and Novikov additivity of sign W and $\varepsilon_r(\tilde{W})$ (valid for the latter because they are linear combinations of τ^s-signatures) that $\sigma_r(M, \varphi)$ depends only on (M, φ) and r.

As we shall see in Lemma 2.2 below, it is always possible to take $n = 1$, but the extra generality in the definition will be useful in §4. We shall, however, always be in a situation where either $n = 1$ or $F = \emptyset$.

The following lemma shows that $\sigma_r(M, \varphi)$ is always defined.

LEMMA 2.2. *Given* (M, φ) *as above, suppose* $M = \partial W$ *with* $H_1(W; Z_m) = 0$. *Then* $\tilde{M} \to M$ *extends to an m-fold cyclic branched covering* $\tilde{W} \to W$ *over a surface* $F \subset$ int W, *such that the canonical covering translation of* $\tilde{M}$ *corresponds to rotation through* $2\pi/m$ *on each fibre of the normal bundle of* $\tilde{F} \subset$ int $\tilde{W}$.

PROOF. $\varphi \in \text{Hom}(H_1(M), Z_m) \cong H^1(M; Z_m)$. Since $H_1(W; Z_m) = 0$, there is a surface $F \subset$ int W such that the image in $H_2(W; Z_m)$ of $[F] \in H_2(W)$ is the Lefschetz dual of $\delta\varphi \in H^2(W, M; Z_m) \cong \text{Hom}(H_2(W, M), Z_m)$. Thus, in terms of intersections,

$$[F] \cdot x \,(\text{mod } m) = \delta\varphi(x) = \varphi(\partial x) \quad \text{for all } x \in H_2(W, M).$$

Let $\rho \in H^2(W, W - F; Z_m) \cong \text{Hom}(H_2(W, W - F), Z_m)$ be dual to the fundamental class in $H_2(F; Z_m)$. Comparing the cohomology exact sequences of the pairs (W, M), $(W, W - F)$, with Z_m coefficients, we see that $\rho = \delta\psi$ for some $\psi \in H^1(W - F; Z_m) \cong \text{Hom}(H_1(W - F), Z_m)$ which extends φ. Note also that since

$$[F] \cdot y \,(\text{mod } m) = \delta\psi(y) = \psi(\partial y) \quad \text{for all } y \in H_2(W, W - F),$$

ψ evaluates to $1 \in Z_m$ on a meridian of F. Then ψ determines the desired branched covering $\tilde{W} \to W$.

3. Surgery descriptions. We now describe a method for computing $\sigma_r(M, \varphi)$. A framed oriented link L, with components $L_1, \cdots, L_n$, in S^3, is a *surgery description* of (M, φ) if

(i) M is obtained by surgery on L (according to its framing), and

(ii) if $\bar{\mu}_i \in H_1(M)$ is the image of the class of a meridian μ_i of L_i, then $\varphi(\bar{\mu}_i) = 1 \in Z_m$ for each $i = 1, \cdots, n$.

(Note that the orientation of L is irrelevant to (i), but not to (ii).)

Surgery descriptions in this sense always exist, for it is known that, given M, there exists a link L satisfying (i) **[16]**, [7], which may now be modified by moves corresponding to handle additions and handle slides until (ii) is also satisfied.

The invariants $\sigma_r(M, \varphi)$ can be expressed in terms of certain invariants of L, as follows. Let $A = (a_{ij})$ be the matrix of linking numbers of L, that is, $a_{ij} = lk(L_i, L_j)$, $i \neq j$, and a_{ii} is the framing integer associated with L_i. Choose a Seifert surface V spanning L, let S be the corresponding Seifert matrix, and let S^T denote the transpose of S. Recall that $\omega = e^{2\pi i/m}$.

LEMMA 3.1. *Let the framed oriented link* L *be a surgery description of* (M, φ). *Then, for* $0 < r < m$,

$$\sigma_r(M, \varphi) = \text{sign } A - \text{sign}((1 - \omega^{-r})S + (1 - \omega^r)S^T) - \frac{2(\sum_{i,j} a_{ij})r(m-r)}{m^2}.$$

PROOF. Let W be the 4-manifold obtained by attaching n 2-handles to the 4-ball B^4 along disjoint tubular neighbourhoods of the components of L, according to the framing of L. Then W is 1-connected and $\partial W = M$. Recalling the proof of Lemma 2.2, we seek $F \subset \operatorname{int} W$ such that

$$[F]\cdot x \pmod{m} = \delta\varphi(x) = \varphi(\partial x) \quad \text{for all } x \in H_2(W, M).$$

For $i = 1, \cdots, n$, let $c_i \in H_2(W)$ be the class represented by the core of the ith 2-handle together with (say) the cone (in B^4) on L_i. Then $H_2(W)$ is free abelian on $c_1, \cdots, c_n$. Also, $H_2(W, M)$ is free abelian on $c_1^*, \cdots, c_n^*$, where c_i^* is the class of the co-core of the ith 2-handle. Let $f = \sum_{i=1}^n c_i$. Then $f\cdot c_j^* = 1, j = 1, \cdots, n$, and by hypothesis $\varphi(\partial c_j^*) = \varphi(\bar{\mu}_j) = 1 \in \boldsymbol{Z}_m$. Hence $f\cdot x \pmod{m} = \varphi(\partial x)$ for all $x \in H_2(W, M)$. Let V' be obtained by pushing the interior of V, the Seifert surface for L, into the interior of B^4, in the obvious way, using a collar of S^3 in B^4. The union of V' with the cores of all the 2-handles is then a surface $F \subset \operatorname{int} W$ representing f. By the proof of Lemma 2.2, we then have an m-fold cyclic branched cover $\tilde{W}$ of (W, F) such that $\partial(\tilde{W} \to W)$ is the covering $\tilde{M} \to M$ determined by φ.

The intersection form on $H_2(W)$ is given, with respect to the basis $c_1, \cdots, c_n$, by the matrix A of linking numbers of L; hence sign W = sign A. Also, $[F]^2 = (\sum_{i=1}^n c_i)^2 = \sum_{i,j} a_{ij}$. This accounts for the first and last terms on the right-hand side of the assertion of the lemma; it remains to identify the middle term as $\varepsilon_r(\tilde{W})$.

Now $\tilde{W} = \tilde{B} \cup \tilde{H}$, where $\tilde{B}$ is the m-fold cyclic branched cover of (B^4, V'), and $\tilde{H}$ is the m-fold cyclic branched cover of ($\bigcup$ 2-handles, $\bigcup$ cores). Thus $\partial\tilde{B}$ is the m-fold cyclic branched cover of (S^3, L), and $\tilde{H}$ is a disjoint union of n 2-handles, attached to $\tilde{B}$ along a tubular neighbourhood of $\tilde{L} \subset \partial\tilde{B}$, where $\tilde{L}$ is the inverse image of L. Since $H_2(\tilde{L}) = 0 = H_2(\tilde{H})$, there is a Mayer-Vietoris exact sequence $0 \to H_2(\tilde{B}) \to H_2(\tilde{W}) \to H_1(\tilde{L})$ which is equivariant with respect to the action of the group of covering translations. Now tensor with $\boldsymbol{C}$, and observe that the resulting exact sequence induces a corresponding exact sequence of eigenspaces. In particular, since the covering translations act trivially on $\tilde{L}$, we have $\varepsilon_r(\tilde{B}) = \varepsilon_r(\tilde{W})$ for $0 < r < m$.

For calculating intersections, it turns out to be more convenient to use, instead of $\tilde{B}$, the corresponding unbranched cover. So consider a tubular neighbourhood $V' \times D^2$ of V' in B^4, and let $\hat{B}$ be the m-fold cyclic cover of $B^4 - V' \times \operatorname{int} D^2$. Then $\tilde{B} \cong \hat{B} \cup V' \times D^2$, and, since $H_2(V') = 0$, we have an equivariant Mayer-Vietoris exact sequence

$$H_2(V' \times S^1) \longrightarrow H_2(\hat{B}) \longrightarrow H_2(\tilde{B}) \longrightarrow H_1(V' \times S^1).$$

Since the covering translations induce the identity on $H_*(V' \times S^1)$, an elementary argument shows that inclusion induces an isomorphism of eigenspaces $E_r(\hat{B}) \cong E_r(\tilde{B})$ for $0 < r < m$, and hence $\varepsilon_r(\hat{B}) = \varepsilon_r(\tilde{B}), 0 < r < m$.

$\hat{B}$ may be described as follows. First let C be obtained by cutting $B^4 - V' \times \operatorname{int} D^2$ along the trace T of the isotopy which pushed the interior of V into the interior of B^4. Observe that $T \cong V \times I$, that $C \cong B^4$, and that C contains two copies $T^\pm$ of T in its boundary. Now take m copies C_s of C, $s \in \boldsymbol{Z}_m$, and identify T_s^+ with T_{s+1}^- for each s. The result is $\hat{B}$. Let $z_1, \cdots, z_k$ be cycles in V representing a basis for $H_1(V)$. These determine cycles $z_1^+, \cdots, z_k^+$, say, in T_0^+ and $z_1^-, \cdots, z_k^-$ in T_1^-, and for each $i = 1, \cdots, k$, the union of the cone on z_i^+ in C_0 and the cone on z_i^- in C_1 determines a class $x_i \in H_2(\hat{B})$. A Mayer-Vietoris argument shows that $x_1, \cdots, x_k$ is a $\boldsymbol{Z}[\boldsymbol{Z}_m]$-basis for $H_2(\hat{B})$. Letting τ as usual denote the automorphism of $H_2(\hat{B})$ induced by the canonical covering translation, which takes each C_s to C_{s+1}, a basis for $H_2(\hat{B})$ over $\boldsymbol{Z}$ is $\{\tau^s x_i : 0 \leqq r < m, 1 \leqq i \leqq k\}$. The intersection form on $H_2(\hat{B})$ with respect to this basis can be readily described. Recall that S is the Seifert matrix of L with respect to V; write $S = (v_{ij})$. Then (with an appropriate modification if $m = 2$)

$$\tau^s x_i \cdot \tau^t x_j = \begin{cases} v_{ij} + v_{ji}, & s = t, \\ -v_{ij}, & s = t + 1, \\ -v_{ji}, & s = t - 1, \\ 0, & \text{otherwise.} \end{cases}$$

Now pass to $H_2(\hat{B}) \otimes \boldsymbol{C}$, but continue to write τ, x_i instead of $\tau \otimes \mathrm{id}$, $x_i \otimes 1$. Let $y_{i,r} = \sum_{s=0}^{m-1} \omega^{-rs} \tau^s x_i$, $0 \leqq r < m$, $1 \leqq i \leqq k$. Then $\{y_{i,r} : 1 \leqq i \leqq k\}$ is a linearly independent set of elements of E_r, $0 \leqq r < m$. Since $E_0 \oplus E_1 \oplus \cdots \oplus E_{m-1} = H_2(\hat{B}) \otimes \boldsymbol{C}$ has dimension mk, it follows that in fact it is a basis for E_r. With respect to this basis, the Hermitianized intersection form on E_r is given by

$$\begin{aligned} y_{i,r} \cdot y_{j,r} &= \sum_{t=0}^{m-1} \sum_{s=0}^{m-1} (\omega^{-rs} \tau^s x_i) \cdot (\omega^{-rt} \tau^t x_j) \\ &= \sum_{t=0}^{m-1} \sum_{s=0}^{m-1} \omega^{-r(s-t)} (\tau^s x_i \cdot \tau^t x_j) \\ &= \sum_{s=0}^{m-1} (v_{ij} + v_{ji} - \omega^{-r} v_{ij} - \omega^r v_{ji}) \\ &= m((1 - \omega^{-r}) v_{ij} + (1 - \omega^r) v_{ji}). \end{aligned}$$

Hence, for $0 < r < m$, $\varepsilon_r(\tilde{W}) = \varepsilon_r(\hat{B}) = \mathrm{sign}((1 - \omega^{-r})S + (1 - \omega^r)S^T)$, and the proof is complete.

REMARK. The signatures $\mathrm{sign}((1 - \omega^{-r})S + (1 - \omega^r)S^T)$, for m prime and $r = [m/2]$, were used by Tristram in **[13]**. (Compare also **[9]** and **[6]**.) The above interpretation of them as eigenspace signatures associated with an m-fold branched cyclic cover has also been given, in somewhat greater generality, by Viro **[15]**.

4. Slice knots. Let K be a knot in S^3. Fix a prime q, and let M_n denote the q^n-fold branched cyclic cover of (S^3, K), $n = 1, 2, \cdots$. (By an argument analogous to the proof of Lemma 4.2 below, $H_*(M_n; \boldsymbol{Q}) \cong H_*(S^3; \boldsymbol{Q})$.) Suppose we have an epimorphism $\varphi: H_1(M_1) \to \boldsymbol{Z}_m$. It is not hard to show that the branched covering projection $M_n \to M_1$ induces a surjection on π_1, and hence on H_1. Composition with φ then defines epimorphisms $\varphi_n: H_1(M_n) \to \boldsymbol{Z}_m$ for all n.

THEOREM 4.1. *Suppose K is a slice knot. Then there is a constant c, and a subgroup G of $H_1(M_1)$ with $|G|^2 = |H_1(M_1)|$, such that if m is a prime power and $\varphi: H_1(M_1) \to \boldsymbol{Z}_m$ is an epimorphism satisfying $\varphi(G) = 0$, then $|\sigma_r(M_n, \varphi_n)| < c$ for all n.*

We remark that the proof of Theorem 4.1 will apply without essential change to any knot K in a homology 3-sphere M such that $(M, K) = \partial(W, D)$ for some 2-disc D in a homology 4-ball W.

We require some preliminary lemmas.

LEMMA 4.2. *Let D be a 2-disc in B^4, and let V_n be the q^n-fold branched cyclic cover of (B^4, D), q prime. Then $\tilde{H}_*(V_n\colon \boldsymbol{Q}) = 0$.*

PROOF. Let $\tilde{X}$ be the infinite cyclic cover of $B^4 - D$. We then have the exact sequence (see [9])

$$\cdots \longrightarrow \tilde{H}_i(\tilde{X}; \boldsymbol{Z}_q) \xrightarrow{t^{q^n}-1} \tilde{H}_i(\tilde{X}; \boldsymbol{Z}_q) \longrightarrow \tilde{H}_i(V_n; \boldsymbol{Z}_q) \longrightarrow \tilde{H}_{i-1}(\tilde{X}; \boldsymbol{Z}_q) \longrightarrow \cdots$$

where t is the automorphism induced by the canonical covering translation of $\tilde{X}$. Since $V_0 = B^4$, $t - 1$ is an isomorphism. Hence, with $\boldsymbol{Z}_q$ coefficients, $t^{q^n} - 1 = (t-1)^{q^n}$ is also an isomorphism, giving $\tilde{H}_*(V_n; \boldsymbol{Z}_q) = 0$. Since V_n is compact, the result follows.

LEMMA 4.3. *Let V be a $\boldsymbol{Q}$-homology 4-ball. If the image of $H_1(\partial V) \to H_1(V)$ has order l, then $H_1(\partial V)$ has order l^2.*

PROOF. Since $H_2(\partial V) = 0$, we have an exact sequence

$$0 \to H_2(V) \to H_2(V, \partial V) \to H_1(\partial V) \to H_1(V) \to H_1(V, \partial V) \to 0.$$

By duality and universal coefficient theorems, $|H_2(V)| = |H_1(V, \partial V)|$ and $|H_2(V, \partial V)| = |H_1(V)|$; hence the result.

A slight extension of [5, Lemma 4.1] yields

LEMMA 4.4. *Let X be a connected complex with $\pi_1(X)$ finitely generated, $H_1(X)$ finite, and $H_1(X; \boldsymbol{Z}_p)$ cyclic for some prime p. Let $\tilde{X} \to X$ be a regular p^r-fold cyclic covering. Then $H_1(\tilde{X}; \boldsymbol{Q}) = 0$.*

For a proof of the following, see [9].

LEMMA 4.5. *Let X be a finite connected complex and $\tilde{X} \to X$ a regular infinite cyclic covering. Let F be a field. If $H_1(X; F) \cong F$, then* $\dim H_1(\tilde{X}; F)$ *is finite.*

PROOF OF THEOREM 4.1. By hypothesis, $(S^3, K) = \partial(B^4, D)$ for some 2-disc $D \subset B^4$. Let V_n be the q^n-fold branched cyclic cover of (B^4, D); thus $\partial V_n = M_n$. By Lemma 4.2, $\tilde{H}_*(V_n; \boldsymbol{Q}) = 0$. Let $i_{n*}\colon H_1(M_n) \to H_1(V_n)$ be induced by inclusion, and let $G = \ker i_{1*}$. By Lemma 4.3, $|G|^2 = |H_1(M_1)|$.

Suppose $m = p^a$, p prime. Since $\varphi(G) = 0$, there is an epimorphism $\psi\colon H_1(V_1) \to \boldsymbol{Z}_{p^b}$ for some b making the diagram

$$\begin{array}{ccc} H_1(M_1) & \xrightarrow{\ i_{1*}\ } & H_1(V_1) \\ \Big\downarrow{\scriptstyle\varphi} & & \Big\downarrow{\scriptstyle\psi} \\ \boldsymbol{Z}_{p^a} & \longrightarrow & \boldsymbol{Z}_{p^b} \end{array}$$

commute, where $\boldsymbol{Z}_{p^a} \to \boldsymbol{Z}_{p^b}$ is multiplication by p^{b-a}. Composing ψ with the epimorphism $H_1(V_n) \to H_1(V_1)$ induced by the branched covering projection gives a commutative diagrm

$$\begin{array}{ccc} H_1(M_n) & \xrightarrow{\ i_{n*}\ } & H_1(V_n) \\ \Big\downarrow{\scriptstyle\varphi_n} & & \Big\downarrow{\scriptstyle\psi_n} \\ \boldsymbol{Z}_{p^a} & \longrightarrow & \boldsymbol{Z}_{p^b} \end{array}$$

for all n.

Let $d_n = \dim H_1(V_n; Z_p)$. By doing surgery on $d_n - 1$ circles in int V_n we may obtain W_n with $H_1(W_n; Z_p)$ cyclic and a commutative diagram

$$\begin{array}{ccc} H_1(M_n) & \xrightarrow{i'_{n*}} & H_1(W_n) \\ \downarrow{\scriptstyle \varphi_n} & & \downarrow{\scriptstyle \psi'_n} \\ Z_{p^a} & \longrightarrow & Z_{p^b} \end{array}$$

where i'_n is inclusion and ψ'_n is surjective. Let $\tilde{W}_n \to W_n$ be the p^b-fold cyclic covering induced by ψ'_n; then $\partial(\tilde{W}_n \to W_n)$ consists of p^{b-a} copies of the p^a-fold cyclic covering $\tilde{M}_n \to M_n$ induced by φ_n.

Since $\tilde{H}_*(V_n; Q) = 0$, the euler characteristic $\chi(V_n) = 1$. Hence $\chi(W_n) = \chi(V_n) + 2(d_n - 1) = 2d_n - 1$, giving $\chi(\tilde{W}_n) = p^b(2d_n - 1)$. By Lemma 4.4, $H_1(\tilde{W}_n; Q) = 0$. Therefore $H_3(\tilde{W}_n; Q)$, which is isomorphic to $H_1(\tilde{W}_n, \partial\tilde{W}_n; Q)$ by duality, has dimension $p^{b-a} - 1$. It follows that $\dim H_2(\tilde{W}_n; Q) = p^b(2d_n - 1) + p^{b-a} - 2$. Note also that since signature is unaffected by surgery, sign W_n = sign $V_n = 0$. Hence

$$|\sigma_r(M_n, \varphi_n)| \leq \frac{1}{p^{b-a}}(p^b(2d_n - 1) + p^{b-a} - 2)$$
$$< p^a(2d_n - 1) + 1.$$

Finally, let $\tilde{X}$ denote the infinite cyclic cover of $X = B^4 - D$, and let t: $H_1(\tilde{X}; Z_p) \to H_1(\tilde{X}; Z_p)$ be the automorphism induced by the canonical covering translation. Then (see the proof of Lemma 4.2) $H_1(V_n; Z_p) \cong \operatorname{coker}(t^{q^n} - 1)$. In particular, $d_n \leq d = \dim H_1(\tilde{X}; Z_p)$, which is finite by Lemma 4.5. We may now set $c = |G|\,(2d - 1) + 1$, and the proof is complete.

5. Some calculations. Let us consider the knots K_k ($k \in Z$) illustrated in Figure 1.

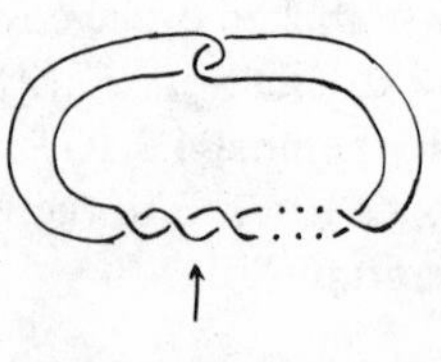

k full positive twists

FIGURE 1

Thus K_k may be described as the k-twisted double of the unknot, or alternatively as the rational (2-bridge) knot corresponding to the rational number $(4k + 1)/2$ **[11]**, **[3]**. Its 2-fold branched cover is the lens space $L(4k + 1, 2)$.

K_k has a Seifert surface of genus 1 with corresponding Seifert matrix

$$\begin{pmatrix} -1 & 1 \\ 0 & k \end{pmatrix}.$$

It follows easily that K_k is algebraically slice precisely when $4k + 1 = l^2$ for some integer l. The first two such values of k, namely 0 and 2, give the unknot and the stevedore's knot respectively, both of which are slice (indeed ribbon) knots. However,

THEOREM 5.1. *K_k is slice only if $k = 0, 2$.*

In fact, the proof shows that if $k \neq 0, 2$, then K_k does not bound a disc in any homology 4-ball (see remark after Theorem 4.1).

PROOF. If K_k is slice, then it is certainly algebraically slice. So for some fixed k such that $4k + 1 = l^2$, let M_n be the 2^n-fold branched cyclic cover of (S^3, K_k). For any divisor m of l we have epimorphisms $\varphi\colon H_1(M_1) \cong Z_{l^2} \to Z_m$, which necessarily satisfy $\varphi(G) = 0$ where $G \subset H_1(M_1)$ has order l. We compute $\sigma_r(M_n, \varphi_n)$ (for suitable φ) using the following surgery description. In Figure 2, surgery with framing $+1$ on the unknotted curve J indicated yields S^3 in such a

FIGURE 2

way that the other unknotted curve, K, becomes K_k. By an isotopy of S^3, Figure 2 may be transformed to Figure 3. Then M_n, the 2^n-fold branched cyclic cover of (S^3, K_k), is obtained by surgery on the link L consisting of the 2^n lifts of J in the

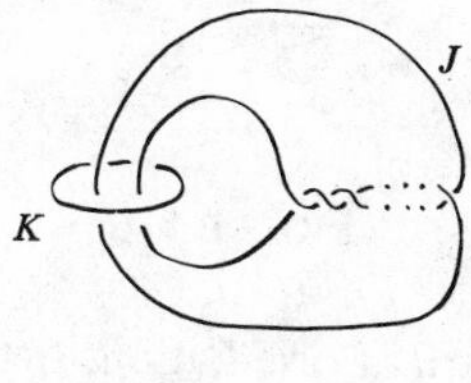

FIGURE 3

2^n-fold branched cyclic cover of (S^3, K). The latter is just S^3, and L is illustrated in Figure 4. To determine the appropriate framing x of a component L_i of L, choose (temporarily) an equivariant orientation of L. Consider a 2-chain C_i whose boundary is a slightly pushed-off copy of L_i determined by the framing of L_i. This projects to a 2-chain C whose boundary is a similarly defined push-off of J. Consideration of the intersections of $\bigcup_i C_i$ and C with L and J respectively gives, for each i,

$$1 = \text{framing of } J = x + \sum_{j \neq i} lk(L_i, L_j) = x - 2k;$$

hence $x = 2k + 1$.

We must now consider $\varphi_n\colon H_1(M_n) \to Z_m$. M_1 is obtained by surgery on the framed link shown in Figure 5, to which, for reasons soon to become apparent,

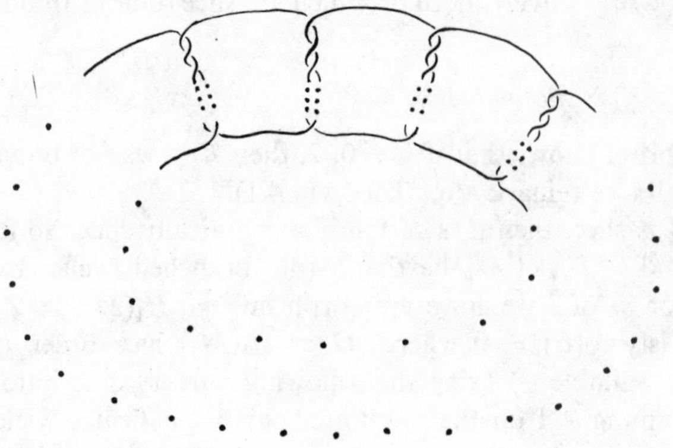

FIGURE 4

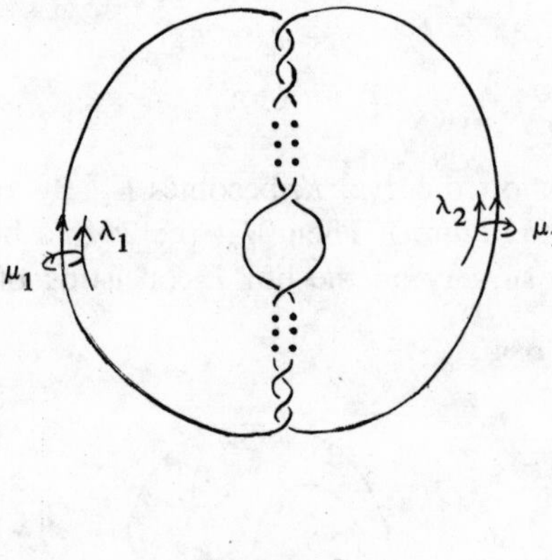

FIGURE 5

we have assigned a nonequivariant orientation. Let λ_1, μ_1 and λ_2, μ_2 be oriented longitude-meridian pairs for the two components, where λ_i is chosen to be null-homologous in the complement of the ith component. The first homology of the complement of the link is free abelian on μ_1, μ_2, and we have $\lambda_1 = 2k\mu_2$, $\lambda_2 = 2k\mu_1$. Surgery has the effect of adding the relations

$$\lambda_1 + (2k + 1)\mu_1 = 0, \qquad \lambda_2 + (2k + 1)\mu_2 = 0.$$

Thus if $\bar{\mu}_i$ denotes the image of μ_i in $H_1(M_1)$, we see that $H_1(M_1)$ is cyclic of order $4k + 1 = l^2$, generated by $\bar{\mu}_1 = \bar{\mu}_2$. Hence $\varphi\colon H_1(M_1) \to \mathbf{Z}_m$ can be chosen to satisfy $\varphi(\bar{\mu}_1) = \varphi(\bar{\mu}_2) = 1$.

More generally, we give the link L which yields M_n the alternating orientation shown in Figure 6. Recalling that φ_n is defined to be the composition $H_1(M_n) \to H_1(M_1) \xrightarrow{\varphi} \mathbf{Z}_m$, it follows that we then have $\varphi(\bar{\mu}_i) = 1$ for each i, where $\bar{\mu}_i \in H_1(M_n)$ corresponds to the meridian of the ith component L_i of L. Thus L is a surgery presentation for (M_n, φ_n) in the sense of §3.

The linking matrix A of L is the $2^n \times 2^n$ matrix

$$\begin{pmatrix} 2k+1 & k & 0 & \cdots & 0 & k \\ k & 2k+1 & k & & & 0 \\ 0 & k & 2k+1 & & & \vdots \\ \vdots & & & \ddots & & 0 \\ 0 & & & & & k \\ k & 0 & \cdots & 0 & k & 2k+1 \end{pmatrix}.$$

Note that $A = kB + I$, where B is

$$\begin{pmatrix} 2 & 1 & 0 & \cdots & 0 & 1 \\ 1 & 2 & 1 & & & 0 \\ 0 & 1 & 2 & & & \vdots \\ \vdots & & & \ddots & & 0 \\ 0 & & & & 2 & 1 \\ 1 & 0 & \cdots & 0 & 1 & 2 \end{pmatrix}.$$

It is easy to verify (for example by inductively computing the principal minors) that B is positive definite. Hence A is also positive definite, that is, $\operatorname{sign} A = 2^n$. Also, $\sum_{i,j} a_{ij} = 2^n(4k+1) = 2^n l^2$.

Let V be the Seifert surface for L illustrated in Figure 6.

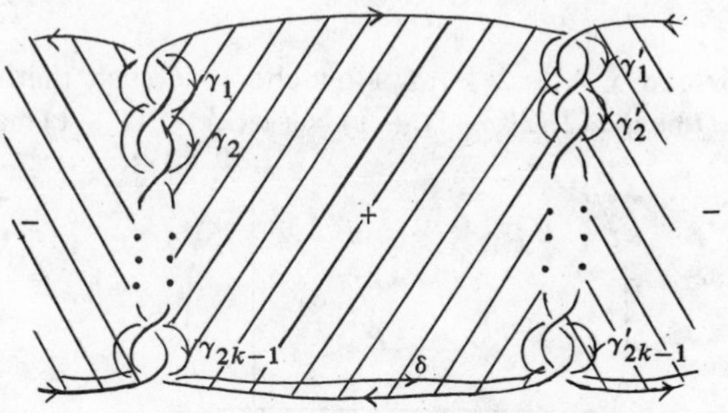

FIGURE 6

The 2^n $(2k-1)$-element sets of which $\{\gamma_1, \cdots, \gamma_{2k-1}\}$, $\{\gamma'_1, \cdots, \gamma'_{2k-1}\}$ is a typical pair, together with δ, determine a basis for $H_1(V)$. In the corresponding Seifert matrix, S_n, say, $\gamma_1, \cdots, \gamma_{2k-1}$ contribute the $(2k-1) \times (2k-1)$ block

$$C = \begin{pmatrix} -1 & 1 & 0 & \cdots & 0 \\ 0 & -1 & 1 & & \vdots \\ \cdot & & -1 & & \\ \cdot & & & \ddots & 0 \\ \cdot & & & & 1 \\ 0 & \cdots & & 0 & -1 \end{pmatrix}$$

and $\gamma_1', \cdots, \gamma_{2k-1}'$ contribute the transpose C^T of C. Thus S_n is the $(2^n(2k-1)+1) \times (2^n(2k-1)+1)$ matrix

$$\left(\begin{array}{cccccc|c}
C & & & & & & 1 \\
& C^T & & & & & \\
& & C & & & & 1 \\
& & & C^T & & & \\
& & & & \ddots & & \\
& & & & & C & 1 \\
& & & & & C^T & \\
\hline
1 & & 1 & & & 1 & -2^{n-1}
\end{array}\right).$$

Let $\omega = e^{2\pi i/m}$. Write $S_{n,r}$ for the Hermitian matrix $(1-\omega^{-r})S_n + (1-\omega^r)S_n^T$, and similarly for C. Then $S_{n,r}$ is

$$\left(\begin{array}{ccccccc|c}
C_r & & & & & & & \bar{\alpha} \\
& C_r^T & & & & & & \alpha \\
& & C_r & & & & & \bar{\alpha} \\
& & & C_r^T & & & & \alpha \\
& & & & \ddots & & & \\
& & & & & C_r & & \bar{\alpha} \\
& & & & & & C_r^T & \alpha \\
\hline
\alpha & \bar{\alpha} & \alpha & \bar{\alpha} & & \alpha & \bar{\alpha} & x
\end{array}\right)$$

where $\alpha = 1 - \omega^r$ and $x = -2^{n-1}\alpha\bar{\alpha}$. Now choose P such that $PC_rP^* = D$ is diagonal, and let Q be the $(2^n(2k-1)+1) \times (2^n(2k-1)+1)$ matrix

$$\left(\begin{array}{ccccc|c}
P & & & & & \\
& \bar{P} & & & & \\
& & \ddots & & & 0 \\
& & & P & & \\
& & & & \bar{P} & \\
\hline
& & 0 & & & 1
\end{array}\right).$$

Then $QS_{n,r}Q^*$ is

$$\left(\begin{array}{ccccc|c}
D & & & & & * \\
& D & & & & * \\
& & \ddots & & & \vdots \\
& & & D & & * \\
& & & & D & * \\
\hline
* & * & \cdots & * & * & x
\end{array}\right)$$

where the entries in the last row (and last column) are periodic with period $2(2k-1)$. We shall see later (in the proof of Lemma 5.2) that C_r is nonsingular. Assuming this, use the diagonal entries of the D's in $QS_{n,r}Q^*$ to clear all the entries (except the last) from the last row and column. Because of the periodicity noted above, this process changes the entry x to $x + 2^{n-1}y = 2^{n-1}(y - \alpha\bar{\alpha})$ for some y independent of n. Hence sign $S_{n,r} = 2^n$ sign $C_r + \eta_r$, where $|\eta_r| \leqq 1$ and η_r is independent of n. By Lemma 3.1, for $0 < r < m$,

$$\sigma_r(M_n, \varphi_n) = 2^n - 2^n \operatorname{sign} C_r - \eta_r - 2^{n+1}r(m - r)(l/m)^2.$$

In particular, note the multiplicative relation

$$\sigma_r(M_n, \varphi_n) + \eta_r = 2^n(\sigma_r(M_1, \varphi) + \eta_r).$$

It follows from Theorem 4.1 that for K_k to be slice we must have $\sigma_r(M_1, \varphi) + \eta_r = 0$, or, equivalently,

$$\operatorname{sign} C_r = 1 - 2r(m - r)(l/m)^2,$$

for every prime power divisor m of l, and every r, $0 < r < m$. (We may remark that the present examples are somewhat deceptive in that the multiplicativity noted above does not hold in general. Nevertheless, a good deal of information can be extracted from a single branched cover; see [**2**, Theorems 2 and 3].)

Since $C_{m-r} = \bar{C}_r$, and m is odd, there is no loss of generality in restricting to $0 < r \leqq (m - 1)/2$. Theorem 5.1 will follow easily from

LEMMA 5.2. *Suppose m is odd and $0 < r \leqq (m - 1)/2$. Then*

$$\operatorname{sign} C_r = -2[2kr/m] - 1.$$

PROOF. Let D_n be the $n \times n$ principal minor of C_r, $n = 1, \cdots, 2k - 1$. Then, writing $\alpha = 1 - \omega^r$ as before, and expanding D_n by (say) the first row, we obtain the difference equation

$$D_n = -(\alpha + \bar{\alpha})D_{n-1} - \alpha\bar{\alpha}D_{n-2}, \qquad n = 2, \cdots, 2k - 1.$$

Since the roots of the corresponding characteristic equation $x^2 + (\alpha + \bar{\alpha})x + \alpha\bar{\alpha} = 0$ are $-\alpha$, $-\bar{\alpha}$, the general solution of this difference equation is $(-1)^n(A\alpha^n + B\bar{\alpha}^n)$, where A and B are arbitrary constants. Our initial values $D_0 = 1$, $D_1 = -(\alpha + \bar{\alpha})$ give $A = \alpha/(\alpha - \bar{\alpha})$, $B = -\bar{\alpha}/(\alpha - \bar{\alpha})$; hence

$$D_n = (-1)^n\left(\frac{\alpha^{n+1} - \bar{\alpha}^{n+1}}{\alpha - \bar{\alpha}}\right).$$

Write $\alpha = \rho e^{i\theta}$, $\rho > 0$. Then $D_n = (-\rho)^n \sin(n + 1)\theta/\sin\theta$. Also, since $\tan\theta = (-\sin 2\pi r/m)/(1 - \cos 2\pi r/m) = -\cot \pi r/m$, and since $-\pi/2 < \theta < 0$, we have $\theta = \pi r/m - \pi/2$. In particular, we see that $D_{2k-1} = \det C_r \neq 0$, a fact which we used earlier. We also see that there are no two consecutive zeros among the D_n, so sign C_r = (number of permanences of sign) − (number of changes of sign) in the sequence $D_0, \cdots, D_{2k-1}$ (where 0's may be assigned either sign). Thus sign $C_r = 2c - (2k - 1)$, where c = number of changes of sign of $\sin n\theta = \sin n(\pi r/m - \pi/2)$, $n = 1, \cdots, 2k$. Write $r = (m - s)/2$, $1 \leqq s \leqq m - 2$, s odd. Then $\sin n(\pi r/m - \pi/2) = -\sin \pi ns/2m$. Hence c = number of changes of sign of $\sin \pi ns/2m$, $n = 1, \cdots, 2k$, $= [2ks/2m] = [k - 2kr/m] = k - 1 - [2kr/m]$. Therefore,

$$\operatorname{sign} C_r = 2\left(k - 1 - \left[\frac{2kr}{m}\right]\right) - (2k - 1) = -2\left[\frac{2kr}{m}\right] - 1$$

as stated.

Returning to the proof of Theorem 5.1, recall that K_k slice implies

$$2r(m - r)(l/m)^2 - 1 + \text{sign } C_r = 0$$

for every prime power divisor m of $l = 4k + 1$, and every r, $0 < r < m$. By Lemma 5.2, this is equivalent to the condition that for every r, $0 < r \leqq (m - 1)/2$,

$$r(m - r)(l/m)^2 - \left[\frac{2kr}{m}\right] - 1 = 0.$$

Replacing $[2kr/m]$ by $(l^2 - 1)r/2m = 2kr/m > [2kr/m]$, it follows that we must have

$$r(m - r)(l/m)^2 - \frac{(l^2 - 1)r}{2m} - 1 < 0.$$

Multiplying by $2/r$, we obtain $(m - 2r)(l/m)^2 + 1/m - 2/r < 0$, and hence, since $m|l$, $m + 1/m < 2(r + 1/r)$. But putting $r = (m - 1)/2$, the value which maximizes $r + 1/r$, gives $m^2 - 4m - 1 < 0$, which is clearly violated by (odd) $m > 3$. Moreover, if $l > 3$, then l has a prime power divisor $m > 3$. Hence K_k can be slice only if $l = 1, 3$, that is, $k = 0, 2$. Indeed we have shown that this fact is detected by the invariants $\sigma_r(M_n, \varphi_n)$ for any r, $0 < r < m$.

REFERENCES

1. M. F. Atiyah and I. M. Singer, *The index of elliptic operators*. III, Ann. of Math. (2) **87** (1968), 546–604.

2. A. J. Casson and C. McA. Gordon, *Cobordism of classical knots*, mimeographed notes, Orsay, France, 1975.

3. J. H. Conway, *An enumeration of knots and links, and some of their algebraic properties*, Computational Problems in Abstract Algebra, Pergamon, New York and Oxford, 1970, pp. 329–358. MR **41** #2661.

4. R. H. Fox and J. W. Milnor, *Singularities of 2-spheres in 4-space and cobordism of knots*, Osaka J. Math. **3** (1966), 257–267. MR **35** #2273.

5. W. C. Hsiang and R. H. Szczarba, *On embedding surfaces in four-manifolds*, Proc. Sympos. Pure Math., vol. 22, Amer. Math. Soc., Providence, R. I., 1971, pp. 97–103. MR **49** #4000.

6. J. Levine, *Knot cobordism groups in codimension two*, Comment. Math. Helv. **44** (1969), 229–244. MR **39** #7618.

7. W. B. R. Lickorish, *A representation of orientable combinatorial 3-manifolds*, Ann. of Math. (2) **76** (1962), 531–540. MR **27** #1929.

8. W. S. Massey, *Proof of a conjecture of Whitney*, Pacific J. Math. **31** (1969), 143–156. MR **40** #3570.

9. J. W. Milnor, *Infinite cyclic coverings*, Conf. on the Topology of Manifolds (Michigan State Univ., E. Lansing, Mich., 1967), Prindle, Weber and Schmidt, Boston, Mass., 1968, pp. 115–133. MR **39** #3497.

10. V. A. Rohlin, *Two-dimensional submanifolds of four-dimensional manifolds*, Funkcional. Anal. i Priložen **5** (1971), 48–60. (Russian) MR **45** #7733.

11. H. Schubert, *Knoten mit zwei Brücken*, Math. Z. **65** (1956), 133–170. MR **18**, 498.

12. H. Seifert, *Über das Geschlecht von Knoten*, Math. Ann. **110** (1934), 57–592.

13. A. G. Tristram, *Some cobordism invariants for links*, Proc. Cambridge Philos. Soc. **66** (1969), 251–264. MR **40** #2104.

14. H. F. Trotter, *On S-equivalence of Seifert matrices*, Invent. Math. **20** (1973), 173–207.

15. O. Ja. Viro, *Branched coverings of manifolds with boundary, and invariants of links*. I, Math. USSR Izvestija **7** (1973), 1239–1256. (Rusian) MR **51** #6832.

16. A. H. Wallace, *Modifications and cobounding manifolds*, Canad. J. Math. **12** (1960), 503–528. MR **23** #A2887.

Cambridge University

The Institute for Advanced Study

Proceedings of Symposia in Pure Mathematics
Volume 32, 1978

CONSTRUCTIONS OF FIBRED KNOTS AND LINKS*

JOHN R. STALLINGS

Introduction. In this paper we consider only polyhedral, that is, nonwild, situations.

In the oriented 3-sphere S^3, let T be a compact, connected, oriented surface with nonempty boundary Bd T. Let T^+ be a copy of T in $S^3 - T$ parallel to T. If the map $\pi_1(T^+) \to \pi_1(S^3 - T)$ is an isomorphism, we call T a *fibre surface*, and its boundary Bd T a *fibred link*. The reason for this language is that, given the condition on the fundamental groups, $S^3 -$ Bd T is the total space of a fibre bundle with base space the circle and fibre the interior of T **[1]**. A fibred link of only one component is called a fibred knot or Neuwirth knot **[2]**.

It is known that the Alexander polynomial $A(t)$ of a fibred knot has degree equal to twice the genus of the corresponding fibre surface, and that it has leading coefficient 1 **[3]**; of course, also, $A(t)$ satisfies a symmetry condition. Every possible such Alexander polynomial occurs as the polynomial of some fibred knot **[4]**. For a fibre surface T, the translation of the fibre around the base-space circle determines an element of the mapping-class group of T, a homeomorphism $h: T \to T$ well defined up to isotopy; this element is called the *holonomy* of the fibre surface; the Alexander polynomial is the characteristic polynomial of the map the holonomy induces on $H_1(T)$. It is also known **[5]** that if the leading coefficient of the Alexander polynomial of an *alternating* knot is 1, then the knot is fibred.

The links which occur as isolated singularities of algebraic surfaces, certain compound torus links, are known to be fibred **[6]**; these are special cases of a closed positive braid, whose Alexander polynomial was found to have leading coefficient 1 **[7]**, and which we shall show are fibred.

This paper discusses several methods of creating fibre surfaces, including plumbing, twisting, and companionization.

AMS (*MOS*) *subject classifications* (1970). Primary 55A25.

*This work was partially supported by NSF contract number MCS 74–03423.

Plumbing. This is a generalization of the technique described by Murasugi [5].

Consider two oriented fibre surfaces T_1 and T_2. On T_i let D_i be 2-cells, and let $h\colon D_1 \to D_2$ be an orientation-preserving homeomorphism such that the union of T_1 and T_2 identifying D_1 with D_2 by h is a 2-manifold T_3. That is to say:

$$h(D_1 \cap \text{Bd } T_1) \cup (D_2 \cap \text{Bd } T_2) = \text{Bd } D_2.$$

We can realize T_3 in S^3 as follows: Thicken D_1 on the positive side of T_1, to get a 3-cell, whose complementary 3-cell E_1 contains T_1 with $T_1 \cap \text{Bd } E_1 = D_1$ and with the negative side of T_1 contained in the interior of E_1. Likewise, there is a 3-cell E_2 containing T_2, with $T_2 \cap \text{Bd } E_2 = D_2$ and with the positive side of T_2 contained in the interior of E_2. The homeomorphism $h\colon D_1 \to D_2$ extends to $h\colon \text{Bd } E_1 \to \text{Bd } E_2$. The union of E_1 and E_2, identifying their boundaries by h—this is S^3—contains T_3 as $T_1 \cup T_2$. We say T_3 is obtained from T_1 and T_2 by *plumbing*.

THEOREM 1. *If T_1 and T_2 are fibre surfaces, so is T_3.*

The proof can be found by examining the map on fundamental groups. We can identify

$$\pi_1(T_3) \approx \pi_1(T_1) * \pi_1(T_2),$$
$$\pi_1(S^3 - T_3) \approx \pi_1(S^3 - T_1) * \pi_1(S^3 - T_2).$$

The map on the second factor is that which we would expect; on the first factor it is slightly different, the image elements of a particular basis of $\pi_1(T_1)$ being multiplied on the left and right by certain elements of $\pi_1(S^3 - T_2)$.

A special interesting case concerns braids [8]. A braid of n strands can be expressed as a word in generators $\sigma_1, \cdots, \sigma_{n-1}$, where σ_i is the braid involving a single

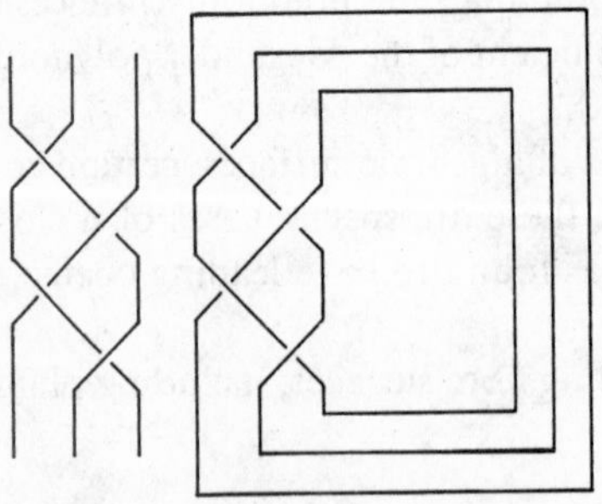

$\sigma_1^{-1}\sigma_2\ \sigma_1^{-1}\sigma_2 = \beta$ $\hat{\beta}$, the knot 4_1

FIGURE 1

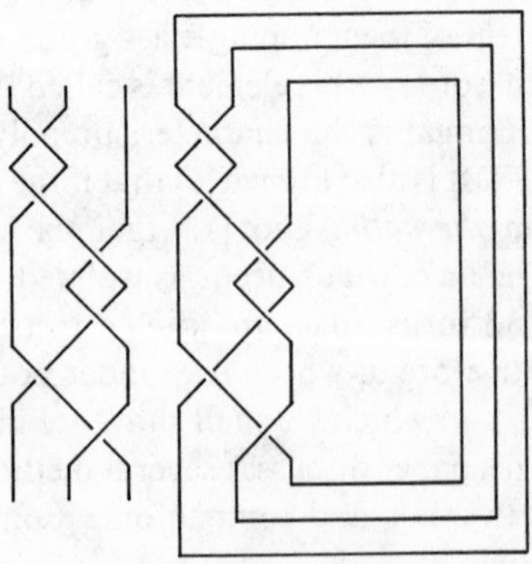

$\sigma_1^2\ \sigma_2^2\ \sigma_1\ \sigma_2^{-1} = \beta$ β, the knot 5_2

FIGURE 2

crossing of the ith and $(i + 1)$st strands. If $\beta = \sigma_{i_1}^{\varepsilon_1} \sigma_{i_2}^{\varepsilon_2} \cdots \sigma_{i_k}^{\varepsilon_k}$, $\varepsilon_j = \pm 1$ has the two properties—(a) every σ_i occurs at least once, (b) for each i, the exponents of all occurrences of σ_i are the same—then we call β *homogeneous*. For example, if all ε_i are $+1$, we have the positive braids studied in [7]. The braid $\sigma_1\sigma_2^{-1}\sigma_1\sigma_2^{-1}$ is homogeneous; the braid $\sigma_1^2\sigma_2^2\sigma_1\sigma_2^{-1}$ is not homogeneous—see Figures 1 and 2.

Given any braid β, we can close it up to obtain a closed braid $\hat{\beta}$. There is an oriented surface T_β whose boundary is $\hat{\beta}$, obtained as the union of n disks, one for each strand, where the ith and $(i + 1)$st disks are joined by a number of half-twisted strips, one for each occurrence of σ_i in β. Then T_β is obtained by plumbing a series of surfaces $T_1, T_2, \cdots, T_{n-1}$, where T_i consists of the ith and $(i + 1)$st disks with the connecting half-twists. If β is homogeneous, the half-twists in T_i are all in the same sense, so that T_i looks like Figure 3 or its mirror image. A direct computation shows that the surfaces in this figure are fibred. Thus

THEOREM 2. *If β is a homogeneous braid, then $\hat{\beta}$ is a fibred link with fibre surface T_β.*

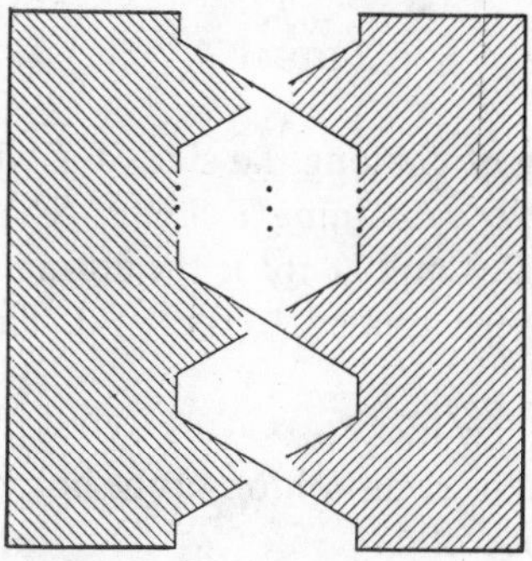

FIGURE 3

This has a curious corollary. If L is any link, it can be represented as $\hat{\beta}$ for some (nonhomogeneous) braid β. Now, by adding to the picture for β several other strands, we can isolate the positive from the negative crossings of β so that they are located on different vertical strata. The new strands can be crossed over each other, so that in the closed form they will represent a single unknot. Furthermore, we can arrange it so that this circle has arbitrarily prescribed linking numbers with the components of L. Figure 4 applies this to the braid of Figure 2.

THEOREM 3. *Given any link L in S^3, there is an unknot K disjoint from L, with arbitrarily prescribed linking numbers with the components of L, such that $K \cup L$ is a fibred link.*

By choosing the linking numbers carefully (making their sum = 1), we can do Dehn surgery [9] on K to obtain the 3-sphere again. This surgery will be compatible with the fibration, and thus *any link can be transformed into a fibred link by a single Dehn surgery.*

Twisting. Suppose T is a fibre surface and C is a simple closed curve on T, such that C is unknotted in S^3, and so that C bounds a disk D which is orthogonal to T along C. This latter condition is equivalent to C and C^+ having linking number

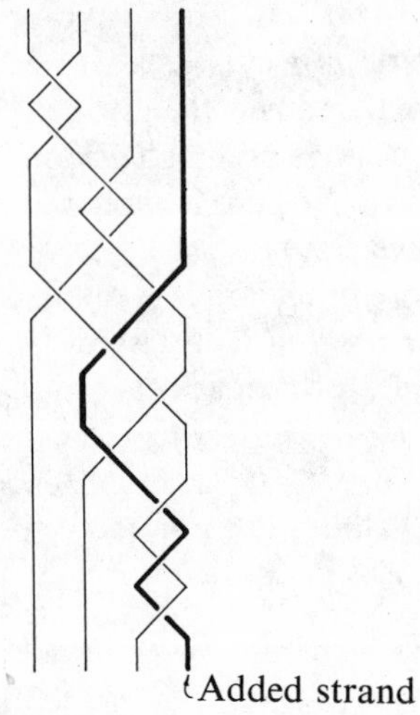

FIGURE 4

zero. Let A be a thickening of C along the side of T where D starts. The complement of A is a donut $S^1 \times D$ containing T. Let $\tau: S^1 \times D \to S^1 \times D$ be a homeomorphism, a twist along D. Look at $\tau(T)$; the fibring of $S^3 - \text{Bd}\, T$ contained in $S^1 \times D$ fits up, after τ has been applied, to that in A. Thus

THEOREM 4. *$\tau(T)$ is a fibre surface.*

The holonomy of $\tau(T)$ is the composition of the holonomy of T with a Lickorish twist [10] in the neighborhood of C.

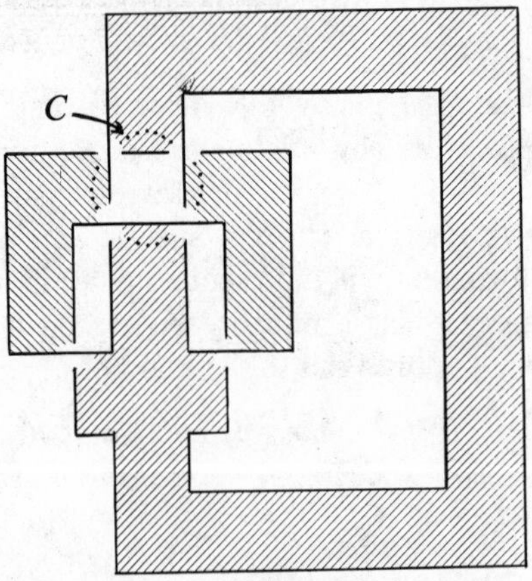

FIGURE 5

As an interesting example, the surface in Figure 5 is a fibre surface, and C is a curve along which such twists are permissible, leading to the fibre surfaces T_n described in Figure 6.

The knots K_n which are the fibred knots of Figure 6 all have the same Alexander polynomial $(t^2 - t + 1)^2$, but they can be distinguished by the fact that if M_n is

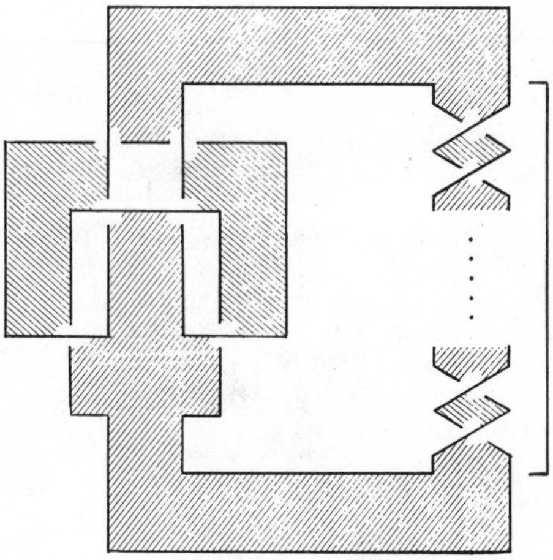

FIGURE 6

the Alexander matrix of K_n, describing the holonomy on the homology of T_n, then $M_n^2 - M_n + I$ has n as an elementary divisor. This shows M_n and M_k for $|n| \neq |k|$ are dissimilar, and so K_n and K_k are inequivalent.

Companionization. Suppose that in $S^1 \times D^2 = A$, there is a link L and an oriented surface T, such that Bd $T = L$, plus n longitudes, $n \geqq 1$, all oriented coherently (of the form $S^1 \times$ boundary point), and such that $A - L$ fibres over S^1 with fibre $T - L$. We would describe T as a fibre surface within $S^1 \times D$. Now, suppose A is embedded in S^3 via a knot K, in such a way that the longitudes in Bd T are "longitudes" of K, i.e., null-homologous in $S^3 - A$.

THEOREM 5. *If K is a fibred knot, $A = S^1 \times D$ is knotted via K, and L is, within $S^1 \times D$, a fibred link whose fibre T as above intersects* Bd A *in n longitudes to K, then L is fibred within S^3. The corresponding fibre surface consists of a fibre T of $A - L$ plus n fibre surfaces of K.*

This is geometrically obvious, but can also be shown from fundamental-group considerations. In Schubert's terminology **[11]**, K is a companion of L. A particular case of this **[12]** is cabling, in which L is a torus knot on a torus parallel to the boundary of A; if we cable a fibred knot K, we obtain a new fibred knot L.

REFERENCES

1. J. Stallings, *On fibering certain 3-manifolds*, Topology of 3-Manifolds and Related Topics, Prentice-Hall, Englewood Cliffs, N.J., 1962. MR **28** #1600.

2. L. Neuwirth, *On Stallings fibrations*, Proc. Amer. Math. Soc. **14** (1963), 380–381. MR **26** #6958.

3. E. S. Rapaport, *On the commutator subgroup of a knot group*, Ann. of Math. (2) **71** (1960), 157–162. MR **22** #6842.

4. G. Burde, *Alexanderpolynome Neuwirtscher Knoten*, Topology **5** (1966), 321–330. MR **33** #7998.

5. K. Murasugi, *On a certain subgroup of the group of an alternating link*, Amer. J. Math. **85** (1963), 544–550. MR **28** #609.

6. J. Milnor, *Singular points of complex hypersurfaces*, Ann. of Math. Studies, no. 61, Princeton Univ. Press, Princeton, N.J., 1968. MR **39** #969.

7. W. Burau, *Über Zopfgruppen und gleichsinnig verdrillte Verkettungen*, Hamburg. Math. Abh. **11** (1935).

8. W. Magnus, *Braid groups: A survey* (Proc. 2nd Internat. Congress on the Theory of Groups), Lecture Notes in Math., vol. 372, Springer-Verlag, Berlin and New York, 1974. MR **50** #5774.

9. M. Dehn, *Die Gruppe der Abbildungsklassen*, Acta Math. **69** (1938). ["Dehn surgery" on an unknot transforms the complementary donut in S^3 by a single twist. In the case under consideration the fibre surfaces are transformed into surfaces orthogonal to the unknot.]

10. W. B. R. Lickorish, *A finite set of generators for the homeotopy group of a 2-manifold*, Proc. Cambridge Philos. Soc. **60** (1964), 769–778. MR **30** #1500.

11. H. Schubert, *Knoten und Vollringe*, Acta Math. **90** (1953), 131–286. MR **17**, 291.

12. Joan Birman, *Cabling fibred knots* (conversation).

UNIVERSITY OF CALIFORNIA, BERKELEY

Proceedings of Symposia in Pure Mathematics
Volume 32, 1978

A METHOD OF CONSTRUCTING 3-MANIFOLDS AND ITS APPLICATION TO THE COMPUTATION OF THE μ-INVARIANT

HUGH M. HILDEN* AND JOSÉ M. MONTESINOS

1. Introduction. There is a representation for any closed, oriented 3-manifold M^3 that, like the representation by framed links (see **[1]**, **[2]**), provides a simply connected 4-manifold W^4 with $\partial W^4 = M^3$. In this paper we shall describe the representation, prove a theorem about branched covering spaces that follows from it, and show how to use it to compute the μ-invariant of a 3-manifold with $H_1(M^3; Z/2Z) = 0$.

2. Representation of 3-manifolds by bands. We orient $S^3 = \{(x, y, z)\} \cup \{\infty\}$ in such a way that the standard frame is positive.

Let F and G be a pair of disjoint, bounded, surfaces in S^3 which should be thought of as "disks with bands". Let $\hat{U}$ and $\hat{V}$ be closed disjoint regular neighborhoods of F and G respectively and let U and V be obtained from them by "pinching to a point" near ∂F and ∂G. There is a natural involution on U and V which we denote by i.

Now let $\bar{D}^4_a$, $\bar{D}^4_b$, and $\bar{D}^4_c$, be three copies of D^4, all oriented so that a positive frame at a point in S^3, followed by the outward normal, is positive. Let $W^4(F, G) = \bar{D}^4_a \cup \bar{D}^4_b \cup \bar{D}^4_c/\sim$, where " $\sim$ " is defined as follows: On $U \cup V$ let $x_a = ix_a$, $x_b = ix_c$ if $x \in U$ and $y_a = iy_b$, $y_c = iy_c$ if $y \in V$. This has the effect of gluing the three balls together along their boundaries, the first and second balls along V, and the second and third balls along U, using the involution i. We let $M^3(F, G) = \partial W^4(F, G)$ oriented to that the inclusion $\bar{S}^3_b - U \to S^3$ preserves orientation, and we say the pair (F, G) is a representation of M^3 by bands.

AMS (MOS) subject classifications (1970). Primary 57A10; Secondary 55A10.

*The first author was supported by an NSF grant.

THEOREM A. *Every closed, oriented 3-manifold has a representation by bands.*

The proof we shall give leads to a more special representation, and a stronger theorem, appropriate to the computation of the μ-invariant.

COROLLARY B. *Given any closed, oriented 3-manifold M^3, there is a simply connected compact 4-manifold W^4, and a 3-fold irregular branched covering space map $p: W^4 \to D^4$, such that $\partial W^4 = M^3$. (See* [4] *or* [5] *for definitions of branched covering space and branch set.)*

PROOF OF COROLLARY B. Let $p: \bar{D}^4_a \cup \bar{D}^4_b \cup \bar{D}^4_c/\sim \;\to D^4/\sim$ be the natural map where $x \sim ix$ if $x \in U \cup V$. Equivalence classes are preserved and this is, in fact, a branched covering space. Note that $D^4/\sim$ is homeomorphic to D^4 and that the branch set is $F \cup G$ "pushed down" into D^4.

PROOF OF THEOREM A. We begin by constructing a particular branched covering space $p: X_g \to D^3$ where X_g is a genus g handlebody. Let $D^3 = E^3_- \cup \{\infty\} = \{x \leqq 0\} \cup \{\infty\}$, let $A_i = \{-1 \leqq x \leqq 0; y = 0, 2i \leqq z \leqq 2i + 1\} \cup \{(x + 1)^2 + (z - 2i - \frac{1}{2})^2 \leqq (\frac{1}{2})^2;\ y = 0\}$ and let $a_i = \overline{\partial A_i - \{x = 0\}}$, $1 \leqq i \leqq g + 2$. (See Figure 1.) Let $x_i \in \pi_1(D^3 - \bigcup \{a_i\}, \infty)$ be as indicated in Figure 1.

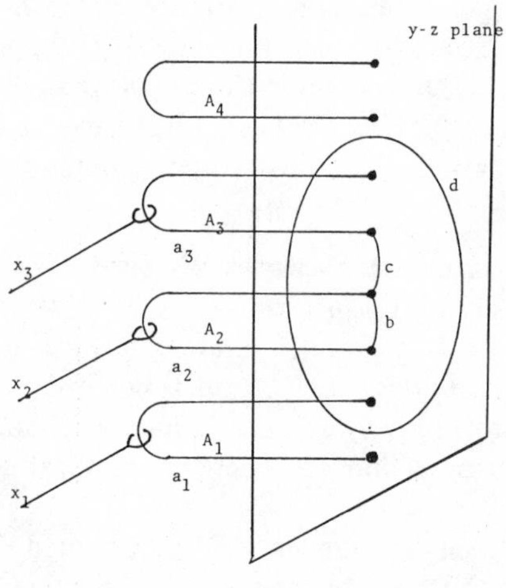

FIGURE 1

Now define a representation of the free group $\rho: \pi_1(D^3 - \bigcup a_i, \infty) \to \Sigma^3$ by $\rho(x_i) = (01)$, $2 \leqq i \leqq g + 2$, and $\rho(x_1) = (02)$. Let $p: X_g \to D$ be the branched covering space associated to the representation (see [5] for directions as to how to construct a branched covering space from a representation). Since X_g is obtained by pasting balls together along disjoint disks it is a handlebody.

Next we define some homeomorphisms of ∂D that lift to ∂X_g. An arc along the z-axis connecting the endpoints of a_i, $2 \leqq i \leqq g + 2$, or an endpoint of a_i to an endpoint of a_{i+1}, $2 \leqq i \leqq g + 1$, such as b or c in Figure 1, has as its inverse image, under $p: \partial X_g \to \partial D$, the union of an arc and a circle. A disk neighborhood on ∂D of such an arc has as its inverse image the disjoint union of a disk and an annulus in ∂X_g. A Dehn disk twist in such a disk lifts to the "disjoint union" of a disk twist and a Dehn annular twist. (See **[13]** for definitions of Dehn twist.) For the moment we call such a homeomorphism a type I homeomorphism. We note that if φ is a type I homeomorphism, the representations ρj_* and $\rho j_*\varphi_*$ of $\pi_1(S^2 - \bigcup_i a_i)$ onto Σ_3 are exactly the same, where j is the inclusion $(S^2 - \bigcup a_i) \to (D^3 - \bigcup a_i)$.

A circle in the yz plane, such as a d in Figure 1, that contains $A_i \cap S^2$ in its interior, $2 \leqq i \leqq m$, $A_j \cap S^2$ in its exterior, $m + 1 \leqq j \leqq g + 2$, and the upper endpoint of a_1 in its interior, has as its preimage under p a pair of circles, one of which double covers it, and one of which is mapped homeomorphically onto it, by p. The square of a Dehn annular twist about such a circle lifts to a Dehn annular twist plus the square of a Dehn annular twist. We shall call the square of such a Dehn annular twist a type II homeomorphism. We note again that the representations ρj_* and $\rho j_*\varphi_*: \pi_1(S^3 - \bigcup_i a_i) \to \Sigma_3$ are exactly the same if φ is type II. It is shown in **[4]** that any orientation preserving homeomorphism of ∂X_g is isotopic to a product of the lifts of type I and type II homeomorphisms.

Next, let $D^{3\prime} = E^3_+ \cup \{\infty\} = \{(x, y, z) \mid x \geqq 0\} \cup \{\infty\}$, let a_i' be the image of a_i under reflection in the y-z plane, and let us construct a branched covering space $p': X_g' \to D^{3\prime}$ exactly as before.

Suppose we have a 3-manifold M^3 represented as a Heegaard splitting: $M^3 = X_g \cup_\varphi X_g'$. We assume, isotoping φ, that φ is a product of lifts of types I and II homeomorphisms. (The two branched covering spaces over $S^2 = \partial D^3 = \partial D^{3\prime}$ are the same.) We can patch the two branched covering space maps together to get a branched covering space, $p \cup p': M^3 = X_g \cup_\varphi X_g' \to D^3 \cup_{\bar\varphi} D^{3\prime}$, branched over $\bigcup_i a_i \cup_{\bar\varphi} \bigcup_i a_i'$ where $\bar\varphi$ is a product of type I and type II homeomorphisms. Moreover, since $\rho j_*\varphi_* = \rho j_*$ and the representations of $\pi_1(D^3 - \bigcup_i a_i)$ and $\pi_1(D^{3\prime} - \bigcup_i a_i')$ agree on $\pi_1(S^2 - \bigcup a_i)$, we obtain a representation of $\pi_1(S^3 - \text{link}) = \pi_1(D^3 \cup_{\bar\varphi} D^{3\prime} - \bigcup_i a_i \cup_{\bar\varphi} \bigcup_i a_i')$ onto Σ_3 from which we could again construct M^3.

Suppose that $\bar\varphi = \prod_{i=1}^n f_i$ where f_i is of type I or II, and suppose that $f_i(t)$ is an isotopy of f_i with the identity, $0 \leqq t \leqq 1$, with $f_i(0) = f_i$ and $f_i(1) = \text{identity}$. (In practice, f_i is a homeomorphism whose restriction to a disk is a "twist," so we choose the isotopy merely to "unwind the twist" and to be supported on the same disk.) Now define $\bar\varphi_t = f_j(n(t - (j-1)/n)) f_{j+1} \cdots f_n$ if $(j-1)/n \leqq t \leqq j/n$; $1 \leqq j \leqq n$. Clearly $\bar\varphi_1$ is the identity, $\bar\varphi_0 = \bar\varphi$, and we can assume $\bar\varphi_t$ has compact support contained in $\{x = 0\}$.

The map $(x, y, z) \to (x, y, z)$ if $x \leqq -1$ or $x > 0$ and $(x, y, z) \to (x, \bar\varphi_{-x}^{-1}(y, z))$ if $-1 \leqq x \leqq 0$ defines a homeomorphism ψ from $D^3 \cup D^{3\prime} = S^3$ onto $D^3 \cup_{\bar\varphi} D^{3\prime}$. The set $\psi^{-1}(\bigcup_i a_i \cup_{\bar\varphi} \bigcup_i a_i')$ is the inverse image of the branch set. Since $f_j(\bigcup_i \partial a_i) = \bigcup_i \partial a_i$, we know from the definition of ψ that $\psi^{-1}(\bigcup_i a_i \cup_{\bar\varphi} \bigcup a_i') \cap \{x = t\} = \bigcup_i a_i \bigcup_i a_i' \cap \{x = t\}$ if $t \leqq -1$, $t \geqq 0$, or $t = -j/n$, $0 \leqq j \leqq n$. The set $\psi^{-1}(\bigcup_i a_i \cup_{\bar\varphi} \bigcup_i a_i') \cap \{x = s\}$ depends only on the isotopy $f_j(t)$ if $j - 1 \leqq -ns \leqq j$. It is not hard to see that the isotopy $f_j(t)$ can be chosen so that the projection of the inverse image of the branch set on the x-z plane appears as in Figure 2.

Next we must determine the representation. For any point P on the link not lying in the shadow of another point when the sun is far away on the positive y-axis, let $x(P)$ be the curve that comes straight down from ∞ to a point very near to P_0, circles the link once and then returns along the same line to ∞. Let $\hat\rho(P) = \hat\rho(x(P))$ where $\hat\rho$ is the induced representation, $\hat\rho = \rho\psi_*: \pi_1(S^3 - \text{link}) \to \Sigma_3$. It can

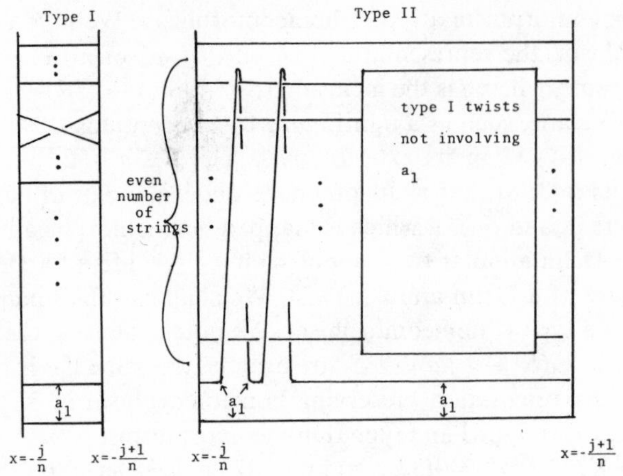

FIGURE 2

be seen that $\hat{\rho}$ is constant on any bridge of the link projected on the x-z plane and that $\hat{\rho}(-1, 0, k) = (01)$ if $4 \leqq k \leqq 2g + 5$. We now can show, by induction on j, and using the relations of form $a = bcb^{-1}$ that occur at each crossover, working from left to right in Figure 2, that $\hat{\rho}(-j/n, 0, 2) = \hat{\rho}(-j/n, 0, 3) = (02)$ and $\hat{\rho}(-j/n, 0, k) = (01)$ for $0 \leqq j \leqq n$ and $4 \leqq k \leqq 2g + 5$.

Next we choose a suitable splitting complex. We can assume that the part of the branch set not containing a_1' lies in the x-z plane except for small arcs near the crossing points where the overcrossing arc has small positive y-coordinate and the undercrossing arc has small negative y-coordinate (refer to Figure 2), and that the part of the branch set containing a_1' lies in the x-z plane except for the "double twists". Let the splitting complex K be the union of the following 2-complexes: First, a disk in the x-z plane large enough to contain the projection of the branch set; second, the part between the overcrossing and undercrossing arcs and the x-z plane; third, a "twisted ribbon" consisting of a 1-parameter family of line segments parallel to the x-axis, as in Figure 3; fourth, the points between the left edge of the twisted ribbon and the x-z plane (again see Figure 3).

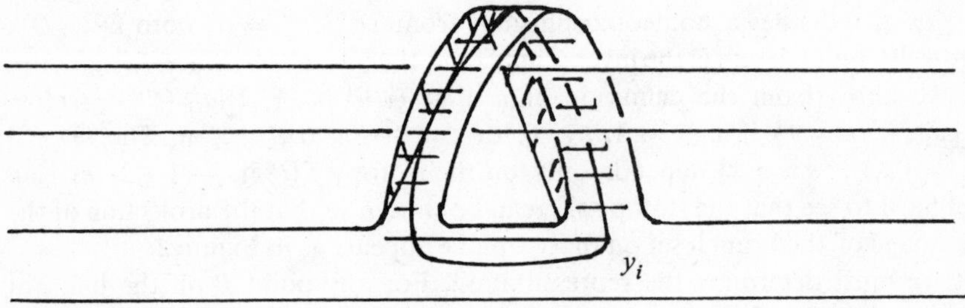

FIGURE 3

Thus K is a closed 2-complex, $S^3 - K$ is connected, and $\pi_1(S^3 - K) = 0$ so that K is a splitting complex. As in [6] we assign a permutation to each σ^2 in the complex K(the image in Σ_3 under $\hat{\rho}$ of a curve beginning at ∞, traversing K once in σ^2 and then returning to ∞) and then split three different copies of S^3 along K and paste them together according to the permutations. We will obtain exactly the same branched covering space if we delete the 2 simplexes from K that have the identity assigned to them before cutting and pasting.

Using the fact that $\hat{\rho}(-j/n, 0, 2) = \hat{\rho}(-j/n, 0, 3) = (02)$ and $\hat{\rho}(-j/n, 0, k) = (01)$, $0 \leqq j \leqq n$, $4 \leqq k \leqq 2g + 5$ and again referring to Figure 2 we see that the part of K to which permutations other than the identity have been assigned consists of the disjoint union of two bounded surfaces F and G. The surface F is the "checkerboard surface" spanned by the components of the link not containing a_1' lying mostly in the x-z plane with the permutation (01) assigned to it and the surface G consists of a disk with a series of nonorientable ribbons attached to it that loop through the holes in F, with the permutation (02) assigned to it.

Since splitting along F and G and pasting gives rise to the 3-manifold $M^3(F, G) = \partial W^4(F, G)$ the proof of Theorem A is complete.

Actually, we have proved more than Theorem A. We now state the somewhat stronger

THEOREM C. *Every closed, orientable 3-manifold M^3 has a representation by bands, $M^3 = M(F, G)$, such that $F \cup G$ has the following special properties*:

There is a basis $\{x_i\}$, $1 \leqq i \leqq r$, for $H_1(F; Z)$ and a basis $\{y_j\}$, $1 \leqq j \leqq s$ for $H_1(G; Z)$ such that if $z = \sum_{i=1}^{r} \varepsilon_i x_i$ and $w = \sum_{j=1}^{s} \varepsilon_j y_j$, ε_i, $\varepsilon_j = 0$ or 1, then z and w are represented by disjoint unions of directed simple closed curves L and M such that

(a) *L is the boundary of a disjoint union of 2-disks $\bigcup_{i=1}^{k} D_i^2$ lying in the x-z plane.*

(b) *The components of M are the boundaries of disjoint 2-disks lying in planes parallel to the y-z plane with no two disks in the same plane and with each component of M puncturing at most one D_i^2.*

PROOF OF THEOREM C. The projection of F onto the x-z plane is a retract of F. We can choose as basis $\{x_i\}$ for $H_1(F; Z)$ the clockwise boundaries of contiguous disks where each disk contains exactly one "hole" of the projection of F. If $z = \sum \varepsilon_i x_i$, $\varepsilon_i = 0$ or 1, then, by cancelling arcs running in opposite directions we obtain a link L representing z and satisfying (a).

The surface G is a disk with nonorientable bands attached. We choose as basis $\{y_j\}$ for $H_1(G; Z)$ the "centerlines" of these bands. After a slight isotopy of G (see Figure 3) these "centerlines" lie in distinct planes parallel to the y-z plane. If M is any disjoint union of "centerlines" then $L \cup M$ satisfies (a) and (b).

3. Computation of the μ-invariant. In this section we consider a fixed $M^3 = M(F, G) = \partial W^4(F, G)$ where the representation satisfies the conclusion of Theorem C.

Recall $W^4 = \bar{D}_a^4 \cup \bar{D}_b^4 \cup \bar{D}_c^4$, $\bar{D}_a^4 \cap \bar{D}_b^4 = V$, $\bar{D}_b^4 \cap \bar{D}_c^4 = U$, and $\bar{D}_a^4 \cap \bar{D}_c^4 = \varnothing$. If we let $X = \bar{D}_a^4 \cup \bar{D}_c^4$ and $Y = \bar{D}_b^4$ then we obtain an isomorphism $\varphi: H_2(W^4) \to H_1(F \cup G) = H_1(F) \oplus H_1(G)$ by a Mayer-Vietoris argument.

If x is an oriented curve in F, let $S(x)$ be the union of the cone on x in $\bar{D}_b$ and the cone on x in $\bar{D}_c$, oriented in such a way that $\varphi[S(x)] = [x]$. The sphere $S(y)$ is defined similarly for an oriented curve y in G. We easily obtain

PROPOSITION D. *The group $H_2(W^4)$ is a free abelian group generated by $\{S(x_1), \ldots, S(x_r), S(y_1), \ldots, S(y_s)\}$ where $\{x_1, \ldots, x_r\}$ is a basis for $H_1(F)$ and $\{y_1, \ldots, y_s\}$ is a basis for $H_1(G)$.*

If V^4 is an oriented 4-manifold an element z of $H_2(V^4)$ is called *characteristic* if its reduction modulo 2 is dual to the second Stiefel-Whitney class $w_2(V^4)$. A characteristic surface F is an oriented surface smoothly embedded in the interior of V^4 such that $[F]$ in $H_2(V^4)$ is characteristic.

Next we show, for $V^4 = W^4(F, G)$, how to find characteristic elements in $H_2(V^4)$ and we show that it is always possible in this case to find a characteristic surface that is a locally flat sphere. The intersection form of V^4 is the pairing $H_2(V^4) \times H_2(V^4) \to Z$ defined by $(z, w) \to \langle \hat{z} \cup \hat{w}, [V^4] \rangle$ where $\hat{z}$, $\hat{w}$ are the Poincaré duals of z and w and $[V^4]$ is the orientation generator for V^4. More geometrically, one can find oriented surfaces that represent z and w, put them in general position, and compute their algebraic intersection. This second method is particularly suitable for $V^4 = W(F, G)$ because, with the basis $\{x_i\}$, $\{y_j\}$ for $H_1(F \cup G)$ of Theorem C, all the elements of the induced basis $\{S(x_i), S(y_j)\}$ can be represented by locally flat spheres.

If z_1 and z_2 are oriented curves in S^3, let $l(z_1, z_2)$ denote their linking number. If $z_1 \in F \cup G$, let $\bar{z}_1$ be the algebraic number of half-twists of a ribbon R in $F \cup G$ with centerline z_1. Orient ∂R so that $\partial R = 2z_1$. Then $l(z_1, \partial R) = \bar{z}_1$. It follows from the definitions that $[S(z)] \cdot [S(z)] = \bar{z}$, $[S(x)] \cdot [S(y)] = l(x, y)$, $[S(x)] \cdot [S(x')] = \frac{1}{2}(\bar{w} - \bar{x} - \bar{x}')$, for closed curves $z \in F \cup G$, $x, x' \in F$, $y \in G$, and w representing $x + x'$. The intersection matrix for $W^4(F, G)$ can thus be read directly from the surface $F \cup G$.

An element $[z] \in H_2(W)$ is characteristic if $[z] \cdot [x] = [x] \cdot [x]$ modulo 2 for all $x \in H_2(W)$. A characteristic element will always exist because, reducing modulo 2, the intersection matrix defines an element of the dual space of $H_2(W; Z/2Z)$ which is a vector space over the field $Z/2Z$. In fact there is always a characteristic element z such that $\varphi(z) = \sum_{i=1}^{r} \varepsilon_i x_i + \sum_{j=1}^{s} \varepsilon_j y_j$. It is easy to compute z.

Next we construct a characteristic surface for z. By properties (a), (b) of Theorem C there is a disjoint set of disks in $\bar{D}_a^4$ whose boundary is $\sum_{i=1}^{r} \varepsilon_i x_i$, and a disjoint set of disks in $\bar{D}_c^4$ whose boundary is $\sum_{j=1}^{s} \varepsilon_j y_j$.

Let H be the union of these disks with the cone on $\varphi(z)$ in $\bar{D}_b^4$. The 2-complex H is a locally flat manifold except at the center of $\bar{D}_b^4$, and $[H] = z$ in $H_2(W^4)$. Remove from H, the intersection of H with the interior of the 4-ball of radius $\frac{1}{2}$ centered at the center of $\bar{D}_b^4$ and consider the link obtained by intersecting H with the sphere of radius $\frac{1}{2}$. By properties (a), (b) of Theorem C the link is of a particularly simple type.

This link is naturally spanned by a finite number of disjoint surfaces that are homeomorphic to disks with holes. We add this spanning surface to the rest of H to obtain $\hat{H}$, a locally flat characteristic surface. One can see that $\hat{H}$ is a finite union of spheres, since joining the rest of H has the effect of filling the holes in the disks.

Finally we join the various components of $\hat{H}$ with tubes to obtain a locally flat characteristic sphere K that represents z.

Recall that the index of an oriented 4-manifold V^4, with $H_2(V^4)$ free abelian, is the signature of the intersection form $H_2(V^4) \times H_2(V^4) \to Z$. Let $I(V^4)$ denote the index.

THEOREM E. *If* $H_1(M^3(F, G); Z/2Z) = 0$ *then* $\mu(M^3(F, G)) = K \cdot K - I(W^4(F, G))$ (mod 16).

PROOF OF THEOREM E. Let $\hat{W}^4$ be a simply connected parallelizable 4-manifold whose boundary is $M^3(F, G)$. By definition $\mu(M^3(F, G)) = I(\hat{W}^4)$. Let $W = \hat{W}^4 \cup W^4(F, G)$. Then W is a closed oriented 4-manifold with K for a characteristic surface. By the formula of Kervaire and Milnor [7], $K \cdot K -$ index $W \equiv 0$ (mod 16). By Novikov additivity [9], $I(W) = I(\hat{W}^4) + I(W^4(F, G))$. It follows that $\mu(M^3(F \cdot G)) \equiv I(\hat{W}^4) \equiv K \cdot K - I(W^4(F, G))$ (modulo 16). In the next section we shall use the formula of Theorem E to compute $\mu(M^3(F, G))$ in a specific example.

4. An example. To illustrate the preceding methods we shall compute the μ-invariant of a particular homology sphere. Let F and G be as in Figure 4 below, let $\{y_1, x_1, x_2, x_3\}$ be the indicated basis for $H_1(F \cup G)$, and let $\{S(y_1), S(x_1), S(x_2), S(x_3)\}$ be the corresponding basis for $H_2(W^4)$. Using the formulas in §3 for computing intersection numbers from linking numbers we see that the intersection matrix A for $W^4(F, G)$ is:

$$\begin{bmatrix} 1 & -1 & 0 & 0 \\ -1 & 3 & 2 & 1 \\ 0 & 2 & 3 & 2 \\ 0 & 1 & 2 & 1 \end{bmatrix}$$

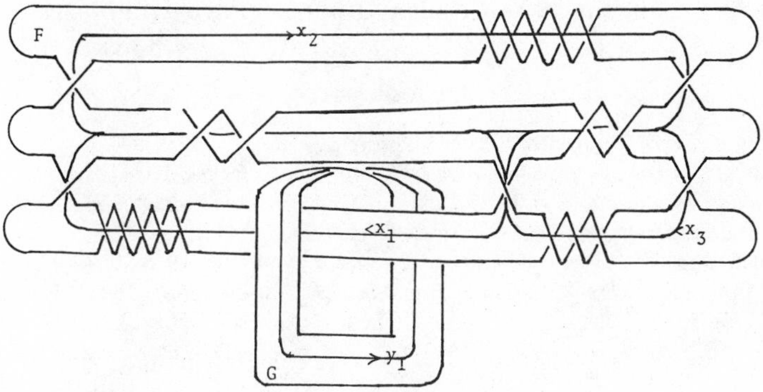

FIGURE 4

We compute determinant $A = -1$. By a Mayer-Vietoris argument this guarantees that $M^3(F, G)$ is a homology sphere.

We next compute a characteristic element directly from the matrix A. Of the sixteen possibilities $z = \varepsilon_1 y_1 + \varepsilon_1 x_1 + \varepsilon_2 x_2 + \varepsilon_3 x_3$, the one that works is $z = x_1 + x_2$. Thus $z \cdot z = x_1 \cdot x_1 + 2x_1 \cdot x_2 + x_2 \cdot x_2 = 3 + 4 + 3 = 10$.

We diagonalize A over the rationals to obtain its set of eigenvalues $\{1, 1, 2, -\frac{1}{2}\}$. Thus index $A = 2$. Hence $\mu(M^3(F, G)) \equiv 8$ (modulo 16).

Problem. Is $M^3(F, G)$ simply connected? One can determine a presentation for the fundamental group using the methods of **[10]**. Note that one is guaranteed $M^3(F, G) \neq S^3$ since $\mu(M^3(G, F)) \neq 0$.

5. Concluding remarks. Cappell and Shaneson **[11]** give an explicit formula for computing the μ-invariant of a homology sphere presented as a dihedral covering space of S^3 branched over a knot. Their formula is fairly complicated but it simplifies considerably if the knot is a ribbon knot. Conversations with Cappell were helpful to the authors in writing up this paper. Kaplan **[2]** shows how to compute the μ-invariant of M^3 if M^3 is presented as a framed link in the sense of Kirby. Also, Birman and Craggs **[12]** have a method for generating homology spheres and computing the μ-invariant, based on the mapping class groups.

All these methods can be used to construct potential counterexamples to the Poincaré conjecture. One constructs (as for example in §4) a homology 3-sphere whose μ-invariant is not zero, computes a presentation for the fundamental group, and attempts to show the group is trivial. Of course, there is no algorithm, in general, to determine if a group is trivial from its presentation, but in practice, with some work one can usually tell this. Using computers, it might be worthwhile to compute simplified presentations of the fundamental groups of, say, several hundred homology spheres with nonzero μ-invariant.

An interesting open question is whether or not every M^3 equals $M^3(F, G)$ with $F \cup G$ orientable. If $F \cup G$ is orientable then $W^4(F, G)$ is parallelizable and one has a simple formula for the μ-invariant. Using the operations defined in **[5]** we can show that either F or G can always be chosen orientable.

ADDED IN PROOF. Correspondence with F. González-Acuña revealed that he already knew a result similar to Corollary B for 2-fold covering spaces.

BIBLIOGRAPHY

1. R. Kirby, *A calculus for framed links in* S^3 (preprint).

2. S. Kaplan, *Constructing framed* 4-*manifolds with almost framed boundaries.*

3. H. Hilden, J. Montesinos and T. Thickstun, *Closed orientable* 3-*manifolds as* 3-*fold branched coverings of* S^3 *of a special type*, Pacific J. Math. **65** (1976), 65–75.

4. H. Hilden, 3-*fold branched coverings of* S^3, Amer. J. Math. **98** (1976), 989–997.

5. J. M. Montesinos, *Three-manifolds as* 3-*fold branched covers of* S^3, Quart. J. Math. Oxford Ser. (2) **27** (1976), 85–94.

6. L. P. Neuwirth, *Knot groups*, Ann. of Math Studies, No. 56, Princeton Univ. Press, Princeton, N. J., 1965.

7. M. Kervaire and J. Milnor, *On* 2-*spheres in* 4-*manifolds*, Proc. Nat. Acad. Sci. U.S.A. **47** (1961), 1651–1657. MR **24** #A2968.

8. F. Hirzebruch, W. D. Neumann and S. S. Koh, *Differentiable manifolds and quadratic forms*, Marcel Dekker, New York, 1971.

9. M. Atiyah and I. Singer, *The index of elliptic operators.* III, Ann. of Math. (2) **87** (1968), 546–604. MR **38** #5245.

10. R. H. Fox, *A quick trip through knot theory*, Topology of 3-manifolds and related topics, Prentice-Hall, Englewood Cliffs, N. J., 1962, pp. 120–167. MR **25** #3522.

11. S. Cappell and J. Shaneson, *Invariants of* 3-*manifolds*, Bull. Amer. Math. Soc. **81** (1975), 559–561. MR **51** #4209.

12. J. S. Birman and R. Craggs, *The* μ-*invariant of* 3-*manifolds and certain structural properties of the group of homeomorphisms of a closed oriented* 2-*manifold* (preprint).

13. J. S. Birman, *Braids, links, and mapping class groups*, Ann. of Math. Studies, No. 82, Princeton Univ. Press, Princeton, N. J., 1974. MR **51** #11477.

UNIVERSITY OF HAWAII

INSTITUTE FOR ADVANCED STUDY

Proceedings of Symposia in Pure Mathematics
Volume 32, 1978

A NEW DECOMPOSITON THEOREM FOR IRREDUCIBLE SUFFICIENTLY-LARGE 3-MANIFOLDS

WILLIAM JACO AND PETER B. SHALEN[1]

This paper is an expanded version of the address given by the first author. The address covered most of the material presented here as §3 without any detail toward proof. We have included some proofs where space permitted and given much more attention to applications. The paper is divided into four sections.

In §1 we state the main theorem and discuss the idea of its proof. We then give an outline of the proof in a special case. This theorem gives a classification (up to homotopy) of certain mappings of Seifert fibered spaces and I-bundles into an irreducible, sufficiently-large 3-manifold.

In §2 we discuss the Annulus-Torus Theorem of Waldhausen **[19]**. We then use the Mapping Theorem of §1 to give a description of the process of replacing a singular mapping of an annulus or a torus into a 3-manifold by an embedding of an annulus or a torus. Our work includes the case where the 3-manifold in question is closed; and for this case we completely describe the problem of replacing a singular mapping of a torus by an embedding of a torus.

In §3 we state the Decomposition Theorem of the title. We include a proof of the uniqueness of the decomposition

In §4 we apply the results of the previous sections to describe certain phenomena in the fundamental group of an irreducible, sufficiently-large 3-manifold. This includes a classification of centralizers of elements in the fundamental group, a description of the root structure of elements in the fundamental group, and a

---

AMS (*MOS*) *subject classifications* (1970). Primary 57A10; Secondary 55A10, 55A35.

Key words and phrases. Seifert fibered spaces, centralizers, roots, Dehn Lemma, loop theorem, sphere theorem, annulus theorem, torus theorem.

[1]The research of both authors is partially supported by NSF Grant MPS 71–03072.

description of two generator subgroups of the fundamental group of an irreducible, sufficiently-large, atoridal 3-manifold.

Similar results to those of §3 have been proved independently by K. Johannson **[12]**. While we used the Decomposition Theorem to describe certain phenomena of the algebraic structure of 3-manifold groups, Johannson used the Decomposition Theorem to describe phenomena of homotopy equivalences between irreducible, sufficiently-large 3-manifolds. The results of Johannson are described elsewhere in these PROCEEDINGS by F. Waldhausen.

We work throughout in the PL-category. All 3-manifolds are considered to be orientable; yet, 2-manifolds may or may not be orientable. In particular, we allow Seifert fibered spaces having a nonorientable decomposition surface, as well as I-bundles over a nonorientable surface.

A 3-manifold M is *irreducible* if every 2-sphere in M bounds a 3-cell in M. In light of the Kneser-Milnor theorem **[13]**, **[14]** which allows that any compact, orientable 3-manifold can be written uniquely as a connected sum of irreducible 3-manifolds and copies of $S^2 \times I$, the convenience of working with irreducible 3-manifolds imposes no real restriction to the purpose of our work.

A discussion of sufficiently-large 3-manifolds may be found in **[17]** or **[18]**. We only remind the reader that a compact, orientable and irreducible 3-manifold M is sufficiently-large if $H_1(M)$ is infinite. Hence, our results apply to all compact, orientable and irreducible 3-manifolds with nonempty boundary (these include the classical knot manifolds of the 3-sphere). Also, among compact, orientable and irreducible 3-manifolds, the sufficiently-large ones are precisely those having fundamental group a free product with amalgamation or an HNN-group.

1. The Mapping Theorem. In this section we state our main theorem. It is a classification (up to homotopy) of certain mappings of Seifert fibered spaces and interval bundles over surfaces into an irreducible, sufficiently-large 3-manifold. We shall also give a discussion of the idea of the proof of this theorem and an outline of the proof in a special case.

To begin we give some definitions and notational conventions that are required for the statement of our main theorem. The technical difficulties can be greatly reduced if one is willing to work in the absolute case as opposed to the strong relative version that we have chosen. However, we believe that the full strength of the theorem is in its relative version and the applications of this version to studying the structure of irreducible, sufficiently-large 3-manifolds. (In particular, see the paper reporting the work of B. Evans and W. Jaco also in these PROCEEDINGS.)

A pair $(\mathscr{X}, \mathscr{Y})$ is called a *polyhedral pair* if $\mathscr{X}$ is a polyhedron and $\mathscr{Y}$ is a subpolyhedron of X. A polyhedral pair $(\mathscr{X}, \mathscr{Y})$ is said to be *connected* if $\mathscr{X}$ is connected. A *component* of a polyhedral pair $(\mathscr{X}, \mathscr{Y})$ is a polyhedral pair (X, Y) where X is a component of $\mathscr{X}$ and $Y = X \cap \mathscr{Y}$. A polyhedral pair $(\mathscr{X}, \mathscr{Y})$ is *compact* if both $\mathscr{X}$ and $\mathscr{Y}$ are compact. The polyhedral pair $(\mathscr{X}', \mathscr{Y}')$ is *contained in the polyhedral pair* $(\mathscr{X}, \mathscr{Y})$, written $(\mathscr{X}', \mathscr{Y}') \subset (\mathscr{X}, \mathscr{Y})$ if $\mathscr{X}' \subset \mathscr{X}$ and $\mathscr{Y}' \subset \mathscr{Y}$.

Let $(\mathscr{P}, \mathscr{Q})$ and $(\mathscr{X}, \mathscr{Y})$ be polyhedral pairs. A PL-map $f: \mathscr{P} \to \mathscr{X}$ is a *map of pairs* if $f(\mathscr{Q}) \subset \mathscr{Y}$. We write this $f: (\mathscr{P}, \mathscr{Q}) \to (\mathscr{X}, \mathscr{Y})$. The maps $g: (\mathscr{P}, \mathscr{Q}) \to (\mathscr{X}, \mathscr{Y})$ and $f: (\mathscr{P}, \mathscr{Q}) \to (\mathscr{X}, \mathscr{Y})$ are *homotopic as maps of pairs* if there exists a map of pairs h: $(\mathscr{P} \times I, \mathscr{Q} \times I) \to (\mathscr{X}, \mathscr{Y})$ with $h(p, 0) = g(p)$ and $h(p, 1) = f(p)$ for each $p \in \mathscr{P}$.

An *n-manifold pair* is a polyhedral pair $(\mathscr{M}, \mathscr{T})$ where $\mathscr{M}$ is an n-manifold and $\mathscr{T}$ is an $(n - 1)$-manifold contained in $\partial\mathscr{M}$. A compact 3-manifold pair $(\mathscr{M}, \mathscr{T})$ is said to be *irreducible* if $\mathscr{M}$ is irreducible and $\mathscr{T}$ is incompressible. An irreducible 3-manifold pair $(\mathscr{M}, \mathscr{T})$ is *sufficiently-large* if each component of $\mathscr{M}$ is sufficiently-large.

A connected 3-manifold pair (M, T) is called an *I-pair* if there exists a homeomorphism h of M onto the total space of an I-bundle over a compact surface, not necessarily orientable, such that $h(T)$ is the total space of the associated ∂I-bundle.

Let E be a Seifert fibered 3-manifold over a surface B with projection $p: E \to B$, where B is not necessarily orientable [**16**]. Then we say that a subset T of E is *saturated* if $T = p^{-1}(p(T))$.

A connected 3-manifold pair (M, T) is called an *S^1-pair* if there exists a homeomorphism h of M onto the total space of a Seifert fibered 3-manifold over a compact surface such that $h(T)$ is a saturated subset.

A *Seifert pair* is a 3-manifold pair $(\mathscr{S}, \mathscr{F})$, each component of which is an I-pair or an S^1-pair. We shall call a connected Seifert pair (S, F) *degenerate* if either $\pi_1(S) = 1$ or $\pi_1(S) \approx \boldsymbol{Z}$ and $F = \emptyset$. Then a Seifert pair $(\mathscr{S}, \mathscr{F})$ is degenerate if it has a degenerate component.

Let $(\mathscr{M}, \mathscr{T})$ be an n-manifold pair and let (X, Y) be any connected polyhedral pair. A map of pairs $f: (X, Y) \to (\mathscr{M}, \mathscr{T})$ is *essentjal* if f_* is injective on $\pi_1(X)$ and the map $f: X \to \mathscr{M}$ cannot be homotoped as a map of pairs to a map g such that $g(X) \subset \mathscr{T}$. If $(\mathscr{X}, \mathscr{Y})$ is any polyhedral pair and $f: (\mathscr{X}, \mathscr{Y}) \to (\mathscr{M}, \mathscr{T})$ is a map of pairs, then we say that f is *essential* if for each component (X, Y) of $(\mathscr{X}, \mathscr{Y})$ the map of pairs $f|_X: (X, Y) \to (\mathscr{M}, \mathscr{T})$ is essential. Otherwise, we say that $f: (\mathscr{X}, \mathscr{Y}) \to (\mathscr{M}, \mathscr{T})$ is *inessential.*

Let $(\mathscr{M}, \mathscr{T})$ be an n-manifold pair and let $(\mathscr{X}, \mathscr{Y})$ be a polyhedral pair contained in $(\mathscr{M}, \mathscr{T})$. We say that $(\mathscr{X}, \mathscr{Y})$ is essential in $(\mathscr{M}, \mathscr{T})$ if the inclusion map $i: (\mathscr{X}, \mathscr{Y}) \to (\mathscr{M}, \mathscr{T})$ is an essential map of pairs.

1.1. MAPPING THEOREM. *Let (M, T) be a connected, sufficiently-large* 3-*manifold pair. Then there exists a compact, essential Seifert pair $(\Sigma, \Phi) \subset (M, T)$ such that for any nondegenerate, connected Seifert pair (S, F) and any essential map $f: (S, F) \to (M, T)$, f is homotopic as a map of pairs to a map $g: (S, F) \to (M, T)$ such that $g(S) \subset \Sigma$ and $g(F) \subset \Phi$.*

1.2. REMARK. Notice that each component of the frontier of Σ in the conclusion of Theorem 1.1 is either an incompressible annulus or an incompressible torus in M.

The idea of the proof is simple and rather straightforward. Since the manifold M is irreducible and sufficiently-large, the situation lends itself to an inductive procedure by splitting M along an incompressible surface. This immediately leads to technical problems; namely the procedure of splitting the manifold M forces one to consider a relative version of the theorem (one reason for pairs (M, T)) and at the same time creates a problem to the proper notion of a sufficiently-large 3-manifold pair. The reason for this latter problem is that splitting M along a surface W may result in a manifold that does not have incompressible boundary or the creation of a pair (M', T') where T' is not incompressible. One way around this problem is to develop the idea of 3-manifolds with a distinguished collection of surfaces in their boundary. However, we believe that the technical language and development that such a method calls for is not merited. In fact, the conclusion of the theorem suggests that there is a convenient way to split M to circumvent the particular problems. Our method puts the burden of proof on finding the appropriate surface for splitting M.

If N is a two-sided $(n - 1)$-dimensional submanifold of the n-manifold M, the *manifold obtained by splitting M along N* will be denoted by $\sigma_N(M)$. There is a canonical identification map $r: \sigma_N(M) \to M$ such that $r^{-1}(N)$ consists of two homeomorphic copies of N, each of which is in the boundary of $\sigma_N(M)$, and $r|r^{-1}(M - N)$ is one-one. In cases where we wish to be specific we denote the mapping r by $r_N(M)$. Suppose that $(\mathcal{M}, \mathcal{T})$ is an n-manifold pair and that $\mathcal{W}$ is a two-sided $(n - 1)$-dimensional submanifold of $\mathcal{M}$ such that $\mathcal{W} \cap \mathcal{T} = \emptyset$. If we split $\mathcal{M}$ along $\mathcal{W}$ there is a canonical pair obtained by splitting. Namely, let $\mathcal{M}' = \sigma_{\mathcal{W}}(\mathcal{M})$, let $\mathcal{W}' = r^{-1}(\mathcal{W})$ and define $\mathcal{T}'$ to be $\mathcal{W}' \cup r^{-1}(\mathcal{T})$. The pair $(\mathcal{M}', \mathcal{T}')$ is said to be *obtained by splitting the pair $(\mathcal{M}, \mathcal{T})$ at $\mathcal{W}$*. We denote $(\mathcal{M}', \mathcal{T}')$ by $\sigma_{\mathcal{W}}(\mathcal{M}, \mathcal{T})$. Now suppose that $(\mathcal{M}, \mathcal{T})$ is an n-manifold pair and $(\mathcal{N}, \mathcal{P})$ is a k-manifold pair. Furthermore, suppose that $\mathcal{W}$ is a two-sided $(n - 1)$-dimensional submanifold of $\mathcal{M}$ with $\mathcal{W} \cap \mathcal{T} = \emptyset$ and $f: (\mathcal{N}, \mathcal{P}) \to (\mathcal{M}, \mathcal{T})$ is a map of pairs such that f, regarded as a map of $\mathcal{N}$ into $\mathcal{M}$, is transverse with respect to $\mathcal{W}$. Then $\mathcal{V} = f^{-1}(\mathcal{W})$ is a two-sided $(k - 1)$-dimensional submanifold of $\mathcal{N}$ and $\mathcal{V} \cap \mathcal{P} = \emptyset$. In this situation we get a natural map of pairs $f': \sigma_{\mathcal{V}}(\mathcal{N}, \mathcal{P}) \to \sigma_{\mathcal{W}}(\mathcal{M}, \mathcal{T})$. We shall denote f' by $\sigma_{\mathcal{W}}(f)$ and call it the *map obtained from f by splitting at $\mathcal{W}$*. If f: $(\mathcal{N}, \mathcal{P}) \to (\mathcal{M}, \mathcal{T})$ is a map of pairs and $(\mathcal{M}, \mathcal{T})$ is a sufficiently-large 3-manifold pair, then a two-sided 2-manifold in $\mathcal{M}$ is said to be *suitable for splitting f* if $\mathcal{W}$ is incompressible and $\mathcal{W} \cap \mathcal{T} = \emptyset$.

In this terminology, we consider a connected, sufficiently-large 3-manifold pair (M, T). We let $\mathfrak{A}$ denote the set of all triples (S, F, h) where (S, F) is a nondegenerate compact, connected Seifert pair and h: $(S, F) \to (M, T)$ is an essential map. Now for $u = (S, F, h) \in \mathfrak{A}$, set $S_u = S$, $F_u = F$ and $h_u = h$. Let $\mathcal{S}$ denote the disjoint union of the S_u as u ranges over $\mathfrak{A}$; set $\mathcal{F} = \bigcup_{u \in \mathfrak{A}} F_u \subset \mathcal{S}$. Define $f: \mathcal{S} \to M$ by $f|S_u = h_u$ for all $u \in \mathfrak{A}$. Then $(\mathcal{S}, \mathcal{F})$ is a nondegenerate Seifert pair and f: $(\mathcal{S}, \mathcal{F}) \to (M, T)$ is an essential map of pairs. While $(\mathcal{S}, \mathcal{F})$ may have infinitely many components, each component of $(\mathcal{S}, \mathcal{F})$ is a compact Siefert pair.

We now consider the essential map f: $(\mathcal{S}, \mathcal{F}) \to (M, T)$ of the nondegenerate Seifert pair $(\mathcal{S}, \mathcal{F})$ into the sufficiently-large 3-manifold pair (M, T); and our objective is to find a two-sided surface W in M which is suitable for splitting f so that our induction hypothesis may be applied.

In order to do this latter part, we need a notion of complexity. If the surface W is incompressible and ∂-incompressible, then standard complexity notions (in particular, those in **[18]** made an invariant of the manifold as opposed to an invariant of a handle decomposition) are sufficient. However, since our surface W must be chosen so as not to meet T, it may be the case that W is not ∂-incompressible; in fact, W may be parallel into ∂M (but never parallel into $\partial M - T$). For this reason, we need to define a complexity of a 3-manifold pair (M, T). Then our objective is to find a two-sided surface W in M which is suitable for splitting f and has the property that if (M', T') is a component of $\sigma_W(M, T)$, then the complexity of (M', T') is strictly less than that of (M, T).

We shall outline the proof under the assumption that there exists a surface W which is suitable for splitting f and that the complexity of any component (M', T') of $\sigma^W(M, T)$ is less than the complexity of (M, T).

OUTLINE OF PROOF OF THEOREM 1.1.

Step 1. Consider mappings of $(\mathscr{S}, \mathscr{F})$ into (M, T) which are equivalent (as maps of pairs) to the original map $f: (\mathscr{S}, \mathscr{F}) \to (M, T)$ and have the additional property that the inverse of W is an incompressible 2-manifold. Standard methods allow that this collection of mappings is not empty. Now, among this collection, choose a map (also called f) $f: (\mathscr{S}, \mathscr{F}) \to (M, T)$ so that $V = f^{-1}(W)$ has a minimal number of components.

Step 2. By assumption $W \cap T = \emptyset$, therefore $V \cap \mathscr{F} = \emptyset$. By classifying incompressible surfaces in a compact, connected Seifert pair (S, F) where the surface does not meet F, we are able to show that the split pair $\sigma_V(\mathscr{S}, \mathscr{F})$ is a Seifert pair. Furthermore, by Step 1 we are able to conclude from the minimality on the number of components of $V = f^{-1}(W)$ that the Seifert pair $\sigma_V(\mathscr{S}, \mathscr{F})$ is nondegenerate and that the split map $\sigma_W(f): \sigma_V(\mathscr{S}, \mathscr{F}) \to \sigma_W(M, T)$ is essential.

Step 3. We apply the induction hypothesis to each component of $\sigma_W(M, T)$. In this way we obtain an essential Seifert pair (Σ', Φ') in $\sigma_W(M, T)$ such that the split map $\sigma_W(f)$ is homotopic (as a map of pairs) to a map $g': \sigma_V(\mathscr{S}, \mathscr{F}) \to \sigma_W(M, T)$ having the property that if (S', F') is a component of $\sigma_V(\mathscr{S}, \mathscr{F})$, then $g'(S') \subset \Sigma'$ and $g'(F') \subset \Phi'$.

Step 4. The idea now is to let (Σ, Φ) be the Seifert pair obtained by gluing together the components of (Σ', Φ') via the mapping $r_W(M)$ and letting $g: (\mathscr{S}, \mathscr{F}) \to (M, T)$ be the mapping defined from g', $r_V(\mathscr{S})$ and $r_W(M)$.

The first problem is to have the components of $\Phi' \cap (r_W(M))^{-1}(W)$ to match up under the mapping $r_W(M)$. Similar to this problem is making g well defined; namely, if v is a component of V and v_0 and v_1 are the components of $(r_V(\mathscr{S}))^{-1}(v)$, then $g|v$ should be definable by considering $r_W(M) \cdot g'|_{v0}$ or $r_W(M) \cdot g'|_{v1}$. Both of these conditions are handled by developing some new notions about mappings into 2-manifolds and setting up a more technical but stronger version of Theorem 1.1 to proceed inductively.

Hence, the problem that remains is to prove that the pair (Σ, Φ) is a Seifert pair.

In order to prove that (Σ, Φ) is a Seifert pair, it is sufficient to prove it in the case that $W \subset \Sigma$, both W and Σ are connected, and $W \cap \Phi = \emptyset$. Along with these assumptions we have that $(\Sigma', \Phi') = \sigma_W(\Sigma, \Phi)$ is a Seifert pair.

There are three cases to consider.

Case 1. *W does not separate Σ and $\sigma_W(\Sigma, \Phi)$ is an I-pair.*

If follows immediately that $\Phi = \emptyset$ and that Σ is a surface bundle over S^1 with fibers homeomorphic to W. To prove that Σ is a Seifert fiber space it is sufficient to prove that the monodromy is periodic. The theory of periodic homeomorphisms of surfaces in **[8]** was developed to handle this case.

Case 2. *W separates Σ and each component of $\sigma_W(\Sigma, \Phi)$ is an I-pair.*

There is only one nontrivial consideration in this case. That is when each component of $\sigma_W(\Sigma, \Phi)$ is a twisted I-bundle over a nonorientable surface. There are a number of ways to handle this case. The objective is to prove that Σ admits a Seifert fibration (observe that $\Phi = \emptyset$). In order to prove this we apply Case 1 to the canonical two sheeted covering $\tilde{\Sigma}$ of Σ to give $\tilde{\Sigma}$ the structure of a Seifert fiber space. We then prove that this situation allows one to give Σ a Seifert fibering.

Case 3. *Some component of* $\sigma_W(\Sigma, \Phi)$ *is an* S^1*-pair and not an I-pair.*

Set $r = r_W(\Sigma)$ and $(\Sigma', \Phi') = \sigma_W(\Sigma, \Phi)$. Then $r^{-1}(W) \subset \Phi'$. Since some component of (Σ', Φ') is an S^1-pair, either W is an annulus or W is a torus. It follows that either (Σ', Φ') is connected or (Σ', Φ') has two components and each is an S^1-pair.

If W is an annulus then it follows immediately that (Σ, Φ) is an S^1-pair.

If W is a torus and W_0, W_1, are the components of $r^{-1}(W)$, then both W_0 and W_1 are contained in Φ' and hence are unions of fibers in a Seifert fibration of Σ'. If α_i is the homotopy class of a fiber of the Seifert fibration of Σ' in W_i $(i = 0, 1)$, then we use the fact that each component of $(\mathcal{S}, \mathcal{F})$ must be an S^1-pair to prove that $r(\alpha_0)$ is homotopic to $r(\alpha_1)$ in W. This establishes that (Σ, Φ) is an S^1-pair.

The argument proving that W exists (along with all of the details for the proofs of Steps 1—4) appears in [9].

2. The Torus-Annulus Theorems. The Torus Theorem and the Annulus Theorem were first announced by Waldhausen in [19]. Later, in a series of papers by Feustel [4] and Cannon and Feustel [2], proofs of these theorems, as stated by Waldhausen, were given. In this section we show how these theorems follow immediately as corollaries of our Theorem 1.1. By using this method we are able to give a complete description for singular mappings of tori; thereby giving a Torus Theorem in the case of closed manifolds.

Let (M, T) be a 3-manifold pair. A mapping f: $(S^1 \times I, S^1 \times \partial I) \to (M, T)$ is *essential relative to* T, written f is essential (rel T), if f induces an injection of both $\pi_1(S^1 \times I)$ into $\pi_1(M)$ and $\pi_1(S^1 \times I, S^1 \times \partial I)$ into $\pi_1(M, T)$. If f: $(S^1 \times I, S^1 \times \partial I) \to (M, \partial M)$ and f is essential (rel ∂M), then we say f is *essential*. Observe that if (M, T) is a sufficiently-large 3-manifold pair, then a map f: $(S^1 \times I, S^1 \times \partial I) \to (M, T)$ being essential (rel T) is equivalent to the map f inducing an injection of $\pi_1(S^1 \times I)$ into $\pi_1(M)$ and not being homotopic to a mapping g: $(S^1 \times I, S^1 \times \partial I) \to (M, T)$ with $g(S^1 \times I) \subset T$. In this relative situation the notion corresponds to our notion of an essential mapping of a Seifert pair. A mapping f: $S^1 \times S^1 \to M$ is *essential* if f induces an injection of $\pi_1(S^1 \times S^1)$ into $\pi_1(M)$.

2.1. Annulus Theorem. *Let M be an irreducible* 3-*manifold. If* f: $(S^1 \times I, S^1 \times \partial I) \to (M, \partial M)$ *is essential, then there exists an embedding* g: $(S^1 \times I, S^1 \times \partial I) \to (M, \partial M)$ *which is essential. Furthermore, if* $f \mid S^1 \times \partial I$: $S^1 \times \partial I \to \partial M$ *is an embedding, then the embedding g may be chosen so that* $g \mid S^1 \times \partial I = f \mid S^1 \times \partial I$.

Proof. Since f is essential, there is no loss in generality to assume that ∂M is incompressible. Hence, we may assume that the pair $(M, \partial M)$ is a sufficiently-large 3-manifold pair.

By Theorem 1.1 there exists an essential Seifert pair $(\Sigma, \Phi) \subset (M, \partial M)$ such that any essential map of a Seifert pair (S, F) into $(M, \partial M)$ is homotopic (as a map of pairs) to a map taking S into Σ and F into Φ. In particular, since $(S^1 \times I, S^1 \times \partial I)$ has the homotopy type of the Seifert pair $(S^1 \times I \times I, S^1 \times I \times \partial I)$, the mapping f may be homotoped (as a map of pairs) to a mapping f': $(S^1 \times I, S^1 \times \partial I) \to (M, \partial M)$ with $f'(S^1 \times I) \subset \Sigma$ and $f'(S^1 \times \partial I) \subset \Phi$. Let (S, F) denote the component of (Σ, Φ) containing $f'(S^1 \times I)$.

We may assume that each component of Fr S is an essential embedding of an annulus or torus. Since $F \neq \emptyset$, either some component of Fr S is an essential, embedded annulus in M or we need only show that an essential mapping $f'\colon (S^1 \times I, S^1 \times \partial I) \to (S, F)$ implies that there exists an essential embedding $g\colon (S^1 \times I, S^1 \times \partial I) \to (S, F)$.

If (S, F) is an I-pair, then S is homeomorphic to an I-bundle over a compact surface G and F corresponds to the associated ∂I-bundle. Furthermore, the existence of f' forces the Euler characteristic of G to be less than zero. It is now easy to exhibit essential embeddings of annuli in (S, F).

If (S, F) is an S^1-pair, then S admits the structure of a Seifert fiber space over a compact surface B with projection $p\colon S \to B$ where F is a saturated subset over a submanifold of ∂B. Again the existence of f' implies that if B is a disk, then S has at least two singular fibers ($\partial B \neq 0$). It is now easy to construct essential embeddings of annuli in (S, F) by considering saturated annuli over essential spanning arcs in B.

To obtain the relative version that if $f | S^1 \times \partial I$ is already an embedding, then g may be chosen so that $g | S^1 \times \partial I = f | S^1 \times \partial I$ requires only the observation that in the above argument we may assume that $f' | S^1 \times \partial I$ is an embedding and, in ∂M, $f' | S^1 \times \partial I$ is homotopic to $f | S^1 \times \partial I$.

REMARK. Another relative version of the Annulus Theorem which is useful in practice is that if $f\colon (S^1 \times I, S^1 \times \partial I) \to (M, T)$ is essential (rel T), then the embedding g may be chosen so that $g(S^1 \times \partial I)$ is contained in a neighborhood of $f(S^1 \times \partial I)$.

The Torus Theorem is similar in nature to the Annulus Theorem, the difference being the consideration of essential singular mappings of tori as opposed to essential singular mappings of annuli. However, the statement of the Torus Theorem takes a slightly more complicated form. This is due to the fact that a theorem for replacing essential singular mappings of tori, as was done in the case of annuli, is false. The standard examples of irreducible 3-manifolds with infinite fundamental group which are not sufficiently-large **[17]** admit essential singular mappings of tori into them but contain no embedded injective surface. These examples are certain Seifert fiber spaces having decomposition surface a 2-sphere and having three singular fibers. It turns out that even in the case of a closed sufficiently-large 3-manifold M there may be essential singular mappings of tori into M; yet no essential embeddings of tori into M. Again such examples are Seifert fiber spaces having decomposition surface a 2-sphere and having three singular fibers. (Another way of viewing these latter examples is to consider the closed 3-manifolds obtained by attaching a solid torus to the boundary of a torus knot manifold by identifying the boundary of the meridian disk of the torus to the longitude of the torus knot manifold.)

It turns out, however, that by using our Theorem 1.1 we are able to completely describe the situation.

2.2. TORUS THEOREM. *Let M be a compact, irreducible and sufficiently-large 3-manifold. If $f\colon S^1 \times S^1 \to M$ is essential, then either there exists no embedding $g\colon S^1 \times S^1 \to M$ which is essential or M is a Seifert fiber space with decomposition surface a two-sphere and having three singular fibers.*

PROOF. Let $(M, \emptyset)$ be the sufficiently-large 3-manifold pair of our Theorem 1.1. Then there exists an essential Seifert pair $(\Sigma, \emptyset) \subset (M, \emptyset)$ and a mapping $f'\colon S^1 \times S^1 \to \Sigma$ such that f' is homotopic in M to f.

Each component of $\partial\Sigma$ is an incompressible torus or $\Sigma = M$ is closed. In the former case it is easy to obtain an embedding $g\colon S^1 \times S^1 \to M$ which is essential. In the latter case, M admits a Seifert fibration with decomposition surface a closed surface B, projection map ρ and (say) q singular fibers ($q \geqq 0$).

If B is any surface except S^2 on $\boldsymbol{RP}^2$, real projective two space, then B contains a noncontractible two-sided simple closed curve J and $p^{-1}(J)$ gives the desired embedding. If B is either S^2 or $\boldsymbol{RP}^2$, then by assumption M is sufficiently-large and therefore the Seifert fibration of M has at least three singular fibers in the case of S^2 and at least two singular fibers in the case of $\boldsymbol{RP}^2$. If $B = S^2$ and M has at least four singular fibers, then there exists a simple closed curve J in B such that $p^{-1}(J)$ is the desired embedding. If $B = \boldsymbol{RP}^2$ and M has at least two singular fibers, then if J is the square of the orientation reversing curve in $\boldsymbol{RP}^2$, then $p^{-1}(J)$ is the desired embedding.

The conclusion of the theorem follows.

Let M be a 3-manifold. A subgroup H of $\pi_1(M)$ is *peripheral* if there exists a component C of ∂M such that H is conjugate into $\mathrm{Im}(\pi_1(C) \subsetneq \pi_1(M))$. If H is abelian, then for H to be peripheral it is sufficient that each element of H be peripheral; i.e. for each element $h \in H$ there exists a component C_h of ∂M such that h is conjugate into $\mathrm{Im}(\pi_1(C_h) \subsetneq \pi_1(M))$.

We say that a mapping $f\colon S^1 \times S^1 \to M$ is *peripheral* if f_* takes $\pi_1(S^1 \times S^1)$ into a peripheral subgroup of $\pi_1(M)$. Otherwise, f is *nonperipheral.* In some cases (see for example **[3]** and **[6]**) one needs to consider essential, nonperipheral mappings of $S^1 \times S^1$ into a 3-manifold M and determine when there exists an essential, nonperipheral embedding of $S^1 \times S^1$ into M. In fact, the original statement of the Torus Theorem by Waldhausen was in these terms.

Our next theorem gives a complete description of this situation.

2.3. THEOREM. *Let M be a compact, irreducible and sufficiently-large 3-manifold. If $f\colon S^1 \times S^1 \to M$ is essential and nonperipheral, then either there exists an embedding $g\colon S^1 \times S^1 \to M$ which is essential and nonperipheral or M is a Seifert fiber space with decomposition surface B and one of the following holds:*

(i) *B is a 2-sphere and M has three singular fibers;*

(ii) *B is a disk and M has two singular fibers;*

(iii) *B is an annulus and M has fewer than two singular fibers;*

(iv) *B is a Moebius band and M has fewer than two singular fibers.*

As an immediate corollary to this theorem we have the original Torus Theorem as announced by Waldhausen **[19]**.

2.4. COROLLARY. *Let M be a compact, irreducible 3-manifold with $\partial M \neq \emptyset$. If $f\colon S^1 \times S^1 \to M$ is essential and nonperipheral, then either there exists an embedding $g\colon S^1 \times S^1 \to M$ which is essential and nonperipheral or the conclusion of the Annulus Theorem holds.*

In **[11]** Johannson announces a classification up to homotopy of mappings of tori into 3-manifolds.

3. The Decomposition Theorem. If (M, T) is a sufficiently-large 3-manifold pair, then by Theorem 1.1 it has associated to it a compact Seifert pair (Σ, Φ) which is fixed; and then for any nondegenerate Seifert pair (S, F) and essential mapping $f: (S, F) \to (M, T)$, f can be deformed (as a map of pairs) to a mapping of (S, F) into (Σ, Φ). Immediately one asks to what extent, if any, the Seifert pair (Σ, Φ) is unique.

It is not difficult to see that without any further requirements than those mentioned above for the pair (Σ, Φ), then (Σ, Φ) is not unique. However, by requiring two further properties we have that (Σ, Φ) is unique up to isotopy. In fact, if in addition to these two properties we require that $\Sigma \cap \partial M = \Sigma \cap T = \Phi$, then (Σ, Φ) is unique up to ambient isotopy of M leaving $\partial M - T$ fixed. In particular, when ∂M is incompressible and $T = \partial M$, then (Σ, Φ) can always be chosen to be unique up to ambient isotopy of M.

Let (M, T) be a sufficiently-large 3-manifold pair. Suppose that the Seifert pair $(\Sigma, \Phi) \subset (M, T)$ satisfies the following properties:

(i) *For any nondegenerate Seifert pair* (S, F) *and any essential map* $f: (S, F) \to (M, T)$, f *is homotopic as a map of pairs to a map* $g: (S, F) \to (M, T)$ *such that* $g(S) \subset \Sigma$ *and* $g(F) \subset \Phi$.

(ii) *Each component of* Fr Σ *is either an essential torus or an annulus which is essential* (rel T).

(iii) *No proper subcollection of components of* (Σ, Φ) *satisfies* (i) *and* (ii).

We shall prove in the next theorem that each sufficiently-large 3-manifold pair (M, T) has an associated Seifert pair (Σ, Φ) which satisfies (i)–(iii) above. Furthermore, we shall prove that such a (Σ, Φ) is unique up to isotopy. Such a Seifert pair satisfying (i)–(iii) above is called the *characteristic Seifert pair for* (M, T). In the case that $T = \partial M$ and (Σ, Φ) is the characteristic Seifert pair for $(M, \partial M)$ we say that (Σ, Φ) is the *characteristic Seifert pair for M*.

A result similar to our next theorem for the case $T = \partial M$ was announced by K. Johannson **[11]**.

3.1. THEOREM. *Let* (M, T) *be a sufficiently-large* 3-*manifold pair. Then a characteristic Seifert pair for* (M, T) *exists and is unique up to isotopy.*

PROOF. Suppose that A is an incompressible annulus embedded in M with $\partial A \subset T$. If A is not essential (rel T), then A is parallel through M to an annulus S' in T. Using this fact and Theorem 1.1 we can prove the existence of a Seifert pair $(\Sigma_1, \Phi_1) \subset (M, T)$ and satisfying both conditions (i) and (ii) above. Among all such Seifert pairs let (Σ, Φ) be one having a minimal number of components. Clearly (Σ, Φ) satisfies conditions (i)–(iii) above.

3.2. CLAIM. *If* $(\Sigma, \Phi) \subset (M, T)$ *satisfies conditions* (i)–(iii), *then* (Σ, Φ) *is unique up to isotopy.*

PROOF. Suppose that $(\Sigma', \Phi') \subset (M, T)$ is a Seifert pair and (Σ', Φ') satisfies conditions (i)–(iii) above.

First, we prove that there is a bijective correspondence between the components of (Σ, Φ) and the components of (Σ', Φ') such that a component (S, F) of (Σ, Φ) corresponds to the component (S', F') of (Σ', Φ') if and only if (S, F) can be deformed (as a pair) into (S', F') and conversely. To see this we use condition (i) above for (Σ', Φ') to obtain a deformation between the inclusion mapping of (Σ, Φ) into (M, T) and a mapping of (Σ, Φ) into (M, T) with $\Sigma \subset \Sigma'$ and $\Phi \subset \Phi'$ (this is a deformation of pairs). We use this deformation and assign a correspondence from the components of (Σ, Φ) into the components of (Σ', Φ') such that the component (S, F) of (Σ, Φ) corresponds to the component (S', F') of (Σ', Φ') if (S, F) is deformed into (S', F'). Now, since (Σ, Φ) satisfies condition (i) above and (Σ', Φ') satisfies condition (iii) above, it follows that this correspondence is surjective. We then use the symmetry of the situation to obtain a surjective correspondence from the components of (Σ', Φ') onto the components of (Σ, Φ) such that the component (S', F') of (Σ', Φ') corresponds to the component (S, F) of (Σ, Φ) if (S', F') deforms into (S, F) under some preassigned deformation of (Σ', Φ') into (Σ, Φ). Since the number of components involved is finite, the correspondences are forced to be injective. This establishes the desired bijection.

If $(N, R) \subset (M, T)$ is an irreducible 3-manifold pair and each component of Fr N is incompressible in M, then (N, R) is an *injective pair in* (M, T). We have the following lemma.

3.3. LEMMA. *Let (M, T) be a sufficiently-large 3-manifold pair. Suppose that $(N, R) \subset (M, T)$ and $(N', R') \subset (M, T)$ are connected injective pairs. Furthermore, suppose that the inclusion mapping of (N', R') into (M, T) is homotopic (as a map of pairs) to a mapping $g: (N', R') \to (M, T)$ with $g(N') \subset N$ and $g(R') \subset R$. Then there exists an isotopy taking (N', R') into (N, R).*

PROOF. Since each component of R is incompressible in T and R' homotopes through T into R, it follows from Lemma 4.1 of [7] that there exists an isotopy of M fixed on $\partial M - T$ moving R' into R. Hence, there is no loss in generality to assume that $R' \subset R$ and the homotopy taking (N', R') into (N, R) is constant on R'.

Let $(\tilde{M}, p)$ denote the covering space of M corresponding to $\pi_1(N)$. There is a lifting $\tilde{N}$ of N to $\tilde{M}$ such that $p | \tilde{N}: \tilde{N} \to N$ is a homeomorphism. It follows from [3] that $\tilde{M}$ has a manifold compactification to a manifold Q where Q is obtained from $\tilde{N}$ via compactification of each component of $\tilde{M}$ − Int $\tilde{N}$ as a product $C \times I$ where C is a component of $F_R(\tilde{N})$ and the product structure is such that $C \times \{0\}$ corresponds to C.

The homotopy taking (N', R') into (N, R) lifts to $\tilde{M}$. Let $\tilde{N}'$ denote the lifting of N'. Using the product structure in each component of $\tilde{M}$ − Int $\tilde{N}$, one obtains an isotopy taking $\tilde{N}'$ into $\tilde{N}$. (If $N \cap \partial M = R$ and $N' \cap \partial M = R'$, then there exists an ambient isotopy of $\tilde{M}$ taking $\tilde{N}'$ into $\tilde{N}$, which is fixed on $\partial \tilde{M}$.) However, since p restricts to a homeomorphism on both $\tilde{N}'$ and $\tilde{N}$ and the isotopy through $\tilde{M}$ is by small deformations across products from a surface in Fr $\tilde{N}'$ to a surface in Fr $\tilde{N}$ one can copy the isotopy in $\tilde{M}$ by an isotopy in M. This gives us the desired conclusion of Lemma 3.3.

Now, returning to the proof of Theorem 3.1, we have for each component (S, F) of (Σ, Φ) an isotopy h_t $(0 \leqq t \leqq 1)$ taking (S, F) into a component (S', F') of (Σ', Φ') such that $h_t(F) \subset T$ for $0 \leqq t \leqq 1$. Similarly we have an isotopy g_t $(0 \leqq t \leqq 1)$ taking (S', F') into (S, F) such that $g_t(F') \subset T$ for $0 \leqq t \leqq 1$. Let $(S_1', F_1') = g_1(S', F')$ and let $(S_1, F_1) = g_1 \cdot h_1(S, F)$. If C is a component of Fr S, let $C_1 = g_1 \cdot h_1(C)$.

Again using a covering space argument and the fact that if $(\tilde{M}, p)$ is the covering space of M corresponding to $\pi_1(C)$, then $\tilde{M}$ has a manifold compactification to the manifold $C \times I$ with the lifting $\tilde{C}$ of C corresponding to $C \times \{\frac{1}{2}\}$, it follows that

C and C_1 are parallel in M. For otherwise, some components of Fr S would be an annulus parallel into T. This latter situation would contradict condition (ii) for (Σ, Φ) to be a characteristic Seifert pair for (M, T). It now follows that between C and C_1 there exists a unique component C_1' of Fr S_1' which is parallel in M to C; for otherwise, some component of Fr S_1' would be an annulus parallel into T contradicting condition (ii) for (Σ', Φ') being a characteristic Seifert pair for (M, T).

From the above arguments it follows that there exists an isotopy through M taking (S', F') onto (S, F). The desired conclusion of Theorem 3.1 follows since the homotopies taking a component (S_1, F_1) of (Σ, Φ) distinct from (S, F) into a component (S_1', F_1') of (Σ', Φ') distinct from (S', F') can be moved off of $(S, F) = (S', F')$.

This completes the proof of Theorem 3.1.

Observe that the only reason that we were unable to obtain an ambient isotopy of M (fixed on $\partial M - T$) taking (Σ, Φ) onto (Σ', Φ') was the possibility that either $\Sigma \cap \partial M \neq \Sigma \cap T = \Phi$ or $\Sigma' \cap \partial M \neq \Sigma' \cap T = \Phi'$. We do, however, have

3.4. Corollary. *Let M be an irreducible sufficiently-large 3-manifold with incompressible boundary (possibly empty). Then a characteristic Seifert pair of M exists and is unique up to ambient isotopy of M.*

Recall that a classical knot in S^3 is *simple* if it has no companions. This is equivalent to the associated knot space having no embedded, essential annuli or embedded, essential, nonperipheral tori. We borrow from this terminology in the following way.

Suppose that (N, R) is a sufficiently-large 3-manifold pair. We say that N is *simple modulo R*, written N is simple (mod R), if the following two conditions are satisfied:

(a) *there exist no essential mappings* $f: (S^1 \times I, S^1 \times \partial I) \to (N, \partial N - R)$, *and*

(b) *there exist no essential, nonperipheral mappings* $f: S^1 \times S^1 \to N$.

If $R = \varnothing$ and N is simple (mod R) we say N is *simple*. This is consistent with the situation for a simple knot space.

3.5. Decomposition Theorem. *Suppose that M is a compact, irreducible, sufficiently-large 3-manifold. Furthermore, if $\partial M \neq \varnothing$ suppose that ∂M is incompressible. Then M can be expressed uniquely (up to ambient isotopy) as a union $M = N \cup \Sigma$ of two 3-manifolds N and Σ such that*

(1) *each component of $N \cap \Sigma$ is either an essential annulus or an essential torus,*

(2) *each component of N is simple* (rel $N \cap \Sigma$), *and*

(3) *each component S of Σ has the property that the pair $(S, S \cap \partial M)$ is a Seifert pair and if $\Phi = \Sigma \cap \partial M$, then the pair (Σ, Φ) satisfies conditions* (i)–(iii) *above.*

Proof. Let (Σ, Φ) be the characteristic Seifert pair for M. Then define N to be the closure of $M - \Sigma$.

4. Centralizers, roots and relations. Let S be a sufficiently-large Seibert fiber space and if $\partial S \neq \varnothing$ assume that ∂S is incompressible. Let J be any regular fiber of M and let $C = \operatorname{Im}(\pi_1(J) \to \pi_1(S))$. Then C is an infinite cyclic normal subgroup of $\pi_1(S)$. Furthermore, C is central in the canonical subgroup of $\pi_1(S)$ where the *canonical subgroup* **[10]** is equal to $\pi_1(S)$ if S has orientable decomposition surface and otherwise has index two in $\pi_1(S)$.

If α is an element of $\pi_1(S)$, where S is as above, and α is conjugate into C, then the centralizer of α in $\pi_1(S)$ is the canonical subgroup of $\pi_1(S)$ and therefore the centralizer of α in $\pi_1(S)$ has index at most two in $\pi_1(S)$. If α is not conjugate into C, then the centralizer of α in $\pi_1(S)$ has an abelian subgroup of index $\leqq 2$; and if α is not in the canonical subgroup, then the centralizer of α in $\pi_1(S)$ is cyclic. Hence, if α is not conjugate into C, the centralizer of α in $\pi_1(S)$ is either $\boldsymbol{Z}$, $\boldsymbol{Z} + \boldsymbol{Z}$, or $\mathscr{K}$, the fundamental group of the Klein bottle.

We see from the preceding discussion that Seifert fibered manifolds have elements in their fundamental groups that have centralizers which are in some sense nontrivial. On the other hand, let N be any compact, irreducible 3-manifold with nonempty boundary. Let S be a Seifert fibered 3-manifold with an orientable decomposition surface and nonempty, incompressible boundary. Then the center of $\pi_1(S)$ is carried by the fundamental group of a regular fiber J in ∂S. Let $f\colon S^1 \times I \to \partial N$ be an embedding which induces an injection of $\pi_1(S^1 \times I)$ into $\pi_1(N)$ and let $g\colon S^1 \times I \to \partial S$ be an embedding onto a neighborhood of J in ∂S. Then the 3-manifold $M = N\,{}_f\!\cup_g S$ is irreducible and $\pi_1(M)$ contains an element α such that the centralizer of α contains a subgroup isomorphic to $\pi_1(S)$. In general M need not not be Seifert fibered. However, this is typical of the phenomenon of an element of a 3-manifold group which does not have a trivial centralizer in the 3-manifold group. The situation is completely described in our next theorem on centralizers.

4.1. THEOREM. *Let M be a compact, irreducible, sufficiently-large* 3-*manifold. Then there exist disjoint, compact Seifert-fibered submanifolds $N_1, \cdots, N_r$ of M having incompressible boundaries such that if $\bar{G}_i$ is the subgroup of $\pi_1(M)$ defined up to conjugacy as* $\operatorname{Im}(\pi_1(N_i) \to \pi_1(M))$, *$G_i$ is the image in $\bar{G}_i$ of the canonical subgroup of $\bar{G}_i$ and C_i is the image in $\bar{G}_i$ of the fundamental group of a regular fiber of N_i $(1 \leqq i \leqq r)$, then*:

(i) *every nontrivial element of $\pi_1(M)$, whose centralizer is noncyclic, is conjugate to an element g of G_i for some $i \leqq r$ and the centralizer of g is contained in $\bar{G}_i$*;

(ii) *any element of $\pi_1(N)$, whose centralizer has no abelian subgroup of index $\leqq 2$ is conjugate to an element of C_i for some $i \leqq r$ where G_i is nonabelian*; *and*

(iii) *the centralizer in $\pi_1(M)$ of any nontrivial element of C_i is precisely G_i.*

The following corollary is an improved statement of the main theorem of **[6]**.

4.2. COROLLARY. *Let M satisfy the hypothesis of Theorem* 4.1. *Then the centralizer of any element of $\pi_1(M)$ is finitely generated. Moreover, among the centralizers of elements of $\pi_1(M)$ there are, up to conjugacy, only finitely many subgroups that contain no abelian subgroup of index $\leqq$* 2.

In the case that M is as above and M contains no embedded Klein bottles, then conclusion (ii) of Theorem 4.1 takes a simplified form. This result is

4.3. ADDENDUM TO THEOREM 4.1. *If M contains no embedded Klein bottles, then*

(ii′) *any element of $\pi_1(M)$ which has nonabelian centralizer is conjugate to an element of C_i for some $i \leqq r$ where G_i is nonabelian.*

In particular, this last statement applies to compact, irreducible proper submanifolds of S^3.

An nth root (n an integer) of an element g in a group G is an element $x \in G$ such that $g = x^n$. Since an nth root of an element $g \in G$ clearly lies in the centralizer of g in G, we can apply the results on centralizers to obtain a description of the set of roots of an element of the fundamental group of a sufficiently-large 3-manifold.

An element of a group is said to have *trivial root structure* if all of its roots lie in a single cyclic subgroup. In addition an element of a group has *nearly trivial root structure* if all of its roots lie in a subgroup which has a finitely generated abelian subgroup of index $\leqq 2$.

4.4. THEOREM. *Let M be a compact, irreducible sufficiently-large 3-manifold. Then there exist disjoint compact Seifert fibered submanifolds $N_1, \cdots, N_r$ of M having incompressible boundaries in M such that if C_i is the image in $\pi_1(M)$ of the fundamental group of a regular fiber of N_i (a cyclic subgroup of $\pi_1(M)$ defined up to conjugacy), then any element of $\pi_1(M)$ that does not have nearly trivial root structure is conjugate to an element of C_i for some $i \leqq r$.*

The following proposition describes the root structure of the elements of the fundamental group of a Seifert fibered space. This proposition together with Theorem 4.4 completely describes the root structure of an element in the fundamental group of a compact, irreducible, sufficiently-large 3-manifold.

4.5. PROPOSITION. *Let N be a Seifert fibered space and let $C = \operatorname{Im}(\pi_1(J) \subsetneq \pi_1(N))$ where J is a regular fiber of N. Let $K_1, \cdots, K_p$ be the singular fibers of N and let $D_1 = \operatorname{Im}(\pi_1(K_i) \subsetneq \pi_1(N))$. An element x of $\pi_1(N)$ is a root of some element of C if and only if x is conjugate in $\pi_1(N)$ to an element of some D_i ($1 \leqq i \leqq p$).*

The next corollary was proved earlier in **[15]**.

4.6. COROLLARY. *Let M be a compact, irreducible sufficiently-large 3-manifold. Let g be a nontrivial element of $\pi_1(M)$. Then there exist at most finitely many integers n such that $g = x^n$ has a solution in $\pi_1(M)$.*

We conclude this section with two results on relations between elements in 3-manifold groups. It is conjectured in **[5]** that the fundamental group of a compact, irreducible, sufficiently-large 3-manifold is Hopfian. The examples of groups which are not Hopfian in **[1]** furnish possible counterexamples to this conjecture. These groups have a presentation $\langle a, b\colon ab^pa^{-1} = b^q\rangle$ for certain integers p and q. It has been known for some time, however, that these groups cannot be 3-manifold groups **[6]**. But an even stronger condition follows from the preceding work.

4.7. THEOREM. *Let M be a compact, irreducible, sufficiently-large 3-manifold. Let p and q be integers and let a and b be elements of $\pi_1(M)$ such that $ab^pa^{-1} = b^q$. Then either $b = 1$ or $p = \pm q$.*

A compact, irreducible, sufficiently-large 3-manifold M is said to be *atoridal* if every essential map $f\colon S^1 \times S^1 \to M$ is peripheral.

4.8. THEOREM. *Let M be an atoridal 3-manifold. Let G be a two generator subgroup of $\pi_1(M)$. Then one of the following holds:*

(i) *G is free;*

(ii) *$G \approx \mathbf{Z} \times \mathbf{Z}$ and is peripheral; or*

(iii) *G is of finite index in $\pi_1(M)$.*

PROOF. Let M and G be as in the hypothesis. Let $(\tilde{M}, p)$ be the covering space of M corresponding to G. Then there exists a compact submanifold N of $\tilde{M}$ such that $\pi_1(N) \subsetneq \pi_1(\tilde{M})$ is an isomorphism.

Using the fact that $\pi_1(N)$ is generated by two elements and a standard Betti number argument, if follows that the only possibilities for ∂N are

(i) ∂N consists of precisely one component which has genus two, or

(ii) ∂N consists of two components each having genus one (hence each a torus), or

(iii) ∂N consists of precisely one component which has genus one (hence ∂N is a torus).

In case (i) it follows from Lemma 5.4 of [7] that $\pi_1(N)$ is free. In case (ii) each component of ∂N is incompressible. It follows that $p | \partial N$ gives rise to essential mappings of tori into M. By hypothesis these mappings are peripheral. Hence, either $G \approx \boldsymbol{Z} + \boldsymbol{Z}$ and is peripheral or G has finite index in $\pi_1(M)$.

In case (iii) if ∂N is not incompressible then $\pi_1(N) \approx \boldsymbol{Z}$ and G is free. If ∂N is incompressible, then $p | \partial N$ is an essential mapping of a torus into M. It follows that G has finite index in $\pi_1(M)$. This completes the proof.

BIBLIOGRAPHY

1. G. Baumslag and D. Solitar, *Some two-generator one-relator non-Hopfian groups*, Bull. Amer. Math. Soc. **68** (1962), 199–201. MR **26** #204.

2. J. W. Cannon and C. D. Feustel, *Essential annuli and Möbius bands in M^3*, Trans. Amer. Math. Soc. **215** (1976), 219–239.

3. B. Evans and W. Jaco, *Covering spaces and manifold compactifications* (manuscript).

4. C. D. Feustel, *On the torus theorem and its applications*, Trans. Amer. Math. Soc. **217** (1976), 1–43.

5. W. Jaco, *The structure of three-manifold groups*, Inst. for Advanced Study, Princeton, N.J., 1971 (manuscript).

6. ———, *Roots, relations and centralizers in three-manifold groups*, Lecture Notes in Math., vol. 438, Springer-Verlag, Berlin and New York, 1974. MR **52** #15469.

7. W. Jaco and P. B. Shalen, *Peripheral structure of* 3-*manifolds*, Invent. Math. **38** (1976), 55–87.

8. ———, *Surface homeomorphisms and periodicity*, Topology (to appear).

9. ———, *Seifert fiber spaces in* 3-*manifolds*. I, *The mapping theorem* (preprint).

10. ———, *Seifert fiber spaces in* 3-*manifolds*. II, *Centralizers, roots and relations* (preprint).

11. K. Johannson, *Equivalences d'homotopie des varietés de dimension* 3, C. R. Acad. Sci. Paris **66** (1975), 1009–1010. MR **52** #11918.

12. ———, *Homotopy equivalences of knot spaces* (preprint).

13. H. Kneser, *Geschlossene Flachen in dreidimensionalen Mannigfaltigkeiten*, Jber. Deutsche Math. Verein. **38** (1929), 248–260.

14. J. Milnor, *A unique factorization theorem for* 3-*manifolds*, Amer. J. Math. **84** (1962), 1–7.

15. P. B. Shalen, *Infinitely divisible elements in* 3-*manifold groups*, Knots, Groups and 3-Manifolds—Papers Dedicated to the Memory of R. H. Fox (edited by L. Neuwirth), Ann. of Math. Studies, no. 84, Princeton Univ. Press, Princeton, N.J., 1975, pp. 293–335.

16. H. Seifert, *Topology of* 3-*dimensional fibered spaces*, Acta Math. **60** (1933), 147–288.

17. F. Waldhausen, *Gruppen mit Zentrum und* 3-*dimensionale Mannigfaltigkeiten*, Topology **6** (1967), 505–517. MR **38** #5223.

18. ———, *On irreducible* 3-*manifolds which are sufficiently large*, Ann. of Math. (2) **87** (1968), 56–88. MR **36** # 7146.

19. ———, *On the determination of some bounded* 3-*manifolds by their fundamental groups alone*, (Proc. Internat.Sympos. Topology, Herceg-Novi, Yugoslavia, 1968) Beograd, 1969, pp. 331–332.

COLUMBIA UNIVERSITY

RICE UNIVERSITY

Proceedings of Symposia in Pure Mathematics
Volume 32, 1978

A GEOMETRIC PROOF OF ROCHLIN'S THEOREM

MICHAEL FREEDMAN* AND ROBION KIRBY**

0. In 1974 Andrew Casson outlined to us a proof of Rochlin's Theorem (stated below) on the index of a smooth, closed 4-manifold M^4. His proof involved th Arf invariant of a certain quadratic form defined on the first homology group of a surface in M^4 which is dual to the second Stiefel-Whitney class of M^4. Our proof was derived from Casson's; it is the same in principle but differs considerably in detail. After this manuscript was written, we discovered that Rochlin had already in 1971 given a short sketch of this proof; it appears in a paper [**R**$_3$] about real algebraic curves in RP^2.

In addition we obtain (Theorem 2) a "stable" converse to the Kervaire-Milnor nonimbedding theorem [**K-M**], and in §2, by relaxing some orientability assumptions, we prove a new (but unspectacular) nonimbedding theorem (Theorem 4) and find an obstruction to approximating unoriented simplicial 3-chains in a 5-manifold by an immersed 3-manifold.

We thank John Morgan for several valuable conversations.

1. Let M^4 be a closed, orientable, PL (hence smooth) 4-manifold. It has an intersection form $H_2(M; Z)/_{\text{torsion}} \times H_2(M; Z)/_{\text{torsion}} \to Z$ which is a symmetric, unimodular, integral bilinear form [**M**$_1$]; denote this pairing by $x \cdot y$ or xy. Its signature is $\sigma(M) =$ index M.

We say that $\omega \in H_2(M; Z)/_{\text{torsion}}$ is characteristic if its mod 2 reduction $[\omega]_2$ is Poincaré dual to the second Stiefel-Whitney class $w_2 \in H^2(M; Z_2)$. This implies that $\omega \cdot x = x \cdot x$ (mod 2) for all $x \in H_2(M; Z)$. For this congruence follows from the equality $[\omega]_2 \cdot y = y \cdot y$ for all $y \in H_2(M; Z_2)$, which is dual to the equality $w_2 \cup \hat{y} = \hat{y} \cup \hat{y}$ ($\hat{}$ denotes Poincaré dual) which is a definition of the second Steifel-Whitney class w_2 of a 4-manifold.

AMS (*MOS*) *subject classifications* (1970). Primary 57D95.
*Partially supported by NSF.
**Partially supported by NSF and the Miller Institute for Basic Research in Science.

It is an easy bit of algebra [**M-H**, p. 24], that $\omega\cdot\omega = \sigma(M)$ (mod 8). Thus if $w_2 = 0$ so that 0 is characteristic, then $\sigma(M) = 0$ (mod 8). Rochlin improved this by a factor of 2.

THEOREM [$\mathbf{R}_2$]. *If M is closed, orientable,* PL *and* $\omega_2 = 0$, *then* $\sigma(M) = 0$ (mod 16).

Rochlin's Theorem is an anomaly in this sense: in dimensions $4k$, $k > 1$, there are closed, orientable, almost parallelizable PL (not smooth) manifolds P^{4k} with $\sigma(P^{4k}) = 8$ [$\mathbf{M}_2$]. These PL manifolds are missing in dimension 4, and this accounts for the counterexamples to existence and uniqueness of PL structures on manifolds [**K-S**], [$\mathbf{S}_2$]. Rochlin's Theorem is not known for topological 4-manifolds; the existence of such a topological 4-manifold of index 8 is equivalent to proving topological transversality in codimension 4 [$\mathbf{S}_1$].

Let θ_3 be the group of homology cobordism classes of homology 3-spheres. Then Rochlin's Theorem provides an epimorphism $\theta_3 \to^R Z_2$. If Σ^3 is a representative of θ_3, it bounds a PL, parallelizable 4-manifold Q^4. Then $\sigma(Q^4)/8$ (mod 2) $\in Z_2$ is easily seen to be an invariant of the homology cobordism class of Σ^3 by use of Rochlin's Theorem [$\mathbf{R}_1$].

The usual proof [$\mathbf{R}_2$], [**M-K**] of Rochlin's Theorem is homotopy theoretic, requiring: the decomposability of Sq^3, the calculation of $J\colon \pi_3(SO) \to \pi_3^s$, Hirzebruch's identity, $\sigma(M^4) = \frac{1}{3}p_1(\tau_{M^4})\,[M^4]$, and the identification of $p_1(\tau_{M^4})$ with $\pm 2\{$obstruction to extending over M a trivialization of $\tau_{M^4}\,|\,(M^4\text{-point})\}$. Our proof is geometric, except for the use of the isomorphism $\Omega_4 \to^\sigma Z$ where Ω_4 is the oriented bordism group of oriented 4-manifolds.

Here is some motivation for our proof. From now on, all 4-manifolds are closed, orientable and smooth. First consider the generalization [**K-M**]: if the characteristic element $\omega \in H_2(M^4; Z)$ is represented by a smooth, imbedded S^2, then $\omega\cdot\omega - \sigma(M^4) = 0$ (16).

In general, ω is represented by an orientable (if $H_1(M; Z/2) = 0$) surface K^2, and $\omega\cdot\omega - \sigma(M^4) = 0$ (8), so it is predictable that there is a Z^2 obstruction associated with surgering K^2 to a 2-sphere. Here is the simplest case where that obstruction occurs.

Let $Q^4 = CP^2 \,\#\#^8\, \overline{CP}^2$ be complex projective space connected sum eight copies reversed orientation. Let α_0 generate $H_2(CP^2; Z)$ and let α_i generate the ith copy of $H_2(\overline{CP}^2; Z)$. Then $\omega = 3\alpha_0 + \alpha_1 + \cdots + \alpha_8$ is characteristic and $\omega\cdot\omega = 1$. Since $\omega\cdot\omega - \sigma(Q^4) = 8$, ω cannot be represented by a smooth imbedded S^2. (To see this directly from Rochlin's Theorem, suppose S^2 represents ω; its normal disk bundle N is the Hopf bundle, since $\omega\cdot\omega = 1$, so $\partial N = S^3$; remove N and sew in B^4; the new manifold has $\omega_2 = 0$ and index -8, a contradiction.) In this case α_0 is represented by any nonsingular cubic, all of which are tori $S^1 \times S^1\cdot\alpha_i$ is represented by $CP^1 = S^2$, so ω is represented by an $S^1 \times S^1$, and we cannot reduce the genus.

We can try to surger K inside M^4 to get a 2-sphere. Let $A_1, \cdots, A_{2s}$ be imbedded circles representing the generators of a symplectic basis of $H_1(K; Z) = Z^{2s}$. To surger some A_i, we must smoothly imbed a 2-ball D_i in M with $D_i \cap K = A_i$, D_i and K transverse, and so that the normal vector field v to ∂D_i which is tangent to K extends to a normal vector field V to D_i. We can then replace the normal 1-disk bundle to A_i in K by the normal 0-sphere bundle (the boundary of the 1-disk bundle determined by the vector field V) of D_i, thereby reducing the genus of K by 1.

We can always imbed D_i transversely to K with $\partial D_i = A_i$. The obstruction to extending v over D_i is an integer $x \in Z = \pi_1(SO(2))$. The algebraic intersection of int D_i with K, $\bigcap(\text{int } D_i, K)$, is another integer d. Also we may spin D_i once around A_i, as in Figure 1, changing x to $x \pm 1$ and d to $d \pm 1$, so that $d - x$ is unchanged. By iteration we can make either d or x zero.

Let S be a smooth imbedded 2-sphere in M^4. Since K is characteristic, $K \cdot S = S \cdot S$ (mod 2); $S \cdot S$ is the Euler class of the normal bundle. Under the connected sum $D_1 \# S$, x changes to $x + S \cdot S$ and d changes to $d + K \cdot S$. Thus $d - x = d + x = d + x + K \cdot S + S \cdot S$ (mod 2) is a possible Z_2 obstruction to surgery on A_i. Associating $d + x$ (mod 2) to each A_i, we obtain a quadratic form $\tilde{q}: H_1(K; Z_2) \to Z_2$. The Arf invariant of $\tilde{q}$, $\phi(M, K)$, is shown in Lemmas 3–5 to be an invariant of the pair (M, K) up to cobordism of such pairs.

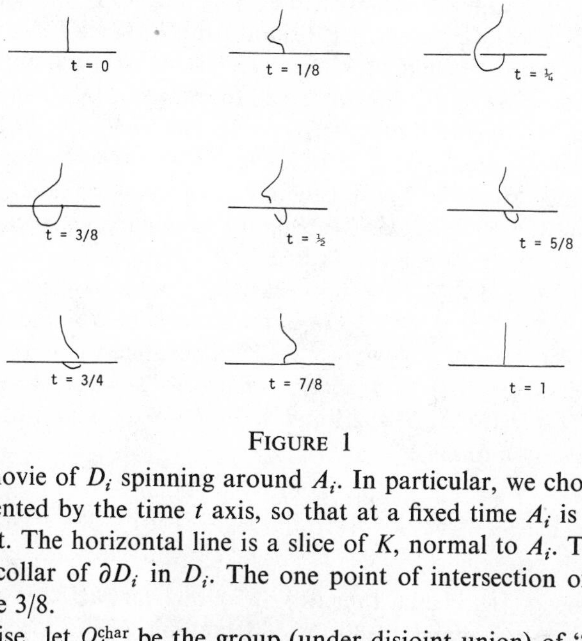

FIGURE 1

This is a movie of D_i spinning around A_i. In particular, we choose an interval of A_i, represented by the time t axis, so that at a fixed time A_i is represented by the center dot. The horizontal line is a slice of K, normal to A_i. The vertical line represents a collar of ∂D_i in D_i. The one point of intersection of K and int D_i occurs at time 3/8.

To be precise, let Ω_4^{char} be the group (under disjoint union) of "characteristic" pairs (M^4, K^2) up to "characteristic" bordism, where M and K are closed and oriented and $[K] \in H_2(M; Z)/_{\text{torsion}}$ is characteristic. Two pairs (M, K) and (M', K') are characteristically bordant if there exist an oriented 5-manifold $\bar{M}$ and an oriented 3-submanifold $\bar{K}^3$ with $[\bar{K}]_2$ dual to $\omega_2(\tau_{\bar{M}})$ and $\partial(\bar{M}, \bar{K}) = (M, K) \cup -(M', K')$.

Thus we have $\phi: \Omega_4^{\text{char}} \to Z_2$.

We show (Lemmas 1 and 2) that $\alpha: \Omega_4^{\text{char}} \to Z \oplus Z$ is an isomorphism where $\alpha(M, K) = (\sigma(M), (K \cdot K - \sigma(M))/8)$, and exhibit generators of Ω_4^{char}. Only here do we need the fact that $\Omega_4 = Z$.

Finally we show that the Z_2-invariant $\theta: \Omega_4^{\text{char}} \to Z_2$, defined by $\theta(M, K) = ((K \cdot K - \sigma(M))/8)(2)$, is equal to ϕ, by showing they are equal on the generators of Ω_4^{char} (Lemma 6, Theorem 1).

LEMMA 1. *$\alpha: \Omega_4^{\text{char}} \to Z \oplus Z$ is an injection.*

PROOF. $\alpha(M, K) = (\sigma(M), (K\cdot K - \sigma(M))/8)$. Signature and intersection are both additive with respect to disjoint sum. σ is well known to be an oriented bordism invariant. Also, $K\cdot K$ is invariant, for if $(M, K) = \partial(\bar{M}, \bar{K})$, then

$$K\cdot K = \hat{\bar{K}} \cup \hat{\bar{K}}[M] = \hat{\bar{K}} \cup \hat{\bar{K}}(i_*[M]) = \hat{\bar{K}} \cup \hat{\bar{K}}(0) = 0$$

(here we denote Poincaré duals by "^" and $i: M \to \bar{M}$ is inclusion). Therefore α is a well-defined homomorphism.

To show α is injective, suppose a characteristic pair (M, K) satisfies $\sigma(M) = K\cdot K = 0$. M and K can be assumed connected. It is tempting to let $\bar{K}$ be an oriented 3-manifold with $\partial\bar{K} = K$ and $\bar{M} = M \times I \cup_{K\times B^2} \bar{K} \times B^2 \cup W$, where W is a spin 5-manifold whose boundary is the component of $\partial(M \times I \cup \bar{K} \times B^2)$ other than M. Unfortunately $\bar{K}$ may not be dual to $\omega_2(\bar{M})$.

Given a Morse function on M, we can assume any extra critical points of index 0 or 4 have been cancelled. The critical points of index 1 determine descending 1-manifolds which in turn determine a family of disjointly imbedded, oriented circles, which, by general position, are disjoint from K. Since K is dual to $\omega_2(M)$, we can frame the tangent bundle of $M - K$. This determines a framing of the normal bundles of the circles. Let N^5 be obtained from $M \times I$ by adding 2-handles to the circles in $M \times 1$ according to the framing. Then $\partial N = M \times 0 \cup \partial_1 M$ and $\partial_1 N$ is 1-connected. Furthermore $K \times I$ in N is dual to $\omega_2(N)$ because the complement is still framed and $K \times I$ is still the obstruction to extending this framing. The same is true for $\tilde{N} = N \natural k(S^2 \times B^3)$, the boundary connected sum of N along $\partial_1 N - K \times 1$ with k copies of $S^2 \times B^3$.

Since $\sigma(\partial_1\tilde{N}) = 0$ and $\pi_1(\partial_1\tilde{N}) = 0$, $\tilde{N}$ has the same intersection form as $X = \#r(S^2 \times S^2)\,\#\varepsilon(S^2 \tilde{\times} S^2)$ where $S^2 \tilde{\times} S^2$ is the nontrivial S^2 bundle over S^2 and $\varepsilon = 0$ or 1. By Theorem W_1 below, $\partial_1\tilde{N}$ is diffeomorphic, via h, to X for some k. We must choose h correctly.

In X, let $\alpha_i, \beta_i \in H_2((S^2 \times S^2)_i; Z)$ and $\gamma, \delta \in H_2(S^2 \tilde{\times} S^2; Z)$ be the standard bases with intersection forms

$$\begin{array}{c|cc} & \alpha_i & \beta_i \\ \hline \alpha_i & 0 & 1 \\ \beta_i & 1 & 0 \end{array} \quad \text{and} \quad \begin{array}{c|cc} & \gamma & \delta \\ \hline \gamma & 0 & 1 \\ \delta & 1 & 1 \end{array},$$

respectively. Since $k_*[K]$ is dual to $\omega_2(X)$, it follows that $h_*[K]$ is a sum of even multiples of the α_i's, β_i's and δ and an odd multiple of γ. Since $K\cdot K = 0$, by Theorem W_2 below [W_2], we can find an orthogonal automorphism of $H_2(X; Z)$ taking $h_*[K]$ to $2n\alpha_1$ (if $\varepsilon = 0$) or to $(2n + 1)\gamma$ (if $\varepsilon = 1$). By Theorem W_3 below [W_3], this automorphism is realized by a diffeomorphism $h': X \to X$.

Let $Y = \#r(S^2 \times B^3)\#\varepsilon(S^2 \tilde{\times} B^3)$, $\partial Y = X$. Let $\bar{M} = \tilde{N} \cup Y$ where we identify ∂Y with $\partial_1\tilde{N}$ by $h'h$. Let $\bar{\alpha}_1$ and $\bar{\gamma} \in H_3(Y, \partial; Z)$ be the classes represented by B^3 fibers so that $\partial\bar{\alpha}_1 = \alpha_1$ and $\partial\bar{\gamma} = \gamma$. There is an oriented 3-manifold $(J, \partial) \subset (Y, X)$ with $\partial J = (h'h)(K \times 1)$ and $[J, \partial] = 2n\bar{\alpha}_1$ or $(2n + 1)\bar{\gamma}$ as $\varepsilon = 0$ or 1. (This follows from a relative version of the representability of codimension two integral homology classes by oriented submanifolds.) Clearly $[J, \partial]$ is dual to $\omega_2(Y)$.

Now let $\bar{K} = K \times I \cup J$, identified along $\partial J = K \times 1$. $\bar{K}$ is dual to $\omega_2(\bar{M})$ (use the Mayer-Vietoris sequence on $\bar{M} = \tilde{N} \cup Y$), $\bar{K}$ is oriented, and $\partial\bar{K} = K$. Thus we have constructed a null bordism of (M, K) in Ω_4^{char}. □

Note. $B\,\mathrm{spin}^c$ is defined as the pullback

$$\begin{array}{ccc} B\,\mathrm{Spin}^c & \longrightarrow & * \\ \downarrow & & \downarrow \\ BSO & \xrightarrow{\delta\omega_2} & K(Z,3) \end{array}$$

A spin^c-structure on an oriented 4-manifold is a lifting of the tangent bundle to $B\,\mathrm{spin}^c$. It is known that every oriented smooth manifold has a spin^c-structure, [H, H]. It follows from Lemma 1 that every oriented smooth 4-manifold, M, with $\sigma(M) = 0$, admits a spin^c-structure which is a spin^c boundary. (Proof. If M is spin let $K \subsetneq M$ satisfy $[K] = 0 \in H_2(M; Z)$. Now $K \cdot K = 0$ and $(M, K) = \partial(\bar{M}, \bar{K})$ with $[\bar{K}] \in H_3(\bar{M}, M; Z)$ dual to $w_2(\tau(\bar{M}))$. So M bounds $\bar{M}$ with $w_2(\tau(M))$ integral; hence there is a spin^c-structure on $\bar{M}$; its restriction is the desired spin^c-structure on M. If M is not spin, by the classification of symmetric-unimodular-odd-bilinear forms there is a basis $\alpha_1, \cdots, \alpha_n, \alpha_{n+1}, \cdots, \alpha_{2n}$ for $H_2(M; Z)$ satisfying $\alpha_i \cdot \alpha_i = +1$ for $1 \leqq i \leqq n$ and $\alpha_i \cdot \alpha_i = -1$ for $n + 1 \leqq i \leqq 2n$, and $\alpha_i \cdot \alpha_j = 0$ for $i \neq j$. Let $K \subsetneq M$ represent $\sum_{i=1}^{2n} \alpha_i$. K is characteristic and $K \cdot K = 0$. Now, as above, M bounds $\bar{M}$ and can be a spin^c-boundary.)

We have used:

THEOREM W_1 [W_1, P. 147]. *If two simply connected, smooth, closed, 4-manifolds, M_1 and M_2, have isomorphic intersection forms, then $M_1 \# k(S^2 \times S^2)$ is diffeomorphic to $M_2 \# k(S^2 \times S^2)$ for some k.*

REMARK. The proof relies on the fact that $\Omega_4^{SO} \to^{\sigma} Z$ is injective.
Only here do we need the calculation of Ω_4^{SO}.

THEOREM W_2 [W_2, THEOREM 4]. *The group of automorphisms of*

$$H_2(M \# 2(S^2 \times S^2) \colon Z)/_{\text{torsion}}$$

which are orthogonal (preserve the intersection form) is transitive on primitive characteristic elements of a given square.

THEOREM W_3 [W_3, THEOREM 2]. *If M is a smooth, simply connected, closed 4-manifold with an indefinite intersection form, then any automorphism of $H_2(M \# S^2 \times S^2, Z)$ can be represented by a diffeomorphism of $M \# S^2 \times S^2$.*

LEMMA 2. *$\alpha \colon \Omega_4^{\mathrm{char}} \to Z \oplus Z$ is onto.*

PROOF. $\alpha\,(CP^2, CP^1) = (1, 0)$ and $\alpha(CP^2 \# \overline{CP}^2, 3\gamma \# \overline{CP}^1) = (0, 1)$, where $\overline{CP}^2$ has the orientation (opposite to CP^2) for which $\overline{CP}^1 \cdot \overline{CP}^1 = -1$ and 3γ is the non-singular complex (elliptic) curve of degree 3 in CP^2 (homologically it is 3 times the generator $[CP^1]$). Since α is additive under disjoint union (or pairwise connected sum), the proof is finished. □

We define a characteristic 2-ad,

$$\begin{array}{ccccc} & \nearrow & B & \searrow & \\ A & & & & M \\ & \searrow & K & \nearrow & \end{array}$$

to be a characteristic pair (M, K) together with an imbedding of an unoriented family of circles, A, in K, and an imbedding of an unoriented surface, B, in M, which meets K normally at $A = \partial B$ and transversally at isolated points of $K - A$. (Generally we will suppress inclusions from our notation.) We say that two char-2-ads,

$$\begin{array}{ccc} & B & \\ \nearrow & & \searrow \\ A & & M \\ \searrow & & \nearrow_{i} \\ & K & \end{array} \qquad \text{and} \qquad \begin{array}{ccc} & B' & \\ \nearrow & & \searrow \\ A' & & M' \\ \searrow & & \nearrow_{i'} \\ & K' & \end{array}$$

are equivalent (written $\sim$) if (M, K) and (M', K') are characteristically bordant, via $(\bar{M}, \bar{K})$, and $A \cup A' \subset \partial(\bar{K})$ bounds an unoriented surface, $\bar{A} \subset \bar{K}$.

We now define a Z_2 valued invariant, q, of characteristic 2-ads. Let $\nu_{B \subset M}$ and $\nu_{\partial B \subset K}$ denote the normal bundles. We have $\nu_{B \subset M} \mid \partial B = \nu_{\partial B \subset K} \oplus \xi$. Since K is orientable the first summand is trivial; so ξ is also trivial (as $\nu_{\partial B \subset K} \oplus \xi$ extends over B). Let F be a framing $\nu_{B \subset M} \mid_{\partial B}$, restricting to a framing on each factor. Let $w_2(\nu_{B \subset M} \mid_{\partial B}, F) = \chi_2 \in H^2(B, \partial B; Z_2)$ be the relative Stiefel-Whitney class.

(χ_2 is the mod 2 reduction of the obstruction in $H^2(B, \partial B, Z_{\text{twisted}})$ to extending $\nu_{\partial B \subset K}$ to a section of $\nu_{B \subset M}$.) Let $X = \chi_2[B, \partial]_2 = 0$ or 1. We define

$$q\left(\begin{array}{ccc} & B & \\ \nearrow & & \searrow \\ A & & M \\ \searrow & & \nearrow \\ & K & \end{array}\right) = \pitchfork_M(\text{int } B, K) + X \pmod 2.$$

By $\pitchfork(\ ,\ )$ we mean the number (mod 2) of transverse intersections.

LEMMA 3. *If*

$$\left(\begin{array}{ccc} & B & \\ \nearrow & & \searrow \\ A & & M \\ \searrow & & \nearrow \\ & K & \end{array}\right) \sim \left(\begin{array}{ccc} & B' & \\ \nearrow & & \searrow \\ A' & & M' \\ \searrow & & \nearrow \\ & K' & \end{array}\right)$$

then

$$q\left(\begin{array}{ccc} & B & \\ \nearrow & & \searrow \\ A & & M \\ \searrow & & \nearrow \\ & K & \end{array}\right) = q\left(\begin{array}{ccc} & B' & \\ \nearrow & & \searrow \\ A' & & M' \\ \searrow & & \nearrow \\ & K' & \end{array}\right).$$

PROOF. $Y = \bar{A} \cup B \cup B'$ is a closed unoriented surface.

Diagram:

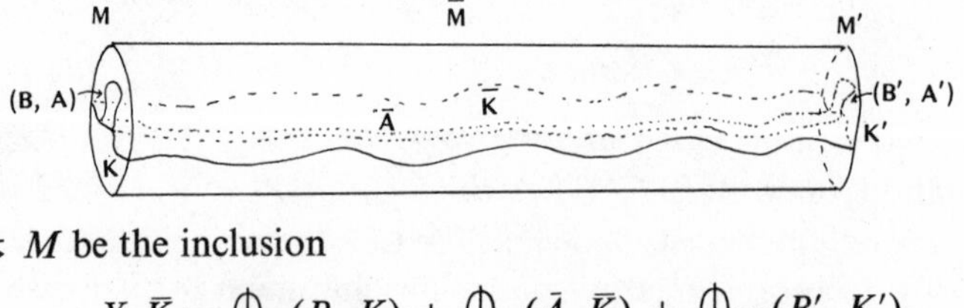

Let $h\colon Y \subset M$ be the inclusion

$$Y \cdot \bar{K} = \pitchfork_M(B, K) + \pitchfork_M(\bar{A}, \bar{K}) + \pitchfork_{M'}(B', K').$$

The middle term is to be interpreted as the obstruction (mod 2) to extending $\nu_{A \amalg A' \subset B \amalg B'}$ to a section of $\nu_{\bar{K} \subset \bar{M}/\bar{A}}$.

On the other hand,

$$\begin{aligned} Y \cdot \bar{K} &= h^*(w_2(\tau(\bar{M})))[Y] = w_2(\tau(\bar{M}) \mid_Y)[Y] \\ &= w_2(\nu_{Y \subset \bar{M}}) + w_2(\tau(Y)) + w_1(\nu_{Y \subset \bar{M}}) \cdot w_1(\tau(Y)). \end{aligned}$$

But $\bar{M}$ is oriented, so $0 = w_1(\tau\bar{M}/Y) = w_1(\nu_{Y\subset\bar{M}}) + w_1(\tau(Y))$, so $w_1(\nu_{Y\subset\bar{M}})\cdot w_1(\tau(Y)) = w_1(\tau(Y))^2 = w_2(\tau(Y))$ by the Wu formula, so $w_2(\tau(\bar{M})|_Y)[Y] = w_2(\nu_{Y\subset\bar{M}})[Y]$. (Note that if Y is oriented and $\bar{M}$ unoriented the preceding assertion is still true.)

Rounding corners, $\nu_{Y\subset\bar{M}/\bar{A}\amalg A'}$ has a framing, $\bar{F}$, which extends F (restricted to A this is (F, inward normal to B)). We can use this framing to break $w_2(\nu_{Y\subset\bar{M}})$ up into three relative Stiefel-Whitney classes

$$w_2(\nu_{Y\subset\bar{M}})[Y] = \underset{\substack{\| \text{ def}\\ x}}{w_2(\nu_{B\subset M}, F)[B, \partial]} + w_2(\nu_{A\subset\bar{M}}, \bar{F})[\bar{A}, \partial] + \underset{\substack{\| \text{ def}\\ x'}}{w_2(\nu_{B'\subset M'}, F')[B', \partial']}$$

$$w_2(\nu_{A\subset\bar{M}}, \bar{F}) = w_2(\nu_{\bar{K}\subset\bar{M}}\,|\,_A, \bar{F}_{2,3}) + w_1(\nu_{\bar{K}\subset\bar{M}}, \bar{F}_{2,3})\cdot w_1(\nu_{A\subset\bar{K}}, F_1)$$

where $\bar{F}_{2,3}$ (and $\bar{F}_1$) denote the last two (and the first) vectors of $\bar{F}$. Let $w_1(\nu_{\bar{K}\subset\bar{M}}, F_{2,3}) = u$ and $w_1(\nu_{A\subset\bar{K}}, F_1) = v$.

$$u + v = w_1(\nu_{\bar{A}\subset\bar{M}}, \bar{F}) = w_1(\tau(\bar{A}), \text{framing of } \tau(A \amalg A'))$$

(because $\bar{M}$ is orientable). As in the closed case, $(u + v) \cup x = \mathrm{Sq}^1 x = x^2$ for all $x \in H^1(\bar{A}, \partial; Z_2)$. $v^2 = (u + v) \cup v = uv + v^2$, so $uv = 0$. Therefore, $w_2(\nu_{Y\subset\bar{M}})[Y] = + \bigcap_M(\bar{A}, \bar{K})$, so

$$\bigcap_M(B, K) + x = \bigcap_{M'}(B', K') + x' \quad (\text{mod } 2). \quad \square$$

Note. The above result holds if the hypothesis that $(\bar{M}; M, M')$ is oriented is replaced by: Y and $\bar{K}$ are oriented.

COROLLARY 1. *If (M, K) is a characteristic pair, q determines a well-defined function, $\tilde{q}\colon H_1(K; Z_2) \to Z_2$.*

PROOF. We kill $H_1(M; Z_2)$ with a finite number of framed 1-surgeries in the complement of K (call the trace N). To $\beta \in H_1(K; Z_2)$ we associate an unoriented surface $(B, \partial) \subset (M, K)$ with $[\partial B] = \beta$. Define

$$\tilde{q}(\beta) = q\left(\begin{array}{ccc} & B & \\ \partial B \nearrow & & \searrow M \\ \searrow & K & \nearrow \end{array}\right).$$

To check that this procedure is well defined let $\tilde{N}$ be an alternative trace. Let

$$(\bar{M}, \bar{K}) = (N, K\times I)\Big/\bigcup_{M,\, K\times 0}(\tilde{N}, K\times I), \qquad \partial(\bar{M}, \bar{K}) = (M_0, K_0) \cup -(M_1, K_1).$$

Let (B_0, ∂) and (B_1, ∂) denote unoriented surfaces in (M_0, K_0) and (M_1, K_1) with $[\partial B_0] = \beta \in H_1(K_0; Z_2)$, $[\partial B_1] = \beta \in H_1(K_1; Z_2)$. There is an unoriented bordism $\bar{A} \subset \bar{K}$ with $\partial\bar{A} = \partial B_0 \cup \partial B_1$, so by definition

$$\left(\begin{array}{ccc} & B_0 & \\ \partial B_0 \nearrow & & \searrow M_0 \\ \searrow & K_0 & \nearrow \end{array}\right) \sim \left(\begin{array}{ccc} & B_1 & \\ \partial B_1 \nearrow & & \searrow M_1 \\ \searrow & K_1 & \nearrow \end{array}\right).$$

By Lemma 3, $\tilde{q}$ is well defined.

LEMMA 4. *$\tilde{q}\colon H_1(K; Z_2) \to Z_2$ is quadratic, i.e. $q(\varepsilon + \delta) - q(\varepsilon) - q(\delta) = \varepsilon\cdot\delta$ for all $\varepsilon, \delta \in H_1(K; Z_2)$.*

PROOF. Let A_ε and A_δ represent ε and δ. Let B_ε and B_δ be as before. Suppose A_ε and A_δ intersect transversally at one point p. Let A_γ be the connected sum (representing $\varepsilon + \delta$) as in Figure 2. Piecewise linearly, we get B_γ from $B_\varepsilon \cup B_\delta \cup T_1 \cup T_2$ where T_1 and T_2 are two curved triangles as shaded in Figure 2; note that $\partial B_\gamma = A_\gamma$. The normal vector fields on B_ε and B_δ must be extended to the new part of A_γ as drawn. Consider the boundary (a circle) of a neighborhood of B in B_γ and push it off itself using the vector field. The two circles link, indicating that the obstruction to extending the vector field over the neighborhood is one. We have verified that $\chi_{\varepsilon+\delta} - \chi_\varepsilon - \chi_\delta = \varepsilon \cdot \delta = 1$.

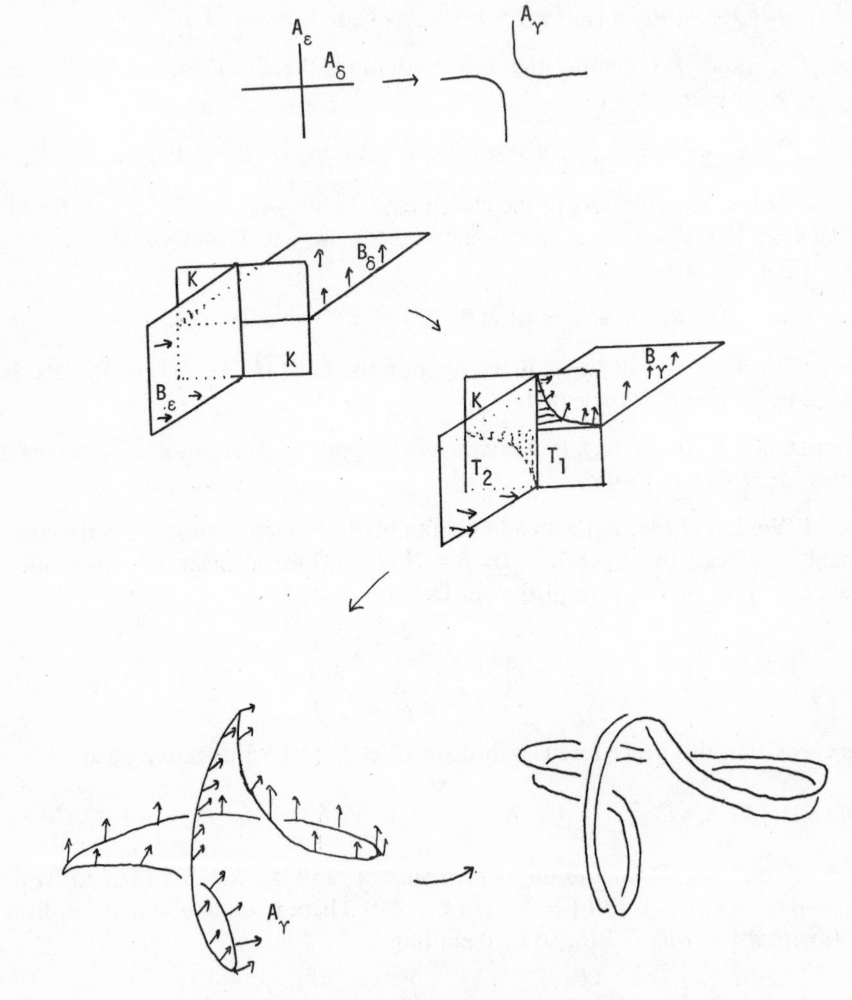

FIGURE 2

If int $B_\varepsilon \cap$ int $B_\delta \neq \varnothing$, then we may push these double points of B_γ off the boundary, adding two points to int $B_\gamma \cap K$. Thus $\pitchfork(B_\varepsilon, K) + \pitchfork(B_\delta, K) = \pitchfork(B_\gamma, K)$ (mod 2). We have shown in this special case that $q(\varepsilon + \delta) - q(\varepsilon) - q(\delta) = \varepsilon \cdot \delta$; the other (easier) cases are left to the reader. □

Let $\phi(M, K)$ be the Arf invariant of $\tilde{q}\colon H_1(K; Z_2) \to Z_2$. (See the Appendix of **[R-S]** for a short presentation of the Arf invariant.)

LEMMA 5. *ϕ determines a well-defined homomorphism:* $\Omega_4^{\mathrm{char}} \to^{\phi} Z_2$.

PROOF. Assume that (M, K) and (M', K') are characteristically bordant via $(\bar{M}, \bar{K})$. For the usual reason

$$\dim(V = \mathrm{Ker}(H_1(K \cup K', Z_2) \to H_1(\bar{K}; Z_2))) = \tfrac{1}{2}\dim(H_1(\partial\bar{K}; Z_2)).$$

The intersection pairing on $H_1(K \cup K'; Z_2)$ is identically zero when restricted to V. $\tilde{q}(K \cup K') = \tilde{q}(K) \oplus \tilde{q}(K')$. If viewed properly, Lemma 3 implies that $\tilde{q}(K \cup K') \mid V$ is identically zero. (If $A \in V$ and $\partial\bar{A} = A$, then $\bar{A} \in H_2(\bar{K}, K \cup K'; Z_2)$; one should regard $\partial(\bar{M}, \bar{K}, \bar{A})$ as $(M \cup -M', K \cup -K', A) \cup (\phi, \phi, \phi)$.) Hence $\phi(\tilde{q}(K \cup K')) = 0$, so $\phi(\tilde{q}(K)) = \phi(\tilde{q}(K'))$. $\square$

LEMMA 6. $\phi(CP^2, CP^1) = 0$ *and* $\phi(CP^2 \# \overline{CP}^2, 3\gamma \# \overline{CP}^1) = 1$.

PROOF. CP^1 is S^2, so H_1 is zero and thus $\phi(CP^2, CP^1) = 0$.

$\phi(CP^2 \# \overline{CP}^2, 3\gamma \# \overline{CP}^1) = \phi(CP^2, 3\gamma)$. In CP^2, 3γ is represented by any cubic; it is convenient to pick $x^3 = y^2 z$. In the coordinate charts $x = 1$ or $y = 1$, the solution is nonsingular, but for $z = 1$, $y^2 = x^3$ is the cone on the (2, 3)-torus knot (= trefoil knot). So 3γ is represented by a smooth 2-sphere except for the cone point. If B^4 is centered at the cone point, replace the cone in B^4 by the Seifert surface (Figure 3) of the trefoil knot in $S^3 = \partial B^4$, obtaining a 2-torus T^2 as a representative of 3γ. The circles A_1, A_2 in the Seifert surface generate $H_1(T^2; Z)$. Each A_i is a trivial knot and bounds a 2-ball B_i in B^4. The obstruction to extending $\nu_{A_i \subseteq T^2}$ to a section of $\nu_{B_i \subseteq B^4}$ is the linking number of A_i and $s(A_i)$ in S^3, where s is a nonzero section of $\nu_{A_i \subseteq T^2}$. $L(A_i, s(A_i)) = \pm 1$. Since $\pitchfork(\mathrm{int}\, B_i, T^2) = 0$, it follows that $q(A_i) = 1$. Thus $\phi(CP^2, T^2) = 1$ (see Appendix **[R-S]**). $\square$

Let $\theta\colon \Omega_4^{\mathrm{char}} \to Z_2$ be the homomorphism given by $\theta(M, K) = (K \cdot K - \sigma(M))/8$ (mod 2).

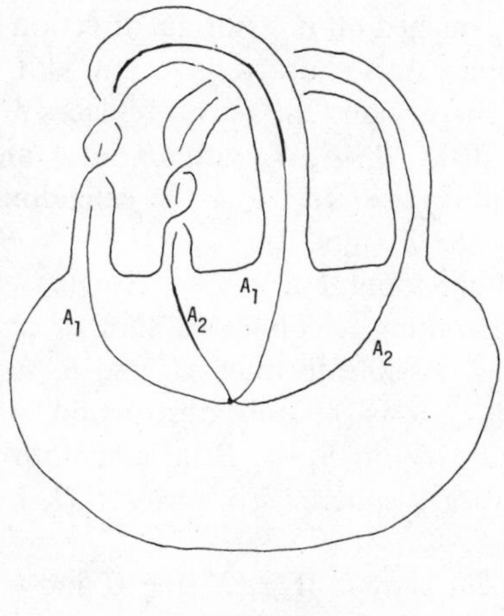

FIGURE 3

THEOREM 1. $\theta(M, K) = \phi(M, K)$.

PROOF. Since both θ and ϕ are homomorphisms (Lemma 5), it is sufficient to check the equality on the generators of Ω_4^{char}, (CP^2, CP^1) and $(CP^2 \# \overline{CP}^2, 3\gamma \# \overline{CP}^1)$. But we have seen (Lemmas 2 and 6), that θ and ϕ are both zero on the first generator, as on the second. □

COROLLARY 2 [**K-M**, THEOREM 1]). *If (M, K) is a characteristic pair, and K is a 2-sphere, then $\theta(M, K) = 0$.*

COROLLARY 3 (ROCHLIN'S THEOREM [**R**$_2$]). *If (M, ϕ) is a characteristic pair, then $\theta(M, \phi) = 0$.*

PROOFS. $H_1(K, Z_2) = 0$ for $K = S^2$ or $\varnothing$, therefore $\phi(M, K) = 0$. By Theorem 1, $\theta(M, K) = 0$.

Note. Suppose $K \subsetneq M$ is a PL imbedding with nonlocally flat points $p_1, \cdots, p_r$ at which K is the cone on knots $S_1, \cdots, S_r$, Let $\mathrm{Arf}(S_i)$ be the Arf invariant of S_i (see [**R**$_2$]). Then if we define

$$\phi(M,K) = \phi(\tilde{q}(H_1(K; Z_2))) + \sum_{i=1}^{r} \mathrm{Arf}(S_i) \pmod 2,$$

we may still conclude that $\theta(M, K) = \phi(M, K)$.

We now show that "stably" ϕ is the only obstruction to surgering K to a 2-sphere. Let $M_s = M \# s(S^2 \times S^2)$, and let j_s be the composition

$$H_2(M; Z) \xrightarrow{\text{inc}_*^{-1}} H_2(M - D^4; Z) \xrightarrow{\text{inc}_*} H_2(M_s; Z).$$

THEOREM 2. *Suppose $\pi_1(M^4) = 0$. Then $\phi(M,K) = 0$ iff for some s, $\exists\, (M_s, K')$ such that K' is a 2-sphere and $i'_*[K'] = j(i_*[K])$.*

Note. Larry Taylor has independently obtained this result.

PROOF. The if direction follows from Theorem 1 [**K-M**] and our Theorem 1.

The argument for the only if direction will be quite liberal with copies of $S^2 \times S^2$. Since $\phi(M, K) = 0$, there is a subspace $V \subset H_1(K; Z_2)$ such that (1) $\dim(V) = \frac{1}{2}\dim(H_1(K; Z_2))$, (2) $v_1 \cdot v_2 = 0$ for $v_1, v_2 \in V$, and (3) $q(v) = 0$ for $v \in V$. Let $A_1, \cdots, A_p$ be circles disjointly imbedded in K representing a basis for V. Let $\tilde{A}_1, \cdots, \tilde{A}_p$ denote copies of $A_1, \cdots, A_p$ pushed off in a normal direction to K so that the linking numbers $L(K, \tilde{A}_i) = 0$. Since K is characteristic, $\tau(M) \mid \tilde{A}_i$ is canonically framed. Since $\pi_1(M) = 0$, framed surgery on $\{\tilde{A}_1, \cdots, \tilde{A}_p\}$ replaces M with M_p. In M_p, there are disjointly imbedded 2-disks, $B_1, \cdots, B_p$ with $\partial B_i = A_i$ and $B_i \cap K = A_i$. There is an Euler class obstruction $\chi_i \in H^2(B_i, \partial; Z)$ to extending $\nu_{A_i \subsetneq K}$ to a section of $\nu_{B_i \subsetneq M_p}$, but $0 = q[A_i] \equiv \chi_i[B_i, \partial] \pmod 2$.

Consider the diagonal imbedding $\Delta\colon S^2 \subsetneq S^2 \times S^2$. $\chi(\nu_{S^2 \subsetneq \Delta S^2 \times S^2})[S^2] = 2$. $\chi_i[B_i, \partial]$ may be altered by ± 2 by taking a connected sum of pairs $(M_p, B_i)\#(S^2 \times S^2, \pm \Delta(S^2))$. In this way it is possible to alter M_p and B_i so that $\chi_i[B_i, \partial]$ becomes zero and M_p becomes M_{p+r}. χ_i is the only obstruction to ambient surgery on K along B_i. The result of surgery on $B_1, \cdots, B_p$ is a smooth submanifold, K', with $H_1(K'; Z_2) = 0$; so K' is a 2-sphere. To verify $i'_*[K'] = j(i_*[K])$, recall that $L(K, \tilde{A}_i) = 0$. □

Note. A similar argument shows: If $\pi_1(M^4) = 0$ and $\xi \in H_2(M^4; Z)$, then for some s, $j_s(\xi)$ is represented by an imbedded torus.

2. In §1, M, $\bar{M}$, K, $\bar{K}$ were taken to be oriented, while B and Y were unoriented (and possibly unorientable). Orienting M, $\bar{M}$, K, $\bar{K}$ was convenient in that it enabled us to calculate the corresponding bordism group, Ω_4^{char}. We chose B (and Y) unoriented because we were defining a quadratic form on $H_1(K; Z_2)$ and orientations would be superfluous.

In the following application we unorient K.

THEOREM 3. *Although* $(CP^2, \gamma) - (CP^2, 3\gamma) = \partial(CP^2 \times I$, *unoriented simplicial 3-chain*), $(CP^2, \gamma) - (CP^2, 3\gamma)$ *cannot be written as* $\partial(CP \times I$, *immersed unoriented 3-manifold*).

PROOF. This is an exercise from the proof of Lemma 3. $\phi(CP^2, \gamma) \neq \phi(CP^2, 3\gamma)$, but a manifold bordism (even an immersed one) $\bar{K}$ from γ to 3γ would force ϕ to assume equal values at each end. □

REMARK 1. There is an old and elegant procedure for approximating a Z_2-simplicial-cycle of dimension 2 in a triangulated 3-manifold by an unoriented, triangulated submanifold. This procedure generalizes for 2-dimensional cycles in any manifold, and for $(n-1)$-dimensional cycles in any n-manifold. The first open case would be: Can you approximate a 3-dimensional Z_2-simplicial-chain in a triangulated 5-manifold by an unoriented, triangulated submanifold? Theorem 3 provides an example (in the relative case) where the answer is no.

REMARK 2. A key calculation occurs in the proof of Lemma 3 in which we show

$$h^* w_2(\tau(\bar{M}))[Y] = w_2(\nu_{Y \subsetneq^h \bar{M}})[Y].$$

We observe that this equality still holds if we replace the assumption (1) $\bar{M}$ is oriented and Y is unoriented, with (2) $\bar{M}$ is unoriented and Y is oriented.

Suppose M, M' are unoriented with $K \subsetneq M$ and $K' \subsetneq M'$ oriented surfaces dual to $w_2(\tau(M))$ and $w_2(\tau(M'))$, respectively, and that image$(H_1(K; Z)) \subset H_1(M; Z)$ and image$(H_1(K'; Z)) \subset H_1(M'; Z)$ are zero. Then if $\bar{M}$ is unoriented and $\bar{K}$ is oriented, with $\bar{K}$ dual to $w_2(\tau(\bar{M}))$, $\tilde{q}(H_1(K; Z_2))$ and $\tilde{q}(H_1(K'; Z_2))$ will be defined and

$$\phi(M, K) = \mathrm{Arf}(\tilde{q}(H_1(K; Z_2))) = \mathrm{Arf}(\tilde{q}(H_1(K'; Z_2))) = \phi(M, K')$$

This is because, with the above assumptions, we will be able to choose B and $\bar{A}$ (see notation preceding Lemma 3) to be oriented compatibly so that Y also is oriented. Now the proof of Lemma 5 goes through using our second set of orientability assumptions. This leads to the following nonimbedding theorem.

THEOREM 4. *Let* M *be an orientable* PL 4-*manifold and let* $\alpha \in H_2(M; Z)$ *satisfy* $(\alpha \cdot \alpha - \sigma(M))/8 \equiv 1 \pmod 2$. *Let* N *be an unorientable* PL 4-*manifold with* $w_2(\tau(N)) = 0$ (*or it is sufficient to assume* $w_2(\tau(N))$ *is represented by a smoothly imbedded oriented surface* $A_2 \subsetneq^{\mathrm{inc}} N$ *with* $\mathrm{inc}_*(H_1(A; Z)) = 0 \in H_1(N; Z)$ *and* $\phi(N, A) = 0$; *note that* ϕ *is defined for the pair despite the fact that* N *is unoriented*). *Then* $\alpha' =$ *image*$(\alpha) \in H_2(M \# N; Z)$ *is not represented by a smooth imbedding* $S^2 \subsetneq^h M \# N$.

LEMMA 7. *A Z-oriented simplicial cycle,* β^3, *of dimension* 3 *in an unoriented triangulated* 5-*manifold may be approximated by a Z-oriented imbedded, triangulated submanifold.*

PROOF. By a modification in a spindle neighborhood of the 2-simplexes of β^3 (see Figure 4), we may assume β^3 is a manifold away from its 1 skeleton.

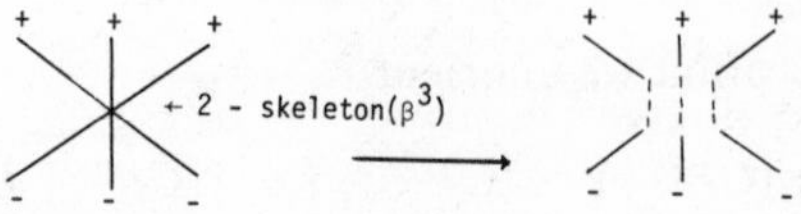

FIGURE 4

Let V be a tubular neighborhood of the 1-skeleton $\partial V = j(S^1 \times S^3) \# K(S^1 \tilde{\times} S^3)$ ($S^1 \tilde{\times} S^3$ denotes the twisted product). $\partial V \cap \beta^3 \overset{\text{def}}{=} L$ is an oriented surface imbedded in ∂V (oriented, because L is imbedded with trivial normal bundle in the $(\beta^3 - 1)$-skeleton). We will show that there is an oriented 3-manifold $\bar{L} \subsetneq V$ with $\partial \bar{L} = L$.

There are $j + k$ normal 4-disks $(D^4, \partial)_i \subset (V, \partial)$ such that V_j "cut" along $\bigcup_i D_i^4$ is a closed 4-disk, X. For each i, there are 2-copies of D_i^4, $D_{i,1}^4$ and $D_{i,2}^4$ included in ∂X. We may assume L meets each D_i in a link L_i. Because L was oriented, the associated links $L_{i,1} \subsetneq \partial D_{i,1}^4$ and $L_{i,2} \subsetneq \partial D_{i,2}^4$ are oriented oppositely (comparing orientations by regluing $D_{i,1}^4$ and $D_{i,2}^4$ to form D_i^4). Let $J_i \subsetneq D_i^4$ be an orientable surface with $\partial J_i = L_i$. Again let $J_{i,1} \subsetneq D_{i,1}^4$ and $J_{i,2} \subsetneq D_{i,2}^4$ be the corresponding copies of J_i in ∂X oriented so that $\partial(J_{i,1}) = - L_{i,1}$ and $\partial(J_{i,2}) = - L_{i,2}$.

$$W \overset{\text{def}}{=} \Big(L - \bigcup_i L_i\Big) \cup_i J_{i,1} \cup_i J_{i,2} \subsetneq \partial X$$

is an oriented surface. Let $Z \subsetneq X$ be an oriented 3-manifold with $\partial Z = W$. Because the orientations on $L_{i,1}$ and $L_{i,2}$ are opposite, so are the orientations on $J_{i,1}$ and $J_{i,2}$. As a result, if we reglue X to form V, the image of Z (which we will call $\bar{L}$) is an orientable 3-manifold contained in V with $\partial \bar{L} = L$.

Now we can approximate β^3 by $(\beta^3 - V) \cup - \bar{L}$, an oriented submanifold. □

PROOF OF THEOREM 4. We only consider the case: $w_2(\tau(N)) = 0$. (As an example N might be $S^1 \tilde{\times} S^3$.) By Theorem 1, α is represented by a smoothly imbedded surface $K \subsetneq^i M$ with $\phi(M, K) = 1$. If we consider $K \subsetneq^{i'} M \# N$ representing α', $\phi(M \# N, K)$ is defined and equal to 1. If α' were represented by a smooth imbedding, $S^2 \subsetneq^h M \# N$, $\phi(M \# N, S^2)$ would be defined and equal 0.

But by a relative form of Lemma 7, there is a smooth oriented 3-manifold, T, with $\partial T = K - S^2$ and a smooth imbedding

$$(T; K, S^2) \xrightarrow{\ j\ } (M \# N \times I, N \# N \times 0, M \# N \times 1)$$

restricting to the imbeddings $i' \times 0$ and $h \times 1$ on the boundary.

Remark 2 shows that the existence of $j : T \subsetneq M \# N \times I$ implies $\phi(M \# N, K) = \phi(M \# N, S^2)$, contradicting the above. □

REFERENCES

[K-M] M. A. Kervaire and J. Milnor, *On 2-spheres in 4-manifolds*, Proc. Nat. Acad. Sci. U.S.A. **47** (1961), 1651–1657. MR **24** #A2968.

[K-S] R. C. Kirby and L. C. Siebenmann, *On the triangulation of manifolds and the Hauptvermutung*, Bull. Amer. Math. Soc. **75** (1969), 742–749. MR **39** #3500.

[M-K] J. Milnor and M. A. Kervaire, *Bernoulli numbers, homotopy groups, and a theorem of Rohlin*, Proc. Internat. Cong. Math. (Edinburgh, 1958), Cambridge Univ. Press, London and New York, 1960, pp. 454–458. MR **22** #12531.

[M-H] J. Milnor and D. Husemoller, *Symmetric bilinear forms*, Springer-Verlag, Berlin and New York, 1973.

[M_1] J. Milnor, *On simply connected 4-manifolds*, Internat. Sympos. on Algebraic Topology, Univ. Nacional Autonoma, Mexico City, 1958, pp. 122–128. MR **21** #2240.

[M_2] ———, *Differential topology*, Lectures on Modern Math., vol. 2, Wiley, New York, 1964, pp. 165–185. MR **31** #2731.

[R_1] R. A. Robertello, *An invariant of knot cobordism*, Comm. Pure Appl. Math. **18** (1965), 543–555. MR **32** #447.

[R_2] V. A. Rochlin, *New results in the theory of 4-dimensional manifolds*, Dokl. Akad. Nauk SSSR **84** (1952), 221–224. (Russian)

[R_3] ———, *Proof of Gudkov's hypothesis*, Functional Anal. Appl. **6** (1972), 136–138.

[R-S] C. P. Rourke and D. P. Sullivan, *On the Kervaire obstruction*, Ann. of Math. (2) **94** (1971), 397–413. MR **46** #4546.

[S_1] M. G. Scharlemann, *Three codimension four transversality theories* (to appear).

[S_2] L. C. Siebenmann, *Disruption of low-dimensional handlebody theory by Rohlin's theorem*, Topology of Manifolds, Markham, Chicago, Ill., 1970, pp. 57–76. MR **42** #6836.

[W_1] C. T. C. Wall, *On simply connected 4-manifolds*, J. London Math. Soc. **39** (1964), 141–149. MR **29** #627.

[W_2] ———, *On the orthogonal groups of unimodular quadratic forms*, Math. Ann. **147** (1962), 328–338.

[W_3] ———, *Diffeomorphisms of 4-manifolds*, J. London Math. Soc. **39** (1964), 131–140. MR **29** #626.

[H-H] F. Hirzebruch and H. Hopf, *Felder von Flächenelementen in 4-dimensionalen Mannigfaltigkeiten*, Math. Ann. **136** (1956), 156–172. MR **20** #7272.

INSTITUTE FOR ADVANCED STUDY

UNIVERSITY OF CALIFORNIA, BERKELEY

Proceedings of Symposia in Pure Mathematics
Volume 32, 1978

SECONDARY INTERSECTIONAL PROPERTIES OF 4-MANIFOLDS AND WHITNEY'S TRICK

YUKIO MATSUMOTO*

In 1961 Kervaire and Milnor **[5]** pointed out that Whitney's trick **[13]** fails in dimension 4. Thus a 4-dimensional manifold certainly contains richer intersectional structure than the ordinary intersection number describes. In fact, the failure itself seems to define certain 'secondary intersection numbers' of 2-cycles in a 4-manifold.

In this note we shall make some elementary observations, from the above point of view, concerning the Arf invariant of characteristic cycles, failure of PL or homotopic versions of Whitney's lemma, a secondary intersection triple and some related examples.

It should be noted that M. Freedman **[1]** independently describes a philosophy of the secondary intersection (the synergistic linking in his terminology). Also in a conversation with the author, he suggested the possibility of defining 'secondary *self*-intersection numbers'. The author is grateful to him for this suggestion which was very useful in writing §1 of the present note.

1. Let M^4 be a compact 4-manifold with $H_1(M^4) = 0$. An integral class $\xi \in H_2(M^4)$ is *characteristic* if $\xi \cdot x + x \cdot x \equiv 0 \pmod 2$ for any $x \in H_2(M^4)$. (If M^4 is closed, it is equivalent to saying that the mod 2 reduction of ξ is dual to $w_2(M^4)$.) By **[11]** or **[2]**, one can associate the Arf invariant, $\operatorname{Arf}(\xi) \in \boldsymbol{Z}_2$, with any characteristic class ξ.

We shall show that if ξ is spherical and represented by an immersed 2-sphere, then the invariant $\operatorname{Arf}(\xi)$ essentially coincides with the number (mod 2) of the intersection points of "Whitney's disks" and the immersed sphere. To be more precise, let $f\colon S^2 \to M^4$ be the immersion and suppose that

AMS (*MOS*) *subject classifications* (1970). Primary 55A99, 55B45; Secondary 55A25, 55G30.
*Supported in part by the National Science Foundation.

(i) all the self-intersection points are transversal double points, and

(ii) there are the same number of positive double points (say $p_1, \cdots, p_r$) as that of negative ones (say $q_1, \cdots, q_r$).

For each p_i (or q_j) we have two preimages p_i', p_i'' (or q_j', q_j'') on S^2. Draw smooth arcs γ_i', γ_i'' on S^2 connecting p_i' and q_i', p_i'' and q_i'', respectively. We may assume that all these curves have no self- or mutual intersections. Since $H_1(M^4) = 0$, the circle $C_i = f(\gamma_i') \cup f(\gamma_i'')$ bounds a smoothly embedded orientable 2-surface Δ_i^2. (We shall refer to Δ_i^2 as a "Whitney's surface," because if M^4 is 1-connected, Δ_i^2 can be taken as a Whitney's disk.) We assume that Δ_i^2 meets $f(S^2)$ normally. Count up the number of intersection points, $\#(\mathrm{Int}\Delta_i^2 \cap f(S^2))$, and denote it by $L(\Delta_i)$.

Next construct a nonzero normal vector field ϕ_i on C_i such that when restricted to $f(\gamma_i')$ it gives a cross-section of a normal 1-vector bundle $\nu(f(\gamma_i') \to f(S^2))$ and when restricted to $f(\gamma_i'')$ a unique extension of $\phi_i | \{p_i, q_i\}$ over $f(\gamma_i'')$ which is normal to both $f(S^2)$ and Δ_i^2. See [**9**, pp. 81–82]. Let $\mathcal{O}(\Delta_i) \in \mathbf{Z} = \pi_1(SO_2)$ be the obstruction to extending ϕ_i over Δ_i.

Then we have

PROPOSITION 1. $\mathrm{Arf}(\xi) = \sum_{i=1}^r L(\Delta_i) + \sum_{i=1}^r \mathcal{O}(\Delta_i) \pmod 2$.

PROOF. Using the above notation, we take a thin tubular neighborhood R_i of $f(\gamma_i'')$ in M^4. Let ν_i be a nonzero cross-section of R_i such that each vector of ν_i is tangent to $f(S^2)$. Take the orthogonal complement T_i of ν_i in R_i. Then T_i is regarded as a 1-handle attached to $f(S^2)$. Surger $f(S^2)$ by these handles T_i, $i = 1, \cdots, r$. We obtain an embedded surface X^2 of genus r. Let $\Delta_i' = \mathrm{closure}(\Delta_i^2 - R_i)$. Δ_i''s are attached 2-surfaces to X^2. Let D_i be a 2-disk fiber of T_i on a point $\in \mathrm{Int} f(\gamma_i'')$. Then D_i's are also attached to X^2. For an attached 2-surface P^2 (to X^2), define $q(\partial P^2) \in \mathbf{Z}_2$ by $q(\partial P^2) = \mathcal{O}(P^2) + L(P^2) \pmod 2$. Then, under the hypothesis that $\xi\ (= [X^2])$ is characteristic, q gives a well-defined quadratic function $q\colon H_1(X^2; \mathbf{Z}_2) \to \mathbf{Z}_2$. The Arf invariant of q depends only on the homology class $[X^2]$. See [**2**], [**8**]. This was the definition of $\mathrm{Arf}(\xi)$; see [**2**], [**11**].

In our situation, the boundaries $\partial\Delta_1', \partial D_1, \cdots, \partial\Delta_1', \partial D_r$ are 'symplectic basis elements' of X^2. Therefore

$$\begin{aligned}\mathrm{Arf}(\xi) &= \sum_i q(\partial\Delta_1')\cdot q(\partial D_i) = \sum_i (\mathcal{O}(\Delta_i) + L(\Delta_i))\cdot 1 \\ &= \sum_i \mathcal{O}(\Delta_i) + \sum_i L(\Delta_i),\end{aligned}$$

as required.

REMARK. A closely related geometric description of the Arf invariant of knots has been given by S. J. Kaplan [**3**].

2. In this section we shall give certain necessary conditions (in terms of the Arf invariant) for 2-homology classes $x_1, \cdots, x_n$ to be represented by pairwise disjoint PL embedded 2-spheres or by continuous maps of S^2 whose images are pairwise disjoint. We assume that $x_1, \cdots, x_n$ are characteristic, and so are the sums $x_i + x_j$ $(i < j), \sum_{i=1}^n x_i$. For example, if M^4 is a spin manifold, one can take classes $x_1, \cdots, x_n$ divisible by 2, or if M^4 has the property that $x \cdot y \equiv 0 \pmod 2$ for any $x, y \in H_2(M)$ (M necessarily has boundary), then every class $\in H_2(M)$ is characteristic.

PROPOSITION 2. *If* $x_1, \cdots, x_n$ *are represented by disjointly embedded* PL *2-spheres, then*

$$\mathrm{Arf}(x_1 + \cdots + x_n) = \mathrm{Arf}(x_1) + \cdots + \mathrm{Arf}(x_n).$$

PROOF. Let $\Sigma_1, \cdots, \Sigma_n$ be the disjoint PL spheres. We may assume that each Σ_i has only one (possibly) nonlocally-flat point p_i. Take a small 4-ball B_i centered at p_i and replace the cone $\Sigma_i \cap B_i$ on the knot $k_i = \Sigma_i \cap \partial B_i$ by a smoothly embedded 2-surface P_i (in B_i) bounded by k_i. Then we have an embedded surface F_i representing x_i. The calculations of $\mathrm{Arf}(x_i)$, $\mathrm{Arf}\,(x_1 + \cdots + x_n)$ can be done using F_i, $F_1 \# \cdots \# F_n$. However, every setting needed (attached 2-surfaces, etc.) can be taken within the pairwise disjoint 4-balls $B_1, \cdots, B_n$. Thus clearly $\mathrm{Arf}\,(x_1 + \cdots + x_n) = \mathrm{Arf}(x_1) + \cdots + \mathrm{Arf}(x_n)$ as required.

EXAMPLE. Following Kirby [6] we use framed links to represent 1-connected 4-manifolds with boundary by attaching 2-handles along them to B^4. Define:

$M_0^4 =$ 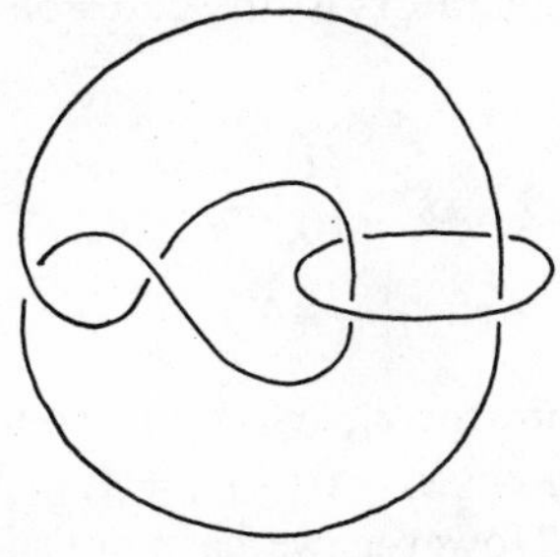

Let x_1, x_2 be the 2-homology classes $\in H_2(M_0^4)$ corresponding to the cores of handles. Then $x_1 \cdot x_2 = 0$. However, we cannot represent them by disjoint PL 2-spheres, because $\mathrm{Arf}(x_1) = \mathrm{Arf}(x_2) = 0$ but $\mathrm{Arf}(x_1 + x_2) = 1$.

REMARK. In his work [4], A. Kawauchi studies the Alexander polynomial associated with an epimorphism $\pi_1(\partial M_0^4) \to \mathbf{Z}$ and shows that any basis of M_0^4 cannot be represented by disjointly embedded PL 2-spheres.

PROPOSITION 3. *If* $x_1, \cdots, x_n \in H_2(M^4)$ *are represented by continuous maps* $f_1, \cdots, f_n\colon S^2 \to M^4$ *with pairwise disjoint images, then*

$$\mathrm{Arf}(x_1 + \cdots + x_n) = \sum_{i<j} \mathrm{Arf}(x_i + x_j) + n\left(\sum_{i=1}^{n} \mathrm{Arf}(x_i)\right).$$

PROOF. Approximating f_i by an immersion closely enough, we can assume that each f_i is an immersion satisfying (i) and (ii) in §1. For each i, let $\Delta^{(i)}$ be the union of Whitney's surfaces attached to the ith sphere $f_i(S^2)$. Let $\mathcal{O}^{(i)} = \mathcal{O}(\Delta^{(i)})$ and let $l_{ij} = \#(\mathrm{Int}\,\Delta^{(i)} \cap f_j(S^2))$. Then by Proposition 1, we have

$$\begin{aligned}\mathrm{Arf}(x_1 + \cdots + x_n) &= \sum_{i=1}^{n} \mathcal{O}^{(i)} + \sum_{i,j=1}^{n} l_{ij} \\ &= \sum_{i<j} (l_{ii} + l_{ij} + l_{ji} + l_{jj} + \mathcal{O}^{(i)} + \mathcal{O}^{(j)}) + n \sum_{i=1}^{n} (l_{ii} + \mathcal{O}^{(i)}) \\ &= \sum_{i<j} \mathrm{Arf}(x_i + x_j) + n \sum_{i=1}^{n} \mathrm{Arf}(x_i)\end{aligned}$$

as asserted.

EXAMPLE. Take the following link which is almost trivial in Milnor's sense [10] and define M^4 by this link

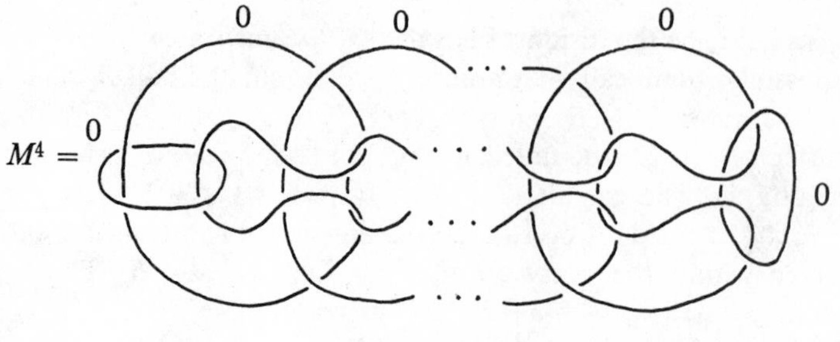

(n components: $n \geqq 3$)

Let $x_1, \cdots, x_n$ be the corresponding generators to the cores of 2-handles. It can be shown that

$$\mathrm{Arf}(x_1 + \cdots + x_n) = \begin{cases} 1, & n = 3, \\ 0, & n \geqq 4, \end{cases}$$

$$\mathrm{Arf}(x_i) = 0 \quad \text{and} \quad \mathrm{Arf}(x_i + x_j) = 0 \quad (i < j).$$

Therefore, if $n = 3$, we cannot represent x_1, x_2, x_3 by continuous maps of S^2 with pairwise disjoint images, though $x_i \cdot x_j = 0\,(\forall\, i, j = 1, 2, 3)$. It is conceivable that the result is the same for $n \geqq 4$. However, we have not succeeded in detecting it.

REMARK. K. Kobayashi [7] proved that if M is a 1-connected 4-manifold, two homology classes x_1, x_2 with $x_1 \cdot x_2 = 0$ can be homotopically separated. Thus $n = 3$ is the minimum number for the above phenomenon.

3. Here is an example of a new numerical homotopy invariant which might be called the 'secondary intersection triple' of 2-dimensional homotopy classes of a 4-manifold. It is closely related to the observations in §§1–2.

Let M^4 be a compact *oriented* 4-manifold with $H_1(M^4) = 0$. Let x_1, x_2, x_3 ($\in$

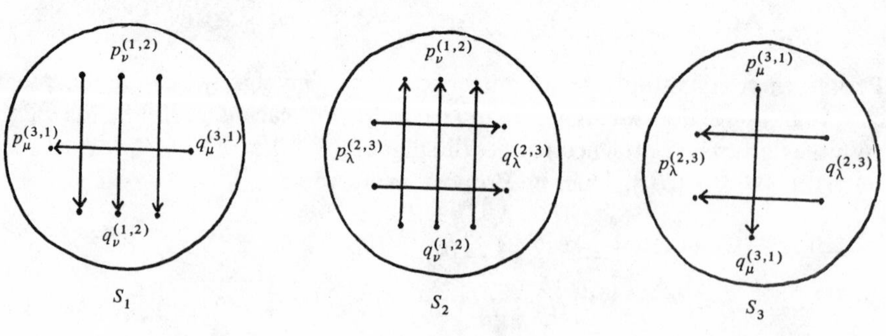

FIGURE 1

$\pi_2(M^4)$) be homotopy classes such that $[x_i] \cdot [x_j] = 0$ ($\forall\, i, j = 1, 2, 3, i \neq j$). Represent them by smooth immersions $f_1, f_2, f_3\colon S^2 \to M^4$. Let S_i be the image $f_i(S^2)$. We denote by $p_\lambda^{(i,j)}$ (or $q_\lambda^{(i,j)}$) the λth positive (or negative) intersection point of S_i and S_j. Draw smooth arcs $\gamma_{\lambda,i}^{(i,j)}$ (or $\gamma_{\lambda,j}^{(i,j)}$) connecting $p_\lambda^{(i,j)}$, $q_\lambda^{(i,j)}$ on the sphere S_i (or S_j). We assume that they have no self-intersection points. As in §1, the circle $\gamma_{\lambda,i}^{(i,j)} \cup \gamma_{\lambda,j}^{(i,j)}$ bounds a smoothly embedded orientable 2-surface $\Delta_\lambda^{(i,j)}$. We orient $\partial\Delta_\lambda^{(i,j)}$ (and thus $\Delta_\lambda^{(i,j)}$) as indicated in Figure 1. Now define an integer $\langle x_1, x_2, x_3\rangle$ as follows:

$$\langle x_1, x_2, x_3\rangle = \sum_\lambda S_1 \cdot \Delta_\lambda^{(2,3)} + \sum_\mu S_2 \cdot \Delta_\mu^{(3,1)} + \sum_\nu S_3 \cdot \Delta_\nu^{(1,2)}$$

$$+ \sum_{\mu,\nu} \frac{\partial\Delta_\mu^{(3,1)} \cdot \partial\Delta_\nu^{(1,2)}}{S_1} + \sum_{\nu,\lambda} \frac{\partial\Delta_\nu^{(1,2)} \cdot \partial\Delta_\lambda^{(2,3)}}{S_2} + \sum_{\lambda,\mu} \frac{\partial\Delta_\lambda^{(2,3)} \cdot \partial\Delta_\mu^{(3,1)}}{S_3},$$

where $S_1 \cdot \Delta_\lambda^{(2,3)}$, etc., denote the intersection number of S_1 and $\Delta_\lambda^{(2,3)}$, etc., and $\partial\Delta_\mu^{(3,1)} \cdot \partial\Delta_\nu^{(1,2)}/S_1$, etc., denote the intersection number of $\partial\Delta_\mu^{(3,1)}$ and $\partial\Delta_\nu^{(1,2)}$ on S_1, etc. Let I ($= I(x_1, x_2, x_3)$) be the ideal of $\boldsymbol{Z}$ defined by

$$I = \left\{\sum_{i=1}^{3} [x_i] \cdot y_i;\, y_j \in H_2(M^4), j = 1,2,3\right\}.$$

PROPOSITION 4. *The residue class of $\langle x_1, x_2, x_3\rangle \in \boldsymbol{Z}/I$ depends only on the homotopy classes x_1, x_2, x_3, and it vanishes if x_1, x_2, x_3 can be represented by continuous maps of S^2 with pairwise disjoint images.*

REMARK. Though the notation $\langle x_1, x_2, x_3\rangle$ is slightly confusing, our triple product is not the same thing with the Massey product. In the sense that the ordinary intersection number is the "4-dimensional counterpart" of the linking number in the 3-dimension, our product corresponds to the higher order linking number defined by Milnor [**10**] and also by Massey [**15**].[1]

PROOF. We shall denote the residue class by the same notation $\langle x_1, x_2, x_3\rangle$. First, the residue class $\langle x_1, x_2, x_3\rangle$ does not depend on the choice of Whitney's surfaces $\Delta_\lambda^{(2,3)}$, etc. This is because if we take another surface $\tilde\Delta_\lambda^{(2,3)}$ with $\partial\tilde\Delta_\lambda^{(2,3)} = \partial\Delta_\lambda^{(2,3)}$, then the difference $S_1 \cdot \tilde\Delta_\lambda^{(2,3)} - S_1 \cdot \Delta_\lambda^{(2,3)}$ is equal to $S_1 \cdot \Sigma_\lambda^{(2,3)}$, where $\Sigma_\lambda^{(2,3)} = \tilde\Delta_\lambda^{(2,3)} \cup -\Delta_\lambda^{(2,3)}$. Since $S_1 \cdot \Sigma_\lambda^{(2,3)} = [x_1] \cdot [\Sigma_{\lambda,j}^{(2,3)}] \in I$, the residue class does not change. Next suppose that we take another $\tilde\gamma_\lambda^{(i,j)}$ on S_i instead of taking $\gamma_{\lambda,i}^{(i,j)}$ but with $\partial\tilde\gamma_{\lambda,i}^{(i,j)} = \partial\gamma_{\lambda,i}^{(i,j)}$. For instance let $\gamma_{\lambda,i}^{(i,j)} = \gamma_{\nu,1}^{(1,2)}$. $\tilde\gamma_{\nu,1}^{(1,2)}$ be regular homotopic (rel. ∂) to $\gamma_{\nu,1}^{(1,2)}$ on S_1. If the regular homotopy misses the intersection points $p_\nu^{(1,2)}, \cdots, q_\nu^{(1,2)}, \cdots, p_\mu^{(3,1)}, \cdots, q_\mu^{(3,1)}, \cdots$, evidently it does not change $\sum_\nu S_3 \cdot \Delta_\nu^{(1,2)}$, $\sum_{\mu,\nu} \partial\Delta_\mu^{(3,1)} \cdot \partial\Delta_\nu^{(1,2)}/S_1$, hence nor $\langle x_1, x_2, x_3\rangle$. In case it hits an intersection point, a typical situation is shown in Figure 2. The movement of Figure 2 increases $\sum_{\mu,\nu} \partial\Delta_\mu^{(3,1)} \cdot \partial\tilde\Delta_\nu^{(1,2)}/S_1$ by ε ($= \pm 1$), the sign of the intersection of $\gamma_{\mu,1}^{(3,1)}$ and $\gamma_{\nu,1}^{(1,2)}$ at p (see Figure 2(ii)). Using this ε, the orientation of S_1 is symbolically described (near p) by

$$(*) \qquad [S_1] = \varepsilon\,[\gamma_{\mu,1}^{(3,1)}] \times [\gamma_{\nu,1}^{(1,2)}].$$

[1]The author thanks Professor J. Milgram who pointed out this article of Massey, in which he defined the higher order linking number applying his triple product to the complement of a link.

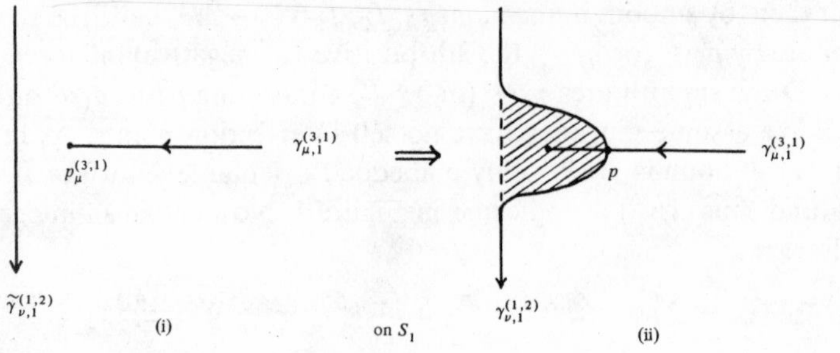

FIGURE 2

Define $\Delta_{\nu}^{(1,2)}$ by attaching the shadowed region (in Figure 2(ii)) to $\tilde{\Delta}_{\nu}^{(1,2)}$. Then the movement of Figure 2 also increases $\sum_{\nu} S_3 \cdot \tilde{\Delta}_{\nu}^{(1,2)}$ by ε' (= the sign of the intersection of S_3 and $\Delta_{\nu}^{(1,2)}$ at the point $p_{\mu}^{(3,1)}$). Here we adopt the orientation convention that the orientation of any 2-surface F^2 and the 'compatible' orientation of the boundary ∂F^2 should be related as

$$[F^2] = [\text{outward vector}] \times [\partial F^2].$$

In particular, $\Delta_{\nu}^{(1,2)}$ (= the shadowed region near $p_{\mu}^{(3,1)}$) is oriented (near p) by $-[\gamma_{\mu,1}^{(3,1)}] \times [\gamma_{\nu,1}^{(1,2)}]$. Then we can calculate ε' symbolically:

$$\begin{aligned}\varepsilon' &= \frac{[S_3] \times [\Delta_{\nu}^{(1,2)}]}{[M^4]} = \frac{[S_3] \times (-[\gamma_{\mu,1}^{(3,1)}]) \times [\gamma_{\nu,1}^{(1,2)}]}{[M^4]} \\ &= -\varepsilon \frac{[S_3] \times [S_1]}{[M^4]} \quad \text{by}(*) \\ &= -\varepsilon,\end{aligned}$$

because $p_{\mu}^{(3,1)}$ is a positive intersection and $[M^4] = [S_3] \times [S_1]$ near $p_{\mu}^{(3,1)}$. Thus the movement of Figure 2 changes $\langle x_1, x_2, x_3 \rangle$ by $\varepsilon + \varepsilon' = 0$, i.e., the triple $\langle x_1, x_2, x_3 \rangle$ does not depend on the particular curve $\gamma_{\nu,1}^{(1,2)}$ etc., once their terminal points $\partial \gamma_{\nu,1}^{(1,2)}$ are fixed. In other words, the order $p_1^{(1,2)}, p_2^{(1,2)}, \cdots; q_1^{(1,2)}, q_2^{(1,2)}, \cdots$, etc. determines the triple product. However, it does not depend even on the order, for if we interchange, say, $p_1^{(3,1)}$ and $p_2^{(3,1)}$, this leaves 6 summations in the definition of $\langle x_1, x_2, x_3 \rangle$ unchanged. See Figure 3. So far, we have proved that $\langle x_1, x_2, x_3 \rangle$ is determined by the geometric figures of S_1, S_2, S_3.

Now we shall show that $\langle x_1, x_2, x_3 \rangle$ depends only on the homotopy classes x_1,

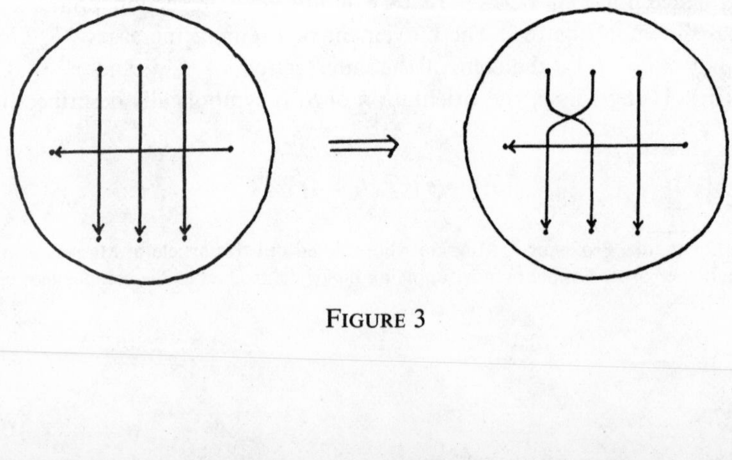

FIGURE 3

x_2, x_3. Suppose the representative of x_i, $f_i: S^2 \to M^4$, is homotopic to another smooth immersion $f_i': S^2 \to M^4$. By Whitney [14], they are regularly homotopic after locally introducing a finite number of Whitney's self-intersection points to f_i. However, such introduction does not change $\langle x_1, x_2, x_3 \rangle$. Thus we may assume that f_i is regularly homotopic to f_i'. If the regular homotopy does not affect the configuration of the curves on $S_1 \cup S_2 \cup S_3$, then it does not change $\langle x_1, x_2, x_3 \rangle$ either. If it affects the configuration, there are two typical cases I and II shown in Figure 4.

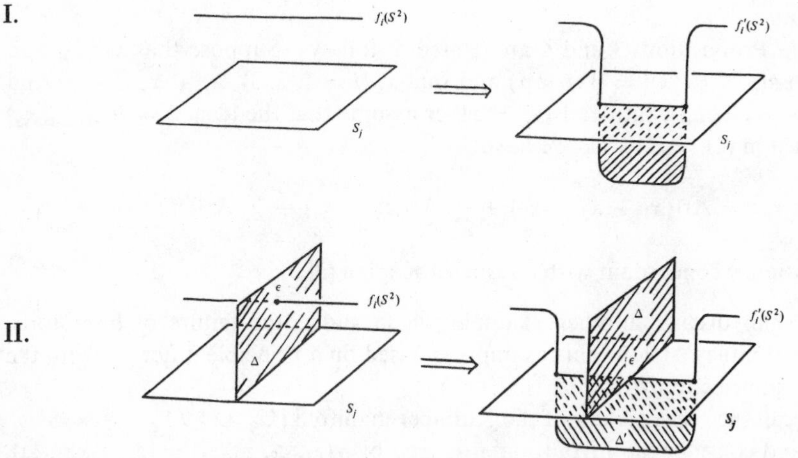

FIGURE 4

Evidently the movement of case I (or its reverse) does not change $\langle x_1, x_2, x_3 \rangle$. In case II, an intersection (of sign ε) of $S_i \cdot \Delta$ vanishes but a new intersection of $\partial\Delta' \cdot \partial\Delta$ (of sign ε') appears. By 'symbolical calculation' of the orientation, we can show that $\varepsilon = \varepsilon'$. Therefore, $\langle x_1, x_2, x_3 \rangle$ remains unchanged again. This completes the proof of homotopy invariance of the residue class $\langle x_1, x_2, x_3 \rangle$.

Finally we have $\langle x_1, x_2, x_3 \rangle = 0$ if $S_1 \cap S_2 = S_1 \cap S_3 = S_2 \cap S_3 = \emptyset$. This completes the proof of Proposition 4.

REMARK. The triple $\langle x_1, x_2, x_3 \rangle$ is skew-symmetric under permutations of x_1, x_2, x_3, and satisfies $\langle \pm x_1, x_2, x_3 \rangle = \pm \langle x_1, x_2, x_3 \rangle$.

EXAMPLE.

$M^4 =$

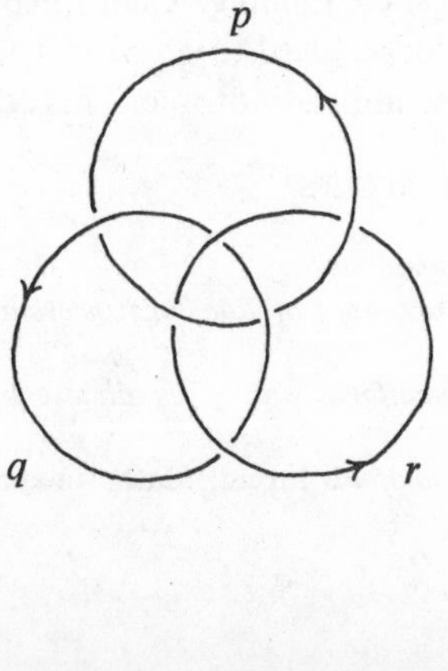

Define M^4 as above and let x_1, x_2, x_3 be the generators of $H_2(M^4)$ corresponding to the handles. Then clearly $x_i \cdot x_j = 0$ $(i < j)$ and we have $\langle x_1, x_2, x_3 \rangle = -1$ mod G.C.D.(p, q, r). (We here adopt the orientation convention: $[M^4] =$ [outward vector] $\times$ $[\partial M^4]$.) Therefore, they cannot be homotopically separated unless G.C.D.$(p, q, r) = 1$.

REMARK. If M^4 is constructed by attaching 2-handles to B^4 along an almost trivial link L of 3-components (with framings p, q, r), then the corresponding generators x_1, x_2, x_3 satisfy $\langle x_1, x_2, x_3 \rangle = -\mu(1, 2, 3)$ mod G.C.D.(p, q, r), where $\mu(1, 2, 3)$ is Milnor's higher linking number **[10]**. (See also **[15]**.) This is proved by the homotopy invariance of $\langle x_1, x_2, x_3 \rangle$ and the homotopy classification of almost trivial links.

REMARK. Propositions 3 and 4 are related as follows: Suppose that $x_1, x_2, x_3 \in \pi_2(M^4)$ satisfy $[x_i] \cdot [x_j] = 0$ $(i < j)$ and that x_i $(i = 1, 2, 3)$, $x_i + x_j$ $(i < j)$ and $x_1 + x_2 + x_3$ are all characteristic. Further assume that the ideal $I = I(x_1, x_2, x_3)$ is contained in $(2) \subset \boldsymbol{Z}$. Then we have

$$\langle x_1, x_2, x_3 \rangle = \operatorname{Arf}(x_1 + x_2 + x_3) + \sum_{i<j} \operatorname{Arf}(x_i + x_j) + \sum_{i=1}^{3} \operatorname{Arf}(x_i) \pmod 2.$$

This equation is concordant with a result of Kaplan **[3]**.

4. We shall discuss another example which shows the failure of homotopic version of Whitney's trick. This example is based on a principle different from the secondary intersection triple.

First recall the construction of the Kummer manifold **[12]**. Let $T^4 = S^1 \times S^1 \times S^1 \times S^1$ and consider the involution σ defined by $\sigma(z_1, z_2, z_3, z_4) = (\bar{z}_1, \bar{z}_2, \bar{z}_3, \bar{z}_4)$, where we consider S^1 as $\{z \in \boldsymbol{C}; |z| = 1\}$. σ has 16 fixed points $p_1, \cdots, p_{16}$. The quotient variety T^4/σ has 16 singular points, each of which locally looks like a cone over a 3-dimensional projective space. Blow up these singularities; in other words, delete small neighborhoods of singular points and glue copies of the total space E of a 2-disk bundle over S^2 with Euler class -2. Thus we obtain a closed smooth 4-manifold M^4 which contains 16 smoothly embedded 2-spheres (as exceptional curves or zero-sections of E's). Denote these spheres by $S_1^2, \cdots, S_{16}^2$. Note that $[S_i^2] \cdot [S_j^2] = -2\delta_{ij}$ (Kronecker's delta). Since $b_2(M^4) = 22$, there is a nonzero homology class $\xi \in H_2(M^4)$ such that $\xi \cdot [S_i^2] = 0$ $(\forall\, i = 1, \cdots, 16)$. Since M^4 is 1-connected **[12]**, ξ is represented by a continuous map $f: S^2 \to M^4$.

ASSERTION. *Although $f(S^2) \cdot [S_i^2] = 0$ $(\forall\, i = 1, \cdots, 16)$, f is not homotopic to a map g such that $g(S^2) \cap (\bigcup_{i=1}^{16} S_i^2) = \emptyset$.*

PROOF. Suppose f were homotopic to such a map g. Then since $M^4 - \bigcup_{i=1}^{16} S_i^2 = T^4/\sigma -$ (the 16 points), the map g would be lifted to $\tilde{g}$: $S^2 \to T^4 -$ (the 16 points). Since $\pi_2(T^4 - 16 \text{ points}) = 0$, $\tilde{g}$ must be null-homotopic. This contradicts $\xi \neq 0$.

REFERENCES

1. M. Freedman, *Knot theory and 4-dimensional surgery.*

2. M. Freedman and R. Kirby, *A geometric proof of Rochlin's theorem*, these PROCEEDINGS, part 2, pp. 85–97.

3. S. J. Kaplan, *Constructing framed 4-manifolds with given almost framed boundaries* (to appear).

4. A. Kawauchi, *On quadratic forms of 3-manifolds* Invent. Math. (to appear).

5. M. Kervaire, and J. Milnor, *On 2-spheres in 4-manifolds*, Proc. Nat. Acad. Sci. U.S.A. **47** (1961), 1651–1657. MR **24** #A2968.

6. R. Kirby, *A calculus for framed links* (to appear).

7. K. Kobayashi, *On a homotopy version of 4-dimensional Whitney's lemma*, Math. Seminar Notes **5** (1977), 109–116, Kobe Univ.

8. Y. Matsumoto, *An elementary proof of Rochlin's signature theorem and its extension by Guillou and Marin*, IAS Seminar Note, March 1977.

9. J. Milnor, *Lectures on the h-cobordism theorem*, Princeton Univ. Press., Princeton, N. J., 1965. MR **32** #8352.

10. ———, *Link groups*, Ann. of Math. (2) **59** (1954), 177–195. MR **17**, 70.

11. V. A. Rochlin, *Proof of Gudkov's hypothesis*, Functional Anal. Appl. **6** (1972), 136–138.

12. E. Spanier, *The homology of Kummer manifolds*, Proc. Amer. Math. Soc. **7** (1956), 155–160. MR **19**, 317.

13. H. Whitney, *The self-intersections of a smooth n-manifold in 2n-space*, Ann. of Math. (2) **45** (1944), 220–246. MR **5**, 273.

14. ———, *The general type of singularity of a set of* $2n - 1$ *smooth functions of n-variables*, Duke Math. J. **10** (1943), 161–172. MR **4**, 193.

15. W. S. Massey, *Higher order linking numbers*, Conf. on Algebraic Topology (Univ. of Illinois at Chicago Circle, Chicago, Ill., 1968), 1969, pp. 174–205. MR **40** #8039.

INSTITUTE FOR ADVANCED STUDY

UNIVERSITY OF TOKYO

GEOMETRY OF DIFFERENTIAL MANIFOLDS AND ALGEBRAIC VARIETIES

Proceedings of Symposia in Pure Mathematics
Volume 32, 1978

SMOOTHINGS OF ISOLATED SINGULARITIES*

E. REES AND E. THOMAS

Consider the following simple example: Let F be a polynomial in n variables with complex coefficients, regarded as a map $F: C^n \to C$. Suppose that $F(0) = 0$; $dF(0) = 0$; $dF(p) \neq 0, p \neq 0$. Set $V = F^{-1}(0) \subset C^n$. Then V is a codimension one algebraic subvariety of C^n with an isolated singularity at 0. However, this singularity can easily be smoothed off by deforming V slightly. That is, if we set $V_t = F^{-1}(t)$, for t a small complex number, then V_t is a nonsingular subvariety for $t \neq 0$, while $V_0 = V$.

Suppose more generally that V is an algebraic subvariety (with arbitrary codimension) of a nonsingular variety W and that V has an isolated singularity at a point $p \in V$, V is said to be algebraically smoothable in W (cf. Hartshorne [3]) if there is an algebraic family $\{V_t\}$, where $t \in$ small complex disk Δ, such that:

(i) for $t \neq 0$, V_t is a nonsingular subvariety of W,

(ii) $V_0 = V$.

See Figure 1.

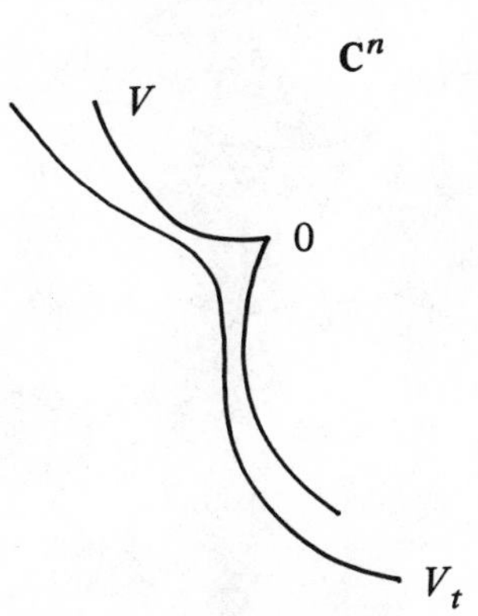

FIGURE 1

AMS (*MOS*) *subject classifications* (1970). Primary 14K05, 57D10, 55E99.

*Research supported by the National Science Foundation.

Here is a way of producing subvarieties with isolated singularities. Suppose that $X \subset P_{n-1}$ is a nonsingular projective variety (P_q = complex projective q-space, $q \geqq 1$). Let ξ denote the canonical complex line bundle over P_{n-1} (ξ is dual to the Hopf bundle [6, p. 59]). Then the total space of ξ, $E(\xi)$, is given by $P_n - \infty$, where $\infty = (0, 0, \cdots, 0, 1)$. Note that $E(\xi|X)$ is an open nonsingular variety in P_n. Define the cone on X by $C(X) = E(\xi|X) \cup \infty \subset P_n$. Clearly $C(X)$ is an algebraic subvariety of P_n with an isolated singularity at ∞. Note that $C(X)$ is defined by precisely the same equations as X, and codim $C(X)$ = codim X. When is $C(X)$ algebraically smoothable in P_n?

By an argument similar to that given in the first paragraph if X is a (nonsingular) complete intersection, then $C(X)$ is algebraically smoothable. On the other hand Thom, in the early 1950's, gave an example of a nonsmoothable cone. Let $P_1 \times P_2 \subset P_5$ denote the Segre embedding, i.e.,

$$((w_0, w_1), (z_0, z_1, z_2)) \to (w_0z_0, w_0z_1, w_0z_2, w_1z_0, w_1z_1, w_1z_2).$$

One then has (see Hartshorne [3]):

THEOREM OF THOM. *$C(P_1 \times P_2)$ is not algebraically smoothable in P_6.*

Thom's proof (which we sketch at the end of the paper) suggests a topological version of the notion of an algebraically smoothable singularity. Suppose that W is now a smooth (nonsingular) manifold and that $V \subset W$ is a smooth submanifold with an isolated singularity at a point p in V (i.e., $V - p$ is a nonsingular submanifold of W). Suppose moreover that $V - p$ has a complex normal bundle in W. We will say that V is topologically smoothable (= t-smoothable) in W if, for any sufficiently small ball B at p, there is a nonsingular submanifold $V' \subset B$, with complex normal bundle, such that the pairs $(\partial B, V \cap \partial B)$ and $(\partial B, V' \cap \partial B)$ are diffeomorphic, with isomorphic complex normal bundles. See Figure 2.

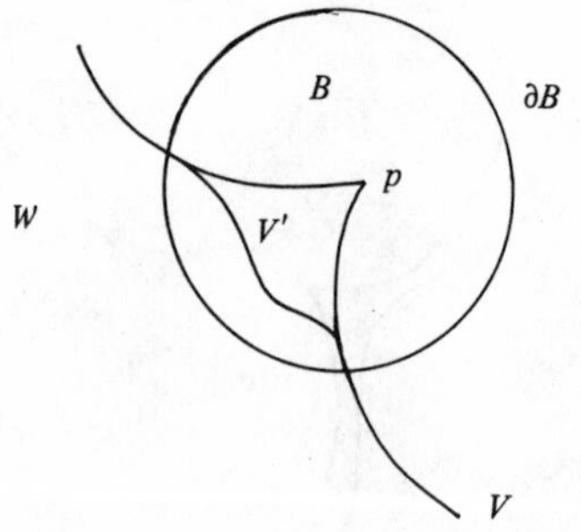

FIGURE 2

Note that we may now form

$$V'' = (V - \text{int } B) \bigcup_{\partial B} V' \subset W;$$

thus V'' is a smooth nonsingular submanifold of W (with stable complex normal bundle) which we regard as the smoothing of V in W.

Given the precise definition of algebraically smoothable (see [3]) it is straightforward to show:

If $V \subset W$ is an algebraic subvariety with an isolated singularity and if V is algebraically smoothable, then it is topologically smoothable.

The point of introducing t-smoothability is that it is measured by an algebraic invariant. To see this, suppose that W is an oriented $(q + 1)$-manifold and that $V \subset W$ has complex codimension n. Then ∂B is an oriented q-sphere S^q; let $L = V \cap \partial B = V \cap S^q$ denote the link of the singularity p. Thus L is a complex codimension n submanifold of S^q, and so by the Pontrjagin-Thom construction [9] we obtain an element $\Lambda(V) \in \pi_q MU(n)$, where $MU(n)$ denotes the Thom complex for the classifying bundle over $BU(n)$. By the Thom transversality theorem [9], one readily shows:

$\Lambda(V) = 0$ if and only if V is t-smoothable in W.

In particular, if V and W are algebraic and if $\Lambda(V) \neq 0$, then V is not algebraically smoothable.

For this theory to be of use we must deal with two problems:

Problem 1. Compute $\pi_* MU(n)$.

Problem 2. Compute $\Lambda(V) \in \pi_* MU(n)$.

Classically one knows that $\pi_{2n+i} MU(n) \approx \Omega_i^U$, for $0 \leqq i \leqq 2n - 2$, where Ω_*^U denotes the complex cobordism ring (see [5]). In particular, since $\Omega_{\text{odd}}^U = 0$, we have:

if $V \subset W^{2q}$ has complex codimension n, with $q \leqq 2n - 1$, then V is t-smoothable in W.

Thus the first unknown group is $\pi_{4n-1} MU(n)$. As usual let $\alpha(n)$ denote the number of ones in the dyadic expansion of the positive integer n. Define

$$\rho(n) = \text{least integer } k \geqq 0 \text{ such that } \alpha(n + k) \leqq 2k.$$

For example, $\rho(n) = 1 \Leftrightarrow n = 2^s + 2^r - 1$.

Our first result is

THEOREM 1. $\pi_{4n-1} MU(n) = Z/2^{\rho(n)-1}$.

We now show how to calculate the element of this group that arises from a cone singularity.

Suppose that $X \subset P_{q-1}$ is a smooth nonsingular submanifold, with a complex normal bundle, say, of dimension n. (Thus, X is a stably almost-complex manifold of (real) dimension $2(q - n - 1)$.) As above, set $C(X) = E(\xi | X) \cup \infty \subset P_q$. Recall that $C(X)$ is simply the Thom complex of $\xi | X$. We write

$$\lambda(X) = \Lambda(C(X)) \in \pi_{2q-1} MU(n).$$

Suppose that $q = 2n$; then $\dim X = 2(n - 1)$. Conversely, suppose we take any stably almost-complex manifold of dimension $2(n - 1)$, and a class $v \in H^2(X; Z)$. Regard v as a map $X \to P_{2n-1}$. By Whitney's theorem, v is homotopic to a smooth embedding $X \subset P_{2n-1}$; also, a simple calculation shows that the normal bundle of this embedding is complex. Thus the pair (X, v) gives a unique element (we display the class v for emphasis): $\lambda(X, v) \in Z/2^{\rho(n)-1}$. How do we compute this integer? Of course, if $\rho(n) = 1$, then $\lambda(X, v) = 0$ and so we obtain:

COROLLARY 1. *Let $X \subset P_{2n-1}$ have complex codimension n. If $n = 2^s + 2^r - 1$, then $C(X)$ is t-smoothable in P_{2n}.*

If $\rho(n) > 1$ we proceed as follows. For $0 \leqq q \leqq 2n - 1$, we have $P_q \subset P_{2n-1}$. Let k be an integer, $0 \leqq k \leqq n - 1$, and let A_k be the element in the complex cobordism ring Ω^U_{2k} defined by:

$$A_0 = X \cap P_n,$$

$$A_k = X \cap P_{n+k} - \sum_{j=1}^{k} A_{k-j} \cdot P_j, \qquad k > 0.$$

Here the dot indicates the product in Ω^U_*. Note that if $[X] \in H_{2(n-1)}(P_{2n-1}; Z)$ denotes the homology class determined by X and if $[X] = d \cdot [P_{n-1}]$, then $A_0 = d \in Z = \Omega^U_0$. (We may think of d as the "degree" of X in P_{2n-1}.)

Our next result is

THEOREM 2. *Let X be a stably almost-complex $2(n-1)$-manifold and let $v \in H^2(X; Z)$. As above, v gives an embedding $X \subset P_{2n-1}$. Suppose that $[X] = d \cdot [P_{n-1}] \in H_{2(n-1)}(P_{2n-1}; Z)$, where $d \in Z$. Then*

$$\lambda(X, v) = \frac{1}{2}\left[\sum_{k=1}^{n-1} c_k[A_k]\binom{n+k}{2k}\right] - \binom{d}{2} \bmod 2^{\rho(n)-1}.$$

Here $c_k[A_k]$ denotes the normal Chern number of $A_k \in \Omega^U_{2k}$.

We can simplify this expression somewhat if we only consider the question of whether $\lambda(X, v)$ generates $\pi_{4n-1}MU(n)$. Since this group is cyclic with order a power of two, $\lambda(X, v)$ is a generator precisely when the integer in Theorem 2 is odd. If we write γ_k for $c_k(X \cap P_{n+k})$, a calculation shows that

$$\gamma_k = ((v(1+v)^{n-k-1}) \cdot c(X))\,[X],$$

where $c(X)$ denotes the total normal Chern class of X. We then have

THEOREM 3. *$\lambda(X, v)$ generates $\pi_{4n-1}MU(n)$ if, and only if,*

$$\frac{1}{2}\sum_{k=1}^{n-1} \gamma_k \cdot \binom{n+k}{2k} + dn + \binom{d}{2} \equiv 1 \bmod 2.$$

EXAMPLE 1. The Segre embedding. By taking products of coordinates as above, one has $P_1 \times P_{n-2} \subset P_{2n-3}$, $n \geqq 3$; by inclusion we obtain $P_1 \times P_{n-2} \subset P_{2n-1}$. Let x, y denote the respective generators for $H^2(P_1)$, $H^2(P_{n-2})$, integer coefficients. Then $v = x + y$, and a calculation gives:

$$\gamma_0\,(=d) = n - 1; \qquad \gamma_k = (-1)^k\, 2\binom{2(k-1)}{k-1}, \qquad k > 0.$$

By Theorem 3 one then has:

For $n \geqq 3$, $\lambda(P_1 \times P_{n-2}, x + y)$ generates $\pi_{4n-1}\, MU(n)$.

EXAMPLE 2. The Veronese embedding. Let x, y denote respectively the canonical generators for $H^2(P_{n-1})$, $H^2(P_{2n-1})$, and let $j_d\colon P_{n-1} \to P_{2n-1}$ be given by $j_d^* y = dx$, $d \in Z$. As remarked above, j_d is homotopic to an embedding, with complex normal bundle. (For $d > 0$, this is the "Veronese embedding" of degree d^{n-1}, which can be algebraically projected down into P_{2n-1}.) Thus $v = dx$. Computing mod 4 one has, for $k > 0$:

$$\gamma_k \equiv \binom{2k}{k} + 2(n-k-1)\binom{2k}{k-1} \mod 4;$$

a calculation then gives, by Theorem 3:

If $n \geqq 3$ and $d \equiv 3 \mod 4$, then $\lambda(P_{n-1}, dx)$ generates $\pi_{4n-1}MU(n)$.

We now consider $\pi_k MU(n)$, with $k > 4n - 1$. It is straightforward to show that $\pi_{4n}MU(n)$ and $\pi_{4n+2}MU(n)$ are free abelian—indeed, each is a direct summand of Ω_*^U. We have calculated $\pi_{4n+1}MU(n)$. It is the direct sum of two cyclic groups, each with order a power of two; and we have also calculated the invariants.

When one considers $\pi_{4n+3}MU(n)$ one sees by the method of [2] that $\pi_{4n+3}MU(n) \otimes Q = Q$, and so a singularity gives rise to a class in this group. This rational class is of course somewhat less delicate than our previous invariants and is essentially the invariant that was considered by Thom. It is defined as follows for cone singularities (cf. [3, Theorem 3.1]). Given an embedding $j: X \subset P_{2n+1}$ of complex codimension n, let $j_*: H^q(X) \to H^{q+2n}(P_{2n+1})$ denote the Gysin homomorphism, define the integer $\mu(X)$ to be

$$[(j_*(c_1(\nu)))^2 - j_*(c_1(\nu)^2) \cdot j_*(1_x)]\,[P_{2n+2}]$$

where $c_1(\nu)$ is the first Chern class of the normal bundle of the embedding j. We identify cohomology classes of P_{2n+1} with those of P_{2n+2} and take cup products in the cohomology of P_{2n+2}.

As an example of this invariant we consider the Segre embedding of $P_r \times P_s$ projected down into $P_{2(r+s)-1}$. A calculation gives that

$$\mu(P_r \times P_s) = \frac{(s-r)^2}{r+s-1}\binom{r+s-1}{s}\binom{r+s-1}{r}.$$

In particular for $r > s > 0$ this is not zero which yields that $C(P_r \times P_s)$ is not t-smoothable in $P_{2(r+s)}$. Thom's example is the case $r = 1, s = 2$.

We conclude by remarking on the role of the integers $\rho(n)$. Much as the divisibility of Chern classes turns up in the case of the unstable homotopy of $BU(n)$ [1], the divisibility of Chern numbers turns up in the unstable homotopy of $MU(n)$. If we let

$$d_i(n) = \text{g.c.d.}(c_i(M)c_{n-i}(M))\,[M] \quad \text{for } 0 \leqq i \leqq n$$

where M varies over Ω_{2n}^U and $c_j(M)$ denotes the (stable) normal Chern class of M, there is a relationship between the integers $d_i(n)$ and the unstable homotopy of $MU(n)$. In fact if we let $\rho_i(n)$ be the least integer $k \geqq 0$ such that $\alpha(n+k) \leqq 2k + i$, so that $\rho(n)$ is $\rho_0(n)$, we have

THEOREM 4. *For $n \geqq 2i \geqq 0$, $d_i(n) = 2^{\rho_i(n)} \cdot N_i(n)$ where $N_i(n)$ is an odd integer. Moreover, $N_0(n) = N_1(n) = N_2(n) = 1$.*

We have no counterexample to the conjecture that $N_i(n) = 1$ for all $i \geqq 0$.

REMARK 1. In our definition of t-smoothable, we have required that the submanifold $V - p$ have complex normal bundle, i.e., have structure group $U(n)$. This was done in order to have the topological definition as close as possible to the algebraic one. By taking a different structure group for the normal bundle (e.g., $O(q)$, TOP (q)), one obtains a number of different definitions of t-smoothable. In general, if $V - p \subset W^{q+1}$ has normal bundle with structure group G, then the obstruction to V being t-smoothable (so that the manifold V'' has normal bundle with structure group G) is a class $\Lambda(V) \in \pi_q MG$.

REMARK 2. Using the method of [2], one can show that the rational homotopy type of $MU(n)$ is given by the (rational) fibration

$$\bigvee_\alpha S_\alpha \xrightarrow{i_n} MU(n) \xrightarrow{U_n} K(Z, 2n).$$

Here α runs over all ordered sets of nonnegative integers $(a_1, a_2, \cdots, a_{n-1})$, with not all a_i zero, and S_α denotes the sphere of dimension $2n + \sum_i 2ia_i$. One can now calculate $\pi_* MU(n) \otimes Q$ by using the Hilton theorem [5]. For example, if $n = 2$ the following groups occur through dimension 17.

i	4	6	8	10	11	12	13	14	15	16	17
$\pi_i MU(2) \otimes Q$	Q	Q	Q	Q	Q	Q	Q	Q	$Q \oplus Q$	Q	$Q \oplus Q$

REMARK 3. We indicate briefly the proof of Theorem 1. Recall that one has a map $S^2 M_n \to M_{n+1}$, where S^2 denotes the double suspension and $M_q = MU(q)$, $q \geqq 1$. Iterating this map and taking the adjoint we obtain a map $M_n \to^{b_n} \Omega^{2N} M_{n+N}$ (N large). Define F_n to be the homotopy fiber of b_n. Since $\pi_{2n+i}\Omega^{2N} M_{n+N} \approx \Omega_i^U$ (for $i \leqq 2N - 2$), we obtain an exact sequence

$$\Omega_i^U \xrightarrow{\partial} \pi_{2n+i-1} F_n \longrightarrow \pi_{2n+i-1} M_n \longrightarrow 0$$

and so

(i) $$\pi_{4n-1} M_n \approx \pi_{4n-1} F_n / \partial \Omega_{2n}^U .$$

We use the method of Milgram [7] to study $H^*(\Omega^{2N} M_{n+N})$ (in the metastable range) and thus obtain $H^*(F_n)$. We find that F_n is $(4n - 2)$-connected, with $\pi_{4n-1} F_n \approx Z$. The generator x_n for $H^{4n-1}(F_n)$ ($\approx Z$) can be chosen so that

(ii) $$\tau(x_n) = \tfrac{1}{2}(Vc_n - V^2) \in H^{4n}(\Omega^{2N} M_{n+N}).$$

Here τ denotes the transgression operator in the fibration $F_n \to M_n \to \Omega^{2N} M_{n+N}$, and $Vc_i = \sigma^{2N}(U_{n+N} c_i)$, with σ the loop homomorphism. Note that if $X^{2q} \in \Omega_{2q}^U$ is regarded as a map $f_X : S^{2q+2N} \to \Omega^{2N} M_{n+N}$, then

$$f_X^*(Vc_q)[S^{2q+2n}] = c_q[X] \in Z.$$

Thus Theorem 1 follows from (i), (ii) and Theorem 4. The calculation of the higher homotopy groups of $MU(n)$ can be carried out in a similar fashion. Finally, Theorem 4 is proved by a careful analysis of the Hazewinkel generators [4] for Ω_*^U, localized at 2.

A more detailed account of this work will be given in a forthcoming paper.

REFERENCES

1. R. Bott, *The space of loops on a Lie group*, Michigan Math. J. **5** (1958), 35–61. MR **21** #1589.

2. O. Burlet, *Cobordismes de plongements et produits homotopiques*, Comment. Math. Helv. **46** (1971), 277–288. MR **45** #4433.

3. R. Hartshorne, *Topological conditions for smoothing algebraic singularities*, Topology **13** (1974), 241–253. MR **50** #2170.

4. M. Hazewinkel, *Constructing formal groups over* $Z_{(p)}$ *algebras*. I, Netherland School of Economics, Report No. 7119, 1971 (mimeographed).

5. P. Hilton, *On the homotopy groups of the union of spheres*, J. London Math. Soc. **30** (1955), 154–172. MR **16**, 847.

6. F. Hirzebruch, *Topological methods in algebraic geometry*, 3rd edition, Springer-Verlag, New York, 1966. MR **34** #2573.

7. R. J. Milgram, *Unstable homotopy from the stable point of view*, Lecture Notes in Math., vol. 368, Springer-Verlag, Berlin and New York, 1974. MR **50** #1235.

8. J. Milnor, *On the cobordism ring* Ω^* *and a complex analogue*. I, Amer. J. Math. **82** (1960), 505–521. MR **22** #9975.

9. R. Thom, *Quelques propriétés globales des variétés differentiables*, Comment. Math. Helv. **28** (1954), 17–86. MR **15**, 890.

ST. CATHERINE'S COLLEGE, OXFORD

UNIVERSITY OF CALIFORNIA, BERKELEY

Proceedings of Symposia in Pure Mathematics
Volume 32, 1978

HODGE THEORY FOR THE ALGEBRAIC TOPOLOGY OF SMOOTH ALGEBRAIC VARIETIES*

JOHN O. MORGAN

The Hodge theory of harmonic forms, when applied to smooth, complex projective varieties, gives deep and important results about the cohomology of such manifolds. In this article I wish to show that the same Hodge theory applies equally well to give nontrivial statements about algebraic topological invariants beyond cohomology. The results will be valid for any smooth, complex, algebraic variety (not just projective ones). That part of the algebraic topology which is amenable to study via Hodge theory is the rational algebraic topology. Since Hodge theory is based on differential forms, all subtle torsion and divisibility questions are ignored. One might be tempted to say that it is possible to study only real (i.e., over $\boldsymbol{R}$) algebraic topology via forms, but since all real algebraic topology of spaces carries an inherent rational structure, the results that we obtain descend automatically from $\boldsymbol{R}$ to $\boldsymbol{Q}$.

The invariants we have in mind are functors which assign to each CW complex an element which has the form of an algebraic structure on a rational vector space. Furthermore, we require that any map $f\colon X \to Y$ which induces an isomorphism on rational homology must induce an isomorphism on the invariants. Examples of such invariants are

(a) the rational cohomology ring,

(b) the rational cohomology ring of the various stages in the Postnikov tower (for simple connected spaces only),

(c) the homotopy groups tensored with $\boldsymbol{Q}$ viewed as a graded Lie algebra under the Whitehead product (again for simply connected spaces only), and

AMS (MOS) subject classifications (1970). Primary 32C10, 14F35.

*This paper has been partially supported by the National Science Foundation under grant number NSF MCS76-08230.

(d) the tower of rational nilpotent Lie algebras associated to the nilpotent quotients of the fundamental group.

(The precise statement of which invariants are included in the theory is given in terms of Sullivan's minimal model [**4**].)

The statement of the main structural result is that all these invariants of smooth algebraic varieties naturally carry *mixed Hodge structures*. These were introduced by Deligne in [**1**] as a generalization of the classical notion of a Hodge structure. Deligne then generalized the Hodge theorem (that the cohomology of a smooth projective variety carries a Hodge structure) by proving that a smooth open variety carries a mixed Hodge structure. This definition, as well as many of the lemmas Deligne proved while carrying out this generalization, plays a central role in the study of the full rational algebraic topology of both projective and open smooth varieties.

The main application of our results is to find restriction on the homotopy types realized as smooth varieties. In the compact case the result agrees with the one obtained by Deligne, Griffiths, Sullivan and the author in [**2**], namely, that the rational homotopy type (including for example the rational homotopy groups if the variety is simply connected) can be algebraically deduced from the cohomology ring. In the open case, we find the first restrictions of any type on the homotopy types. For example, the group $F(x, y)/\{[x, [x, [x, [x, y]]]]\}$ is not the fundamental group of any smooth variety. Likewise, we find restrictions on the possible simply connected homotopy types.

I will describe first the classical results on Hodge theory for the cohomology of smooth projective varieties and Deligne's generalization to open varieties. Then, I will turn to the results on the more general algebraic topology. More detailed statement and proofs can be found in [**3**].

Hodge theory for the cohomology of smooth varieties. If X is a complex manifold and $\mathcal{E}^*(X, \mathbf{C})$ represents the complex valued C^∞ forms on X, then one has the splitting into types $\mathcal{E}^n(X; \mathbf{C}) = \bigoplus_{p+q=n} \mathcal{E}^{p,q}(X)$ where $\mathcal{E}^{p,q}(X)$ are the forms of type (p, q). Type (p, q) means that locally, in a holomorphic coordinate system $(z_1, \cdots, z_n)$ the form is a sum of monomials of the following type:

$$\omega = f(z, \bar{z})\,(dz_{i_1} \wedge \cdots \wedge dz_{i_p} \wedge d\bar{z}_{j_1} \wedge \cdots \wedge d\bar{z}_{j_q}).$$

The differential, d, becomes $\partial + \bar{\partial}$ where $\partial : \mathcal{E}^{p,q}(X) \to \mathcal{E}^{p+1,q}(X)$ and $\bar{\partial} : \mathcal{E}^{p,q}(X) \to \mathcal{E}^{p,q+1}(X)$. The Hodge theorem says that if X is in addition a complex projective variety then this decomposition passes to cohomology.

HODGE THEOREM. *Let X be a smooth projective variety.*

(a) *Define $H^{p,q}(X) \subset H^{p+q}(X; \mathbf{C})$ to be the subspace of classes with closed form representatives of type (p, q). Clearly $H^{p,q}(X) = \overline{H^{q,p}(X)}$. Furthermore, $H^n(X; \mathbf{C}) = \bigoplus_{p+q=n} H^{p,q}(X)$.*

Such a decomposition is called a Hodge structure of weight n.

(b) *If ω is an exact form of type (p, q), then $\omega = d(\eta)$ for some form η of type $(p, q-1)$. (Notice that $\partial\eta$ must be* 0.)

To prove point (a) one introduces the natural metric on $\mathbf{CP}^N$ and takes the induced metric on the projective variety X. Hodge's work shows that for this metric the Laplacian is homogeneous of type (0, 0). Thus a complex form ω is harmonic if and only if all its (p, q) components are. This gives a splitting of the space of complex valued harmonic forms, $\mathscr{H}^n = \bigoplus_{p+q=n} \mathscr{H}^{p,q}$. Applying the Hodge theorem that the space of harmonic forms of dimension n equals H^n we see immediately that $H^n = \bigoplus_{p+q=n} H'^{p,q}$ where $H'^{p,q} \subset H^{p,q}$. ($H'^{p,q}$ is the space of cohomology classes corresponding to $\mathscr{H}^{p,q}$.) To finish the proof of (a), i.e., to show that $H'^{p,q} = H^{p,q}$, one uses (b) and a simple component-by-component argument.

Part (b) is equivalent to the degeneration of the Fröhlicher spectral sequence, $(E_1 = H^q(\Omega^p)) \Rightarrow H^*(X; \boldsymbol{C})$, at E_1. It is proved using a deeper analysis of the Laplacian.

This theorem is appealingly simple to state, but it has remarkable consequences. As one, note that it implies that the odd Betti numbers of a projective variety are even. The reason is that if $n = 2k + 1$ then

$$H^n(X; \boldsymbol{C}) = \Big(\bigoplus_{p+q=n;\, p \geqq k+1} H^{p,q}(X) \Big) \oplus \Big(\bigoplus_{p+q=n;\, q \geqq k+1} H^{p,q}(X) \Big),$$

and the second space is the complex conjugate of the first.

Hodge's study of the cohomology began here but went much further into the ring structure on cohomology. One sample result is that the signature of X is equal to $\sum_{p,q} (-1)^q \dim_{\boldsymbol{C}} H^{p,q}$. Thus we see that the signature which depends on the real cohomology ring of X is determined by the complex cohomology groups together with their extra structure.

To prepare the way for things to come let us give a different formulation of a Hodge structure on a real vector space. (Notice that there was implicitly a real structure used when we observed that $\overline{H^{p,q}(X)} = H^{q,p}(X)$.) If V is a real vector space, then a Hodge structure of weight n on V is a decreasing (finite) filtration on $V \otimes \boldsymbol{C}$:

$$V \otimes \boldsymbol{C} = F^0(V \otimes \boldsymbol{C}) \supset F^1(V \otimes \boldsymbol{C}) \supset \cdots \supset F^{n+1}(V \otimes \boldsymbol{C}) = 0,$$

such that $V \otimes \boldsymbol{C} = \bigoplus_{p+q=n} (F^p(V \otimes \boldsymbol{C}) \cap \bar{F}^q(V \otimes \boldsymbol{C}))$. Were we given such a filtration, we would define $H^{p,q}(V \otimes \boldsymbol{C})$ to be $F^p \cap \bar{F}^q$. Conversely, if we have $V \otimes \boldsymbol{C} = \bigoplus_{p+q=n} H^{p,q}(V \otimes \boldsymbol{C})$ with $\overline{H^{p,q}} = H^{q,p}$, we set $F^p(V \otimes \boldsymbol{C}) = \bigoplus_{p' \geqq p} H^{p',q'}(V \otimes \boldsymbol{C})$, and then it follows that $\bar{F}^q(V \otimes \boldsymbol{C}) = \bigoplus_{q' \geqq q} H^{p,q}(V \otimes \boldsymbol{C})$.

In the classical case of cohomology of a projective variety the filtration equivalent to the Hodge structure is called the Hodge filtration. It is induced from the decreasing filtration on the complex differentiable forms (also called the Hodge filtration) defined by

$$F^p(\mathscr{E}^*(X; \boldsymbol{C})) = \bigoplus_{p' \geqq p} \mathscr{E}^{p',q}(X).$$

Now let us consider the cohomology of open smooth varieties. We begin with the simplest possible case. Suppose X is a projective variety and D is a codimension 1 subvariety (both smooth), and let $\mathfrak{A} = X - D$. The Gysin sequence tells us how to calculate the cohomology of $\mathfrak{A}$ from that of X and D:

$$\cdots \longrightarrow H^i(D) \xrightarrow{\text{Gysin}} H^{i+2}(X) \longrightarrow H^{i+2}(\mathfrak{A}) \longrightarrow H^{i+1}(D) \longrightarrow \cdots.$$

The terms $H^*(D)$ and $H^*(X)$ have Hodge structures: What about $H^*(\mathfrak{A})$? The first problem is that the Gysin map $H^i(D) \to H^{i+2}(X)$ is not a map of Hodge structures since it sends $H^{p,q}(D)$ to $H^{p+1,q+1}(X)$. However, since the map is well behaved with respect to the Hodge decompositions, one is able to remedy this problem in a purely formal manner. Redefine the Hodge structure on $H^i(D)$ by setting $H^{p+1,q+1}(D)$ in the new structure equal to $H^{p,q}(D)$ in the old structure. Now the Gysin map $H^i(D) \to H^{i+2}(X)$ is a map of Hodge structures of weight $(i + 2)$. Consequently the cokernel has a Hodge structure of weight $(i + 2)$. Likewise the kernel of $H^{i+1}(D) \to H^{i+3}(X)$ has a Hodge structure of weight $(i + 3)$. So we have

$$0 \longrightarrow \mathrm{Coker}^{i+2} \longrightarrow H^{i+2}(\mathfrak{A}) \longrightarrow \mathrm{Ker}^{i+3} \longrightarrow 0$$

where the superscripts on the first and last term denote the weight of the Hodge structures they carry. Thus one should not hope that $H^{i+2}(\mathfrak{A})$ has a Hodge structure of weight $(i + 2)$ but rather that it have 2 "pieces" one with a Hodge structure of weight $(i + 2)$ and the other with a Hodge structure of weight $(i + 3)$. It is considerations like this that lead to the definition of a mixed Hodge structure.

DEFINITION (DELIGNE [1]). *Let V be a vector space defined over $K \subset \mathbf{R}$. A mixed Hodge structure on V consists of*

(a) *an increasing filtration of V (the weight filtration)*

$$0 = W_{-1}(V) \subset W_0(V) \subset \cdots \subset W_N(V) = V$$

such that $W_i(V)/W_{i-1}(V)$ has a Hodge structure of weight i, and

(b) *a decreasing filtration on $V \otimes \mathbf{C}$ (the Hodge filtration)*

$$V \otimes \mathbf{C} = F^0(V \otimes \mathbf{C}) \supset F^1(V \otimes \mathbf{C}) \supset \cdots \supset F^{N+1}(V \otimes \mathbf{C}) = 0$$

so that $F^*(V \otimes \mathbf{C})$ induces the Hodge structure on each of the $W_i(V)/W_{i-1}(V)$.

Explanation. Part (a) naively says that V has a collection of "pieces" (the graded objects of the filtration $W_*(V)$) each of which has a Hodge structure. Furthermore the weight of the Hodge structure of a piece is given by the index of the weight filtration of that piece. Part (b) says that the Hodge structures on the various pieces all have a common origin, namely a filtration on $V \otimes \mathbf{C}$. $F^*(V \otimes \mathbf{C})$ automatically defines a filtration $F^*((W_i(V)/W_{i-1}(V)) \otimes \mathbf{C})$, and it is these induced filtrations which define the Hodge structure on $W_i(V)/W_{i-1}(V)$.

In the case of $\mathfrak{A} = X - D$ and $H^{i+2}(\mathfrak{A})$ as above, the weight filtration is

$$0 = W_{i+1}(H^{i+2}(\mathfrak{A})) \subset W_{i+2}(H^{i+2}(\mathfrak{A})) \subset W_{i+3}(H^{i+2}(\mathfrak{A})) = H^{i+2}(\mathfrak{A})$$

$$W_{i+2}(H^{i+2}(\mathfrak{A})) = \mathrm{Im}[H^{i+2}(X) \longrightarrow H^{i+2}(\mathfrak{A})].$$

Thus $W_{i+2}(\mathfrak{A})/W_{i+1}(\mathfrak{A}) = \mathrm{Coker}^{i+2}$ and $W_{i+3}(\mathfrak{A})/W_{i+2}(\mathfrak{A}) = \mathrm{Ker}^{i+3}$. The Hodge structures on these subquotients are the ones induced from the usual Hodge structures on $H^*(D)$ and $H^*(X)$.

THEOREM (DELIGNE [1]). *Let $\mathfrak{A}^n$ be a smooth complex algebraic variety. Then $H^*(\mathfrak{A})$ has, in a natural way, a mixed Hodge structure. It has the following properties:*

(1) *$W_i(H^K(\mathfrak{A})) = 0$ for $i < K$. (The weight filtration starts at weight equal to dimension.)*

(2) *$W_i(H^K(\mathfrak{A})) = H^K(\mathfrak{A})$ for $i \geqq 2K$. (It terminates at weight equal to twice the dimension.)*

(3) *If $\mathfrak{A}^n \hookrightarrow V^n$ with V^n compact then $W_K(H^K(\mathfrak{A})) = \mathrm{Im}(H^K(V) \to H^K(\mathfrak{A}))$.*

(4) *If $\mathfrak{A}$ is compact then the mixed Hodge structure becomes the usual Hodge*

structure (i.e., $W_{K-1}(H^K(\mathfrak{A})) = 0$ *and* $W_K(H^K(\mathfrak{A})) = H^K(\mathfrak{A})$ *and the Hodge structure on* $H^K(\mathfrak{A})$ *is the classical one).*

The proof derives from the classical Hodge theory by relating the cohomology of $\mathfrak{A}$ to the cohomology of various compact smooth varieties. The deus-ex-machina which establishes such a relation is the Hironaka resolution of singularities theorem. Applied in this context it says, given $\mathfrak{A}^n$, there is a compact smooth variety V^n and a collection of smooth divisors $D_1^{n-1}, \cdots, D_l^{n-1}$ all meeting transversally such that $\mathfrak{A} = V^n - \bigcup_{i=1}^{l} D_i^{n-1}$.

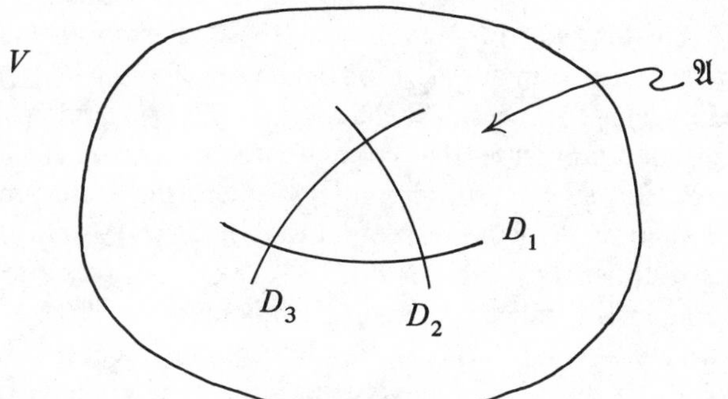

Once we have such a picture, there is a generalization of the Gysin long exact sequence, the Gysin spectral sequence, which tells us how to calculate the cohomology of $\mathfrak{A}$ from the cohomology of the various compact pieces. Namely, there is a spectral sequence converging to $H^*(\mathfrak{A})$ whose E_1 term is

$$\begin{array}{c|c|c}
\ddots \quad \vdots & \vdots & \\
\bigoplus_{i<j} H^0(D_i \cap D_j) \longrightarrow & \bigoplus_i H^2(D_i) \longrightarrow & \vdots \\
 & \bigoplus_i H^1(D_i) \longrightarrow & H^3(V) \\
 & \bigoplus_i H^0(D_i) \longrightarrow & H^2(V) \\
 & & H^1(V) \\
 & & H^0(V) \\
\hline
\end{array}
\qquad \left(E_1^{-p,q} = \bigoplus_{I=i_1<\cdots<i_p} H^{q-2p}(D_I)\right).$$

Thus every term at E_1 has a Hodge structure, and we wish to know that the abutment has a mixed Hodge structure. Deligne constructed this mixed Hodge structure by finding a differential algebra $\mathscr{E}^*(\mathfrak{A})$ (meromorphic forms on V with, at worst, logarithmic singularities along $D = D_1 \cup D_2 \cup \cdots \cup D_l$), which has a weight filtration giving the Gysin spectral sequence. In addition, it has a Hodge filtration. This Hodge filtration induces all the Hodge structures on the E_1-term above. Proceeding through the spectral sequence Deligne showed that $E_2 = E_\infty$, and that the two filtrations on $\mathscr{E}^*(\mathfrak{A})$ induce a mixed Hodge structure on $H^*(\mathfrak{A})$.

This result has many deep consequences. One of the nicest is the invariant cycle theorem. The local version of this theorem supposes that one has a complex space

V^{n+1} and a map $\pi: V^{n+1} \to D$ (= unit disk in $\boldsymbol{C}$) with $V - \pi^{-1}(0)$ smooth, and $\pi|: (V - \pi^{-1}(0)) \to D - \{0\}$ a submersion. One asks which classes $\alpha_t \in H_i(V_t; \boldsymbol{Q})$ come from classes $\tilde{\alpha} \in H_{i+2}(V; \boldsymbol{Q})$ by intersecting with V_t. (Here $V_t = \pi^{-1}(t)$ for some fixed $t \neq 0$.)

An obvious necessary condition is that α_t be invariant under the automorphism describing the fiber bundle with fiber V_t over the circle $|z| = |t|$. The local invariant cycle theorem asserts that this condition is sufficient as well. This result is proved using the above ideas. Instead of proving that, let me consider the global version. It is a direct consequence of the existence of mixed Hodge structures. In the global version one has $\pi: V^{n+1} \to S$ with S compact of dimension 1. There are points $s_0, \cdots, s_l \in S$ such that $\mathfrak{A} = V - \pi^{-1}(s_0 \cup \cdots \cup s_l)$ is smooth and $\pi|: \mathfrak{A} \to (S - \bigcup_i s_i)$ is a submersion. Again one asks for the image of $H_{i+2}(V; \boldsymbol{Q}) \to^{\cap V_t} H_i(V_t; \boldsymbol{Q})$ for some fixed $t \notin \bigcup_i s_i$. The result is that $\alpha_t \in H_i(V_t; \boldsymbol{Q})$ is the image of this map if and only if α_t is invariant under the action of $\pi_1(S - \bigcup_i s_i)$ on $H_i(V_t; \boldsymbol{Q})$. The proof goes as follows. We switch to cohomology where the statement becomes $\alpha_t^* \in H^{2n-i}(V_t)$ is in the image of the restriction map $H^{2n-i}(V) \to H^{2n-i}(V_t)$ if and only if α_t^* is invariant under the action of $\pi_1(S - \bigcup_i s_i)$. The invariance of α_t^* is equivalent to α_t^* being in the image of the restriction map $H^{2n-i}(\mathfrak{A}) \to H^{2n-i}(V_t)$. (This uses the fact that $S - (s_0 \cup \cdots \cup s_l)$ is homotopy equivalent to a wedge of circles.) It is at this point that the mixed Hodge structures enter. We have the mixed Hodge structures on $H^{2n-i}(\mathfrak{A})$ and the usual Hodge structure on $H^{2n-i}(V_t)$. Thus $\alpha_t^* \in H^{2n-i}(V_t)$ has weight $(2n - i)$ and is in the image of $H^{2n-i}(\mathfrak{A})$. Suppose we could conclude that α_t^* was in the image of $W_{2n-i}(H^{2n-i}(\mathfrak{A}))$. Since $\mathrm{Im}[H^{2n-i}(V) \to H^{2n-i}(\mathfrak{A})]$ equals $W_{2n-i}(H^{2n-i}(\mathfrak{A}))$, we could immediately conclude that α_t^* was in the image of $H^{2n-i}(V) \to H^{2n-i}(V_t)$. Such statements are always true for mixed Hodge structures as the following lemma shows.

LEMMA. *Let (A, W_*, F^*) and (B, W_*, F^*) be mixed Hodge structures, and $f: A \to B$ a map compatible with both filtrations, i.e., $f(W_i(A)) \subset W_i(B)$ and $f(F^i(A \otimes \boldsymbol{C})) \subset F^i(B \otimes \boldsymbol{C})$, for all i. If $b \in W_i(B)$ and if $b \in \mathrm{Im} f$, then $b \in \mathrm{Im}(f | W_i(A))$.*

The proof is an exercise in linear algebra.

Hodge theory for the homotopy theory of smooth varieties. The existence of a mixed Hodge structure throughout the algebraic topology of a smooth variety is founded upon the same principles as Deligne's existence result for the cohomology. That is to say, one first finds $\mathfrak{A} = V - D_1 \cup \cdots \cup D_l$, with V and the various D_i being smooth and compact, and the various intersections being transversal. In this situation one considers the complex of meromorphic forms on V with, at worst, logarithmic singularities along the various D_i. This differential algebra calculates the cohomology of $\mathfrak{A}$ and has two filtrations, W_* and F^*. The bridge from this situation to homotopy theory is provided by Sullivan's extension of de Rham's theorem. His results [4] say that given a graded commutative, differential algebra A, it is possible to build a canonical "model" for A, $\mathfrak{M}_A$. $\mathfrak{M}_A$ is a model for A in the sense that $\mathfrak{M}_A$ maps to A inducing an isomorphism on cohomology. What makes $\mathfrak{M}_A$ canonical is its internal structure. $\mathfrak{M}_A$ is required to be built up inductively from the ground field (say $\boldsymbol{C}$) by an infinite inductive procedure:

$$C = \mathfrak{M}_0 \subset \mathfrak{M}_1 \subset \mathfrak{M}_2 \subset \cdots, \qquad \bigcup_i \mathfrak{M}_i = \mathfrak{M}_A$$

where, as an algebra, $\mathfrak{M}_{i+1} = \mathfrak{M}_i \otimes \Lambda(V_{i+1})$. (Here $\Lambda(V)$ means the free graded commutative algebra generated by V in some degree $n > 0$. If n is even, this is the polynomial algebra on V, if n is odd, it is the exterior algebra on V.) In addition the differential of $\mathfrak{M}_{i+1}$ sends V into $\mathfrak{M}_i$ (and, of course, restricts to the differential already given on $\mathfrak{M}_i$). All these properties determine $\mathfrak{M}_A$ up to isomorphism.

We apply these results to the algebra of differential forms on V with logarithmic singularities along $D = D_1 \cup \cdots \cup D_l$, $\mathscr{E}^*(\mathfrak{A})$. (Recall that the differential algebra calculates the cohomology of $\mathfrak{A} = V - D$.) We build the minimal model with special attention to the weight and Hodge filtrations. It turns out that it is possible to build the model so that it has weight and Hodge filtrations compatible with the ones on $\mathscr{E}^*(\mathfrak{A})$, and, furthermore, these filtrations induce a mixed Hodge structure on $\mathfrak{M}$ with both the differential, d, and the wedge product being maps of mixed Hodge structures. Thus a subquotient of $\mathfrak{M}$ formed out of d and wedge product will inherit a mixed Hodge structure.

The second component of Sullivan's theory asserts that such subquotients are exactly the interesting complex homotopy invariants of the space $\mathfrak{A}$. For example let us consider four invariants of spaces.

If $\pi_1 = 0$ we take

(1) the homotopy groups with the Whitehead product,

(2) the cohomology rings of the various stages of the Postnikov tower, and

(3) the k invariants.

Lastly if $\pi_1(X) \neq e$ we consider

(4) the tower of real Lie algebras associated with the real nilpotent completions of $\pi_1(X)$.

Each of these invariants clearly has a complex form and their complex forms occur as subquotients of $\mathfrak{M}$ as follows. Let $\mathfrak{M}$ be the minimal model for $\mathscr{E}^*\mathfrak{A}$ as before; let $I \subset \mathfrak{M}$ be the ideal of elements in degrees > 0; and let $\mathfrak{M}^n \subset \mathfrak{M}$ be the subdifferential algebra generated by all elements in degrees at most n. If $\pi_1(\mathfrak{A}) = 0$, then

(1) $\pi_*(\mathfrak{A}) \otimes C$ is dual to $I/I \wedge I$ as a graded vector space, and the Whitehead product giving a graded Lie algebra structure on $\pi_*(\mathfrak{A}) \otimes C$ is dual to the map $d\colon I/I \wedge I \to I \wedge I/I \wedge I \wedge I$;

(2) the cohomology of $\mathfrak{M}^n$ equals the complex cohomology of the nth stage of the Postnikov tower for $\mathfrak{A}$; and

(3) the $(n+2)$nd k-invariant $k \in H^{n+2}$ (nth stage; $\pi_{n+1}(\mathfrak{A})$) is given by the dual of $d\colon (I/I \wedge I)^{n+1} \to H^{n+2}(\mathfrak{M}^n)$.

Thus the invariants in (1) and (2) have mixed Hodge structure, and the k-invariants of $\mathfrak{A}$ are maps of mixed Hodge structures. Lastly if $\pi_1(\mathfrak{A}) \neq e$, then the tower of nilpotent complex Lie algebras associated to $\pi_1(\mathfrak{A})$ is "dual" to $\mathfrak{M}^1$. $\mathfrak{M}^1$ itself has a canonical presentation $C = \mathfrak{M}_0^1 \subset \mathfrak{M}_1^1 \subset \mathfrak{M}_2^1 \subset \cdots$, $\bigcup_{i=1}^{\infty} \mathfrak{M}_i^1 = \mathfrak{M}^1$ where $\mathfrak{M}_{i+1}^1 = \mathfrak{M}_i^1 \otimes \Lambda(V_{i+1})$ with $d_i V_{i+1} \to \mathfrak{M}_i^1$. Dual to this increasing union of differential algebras each of which is finitely generated in degree 1 is a tower of complex nilpotent Lie algebras. This is the complex form of the tower associated to $\pi_1(\mathfrak{A})$. Thus this tower has a mixed Hodge structure.

The main application of the existence of the mixed Hodge structures on the minimal model is to show the existence of restrictions on the possibly underlying homotopy types of smooth varieties. For this application one needs the following lemma.

LEMMA. *Let V be an algebraic structure on a rational vector space (e.g., ring, differential algebra, Lie algebra) and suppose the underlying vector space has a mixed Hodge structure with the structure maps for the algebraic structure on V being maps of mixed Hodge structures. Then the weight filtration $W_*(V)$ can be split, i.e., $W_n(V) = \bigoplus_{i \leq n} G_i(V)$ so that all the structure maps are homogeneous of weight* 0.

This lemma is proved using the fact that the Hodge filtration can be used to find a complex splitting for $W_*(V) \otimes \boldsymbol{C}$ [1]. The general theory of linear algebraic groups gives the fact that the complex splitting implies the existence of a rational splitting.

Any such graded algebraic object is said to have positive weights if $G_i(V) = 0$ for $i \leqq 0$. Thus we have the following

COROLLARY. *If $\mathfrak{A}$ is a simply connected smooth variety, then the Lie algebra $(\pi_*(\mathfrak{A}) \otimes \boldsymbol{Q}$, Whitehead product) and the cohomology ring of each stage of the Postnikov tower have gradings with positive weights. In addition, we can choose these splittings so that if we view the k-invariants as maps $[\pi_{n+1}(\mathfrak{A}) \otimes \boldsymbol{Q}]^* \to^{k^{n+2}} H^{n+2}$(nth stage) then they preserve the splittings.*

If $\mathfrak{A}$ is not simply connected, then the tower of rational, nilpotent Lie algebras associated to $\pi_1(\mathfrak{A})$ is a tower of graded Lie algebras with positive weights.

Because of the restrictions Deligne found for the possible nontrivial weights in cohomology (parts (b) and (c)) there are restrictions on the nontrivial weights for the further invariants. For $\pi_n(\mathfrak{A}) \otimes \boldsymbol{Q}$ and H^n (kth stage of the Postnikov tower of $\mathfrak{A}$) the only nonzero weights lie between n and $2n$. (This is for $\pi_1(\mathfrak{A}) = e$.) For the Lie algebras associated to $\pi_1(\mathfrak{A})$ they have homogeneous generators of weights 1 and 2 and homogeneous relations of weights 2, 3 and 4. In particular the relations are at most 4th order. This explains why a group such as $F(x, y)/\{[x, [x, [x, [x, y]]]]\}$ cannot occur. Its associated tower of Lie algebras looks like $\{\mathscr{F}^n(x, y)/ [x, [x, [x, [x, y]]]]\}$ where $\mathscr{F}_n(x, y)$ is the free, nilpotent Lie algebra, nilpotent of order n. This algebra does not admit a splitting with generators of weight 1 and 2 and relations of weight 2, 3 and 4 because the only relation is a 5th order commutator and could never have weight $\leqq 4$. Likewise any time the Lie algebra nilpotent of order 5 associated to $\pi_1(\mathfrak{A})$ is free, then the nilpotent Lie algebras of all order must be free. More generally, there is an algebraic process for determining the complete graded tower from the graded term nilpotent of order 5.

In case $\mathfrak{A}$ is compact, i.e., $D = \varnothing$, then the grading on the minimal model for $\mathscr{E}^*(\mathfrak{A})$, $\mathfrak{M} = \bigoplus_{i \geq 0} G_i(\mathfrak{M})$ with d and λ homogeneous of degree 0, has the property that the induced grading on cohomology agrees with the usual grading by degrees. Thus $G_i(H^*(\mathfrak{M})) = H^i(\mathfrak{M})$. The reason for this is that the grading $G_i(H^*(\mathfrak{M}))$ agrees with a splitting for the mixed Hodge structure on $H^*(\mathfrak{A})$. But when $\mathfrak{A}$ is compact $H^i(\mathfrak{A})$ has a Hodge structure of weight i. It is a simple exercise to show that any time a minimal model admits such a grading the model is also the minimal model for its cohomology ring. Thus when $\mathfrak{A}$ is compact to build the minimal model of $\mathscr{E}^*(\mathfrak{A})$ (and hence know the complete rational homotopy type of $\mathfrak{A}$) one needs only build the minimal model for the cohomology ring of $\mathfrak{A}$. Thus we recover the main result of [2] which says that for compact smooth varieties the cohomology ring determines the rational homotopy type. As an example, this says that the tower of nilpotent Lie algebras associated to $\pi_1(\mathfrak{A})$ is obtained as follows. Take $H_1(\mathfrak{A}; Q)$ and let $R = \operatorname{Im}(H_2(\mathfrak{A}) \to^{\Delta} H_1(\mathfrak{A}) \wedge H_1(\mathfrak{A}))$. Form the graded Lie algebra $F(H_1(\mathfrak{A}; Q))/\mathrm{Ideal}(R)$. The nilpotent quotations of this agree with those associated to $\pi_1(\mathfrak{A})$. For example if $\mathfrak{A}$ is a compact Riemann surface of genus g, the nilpotent quotients of $F(x_1, \cdots, x_{2g})/I(\sum_{i=1}^{g}[x_{2i-1}, x_{2i}])$ agree with those associated to $\pi_1(\mathfrak{A})$.

References

1. P. Deligne, *Théorie de Hodge*. II, Inst. Hautes Études Sci. Math. Publ. **40** (1972), 1–38.

2. P. Deligne, P. Griffiths, J. Morgan and D. Sullivan, *Real homotopy theory of Kähler manifolds*, Invent. Math. **29** (1975), 245–274. MR **52** #3584.

3. J. Morgan, *The algebraic topology of smooth algebraic varieties*, Inst. Hautes Études Sci. Math. Publ., 1976 (Preprint).

4. D. Sullivan, *Infinitesimal calculation in topology*, Ann. of Math. (to appear). Preprints available, Institute for Advanced Study, Princeton, N. J.

Columbia University

Proceedings of Symposia in Pure Mathematics
Volume 32, 1978

AN INTRODUCTION TO EMBEDDINGS, IMMERSIONS AND SINGULARITIES IN CODIMENSION TWO

SYLVAIN E. CAPPELL* AND JULIUS L. SHANESON**

Dedicated to the memory of C. D. Papakyriakopolous

"Theaetetus (the mathematician): If it depends upon my zeal, Socrates, the truth will come to light."
from Plato's "Theaetetus"

Introduction. There are many familiar algebraic and geometric constructions which illustrate how frequently embeddings and immersions of manifolds in real codimension two have singularities. Consider, for example, the sets of points V in the complex n-dimensional space $C^n = (z_1, \cdots, z_n)$ which are the solutions of $z_1^2 + z_2^3 = 0$. This set V is a PL manifold embedded in C^n, of real codimension two, which fails to have a linear normal bundle (i.e., is singular or nonlocally flat) along the subvariety K of V where $z_1 = 0$ and $z_2 = 0$. When $n = 2$ the isolated point singularity obtained at the origin is easily described; (C^n, V) is locally homeomorphic to the pair (D^4, D^2) obtained by taking the cone on the nonsingular knot pair $\alpha = (S^3, S^1)$, for α the trefoil knot. This pair (S^3, S^1) which characterizes the singularity of this point is called the link pair of the singularity. For $n > 2$ the singularities are no longer isolated points and the link pairs are the $(n - 2)$ suspension of the trefoil knot. In particular the link pairs of the singular points are themselves now embeddings which have nonlocally flat points and, in fact, stratified singularity sets. Thus, unlike the situation in higher codimension, a PL embedding

AMS (*MOS*) *subject classifications* (1970). Primary 57A35, 57A40, 57A45, 57C35, 57C40, 57C45, 57E10, 57E30, 57D40, 57D65, 18F25, 15A63.

*This research was supported by the National Science Foundation Grant NSF-MCS 76-06974.
**This research was supported by a Sloan Foundation Grant.

or immersion $f\colon M^n \to W^{n+2}$ of PL manifolds is not necessarily locally PL homeomorphic, or even homeomorphic, to a pair (U, V), V a vector subspace of the vector space U.

Reflection on the many examples of PL embeddings and immersions and their singularities leads to consideration of some key problems. Among these are questions concerning: (1) the existence and construction, for fixed X^n and Y^{n+2} of embeddings, and immersions, of X in Y; (2) the structure of the singularity sets (sets of nonlocally flat points) of such immersions and embeddings as stratified spaces; (3) the relation between the singularity sets and such global invariants of X and Y as the Pontrjagin classes; and (4) the relationship of these general questions to our previous global approach to knots and nonsingular codimension two embeddings. In particular, why, for given X^n and Y^{n+2}, is one vastly more likely to find PL embeddings or immersions of X^n in Y^{n+2} if we permit singularities and does this remain true if we restrict the dimension of the singularities? There are further problems concerning the relation between geometry in codimension two and questions about branched covering spaces, group action, spines and spinelessness, etc.

The present paper is an introduction to our work in codimension two geometry and to the global and local study of PL embeddings and immersions in codimension two and their singularities. In this paper a PL embedding (respectively, immersion) is a (respectively, locally), one-to-one PL homeomorphism of M to its image. Here, as no local flatness (linearity) conditions are presupposed, the usual "linearization" techniques for studying PL submanifolds, often modeled or adapted from smooth techniques involving linearization, and likewise the usual surgery methods (**[B1]**, **[No]** in the simply connected case, **[W]** in general) which are used systematically in the study of embedded submanifolds in codimension greater than two **[B2]**, **[B3]**, **[W]** do not directly apply.[1] Further results, complete proofs, more examples, and references to related works are in **[CS2]**, **[CS6]**, **[CS7]**, **[CS9]**, **[CS12]**, **[CS17]**, **[CS11]**; see also the announcements **[CS1]**, **[CS5]**, **[CS9]**.

A key feature of codimension two embeddings is that the group homomorphism $\pi_1(Y - X) \to \pi_1(Y)$, produced from an inclusion of the complement $Y - X$ of a codimension two submanifold $X^n \subset Y^{n+2}$ is surjective but not usually injective. The approach outlined here explores systematically consequences of this.

1. Some global results on embeddings and immersions. This section summarizes some of our results on the study, for fixed PL manifolds M^n and W^{n+2}, of immersions and embeddings of M in W. In most of the results and examples to be considered here, it is easy to see that locally flat (nonsingular) immersions and embeddings do not exist. For example, the rational or real Pontrjagin classes give rise to obstructions to existence of such locally flat immersions. In the more general setting to be considered here these characteristic classes provide global information on the geometric structure of the singularity sets of nonlocally flat points as a stratified space (see §2 below).

Here we will concentrate on existence results and begin with the problem of which PL manifolds immerse in codimension two Euclidean space. Our criterion is stated in terms of stable homotopy theory.

[1]See **[C1]**—**[C5]** for a study of submanifolds in codimension one.

COROLLARY **[CS17]**. *Let M^n be an orientable closed connected* PL *manifold (resp. and with $H^2(M; Z) = 0$). Then M immerses* PL *in S^{n+2} if (resp. and only if) $\pi_{n+k}(\Sigma^k M) \xrightarrow{h} H_{n+k}(\Sigma^k M)$ is surjective, for some k.*

Here $\Sigma^k M$ denotes the kth suspension of M and h is the Hurewicz homomorphism. From this reduction to stable homotopy theory, it follows for example that every manifold of the homotopy type of an almost parallelizable manifold immerses in codimension two Euclidean space, although they do not usually embed. The corollary is a special case of the following result.

THEOREM **[CS17]**. *Let M^n be an orientable connected closed* PL *manifold. Then M hes a* PL *immersion in S^{n+2} if and only if there is an orientable 2-plane bundle ξ over M such that $\pi_{n+k}(\Sigma^k T(\xi)) \xrightarrow{h} H_{n+k}(\Sigma^k T(\xi))$ is surjective, $k > n$.*

Here $\Sigma^k T(\xi)$ denotes the kth suspension of the Thom complex,

$$T(\xi) = \{\text{the total space of } \xi\}/\{\text{the vectors of length} \geqq 1\}.$$

We present an equivalent bundle-theoretic criterion for the existence of immersions in codimension two Euclidean space in **[CS17]**; this is also applicable to nonclosed manifolds.

An important consequence of the above results is that the problem of the existence of general PL immersions in codimension two depends only on the homotopy types of the manifolds. This, as will be seen shortly, is true in a very general setting for immersions. A somewhat analogous principle, implying a reduction to homotopy theory, is also valid, except in the nonsimply connected even dimensional cases, for the problem of the existence of PL embeddings in codimension two. Of course, there is no such principle for locally flat immersions or embeddings.

It is instructive to compare the above results on immersions with our criteria for embedding closed manifolds in codimension two Euclidean space.

THEOREM **[CS12]**. *Let M^n, $n \geqq 3$, be a closed connected orientable* PL *manifold. If the Hurewicz homomorphism $\pi_{n+1}(\Sigma M) \xrightarrow{h} H_{n+1}(\Sigma M)$ is surjective, then there is a* PL *embedding of M in S^{n+2}.*

This is a near converse to the observation that, as a consequence of the Thom-Pontrjagin construction, if an orientable closed manifold M^n PL embeds in S^{n+2}, then $\pi_{n+2}(\Sigma^2 M) \to H_{n+2}(\Sigma^2 M)$ is surjective. This result applies, for example, to show that any manifold which has a homotopy (or just integral homology) equivalence to any products of orientable 1, 2 or 3 manifolds, or spin 4 manifolds, PL embeds in codimension two Euclidean space **[CS10]**. Many manifolds satisfying these hypotheses have nontrivial Pontrjagin classes and hence do not have locally flat embeddings in S^{n+2}. More precisely, results of §2 show how lower bounds on the dimension of the singularity sets can be obtained from characteristic classes. (For example, Browder-Novikov surgery theory can be readily used to construct infinitely many different smooth manifolds M^{q+4} homotopy equivalent to the product of spheres $S^4 \times S^q$, for any $q > 1$, all with nonzero first Pontrjagin class. The above theorem shows that all these M embed PL in S^{q+6}; however, the results of §2 show that for any embedding of M in S^{q+6} the singularity subset of M consisting of points at which the embedding is not locally fiat will have dimension at least q. In this case, this is actually the best possible lower bound on the size of the singularity sets of the embeddings.)

These results and examples illustrate the principle that the existence of PL immersions, and often of PL embeddings, can be reduced to homotopy theory. However, the more delicate questions involving the structure of nonlocally flat points depends on geometrical considerations, such as the Pontrjagin classes. The next few results are applications and actualizations of the reduction of the existence problem. First, a result on immersions:

THEOREM [CS17]. *Let* $f\colon M^n \to W^{n+2}$ *be a* PL *immersion of orientable manifolds with* M *compact.*

(i) *Suppose* $g\colon (M', \partial M') \to (M, \partial M)$ *is a homotopy (or just integral homology) equivalence of* PL *manifolds, with* M' *compact. Then* $f \circ g$ *is homotopic to a* PL *immersion.*

(ii) *Suppose* $h\colon (W, \partial W) \to (W', \partial W')$ *is a homotopy (or just integral homology) equivalence of compact* PL *manifolds. Then* $h \circ f$ *is homotopic to a* PL *immersion.*

And now, an analogous result for embeddings:

THEOREM [CS12]. *Let* $f\colon M^n \to W^{n+2}$ *be a* PL *embedding with* M *closed,* M *and* W *orientable and* $n \geqq 3$. *If* n *is even assume that* $\pi_1 W = 0$.

(i) *Let* $g\colon M' \to M$ *be a homotopy equivalence of closed* PL *manifolds. Then* $f \circ g$ *is homotopic to a* PL *embedding.*

(ii) *Let* $h\colon (W, \partial W) \to (W', \partial W')$ *be a homotopy equivalence of pairs of compact* PL *manifolds. Then* $h \circ f$ *is homotopic to a* PL *embedding.*

In both of these results the assumptions on orientability can be weakened by requiring merely that f be orientable-true, i.e., that $f^*\omega_1(W) = \omega_1(M)$, ω_1 denoting the first Stiefel-Whitney class. However, the simple connectivity hypothesis, which is imposed when n is even to reduce embedding problems to homotopy-theoretic questions, is necessary as the examples of totally spineless manifolds, recalled below, will illustrate.

Let $h\colon M^n \to W^{n+2}$, M closed and connected, W compact be a homotopy equivalence of PL manifolds. It is natural to ask to what extent can W be visualized as being like the total space of a tubular neighborhood of M; that is, when is h homotopic to a PL embedding? When h is homotopic to a PL embedding, we call M, or more precisely its embedded image in W, a spine of W.

THEOREM [CS12]. *Let* $f\colon M^n \to W^{n+2}$ *be an orientable-true homotopy equivalence,* M *and* W *compact connected* PL *manifolds,* M *closed,* $n \geqq 3$. *If* n *is odd, or if* n *is even and* $\pi_1 W = 0$, *then* f *is homotopic to a* PL *embedding.*

The spines produced by this result will often necessarily have very high dimensional sets of singular (nonlocally flat) points. The size of these singularity sets of a spine can sometimes be reduced or even eliminated entirely by replacing M by another closed PL manifold of the same homotopy type. This happens particularly when M is simply connected; see [CS2], [CS12]. An extreme example where this does not occur is for M the n torus $S^1 \times \cdots \times S^1$, $n \geqq 3$. There are PL manifolds W^{n+2} which have the homotopy type of T^n but for which the first Pontrjagin class is not zero.[2] It can be shown using results of §2 below that for such a W any PL manifold spine has a nonlocally flat set of dimension at least $n - 4$. More delicate methods lead to examples where the nonlocally flat set must have dimension precisely $n - 2$. This is best possible as the set of nonlocally flat points of any PL embedding (or immersion) is a subcomplex of the manifold being embedded of dimension at most $n - 2$.

A yet more exotic example, though one of a different kind, is given by the construction in even dimensions of totally spineless manifolds **[CS7]**. This showed in a strong way that in the even-dimensional nonsimply connected case the existence problem for PL embeddings cannot be directly reduced to homotopy theory.

Existence of totally spineless manifolds **[CS7]**. *Let M^n be a closed connected* PL *manifold with $n \geqq 4$, n even, and $H_1(M) \neq 0$. Then there exists a compact* PL *manifold W^{n+2} which is* (PL *tangentially and simply*) *homotopy equivalent to M but which has no* PL *manifold spine. In particular, the given homotopy equivalence of M to W is not homotopic to a* PL *embedding.*

When M^n is smooth (respectively, parallelizable, orientable) our construction of spineless manifolds produces a W^{n+2} which enjoys the same property. The hypothesis $H_1(M) \neq 0$ can be considerably weakened, see **[CS7]**; it is natural in view of the above results to conjecture that it can be replaced by $\pi_1(M) \neq 0$. When $n = 2$, an example of a spineless manifold has been given by Matsumoto **[Ma]**.

On the other hand, *immersed* spines exist quite generally.

THEOREM **[CS17]**. *Let $h: (M^n, \partial M) \to (W^{n+2}, \partial W)$ be an orientation-true homotopy equivalence of pairs of compact* PL *manifolds. Then h is homotopic to a* PL *immersion.*

Of course, these immersed spines will usually have high dimensional complexes of nonlocally flat points. Interestingly, in many cases replacing the PL embedded spines by immersed spines can be used to lower somewhat the dimension of the set of nonlocally flat points. For example, we referred above to the existence of manifolds W^{n+2} of the homotopy type of the n-torus which have embedded spines but for which every embedded spine has a singularity set of nonlocally flat points of codimension 2 in the torus. Any such W can be shown to have an immersed PL spine with singularity set of codimension 8. This interplay between singularity points and double points of PL immersions merits further study **[CS17]**. We do not, in general, know what is the minimum dimension of the set of double points of the immersed spines in the totally spineless manifolds.

We conclude this discussion of spines by remarking that even in the more difficult even-dimensional nonsimply connected case there are certain special but naturally arising classes of compact PL manifolds which always have spines. For $h: M^n \to W^{n+2}$ a homotopy equivalence with M a closed oriented PL manifold and W a compact oriented PL manifold, let $\chi(W) \in H^2(M)$, the Euler class of W, denote the image of $[M]$ under the maps $H_n(M) \xrightarrow{h_*} H_n(W) \to H_n(W, \partial W) \xrightarrow{\alpha} H^2(W) \xrightarrow{h^*} H^2(M)$, where α is Poincaré duality.

PROPOSITION **[CS9]**. *$h: M \to W$ and $\chi(W)$ as above. Assume that $H_2(\pi_1 W; Z) = 0$ and that $\chi(W) \in H^2(M)$ is a primitive generator. Then h is homotopic to a* PL *embedding.*

[2]These exotic manifolds are characterized by this together with the vanishing of the associated Euler class described below.

For example, this happens when W is any simply connected compact $2n + 2$ manifold which has a boundary preserving integral homology equivalence to $\overline{CP^{n+1} - D^{n+1}}$, a punctured complex projective space. (We do not assume $\pi_1(\partial W) = 0$.) It is easy to construct many such W. Then every closed manifold which is homotopy equivalent to CP^n occurs as a spine in W and almost all of these spines will not be locally flat. In this example $\chi(W)$ is a primitive generator; by contrast, the totally spineless manifolds which we construct in **[CS7]** have $\chi(W) = 0$.

We will now consider the general problem of the existence of general PL immersions and embeddings in codimension two. Further results, in which the codimension of the singularity sets is prescribed are presented in **[CS12]**, **[CS17]**. We begin with immersions.

THEOREM **[CS17]**. *Let f: $M^n \to W^{n+2}$ be a continuous map of* PL *manifolds, M compact. Suppose that the underlying spherical fibre space $S(\nu_M \oplus f^*\tau_M)$ of the stable* PL *bundle $\nu_M \oplus f^*\tau_W$ reduces to an oriented circle bundle over M. Then f is homotopic to a* PL *immersion.*

Conversely, if f: $M \to W$ is an orientation true PL *immersion, then $S(\nu_M \oplus f^*\tau_W)$ reduces to an oriented circle bundle.*

Here τ_W denotes the normal bundle of W and ν_M denotes the stable normal bundle of M. The hypothesis that a spherical fibre space reduces to a circle bundle, i.e., an $SO(2)$ bundle, is equivalent to the requirement that it reduces to a spherical fiber space with fiber the circle S^1. When M is closed and orientable with $H^2(M) = 0$ and W is orientable the criteria of the theorem for the existence of immersions is equivalent to the surjectivity of

$$\pi_{n+k}(T(f^*\nu_W)) \xrightarrow{h} H_{n+k}(T(f^*\nu_W)), \qquad k = \dim \nu_W > n,$$

$T(f^*\nu_W)$ denoting the Thom complex of $f^*\nu_W$.

There are many interesting examples which satisfy the hypothesis of the above existence theorem. This abundance of geometrical examples of general codimension two PL immersions can be explicated by viewing the above result using Eilenberg obstruction theory. It implies that the obstruction to the existence of general PL immersions always take their values in finite groups. In particular, there are no rational or real obstructions to existence of such immersions.

This is in sharp contrast with the smooth or PL locally flat immersions in codimension two. Differentiable immersions of spheres in Euclidean space were classified by Smale **[Sm]** and the general theory for the smooth case was given by Hirsch **[Hi]**. Haefliger-Poenaru and Haefliger used microbundles and PL bundles to classify PL immersions that are locally flat. In contrast to the above theorem, it follows from **[HP]**, **[H]** that an orientation true map f: $M^n \to W^{n+2}$ will be homotopic to a locally smoothable immersion if and only if the stable PL bundle $\nu_M \oplus f^*\tau_W$, rather than its underlying spherical fiber space, has the form $\eta \oplus \varepsilon$, η an $SO(2)$ bundle, ε trivial.

We will now consider the problem of the existence of general PL embeddings of a given manifold M^n in a given manifold W^{n+2}. The general criteria provided by **[CS12]**, **[CS9]** shows that the existence of a homotopy-theoretic template is a necessary and sufficient condition for the existence of actual geometric embeddings. The homotopy-theoretic template used is an adaptation of the notion of an h-Poincaré embedding; this consists of a homotopy-theoretic model $E(\xi)$ for the proposed regular neighborhood as well as a candidate homotopy type E for the complement of M in W. These are glued together along a circle bundle $S(\xi)$. Precisely, an h-Poincaré embedding of M in W consists, when M and W are closed, of

(i) an oriented 2-plane bundle ξ over M,

(ii) a CW complex E and an inclusion of the sphere bundle $S(\xi) \subset E$,

(iii) a homotopy equivalence $h\colon W \to E(\xi) \cup_{S(\xi)} E$ with the composite map $W \to E(\xi)/S(\xi) = T(\xi)$ having degree 1.

REMARKS. (1) We have formulated this notion here only for closed manifolds. It can also be done in the case of manifolds with boundary and for results in that case see **[CS12]**.

(2) In view of the examples of totally spineless manifolds **[CS7]**, some restriction must be placed on fundamental groups in the nonsimply connected even dimensional case to obtain geometrical embeddings as consequences of the existence of a Poincaré embedding.

(3) The actual regular neighborhoods of M to be constructed in W will usually necessarily have far more complicated normal structure than ξ (see **[CS12**, §5**]**). In particular the boundary of such a regular neighborhood will be homology equivalent, with suitable local coefficients, to $S(\xi)$, but it will rarely be homotopy equivalent to $S(\xi)$. As an exercise, compute the homotopy type of the boundary of the regular neighborhood of the embedded S^2 in S^4 arising as the suspension of a knot $S^1 \subset S^3$.

(4) Related to (3) is the fact that instead of attempting to prescribe the precise homotopy type of the complement of the geometrical embedding of M in W, it is more natural in codimension two problems to attempt to construct embeddings of M in W for which the complement is homology equivalent (with coefficients in any local coefficient system over $\pi_1 W$) to E. This is the key idea of homology surgery **[CS2]** which underlies all these results. The point is that in codimension 2, the natural map $\pi_1 E \to \pi_1 W$ induced from the inclusion of the complement of a submanifold may be far from being an injection, but is always surjective. We use this principle in constructing embeddings by first producing complements which have the "right" homotopy type only after the fundamental group is "corrected" by adding on the relations adjoined when the regular neighborhood of M is attached to reassemble W. This idea is described in more detail in §4 below.

For simplicity, here we record only some results on when the composed map f,

$$\underbrace{M \hookrightarrow E(\xi) \hookrightarrow E(\xi) \cup_{S(\xi)} E \xrightarrow{h^{-1}} W}_{f}$$

produced from θ and defined up to homotopy, is homotopic to a PL embedding. The map f is called the underlying map of the h-Poincaré embedding θ **[CS12]**.

THEOREM **[CS12]**. *Let M^n and W^{n-2} be oriented closed* PL *manifolds, $n \geqq 3$. If n is even assume that W is simply connected. Let $f\colon M \to W$ be a continuous map. Then f is homotopic to a* PL *embedding if and only if f is the underlying map of an h-Poincaré embedding.*

The result can be improved in a number of ways, some referred to in the remarks above. For example, the orientability assumption on M and W can be replaced by assuming that f is orientation true. That this is useful is indicated by the following example.

EXAMPLE. Let X^n and Y^{n+2}, $n > 2$, be the quotient spaces on fixed point free involutions on S^n and S^{n+2} respectively; i.e., X and Y are homotopy real projective spaces. Then there is a PL embedding of X in Y.

This and related examples can be interpreted in terms of group actions; see §5 below, **[CS4]**, **[CS12]**, **[Ho]**.

Another way in which the above theorem can be strengthened is by studying what can be concluded about the size and structure of the set of nonlocally flat points. For example, in many cases, the examples constructed above must have nonlocally flat sets S large enough so that $H_i(S; Z_2) \to H_i(X; Z_2)$ is surjective for $i < n$. Results involving existence of embeddings with prescribed singularities are presented in **[CS12]**.

In this section, we have concentrated upon existence results. A general framework for discussing classification results for general PL embeddin gs in codimension two is provided by **[CS2]**, **[CS12]**; there are, however, many problems in studying some of the geometrically arising situations. A discussion of classification results for general PL immersions is presented in **[CS11]**. Here we will present only a single rather simple special result.

Let C_{n-1} denote the knot cobordism group of (smooth or) locally flat PL knotted (homotopy) $(n-1)$ spheres in S^{n+1}. Let $\tilde{C}_{n-1} \subset C_{n-1}$ be the subgroup consisting of knots with vanishing index for $n \equiv 0 \pmod 4$ and vanishing Arf-Kervaire invariant for $n \equiv 2 \pmod 4$ and all knots for n odd. C_{n-1} and $\tilde{C}_{n-1}$ are zero for n odd and not finitely generated for n even.

Finally, let π_n^s denote the stable n-stem of the homotopy groups of spheres, $\pi_{n+k}(S^k)$ for k large.

THEOREM **[CS11]**. *The group of concordance classes of* PL *immersions of the n-sphere S^n in S^{n+2} is isomorphic, for $n > 1$, to $\pi_n^s \oplus \tilde{C}_{n-1}$.*

Actually, for this simple special case of **[CS11]**, each PL immersion is concordant to one with at most one isolated singularity. This is, however, rarely true in general as the results of the next section show.

2. Singularities and PL characteristic classes. There is a close relation between the rational or real characteristic classes of manifolds and the structure of the singularity sets, consisting of the nonlocally flat points, of immersions and embeddings of manifolds in codimension two. These results are presented in a Riemann-Roch type result in terms of the rational PL Pontrjagin classes. Further results, including many delicate torsion considerations, are implicit in the discussion of the relation of normal invariants to singularities in **[CS12]**; those lead to a general reduction of questions concerning the dimension of singularity sets to a problem in homotopy theory.

Let $f: M^n \to W^{n+2}$ be a PL immersion. Let $S(f) \subset M$ be the set of points where f is not locally flat, i.e., the set of points $x \in M$ at which for D^n a neighborhood of x in M, the pair $(f(D^n), W^{n+2})$ is not locally homeomorphic to (U, V), U a linear subspace or half-space of $V \cong R^{n+2}$. Then $S(f)$ is a subcomplex of M^n. It is not difficult to see that $\dim(S(f)) \leq n-2$; this is the dual of the observation that every embedding of S^0 in S^2 is unknotted.

When $H^2(M; Q)$ is not zero we will need to begin by associating to an immersion

$f: M^n \to W^{n+2}$ of oriented manifolds (or to an orientation-true immersion) an Euler class $E(f) \in H^2(M; Z)$, M a closed manifold, defined as follows. Let $(V^{n+2}, \partial V)$ be a regular neighborhood **[CS17]** of the immersion f so that $M \subset V$ is a homotopy equivalence. Then the image of the fundamental class $[M] \in H_n(M)$ under the map

$$H_n(M) \to H_n(V) \to H_n(V, V) \xrightarrow[\text{duality}]{\text{Poincaré}} H^2(V) \to H^2(M)$$

is called the Euler class of f. From this definition, it is clear that when f is an embedding the Euler class is just the image of $[M]$ under the map which depends only on the homotopy class of f,

$$H_n(M) \to H_n(W) \to H_n(W^{n+2}, \partial W) \to H^2(W) \to H^2(M).$$

Let $x = E(f)$ and let $L(x)$ denote the total Thom-Hirzebruch L-polynomial in this 2-dimensional class,[3]

$$L(x) = 1 + x^3/3 - x^4/45 + 2x^6/945 - \cdots.$$

For M a PL manifold, let $L(M)$ denote the total Thom-Hirzebruch L-genus. This is a polynomial in the rational Pontrjagin classes of M and determines the Pontrjagin classes; thus

$$L(M) = 1 + L_1(M) + L_2(M) + \cdots, \qquad L_i(M) \in H^{4i}(M).$$

Last, set $L(f) = L(M)L(x) - f^*L(W)$. The Poincaré dual of this class, $L(f) \cap [M]$, is a total homology class in $H_{n-4}(M) \oplus H_{n-8}(M) \oplus H_{n-12}(M) + \cdots$.

THEOREM **[CS17]** (SEE ALSO **[CS12]**). *Let $f: M^n \to W^{n+2}$ be a* PL *immersion. Then $L(f) \cap [M]$ is in the image of the map induced by the inclusion of the singular points of f, $H_*(S(f)) \to H_*(M)$. In particular, if $L(f)$ has a nonzero component in dimension j, then* dim $S(f) \geqq (n - j)$.

This result can be strengthened by considering a natural stratification of $S(f)$ and relating the components of $L(f)$ in different dimensions to this stratification. The points of $S(f)$ are stratified by measuring how far away they are from being nonsingular; this is accomplished by assigning to each point p the number $s(p) =$ the maximum number of times its link pair desuspends.

It is easy to find many examples of manifolds to which this theorem applies to get very strong consequences. For example, let M^n, $n > 4$, be any simply connected closed PL manifold with $H^{4i}(M; Q) \neq 0$, $4i < n$, and $\pi_{n+1}(\Sigma M) \to H_{n+1}(\Sigma M)$ (respectively, $\pi_{n+k}(\Sigma^k M) \to H_{n+k}(\Sigma^k M)$, some k) surjective. It is an exercise in surgery theory to show that in the homotopy type of M, when $n > 4$, there exist infinitely many manifolds N with $p_i(M)$, the ith rational Pontrjagin class, not zero. Then in view of the above results, such manifolds N will PL embed (respectively, immerse) in S^{n+2}, but always with the dimension of the singularity sets satisfying (respectively, when $H^2(M; Q) = 0$) $n - 4i \leqq \dim S(f) \leqq n - 2$.

Another example is provided by the manifolds homotopy equivalent to complex projective space CP^n, $n > 1$. In view of results on embeddings presented above, all of these PL embed in any given homotopy CP^{n+1}; however, by choosing examples of manifolds homotopy equivalent with CP^n but with different first Pontrjagin

[3]Alternatively, as $BSO(2) \cong K(Z, 2)$, x determines a 2-plane bundle and $L(x)$ is the total Thom-Hirzebruch class of that bundle.

class we can construct examples which embed in CP^{n+1} but only with singularity sets $S(f)$ satisfying $H_i(S(f)) \to H_i(CP^n)$ is onto for $i \leqq 2n - 4$. In fact, for some of these examples, necessarily dim $S(f) = 2n - 2$. For an analysis of the dimensions of the singularity sets of embeddings of a homotopy CP^n (or RP^n) in a homotopy CP^{n+1} (resp. RP^{n+2}) see **[CS12]**, **[CS4]**; see also some related examples involving projective spaces in **[CS]**. A complete analysis of the dimensions of the singularity sets arising from embedding homotopy lens spaces in homotopy lens spaces is in the thesis of William Homer **[Ho]**.

Sometimes, when there is not too much torsion in the cohomology of M, the converses to the above theorem can be stated in terms of rational characteristic classes.

THEOREM **[CS17]** (**[CS12]**). *Let M^n be a compact* (*resp. closed*) PL *manifold, k an integer, $k \leqq n$. Suppose that $H^{4j}(M; Z)$ has no torsion for $4j < k$ and $H^{4j+2}(M; Z_2) = 0$ for $4j + 2 < k$. When $k \equiv 3$ (mod 4), assume also that $H^{k-1}(M^{(k-1)}; Z_2) = 0$, $M^{(k-1)}$ a $(k - 1)$ skeleton of M. Let f: $M^n \to W^{n+2}$ be a continuous map of oriented* PL *manifolds* (*resp., and assume if n is even that $\pi_1 W = 0$*). *Then the following are equivalent.*

(i) *f is homotopic to a* PL *immersion with associated $SO(2)$ fibre space ξ and Euler class $x \in H^2(M; Z)$* (*resp., a* PL *embedding with x denoting the characteristic class associated to f: $M^n \to W^{n+2}$*) *and this immersion* (*resp., embedding*) *has singularities of intrinsic codimension at least k; and*

(ii) *$S(\nu_M \oplus f^*\tau_W)$ is fibre homotopy equivalent to $S(\xi \oplus$ trivial bundle$)$* (*resp., f is the underlying map of an h-Poincaré embedding*) *and $L(f) = L(M)\, L(x) - f^*L(W)$ vanishes below dimension k.*

The most general results, which take torsion considerations into account, relate the dimension of singularity sets of immersions or embeddings of M to questions concerning existence of lifts to connective fibres of certain associated maps of M to the classifying space G/PL **[CS12]**, **[CS17]**.

The above results suggest a central problem: Find a precise formula for the Pontrjagin classes in terms of the singularities of embeddings or immersions. The results above state that the singularities carry the Poincaré dual of these characteristic classes. What is now wanted is an explicit description of a cycle describing $L(f) \cap [M]$ as a sum of simplices in $S(f)$ with coefficients determined by the link pairs of the simplices. For the higher Pontrjagin classes this might involve studying invariants, perhaps index-type invariants, of nonlocally flat knots. This suggests introducing certain exotic cobordism theories of nonlocally flat knots where the dimension of the singularities of the knots and cobordisms is prescribed.

3. Global methods of constructing PL **immersions and embeddings.** Many of the results of §1 above asserted the existence of large classes of embeddings and immersions of manifolds in codimension two. In view of the relations between PL singularities and characteristic classes discussed above, these embeddings and immersions can rarely be made locally flat and, in fact, have a rich associated structure of singularities. In this section we outline roughly how these embeddings and immersions are constructed.

The construction of an immersion or an embedding in the homotopy class of a map $f\colon M^n \to Y^{n+2}$ can be broken down into two steps. First, we construct a PL manifold regular neighborhood R of M, $j\colon M \to R$, satisfying $j^*\tau_R = f^*\tau_Y$; here τ_R (τ_Y) denotes the stable PL tangent bundle of R (respectively, Y).[4] This regular neighborhood $j\colon M \subset R$ cannot usually be chosen to have a block bundle structure and, recalling the results on characteristic classes above, j will often necessarily have high dimensional sets of singular points. The construction of R will be discussed shortly. To produce immersions, in the second step we construct, by applying results of Haefliger-Poenaru on immersions in codimension zero, an immersion $g\colon R \to Y$. Now the desired immersion is formed by taking the composite $gj\colon M \to Y$.

The second stage, in the construction of embeddings in codimension two, involves finding a candidate for the complement of the submanifold: that is, an $(n + 2)$-dimensional manifold V such that $\partial V \cong \partial R$ and $V \cup_{\partial R} R \cong W$. However, as $\pi_1(V) \to \pi_1(W)$ cannot generally be taken to be an isomorphism, as M is to be a submanifold of codimension two, the usual surgery theory of Browder **[B1]** and of Novikov **[No]** and its extension by Wall **[W]** cannot be applied to construct such a complement V. We appeal instead to the methods of our global theory of locally flat submanifolds of codimension two and homology equivalent manifolds **[CS2]**, recalled in the next section. This approach takes advantage of the surjectivity of $\pi_1 V \to \pi_1 W$, for complements of codimension two submanifolds in attempting to construct candidate complements with an appropriate fundamental group and homology type with local coefficients over $\pi_1 W$; we do not, however, attempt to prescribe the precise homotopy type of the complement V of the embedding to be constructed. When n is odd, we show that this always leads to an embedding result. When n is even, a kind of homology-surgery obstruction arises; moreover if $\pi_1 W = 0$, we show how this surgery obstruction can be identified with a subgroup of the knot cobordism groups. Then, we can kill it by changing the choice of a regular neighborhood R by introducing one more isolated singular point. This leads to the embedding theorems in the odd dimensional and simply connected even dimensional cases **[CS12**, §5]. In the nonsimply connected even dimensional case, the homology surgery obstruction encountered takes values in groups that are frequently far larger than knot theory; in particular, they are not finitely generated. By careful analysis of this situation in **[CS7]** we construct examples where the obstructions cannot be killed by singularities, isolated or not. This leads to the results on spineless manifolds in the even dimensional nonsimply connected case. An introduction to the relation between codimension two geometry and homology surgery is given in §4 below.

Last, we outline the construction of the regular neighborhood R which was used above. This part of the argument involves local considerations in PL singularity theory as well as the general point of view and many of the results of our work on homology surgery and nonsingular codimension two embeddings. We used, in particular, the point of view on knot cobordism groups as the natural "coefficient" groups for codimension two smooth embeddings, outlined in §4 below. Now, we recall the construction of the classifying space for (respectively, oriented) codimen-

[4]This is, as we said, roughly what R must satisfy. For a precise discussion see **[CS17]** for immersions and **[CS12]** for embeddings.

sion two regular neighborhoods BRN_2 (respectively, $BSRN_2$). See **[CS12]** for detailed constructions and proofs and for references.

Some care is required in showing the existence of such a space; note, for example, the nonuniformity of a general regular neighborhood. An important step in the construction of this space is the following transversality method used in producing induced regular neighborhoods.

Let $M^n \subset W^{n+2}$ be an (resp. oriented) abstract regular neighborhood, i.e., a (resp. oriented) codimension two thickening, proper (i.e., $\partial M = M \cap \partial W$) if ∂M is not empty. This manifold thickening W determines a stratification of M, $\chi = (\chi_0, \cdots, \chi_n)$ with $\chi_n =$ locally flat points and $\chi_i =$ set of points of $M \subset W$ whose link pairs desuspend precisely i times. Now to define induced regular neighborhoods, consider first $N \subset M$, a locally flat submanifold. Then using results of Stone, we first deform N in M to be transverse to χ. Using further results of Stone, we then restrict $M \subset W$, proceeding inductively along strata of χ and its "regular neighborhood system", to obtain a codimension two regular neighborhood of N^m with a stratification $\chi \cap N = \{\chi_{n-m} \cap N, \cdots, \chi_n \cap N\}$. Having defined appropriately induced regular neighborhoods in this case, it follows that the functor which assigns to a manifold the concordance classes of (resp. oriented) regular neighborhoods is representable by a classifying space BRN_2 (resp. $BSRN_2$).

A subsequent reformulation of the definition of BRN_2 and $BSRN_2$ which we give in **[CS12]** has been given by McCrory **[Mc]** using his improved and simplified approach to PL transversality in place of Stone's treatment. McCrory also reproves in his context the local geometric results (e.g., **[CS12**, Proposition 1.3]) which we needed.

To obtain the results on embeddings and immersions, as noted above we must be able to construct regular neighborhoods $j: M \subset R$ with fairly general stable PL tangent bundles, e.g., with general rational PL Pontrjagin classes. This involves showing that $BSRN_2$ is quite large in a sense which will shortly be explained. To begin with, the homotopy groups of $BSRN_2$ are identified, by definition, with regular neighborhoods of spheres. As these can be taken, up to concordance, to have only a single isolated singularity, which looks like the cone on a knot, we may identify these homotopy groups, except in dimension 2, with knot cobordism. For $\pi_2(BSRN_2)$, there is an additional Z corresponding to the Euler class of the regular neighborhood. In particular, it follows that the even dimensional homotopy groups of $BSRN_2$ are not finitely generated while the odd dimensional groups are zero **[CS12**, §3].

Now, by assigning to each regular neighborhood R its Euler class $\chi(R)$ and also a (normal) map $\eta(R)$ comparing the regular neighborhood with a linear 2-plane bundle with that same Euler class we obtain a canonical map **[CS12**, §4]

$$(\chi, \eta) : BSRN_2 \to BSO_2 \times G/\mathrm{PL}.$$

Recall that rationally G/PL is just the same as BSO or $BSPL$; that is, it rationally classifies the PL Pontrjagin classes. In fact **[CS12]**, the natural map given by stabilization $BSRN_2 \to BSPL$ factors through (χ, η). On the homotopy group level η is the map which assigns to a knot its Arf-invariant, index or zero, depending upon the dimension.

The principle that there are enough regular neighborhoods to begin the construction of all of our embeddings and immersions is the following result.

COROLLARY 4.9 OF **[CS12]**. $(\chi, \eta) : BSRN_2 \to BSO_2 \times G/\mathrm{PL}$ *has a cross-section.*

To prove this result a complete description of the homotopy type of $BSRN_2$ was given in **[CS12]**. This description is related to Sullivan's **[Su2]** description of the homotopy type of G/PL using Browder-Novikov surgery theory. For $BSRN_2$ we used the homology surgery theory of **[CS2]** to define and study in **[CS12**, §2] splitting invariants

$$s\colon \Omega_*(BSRN_2) \to \Gamma_{*+2}(\phi_0),$$
$$s\colon \Omega_*(BSRN_2\,;Z_n) \to \Gamma_{*+2}(\phi_0) \otimes Z_n.$$

Here $\Gamma_{*+2}(\phi_0)$ is as in **[CS2**, §3], see §4 below, a homology surgery obstruction group and

$$\Gamma_{n+2}(\phi_0) \cong \begin{cases} 0, & n \text{ odd} \\ C_{n-1}, & n \text{ even} \end{cases} \qquad (n \geqq 5),$$

C_{n-1} the knot cobordism group of (smooth or) locally flat PL knotted (homotopy) $(n-1)$-spheres in S^{n+1}. The needed definition of s and of s_n for general regular neighborhoods is rather subtle and requires the delicate piecewise-linear geometry of **[CS12**, §1].

The homology surgery theory of **[CS2]**, outlined in §4 below, gives an identification $\Gamma_* = \Gamma_{*+4}$ and leads to needed multiplicative properties for s **[CS12**, Appendix to §2]. The geometric periodicity of knot cobordism of **[CS2]** here translates into a homotopy-theoretic periodicity for $BSRN_2$ modulo BSO_2.

The information obtained from these geometric invariants is assembled into a homotopy-theoretic description of $BSRN_2$ **[CS12**, Theorem 4.6]. This procedure is in the pattern of Sullivan's characteristic variety theorem for G/PL **[Su1]**, **[Su2]**; here, as in **[Su2]**, this assembly procedure is much simplified by the special fact that the homotopy groups of the classifying space are zero in odd dimensions **[CS12**, §3]. We show from our description of $BSNR_2$ that, in particular, the map χ gives a product fibration.

Corollary 4.9 was then proved by comparing the description of $BSRN_2$ with that of G/PL. A cross-section of η is induced by splitting the map η_* on homotopy groups. Recall η_* was identified with the map that assigns to a knot $S^{k-1} \subset S^{k+1}$ its Arf-invariant when $k \equiv 2 \pmod 4$, or index when $k \equiv 0 \pmod 4$. Splitting η_2 uses the existence of an invertible, amphicheiral knot in S^2 with nontrivial Arf-invariant (the "figure-eight knot").

We observe that this triviality of χ does not follow a priori from the existence of the obvious cross-section $BSO_2 \to BSRN_2$ obtained by viewing a bundle as a regular neighborhood. Rather, the proof of this fact makes use of our definition of splitting invariants when the associated bundle is nontrivial; in particular this is used in showing that $BSRN_2$ is an H-space. It would be interesting to have a geometric H-space structure on $BSRN_2$ (extending tensor product of bundles) that would imply the triviality of χ, a priori.

A generalized characteristic variety theorem, applicable to any space for which there are generalized splitting invariants, which satisfy necessary properties, is given by L. Jones in these PROCEEDINGS, part 1, pp. 131–140.

Classifying spaces $BSRN_{2,k}$ for oriented regular neighborhoods with nonlocally flat points of intrinsic codimension $\geq k$ are also defined in **[CS12]**. These are needed for the results involving the structure of the nonlocally flat points. There are natural maps

$$BSO_2 = BSRN_{2,\infty} \to \cdots \to BSRN_{2,k+1} \to BSRN_{2,k} \to \cdots \to BSRN_2.$$

$BSRN_{2,q}$ is the $(q-1)$-connective fibering of $BSRN_2$, mod BSO_2. Thus, killing nonlocally flat points in codimension less than q kills homotopy groups of $BSRN_2$ below dimension q, except for $\pi_2(BSO_2)$.

Using these spaces, questions concerning the structure and the dimension of the nonlocally flat points are translated into homotopy-theoretic questions about lifting normal invariants to connective fibers of G/PL. This leads in particular to lower bounds on the dimensions of the nonlocally flat points. See §2 above and **[CS12]**, **[CS17]**, **[CS4]**, **[Ho]**.

A further family of classifying spaces, which play a key role in the study of branched covers and nonlocally smoothable Z_p actions with codimension two fixed points, is discussed in §5 below and **[CS11]**.

4. Smooth and nonsingular PL **embeddings in codimension two.** The theory of smooth or locally smoothable (i.e., locally flat) submanifolds, besides its independent interest, plays an important role in the general theory of PL embeddings and immersions. Knot theory describes the isolated singularity of a PL embedding or immersion; this provided part of the motivation for the first study of knot cobordism **[FM]**, **[N]**. In the global theory of submanifolds to be outlined in this section, knots will be seen to serve as the "coefficients" of the theory. The methods and results of the previous sections use this global approach to smooth or locally flat embeddings.

Let $M^n \subset W^{n+k}$ be a smooth or PL locally smoothable embedding. For $k > 2$, $\pi_1(W - M) \cong \pi_1 W$, and so Poincaré-Lefschetz duality over $Z\pi_1 W$ gives information on the homotopy theory of $W - M$. For $k = 2$, we still have duality over $Z\pi_1 W$, but the inclusion induces only a surjection $\pi_1(W - M) \to \pi_1 W$ that is only rarely an isomorphism. Thus it becomes necessary and appropriate to study manifolds with a fundamental group π (e.g., $\pi_1(W - M)$), and of a given (simple) homology type over a quotient ring Λ of $Z\pi$ (e.g., $Z\pi_1 W$).

As an example of this point of view, let $G_n(M)$ denote the cobordism classes of embeddings of $S^n \times M^k$ in $S^{n+2} \times M$, which commute up to homotopy with projection on M; see **[CS2]** for the exact definition. $G_n(\mathrm{pt})$ is the knot cobordism group. The technique of homology equivalent manifolds shows that classification up to concordance of codimension two embeddings depends primarily upon $\pi_1 M$. Thus we have:

THEOREM A. *If $\pi_1 M = \{e\}$, the map $G_{n+k}(e) \to G_n(M)$ given by connected sum with $i_0 \times \mathrm{id}_M$, $i_0\colon S^n \to S^{n+2}$ the unknot, is a bijection.*

Thus, for $\pi_1 M$ trivial, the knottedness of an element of $G_n(M)$ can be pushed into the neighborhood of a point, in a unique way up to cobordism.

The theory of homology equivalent manifolds is also periodic of period 4; in particular one has:

THEOREM B. *For* $\pi_1 M = \{e\}$, $\delta_{CP^2} : G_n(M) \to G_{n+4}(M \times CP^2)$ *given by taking products with complex projective space is a bijection.*

These two results taken together give a geometric proof and interpretation of the periodicity $G_n(e) \cong G_{n+4}(e)$ of knot cobordism groups **[CS2]**.[5] $G_n(M)$ for M non-simply connected has been studied from this point of view of homology equivalent manifolds by S. Ocken **[O]**.

The same point of view applies to such problems as equivariant cobordism theory **[CS2]**, the theory of semilocal knots **[CS2]**, close embeddings **[CS2]**, **[CS6]** and codimension two splitting **[CS2]**, **[CS9]**. To study homology equivalent manifolds consider a normal map

$$\begin{array}{ccc} \nu_M & \xrightarrow{\ b\ } & \xi \\ \downarrow & & \downarrow \\ M^n & \xrightarrow{\ f\ } & X^n \end{array}$$

$\pi_1 X = \pi$ and suppose $\mathscr{F}: Z\pi \to \Lambda$ is a surjective homomorphism of rings with involution. Here $Z\pi$ has the involution given by $\bar{g} = w(g)g^{-1}$, $g \in \pi$, $w: \pi \to \{\pm 1\}$ the orientation character. Then there is an algebraically defined functor $\Gamma_n(\mathscr{F})$, of maps $\mathscr{F}$ to abelian groups, and an obstruction $\sigma(f, b) \in \Gamma_n(\mathscr{F})$ that vanishes if and only if (f, b) is normally cobordant to a (simple) homology equivalence (with local coefficients) over $Z\pi_1 X$. For $n = 2k$, $\Gamma_n(\mathscr{F})$ is a reduced Grothendieck group of $(-1)^k$ Hermitian quadratic forms over Z that become unimodular forms on stably free modules over Λ. The reduction is accomplished by declaring to be zero forms that contain an anisotropic subspace that, over Λ, becomes a free summand of $\frac{1}{2}$ the rank **[CS2**, Chapter I**]**. For n odd, $\Gamma_n(\mathscr{F})$ is obtained from a subgroup of the stabilized automorphism s of trivial hyperbolic $(-1)^{n-1}$ -forms. See **[CS2]** for the details. Recall in particular that $\Gamma_{\text{odd}}(\mathscr{F}) \subset L_{\text{odd}}(\Lambda)$, but that $\Gamma_{\text{even}}(\mathscr{F})$ is much large than $L_{\text{even}}(\Lambda)$ or $L_{\text{even}}(\pi)$; in fact, in the cases of interest here it is usually not finitely generated. Here $L_n(\Lambda)$ denotes the Wall surgery group of Λ and $\Gamma_n(id_\Lambda)$ coincides with $L_n(\Lambda)$.

When $f(M, \partial M) \to (X, \partial X)$ is actually a normal map of a manifold with boundary, and $f|\partial M: \partial M \to \partial X$ is already a Λ-homology equivalence, the same theory applies relative to the boundary. But if not, and one wishes to modify f on the boundary (or on some part thereof), one considers a commutative diagram:

$$\begin{array}{rcl} \Phi\colon \pi_1 \partial X & \xrightarrow{\ \mathscr{G}\ } & \Lambda' \\ \downarrow{\scriptstyle i_*} & & \downarrow{\scriptstyle j} \\ \pi_1 X & \xrightarrow{\ \mathscr{F}\ } & \Lambda \end{array}$$

Then there exists a relative obstruction $\sigma(f, b) \in \Gamma_n(\Phi)$ that vanishes if and only if (f, b) is normally cobordant to a Λ-homology equivalence that restricts to a Λ'-equivalence on the boundary. All the obstructions can be realized, and there is an exact sequence

[5] Bredon **[Br]** gave an interpretation of knot periodicity in terms of transformation groups. Kauffmann **[Ka]** expressed this in terms of his product operations. Matsumoto **[Ma]** gave an interpretation of periodicity in terms of locally flat spines, using certain Seifert form obstructions to *ambient* codimension two surgery that apply in some special cases.

$$\Gamma_n(\mathscr{G}) \to \Gamma_n(\mathscr{F}) \to \Gamma_n(\Phi) \to \Gamma_{n-1}(\mathscr{G}).$$

See **[CS2]** for more details.

Let $\alpha\colon S^n \subset S^{n+2}$ be a PL locally flat knot, and let α_0 be the unknot. Then it is not hard to show that there exists a degree one map $f\colon S^{n+2} \to S^{n+2}$, transverse to α_0, with $f^{-1}(\alpha_0) = \alpha$. Let $H\colon S^{n+2} \times I \circlearrowleft$ be a homotopy of f to the identity and consider V the closure of the complement in $S^{n+2} \times I$ of a tubular neighborhood of $H^{-1}(\alpha_0 \times I)$ that meets the boundary regularly. Now $\partial V = A \cup B \cup (S' \times D^{n+1} \times 1)$ and H induces

$$\bar{H}\colon (V; A, B, S^1 \times D^{n+1} \times 1) \to (S^1 \times D^{n+1} \times I, S^1 \times D^{n+1} \times 0, S^1 \times S^n \times I, S^1 \times D^{n+1} \times 1),$$
$$\bar{H}|S^1 \times D^{n+1} \times 1 = \text{identity} \quad \text{and} \quad (\bar{H}|A)\colon A \to S^1 \times D^{n+1} \times 0$$

is a homology equivalence over Z by Alexander duality. (Recall A = closed knot complement of α, a homology circle.) Let

$$\begin{array}{ccc} \Phi_0\colon Z[Z] & \xrightarrow{\text{id}} & Z[Z] \\ \downarrow{\scriptstyle \text{id}} & & \downarrow{\scriptstyle a} \\ Z[Z] & \xrightarrow{a} & Z \end{array} \qquad a \text{ the augmentation.}$$

Then the obstruction $\sigma(\bar{H}) \in \Gamma_{n+3}(\Phi_0)$ is defined; it is the obstruction to deforming $\bar{H}$ by a normal cobordism relative to $A \cup S^1 \times D^{n+1} \times 1$ to an integral homology equivalence that restricts to a homotopy equivalence of B to $S^1 \times S^n \times I$. The precise basic result that shows how knot theory forms the coefficients of the general theory is the following **[CS2]**

THEOREM.[6] *The assignment $\alpha \to \sigma(\bar{H})$ induces an isomorphism τ_0 of the knot cobordism group G_n and $\Gamma_{n+3}(\Phi_0)$, for $n \neq 1, 3$.*

Of course, the first complete algebraic description of high dimensional knot cobordism was given by Levine **[L1]**, **[L2]**, extending results of Kervaire **[K]** in terms of Seifert matrices. The first complete algebraic description of equivariant knot cobordism theorems was given in **[CS2]**. In the special case in which the knot is fixed point set, one obtains the cobordism groups of counterexamples to the generalized Smith conjecture as a Γ-group. This in turn enables one to study equivariant PL embeddings and to obtain the codimension two converse to the Smith theorem described in **[CS11]**; see also §5 below.

As an example illustrating the role of knots in the general theory of locally flat embeddings, recall the notion of a local knot, i.e., a smooth PL or topological locally flat embedding $i\colon M^n \subset E(\xi)$, ξ an $SO(2)$ bundle over M, homotopic to the zero section. In **[CS2]** it is shown that for $n \geq 4$, the cobordism classes of Euclidean embeddings form a group, denoted $C_H(M, \xi)$. Let Φ_ξ be the diagram:

$$\begin{array}{ccc} Z(\pi_1(S(\xi))) & \xrightarrow{\text{id}} & Z(\pi_1(S(\xi))) \\ \downarrow{\scriptstyle \text{id}} & & \downarrow{\scriptstyle p_*} \\ Z(\pi_1 S(\xi)) & \xrightarrow{p_*} & Z\pi_1 M \end{array}$$

THEOREM **[CS2]**. *There is an exact sequence*

---

[6]For $n = 3$, $G_n \subset \Gamma_{n+3}(\Phi_0)$ is a subgroup of index 2; see **[CS2]**. For topological locally flat knots, $G_3^{\text{TOP}} \cong \Gamma_6(\Phi_0)$ **[CS3]**. For $n = 1$, $G_n \to \Gamma_{n+3}(\Phi_0)$ is surjective, but **[CG]** not injective.

$$0 \to C_H(M, \xi) \xrightarrow{\Sigma} \Gamma_{n+3}(\Phi_\xi) \to \text{coker } s_H,$$

where $s_H: [\Sigma E(\xi), G/H] \to L_{n+3}(p_*)$, $H = 0$, PL *or* TOP *as appropriate, is the usual surgery obstruction map on normal invariants* **[B1]**, **[N]**, **[W]**.

For example, $C_H(S^n$, trivial) is easily seen to be the knot group again, $n > 1$. There is a natural map $\Gamma_{n+3}(\Phi_0) \xrightarrow{\rho_0} \Gamma_{n+3}(\Phi_\xi)$ induced by the map $Z = \pi_1 S^1 \to \pi_1 S_\xi$ obtained from the inclusion of a fibre. Knots form the coefficient theory in the sense that the action of the element of knot cobordism groups on an element $\alpha \in C_H(M, \xi)$ by connected sums corresponds to the addition of $\rho_0(\tau(\xi))$ to $\Sigma\alpha$, i.e.,

$$\Sigma(\alpha \# \xi) = \Sigma(\alpha) + \rho_0(\tau_0(\xi)).$$

In particular, for $\pi_1 M = \{e\}$ and ξ trivial, $C_H(M, \xi)$ is isomorphic to the appropriate knot cobordism group. However, for $M = T^n$, for example, $C_H(M$; trivial) is much bigger than knot cobordism G_n; see **[CS2]** again.

The method of homology equivalences also applies to group action in spheres. For example, in **[CS2]** the following result is proven.

THEOREM. *Let ξ be a free smooth* (PL *or topological*) *action of the cycle group Z_s on S^{2k+1}. Then, if s and $k \geqq 2$ are not both even, then there exists a smoothly* (PL *or topologically locally flat*) *embedded sphere $S^{2k-1} \subset S^{2k+1}$ that is invariant under the action of ρ.*

In the *general* case one can describe an obstruction for the existence of such a sphere. Furthermore we have

THEOREM **[CS2**, 9.4**]**. *Let τ_1 be the restriction of a free action of ρ on S^{2k+1} to an invariant sphere of codimension two. Let τ_2 be homotopy equivalent to τ_1. Then τ_2 is the restriction to ρ of an invariant sphere if and only if τ_1 is normally cobordant to τ_2.*

This result is in sharp contrast with the general PL case, see §5, where every action arises as invariant (not locally smoothable) spheres of every action two dimensions higher.

An algebraic classification of cobordism classes of invariant spheres is also given in **[CS2]**. We will content ourselves with recalling here the following result for the Z_2 case:

THEOREM. *Let Z_2 act freely on S^{2k}. Then there is at most one cobordism class of smooth,* PL *or topological invariant homotopy spheres.*

5. Equivariant and invariant submanifolds of group actions. Examples of applications of our methods in codimension two geometry and in homology surgery to questions concerning group actions are scattered throughout this paper and **[CS12]**, **[CS17]**. Here we bring together three examples of applications to (1) locally flat or smooth equivariant submanifolds of group actions, (2) nonlocally flat PL equivariant submanifolds of group actions, and (3) the fixed points of group actions which are not locally linear.

(1) The role of characteristic invariant submanifolds of free cyclic actions was brought out in **[LM1]**, **[LM2]**, **[B4]**, **[CS2]**. The restriction of a free action ρ of the cyclic group Z_s on a homotopy sphere Σ^{2k+1} to a PL (or smooth) invariant sub-

sphere Σ^{2j+1} is said to be characteristic if there is an equivariant homotopy equivalence j: $(\Sigma^{2k+1}, \rho) \to (S^{2k+1}, \phi)$, ϕ some linear action on S^{2k+1}, with f transverse to a linear subsphere $S^{2j+1} \subset S^{2k+1}$ and $\Sigma^{2j+1} = f^{-1}(S^{2j+1})$. It can be shown that, provided $(2j + 1) > \frac{1}{2}(2k + 1)$, an invariant subsphere Σ^{2j+1} will be characteristic if and only if the Z_s-normal bundle of Σ^{2j+1} in Σ^{2k+1} splits into a sum of Z_s-plane bundles. The following result describes how characteristic subspheres fit into a tower invariant characteristic spheres.

TOWER THEOREM **[CS2]**. *Let τ be a free action of Z_s on the (homotopy) sphere Σ^{2j-1} that is the restriction of the free action ρ on Σ^{2k+1} to an invariant sphere. Assume s is odd. Then Σ^{2j+1} is a characteristic sphere of ρ (i.e., Σ^{2j+1}/τ is a characteristic submanifold of Σ^{2k+1}/τ) if and only if there is a tower*

$$\Sigma^{2j+1} \subset \Sigma^{2j+3} \subset \cdots \subset \Sigma^{2k-1} \subset \Sigma^{2k+1}$$

of invariant spheres.

(2) Except possibly for some low dimensional cases, every free Z_s action τ on a homotopy sphere Σ^{2k+1} can be obtained by restricting any given free action τ of Z_s on Σ^{2j+1} to appropriate invariant (and characteristic) locally flat subspheres, provided $(2k + 1) - (2j + 1) > 2$. This is far from being true in the category of locally flat submanifolds in codimension 2; this follows readily from standard characteristic class considerations. Some of our results on which actions arise on locally flat subspheres were presented in §4; the conditions are rather restrictive. Thus the following result for general PL embedded subspheres is in sharp contrast with the locally flat case.

THEOREM **[CS12]**. *Let G be finite cyclic or S^1. Let ρ and τ be arbitrary free actions of G on S^n and S^{n-2}, respectively, in the* PL *category. Assume $n \neq 4, 6$, and $n \neq 5$ if $G = S^1$. Then there is an equivariant* PL *embedding of S^{n-2} in S^n; i.e., there is a one-to-one* PL *map f: $S^{n-2} \to S^n$ with*

$$f(\tau(t, x)) = \rho(t, f(x)), \qquad t \in G,\ x \in S^{n-2}.$$

The calculation of the minimal dimension of the singularities among equivariant embeddings of τ in ρ was shown in **[CS12]** to be equivalent to a question in homotopy theory concerning lifting normal invariants. This led to complete results for $G = S^1$ or Z_2 **[CS12]**, **[CS4]**. For general G, there are difficult odd torsion considerations. A complete solution of this problem was obtained by W. Homer in his dissertation **[Ho]**. There are also many interesting examples of possible singularity dimensions discussed there.

For some S^1 actions, even with one of the actions taken to be linear, the singularity set produced by an equivariant embedding necessarily has codimension two in the subsphere **[CS12]**. However, by replacing the embedding by a PL immersion of S^1-actions, one can always reduce the size of the singularity sets to codimension eight **[CS17]**. Here multiple points are being traded off for singularity points.

(3) In **[CS11]** we described some results on the codimension two submanifold fixed points of semifree PL Z_s actions on manifolds. The study of this situation is equivalent, when the viewer's eyeball is moved down to the quotient space, to the study of the manifold branched covers of manifolds obtained by branching along

general PL submanifolds. This makes possible a systematic approach to a question of Fox on when the branched cover of a manifold is again a manifold. In **[CS11]** we introduce a classifying space $BSRN_2^{Z_s}$ for regular neighborhoods with s-fold manifold branched covering space. The homotopy groups of $BSRN_2^{Z_s}$ are, except in dimension two where there is an additional Z, groups of cobordism classes of counterexamples to the generalized P. A. Smith conjecture.

An algebraic description of these groups in high dimensions was given, and some geometric consequences drawn in **[CS2]**. Not much is known about $\pi_2(BSRN_2^{Z_s})$ (see §2 of **[CS11]**); this is essentially the classical Smith conjecture (see **[CS14]** for some recent results on this).

As an application of the use of the spaces $BSRN_p^{Z_p}$ we have the following result which is a kind of converse to P. A. Smith theory. Related results were obtained by L. Jones in high codimensions **[J]**.

THEOREM **[CS11]**. *Let M^n be a Z_p homology sphere with $\pi_{n+1}(\Sigma M) \to H_{n+1}(\Sigma M)$ surjective. Assume that $H^2(M; Z_2) = 0$. Then there exists a semifree* PL *action of Z_p on S^{n+2} with M as fixed points.*

This result should be compared to our related result on embedding in codimension two Euclidean space presented in §1 above.

Many M which satisfy the hypothesis of this theorem do not have locally flat embeddings in S^{n+2}. If the condition in Theorem 3 on the surjectivity of the Hurewicz map is dropped, we can still show, for n odd, that there is a Z_p homology sphere V^{n+2} with a semifree Z_p action and with M as fixed points. The condition on $H^2(M; Z_2)$ arises from the unsettled status of the 3-dimensional P. A. Smith conjecture.

BIBLIOGRAPHY

[Br] G. Bredon, *Regular O(n) manifolds, suspensions of knots and knot periodicity* (to appear).

[B1] W. Browder, *Surgery on simply-connected manifolds*, Springer-Verlag, Berlin and New York, 1972.

[B2] ———, *Embedding 1-connected manifolds*, Bull. Amer. Math. Soc. **72** (1966), 220–231; erratum, ibid. **72** (1966), 736. MR **32** #6467.

[B3] ———, *Embedding smooth manifolds*, Proc. Internat. Congr. Math. (Moscow, 1966), Izdat. "Mir," Moscow, 1968, 712–719. MR **38** #6611.

[B4] ———, *Free Z_p-actions on homotopy spheres*, Topology of Manifolds (Proc. Inst. Univ. of Georgia, Athens, Ga., 1969), pp. 217–226, Markham, Chicago, Ill., 1970. MR **43** #2720.

[BL] W. Browder and G. R. Livesay, *Fixed point free involutions on homotopy spheres*, Tôhoku Math. J. (2) **25** (1973), 69–87. MR **47** #9610.

[BPW] W. Browder, T. Petrie and C. T. C. Wall, *The classification of free actions of cyclic groups of odd order on homotopy spheres*, Bull. Amer. Math. Soc. **77** (1971), 455–459. MR **43** #5547.

[C1] S. E. Cappell, *A splitting theorem for manifolds*, Invent. Math. **33** (1976), 69–170.

[C2] ———, *On connected sums of manifolds*, Topology **13** (1974), 395–400. MR **50** #11252.

[C3] ———, *Manifold with fundamental group a generalized free product*. I, Bull. Amer. Math. Soc. **80** (1974), 1193–1198. MR **50** #8562.

[C4] S. E. Cappell, *Unitary nilpotent groups and Hermitian K-theory*, Bull. Amer. Math. Soc. **80** (1974), 1117–1122. MR **50** #11274.

[C5] ———, *On homotopy invariance of higher signatures*, Invent. Math. **33** (1976), 171–179.

[CS1] S. E. Cappell and J. L. Shaneson, *Submanifolds, group actions, and knots*. I, II, Bull. Amer. Math. Soc. **78** (1972), 1045–1052.

[CS2] S. E. Cappell and J. L. Shaneson *The codimension two placement problem and homology equivalent manifolds*, Ann. of Math. (2) **99** (1974), 277–348. MR **49** #3978.

[CS3] ———, *Topological knots and knot cobordism*, Topology **12** (1973), 33–40. MR **47** #9632.

[CS4] ———, *Nonlocally-flat embeddings, and group actions*, Bull. Amer. Math. Soc. **79** (1973), 577–582. MR **47** #9635.

[CS5] ———, *Submanifolds in codimension two and homology equivalent manifolds*, Ann. Inst. Fourier (Grenoble) **23** (1973), 19–30. MR **49** #11522.

[CS6] ———, *Close codimension two embeddings of even dimensional manifolds*, Amer. J. Math. **97** (1975), 733–740. MR **52** #4314.

[CS7] ———, *Totally spineless manifolds* (to appear).

[CS8] ———, *Branched cyclic covers*, in Knots, Groups and 3-Manifolds, Papers in Memory of R. H. Fox, Ann. of Math. (2) **84** (1975), 165–173.

[CS9] ———, *Fundamental groups, Γ groups, and codimension two embeddings*, Comment. Math. Helv. **39** (1976), 437–446.

[CS10] ———, *Every spin 4-manifold PL embeds in* S^6 (to appear).

[CS11] ———, *Branched covering spaces and PL singularities* (to appear).

[CS12] ———, *Piecewise linear embeddings and their singularities*, Ann. of Math. (2) **103** (1976), 163–228. MR **53** #11629.

[CS13] ———, *There exist inequivalent knots with the same complement*, Ann. of Math. (2) **103** (1976), 349–353.

[CS14] ———, *A note on the Smith conjecture*, Topology (to appear).

[CS15] ———, *Surgery on 4-manifolds and applications*, Comment. Math. Helv. **46** (1971), 500–528.

[CS16] ———, *Some new 4-manifolds*, Ann. of Math. (2) **104** (1976), 61–72.

[CS17] ———, *Immersions and singularities*, Ann. of Math. (2) **105** (1977), 539–552.

[CG] A. Casson and C. Gordon, *Classical knot cobordism* (to appear).

[F] R. H. Fox, *A quick trip through knot theory*, Topology of 3-Manifolds and Related Topics, Prentice-Hall, Englewood Cliffs, N. J., 1962. MR **25** #3522.

[FM] R. H. Fox and T. W. Milnor, *Singularities of 2-spheres in 4-space*, Osaka J. Math. **3** (1966), 257–267. MR **35** #2273.

[H] A. Haefliger, *Lissage des immersions*. II, mimeographed.

[HP] A. Haefliger and V. Poenaru, *La classification des immersions combinatoires*, Inst. Hautes Études Sci. Publ. Math. **23**, 651–667.

[Hi] M. Hirsch, *Immersions of manifolds*, Trans. Amer. Math. Soc. **93** (1959), 242–276. MR **22** #9980.

[Ho] W. Homer, *Group actions and piecewise-linear singularities*, Ph.D. Thesis, Princeton University, 1975.

[Hs] W. C. Hsiang, *A note on free differentiable actions of* S^1 *and* S^3 *on homotopy spheres*, Ann. of Math. (2) **83** (1966), 266–272. MR **33** #731.

[HS] W. C. Hsiang and J. L. Shaneson, *Fake tori*, Topology of Manifolds (Proc. Inst., Univ. of Georgia, Athens, Ga., 1969), Markham, Chicago, Ill., 1970, pp. 19–50. MR **43** #6930.

[Hu] S. T. Hu, *Homotopy theory*, Academic Press, London and New York, 1959.

[H] J. F. P. Hudson, *PL topology*, Benjamin, New York, 1970.

[J] L. Jones, *The converse to the fixed point theorem of P. A. Smith*, Ann. of Math. (2) **94** (1971), 52–68. MR **45** #4427.

[K] M. Kervaire, *Les noeuds de dimension supérieure*, Bull. Soc. Math. France **93** (1965), 225–271. MR **32** #6479.

[Ka] L. Kauffmann (to appear).

[KaMa] K. Kato and Y. Matsumoto, *Simply-connected surgery of submanifolds in codimension two*, J. Math. Soc. Japan **24** (1972), 586–608.

[L1] J. Levine, *Knot cobordism in codimension two*, Comment. Math. Helv. **44** (1968), 229–244. MR **39** #7618.

[L2] ———, *Invariants of knot cobordism*, Invent. Math. **8** (1969), 98–110; Appendum; ibid. **8** (1969), 355. MR **40** #6563.

[L3] ———, *Polynomial invariants of knots in codimension two*, Ann. of Math. (2) **84** (1966), 537–554. MR **34** #808.

[LM1] S. Lopez de Medrano, *Involutions on manifolds*, Springer-Verlag, Berlin and New York, 1970.

[LM2] ———, *Invariant knots and surgery in codimensions two*, Actes Congrès Internat. Math. **2** (1971), 99–112.

[Ma] Y. Matsumoto, *A 4-manifold with no spine*, Bull. Amer. Math. Soc. **81** (1975), 467–470. MR **52** #1711.

[Mc] C. McCrory, *Cone bundles*, Trans. Amer. Math. Soc. **228** (1977), 157–163.

[MM] T. Matsumoto and Y. Matsumoto, *The unstable difference between homology cobordism and piecewise linear bundles* (to appear).

[MS] J. Morgan and D. Sullivan, *The transversality characteristic class and linking cycles in surgery theory*, Ann. of Math. (2) **99** (1974), 463–544. MR **50** #3240.

[N] H. Noguchi, *Obstructions to locally flat embeddings of combinatorial manifolds*, Topology **5** (1966), 203–213. MR **33** #724.

[No] S. P. Novikov, *Homotopically equivalent smooth manifolds*. I, Amer. Math. Soc. Transl. (2) **48** (1965), 271–396.

[RS] C. P. Rourke and B. J. Sanderson, *Block bundles*. I, II, III, Ann. of Math. (2) **87** (1968), 1–28; ibid. (2) **87** (1968), 256–278; ibid. (2) **87** (1968), 431–483. MR **37** #2234a; **37** #2234b; **38** #729.

[O] S. Ocken, *Parametrized knot theory*, Mem. Amer. Math. Soc. No. 170, Amer. Math. Soc., Providence, R. I., 1976.

[Sm] S. Smale, *The classification of immersions of spheres in euclidean spaces*, Ann. of Math. (2) 69 (1959), 327–344. MR **21** #3862.

[St] D. Stone, *Stratified polyhedra*, Lecture Notes in Math., vol. 252, Springer-Verlag, Berlin and New York, 1972.

[Su1] D. Sullivan, *Triangulating homotopy equivalences*, Thesis, Princeton University, 1966.

[Su2] ———, *Triangulating and smoothing homotopy equivalences*, Geometric Topology Seminar Notes, Princeton University, 1967.

[Su3] ———, *Geometric topology*. I, Localization, periodicity, and Galois symmetry, mimeographed notes, M.I.T., 1970.

[Su4] ———, *Genetics of homotopy theory*, Ann. of Math. (2) **100** (1974), 1–79.

[Th] R. Thom, *Les classes caractéristiques de Pontrjagin des variétés triangulées*, Sympos. de Topology Algebra (Mexico, 1956).

[W] C. T. C. Wall, *Surgery on compact manifolds*, Academic Press, New York, 1970.

COURANT INSTITUTE OF MATHEMATICAL SCIENCES

Proceedings of Symposia in Pure Mathematics
Volume 32, 1978

IMMERSIONS OF PROJECTIVE SPACES

DONALD DAVIS AND MARK MAHOWALD

The problem of immersing projective spaces in smallest possible Euclidean space is a classical geometric question which has received much attention from homotopy theorists. Surveys of the history and status of this question were given in 1970 by Gitler **[8]** and in 1971 by James **[12]**. One conjecture discussed in **[8]** was made in 1968. *Once conjectured* **[10]**: If $n \equiv 7\ (8)$, then the smallest Euclidean space in which RP^n can be immersed has dimension

$$2n - 2\alpha(n) + \begin{cases} 0 \\ 1 \\ -1 \end{cases} \quad \text{if } \alpha(n) \equiv \begin{cases} 0 \\ 1, 2\ (4), \\ 3 \end{cases}$$

where $\alpha(n)$ denotes the number of 1's in the binary expansion of n.

The rationale behind this conjecture was that it was known to be true when $n = 2^r - 1$ (**[11]**, **[9]**) and when $\alpha(n) = 4$ (**[1]**, **[15]**), and there was something which looked like obstructions to immersions in smaller Euclidean spaces, and it was felt that anything else was too irregular to be an obstruction.

We announced in **[4]** that we had proved the nonimmersion part of this conjecture, but we found a mistake in our proof **[6]** and then found counterexamples when $\alpha(n) = 6$ or 7 (**[7]**, **[2]**). The things which were thought to be obstructions are in the indeterminacy, so that one must look much harder to find obstructions. It appears that the ultimate answer will be very complicated.

Our results are

THEOREM 1. *If $n \equiv 7\ (8)$, $\alpha(n) = 6$, and $n \neq 63$, then $RP^n \subseteq R^{2n-14}$.*
If $n \equiv 7\ (8)$, $\alpha(n) = 7$, and $\nu(n+1) = 4$ or 5, then $RP^n \subseteq R^{2n-16}$.

Here $\nu(2^a(2b+1)) = a$. These immersions are respectively 3 and 1 better than the once conjectured best possible immersions. They are both within 3 of best possible.

AMS (MOS) subject classifications (1970). Primary 57A35.

THEOREM 2. *If* $n \equiv 7\ (8)$ *and* $\alpha(n) = 6$, *then* $RP^n \not\subseteq R^{2n-18}$.
If $n \equiv 7\ (8)$ *and* $\alpha(n) = 7$, *then* $RP^n \not\subseteq R^{2n-20}$.

The second part of Theorem 2 can be extended to

THEOREM 2′. *If* $\nu(n+1) \geqq \alpha(n) - 4 \geqq 3$, *then* $RP^n \not\subseteq R^{2n-2^{\alpha(n)-3}-4}$.

This is probably not near to best possible for $\alpha(n) \geqq 8$, but it gives the densest set of metastable nonimmersions known to us. It also suggests the possibility that the ultimate answer may involve $2^{\alpha(n)}$.

We also have results for CP^n. In **[5]** we reproved all known nonimmersions (**[14]**, **[16]**) in dimension roughly $2n - 2\alpha(n)$ and showed that for $\alpha(n) \leqq 8$ they are within 1 of best possible. We hope to prove this for all $\alpha(n)$. I will say a bit more about our future goals for this project at the end of the talk.

Our approach is of course obstruction theory. It is well known **[15]** that $RP^n \subseteq R^{n+k}$ if and only if the stable normal bundle has geometric dimension $\leqq k$, i.e., there exists a lifting l in

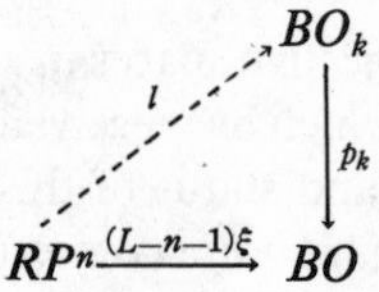

where ξ is the Hopf bundle over RP^n and L is a large power of 2. There is a $2k$-equivalence $RP^\infty/RP^{k-1} = P_k \to V_k$, where $V_k = \text{fibre}(p_k)$, so that the coefficients for the obstructions are $\pi_*^s(P_k)$, which has been tabulated through dimension $k + 29$ in **[13]**. These tables are very complicated, although they do have some regular features. For example if $k \equiv 11\ (16)$, we have

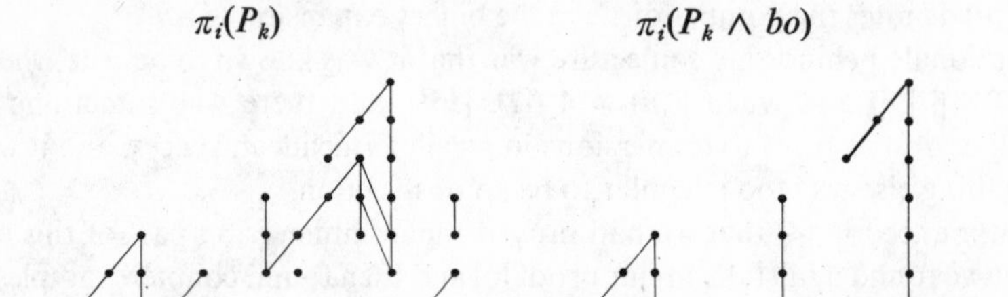

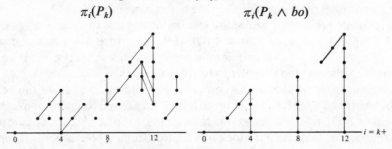

As usual, vertical lines correspond to multiplication by 2 up to elements of higher filtration and diagonal lines of negative slope (\) are differentials in the Adams spectral sequence, which imply that the classes involved do not give rise to homotopy classes. *bo* denotes the spectrum for connective *KO*-theory localized at 2. The map $P_k \to P_k \wedge bo$ sends many regular homotopy classes nontrivially.

Our most useful tool is the main result of **[3]**, which effectively says that the *bo*-primary obstruction for the *p*-fold Whitney sum pH_k of the Hopf bundle over quaternionic projective space QP^k in degree $4i-1$ is the class of height $\nu\binom{p}{i}$ if it exists. These numbers occur because they measure divisibility of symplectic Pontryagin classes.

A simple example of a geometric dimension (gd) result which is immediately implied by this is

PROPOSITION. *If* $i \equiv 3\ (4)$, $\nu\binom{p}{i} \geqq 1$, $\nu\binom{p}{i+1} \geqq 4$, $\nu\binom{p}{i+2} \geqq 5$, *and* $\nu\binom{p}{i+3} \geqq 8$, *then* $\mathrm{gd}(4p\xi_{4i+15}) \leqq 4i-1$.

This result is not very interesting or useful, but it illustrates very simply our method. The map $RP \to^{4p\xi} BO$ is factored through QP^{i+3}, and this map lifts by the result of [3] since the above chart shows that $\pi_{4j-1}(P_{4i-1})$ is *bo*-primary for $j \leqq i+3$.

$$\begin{array}{ccccc} & & & & BO_{4i-1} \\ & & & \nearrow & \downarrow \\ RP^{4i+15} & \longrightarrow & QP^{i+3} & \xrightarrow{pH} & BO \end{array}$$

Finally, I sketch the proof of Theorem 2′. The map $P^n \to^{(L-n-1)\xi} BO$ lifts to $BO[8]$ since $KO(RP^7) \approx Z_8$. Let $\alpha = \alpha(n)$. We show the composite

$$P^{n-2^{\alpha-3}+11} \longrightarrow P^n \xrightarrow{(L-n-1)\xi} BO[8] \longrightarrow C = BO[8]/BO_{n-2^{\alpha-3}-4}[8]$$

is essential. The Adams spectral sequence for $\pi_*(C)$ is closely related to that of $\pi_*(\Sigma P_{n-2^{\alpha-3}-4})$.

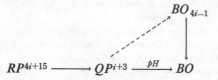

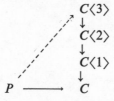

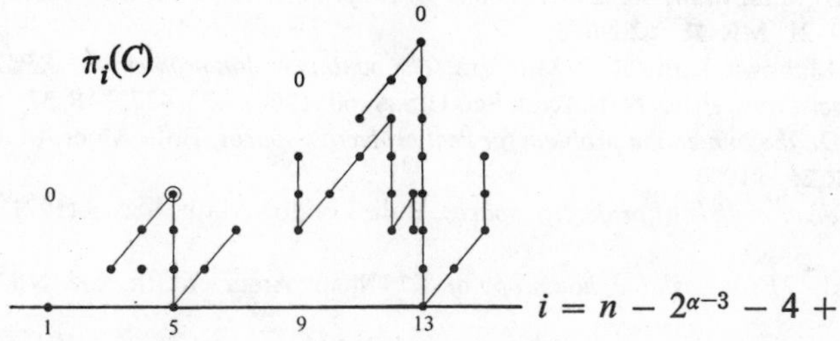

The *bo*-primary obstructions for the corresponding bundle over QP are as indicated. This is where the assumptions on $\alpha(n)$ and $\nu(n+1)$ are used. This enables us to prove that the map $P \to C$ has filtration 3, i.e., there exists a lifting

$$\begin{array}{ccc} & & C\langle 3\rangle \\ & \nearrow & \downarrow \\ & & C\langle 2\rangle \\ & & \downarrow \\ & & C\langle 1\rangle \\ & & \downarrow \\ P & \longrightarrow & C \end{array}$$

where the tower is an Adams resolution of C. The cohomology generators of $C\langle 3\rangle$ correspond to the dots at height 3 in the chart. Our lifting sends only $k_{n-2^{\alpha-3}+1}$ nontrivially. Finally by computing the Adams spectral sequence converging to $[P^{n-2^{\alpha-3}+11}, C]$ we show any such map is nontrivial.

Obviously, such detailed computations cannot be carried much farther. To obtain better collections of nonimmersion results, we feel that a different filtration (such as *bo* or *BP*) must be placed on $\pi_*(P_N)$ so that the obstructions have lower filtration and there are fewer differentials to consider. Our immersion results may possibly be proved for all $\alpha(n)$ if we can iron out some technical difficulties in a certain induction argument. This would prove that the known nonimmersions of CP^n are within 1 of best possible for all n. But so far I guess the most significant piece of information to come out of our work is that the immersion question for RP^n is very difficult.

References

1. J. Adem and S. Gitler, *Non-immersion theorems for real projective spaces*, Bol. Soc. Mat. Mexicana (2) **9** (1964), 37–50. MR **32** #461.

2. D. Davis, *Connective coverings of BO and immersions of projective spaces* (to appear).

3. D. Davis and M. Mahowald, *The geometric dimension of some vector bundles over projective spaces*, Trans. Amer. Math. Soc. **205** (1975), 295–316. MR **51** #9058.

4. ———, *A strong non-immersion theorem for* RP^{8l+7}, Bull. Amer. Math. Soc. **81** (1975), 155–156.

5. ———, *Immersions of complex projective spaces and the generalized vector field problem*, Proc. London Math. Soc. **34** (1977).

6. ———, *Obstruction theory and ko-theory*, Lecture Notes in Math.

7. ———, *The immersion conjecture for* RP^{8l+7} *is false*, Trans. Amer. Math. Soc. **236** (1978), 361–383.

8. S. Gitler, *Immersion and embedding of manifolds*, Proc. Sympos. Pure Math. Vol. 22, Amer. Math. Soc., Providence, R. J., 1971, pp. 87–96. MR **47** #4275.

9. S. Gitler and M. Mahowald, *Some immersions of real projective spaces*, Bol. Soc. Mat. Mexicana (2) **14** (1969), 9–21. MR **41** #2696.

10. S. Gitler, M. Mahowald, and R. J. Milgram, *The nonimmersion problem for* RP^n *and higher order cohomology operations*, Proc. Nat. Acad. Sci. U.S.A. **60** (1968), 432–437. MR **37** #3581.

11. I. M. James, *On the immersion problem for real projective spaces*, Bull. Amer. Math. Soc. **69** (1963), 231–238, MR **26** #1900.

12. ———, *Euclidean models of projective spaces*, Bull. London Math. Soc. **3** (1971), 257–276. MR **45** #7729.

13. M. Mahowald, *The metastable homotopy of* S^n, Mem. Amer. Math. Soc. No. **72** (1967). MR **38** #5216.

14. K. H. Mayer, *Elliptische Differentialoperatoren und Ganzahligkeitssätze für Charakteristische Zahlen*, Topology **4** (1965), 295–313. MR **33** #6650.

15. B. J. Sanderson, *Immersions and embeddings of projective spaces*, Proc. London Math. Soc. (3) **14** (1964), 137–153. MR **29** #2814.

16. B. J. Sanderson and R. L. E. Schwarzenberger, *Non-immersion theorems for differentiable manifolds*, Proc. Cambridge Philos. Soc. **59** (1963), 319–322. MR **26** #5589.

LEHIGH UNIVERSITY

NORTHWESTERN UNIVERSITY

Proceedings of Symposia in Pure Mathematics
Volume 32, 1978

SINGULARITIES IN LENS SPACES

WILLIAM D. HOMER

Let $f\colon M^n \to W^{n+2}$ be a PL embedding of PL manifolds. Then f is said to be locally flat at a point x in M if there is a neighborhood N of $f(x)$ in W and a PL isomorphism $h\colon N \to D^{n+2}$ such that $h(N \cap f(M)) = D^{n+2} \cap \boldsymbol{R}^n = D^n$. Here D^{n+2} is the unit disc in euclidean space $\boldsymbol{R}^{n+2}$ and $\boldsymbol{R}^n$ is regarded as a subset of $\boldsymbol{R}^{n+2}$ in the usual way. A point at which f fails to be locally flat is called singular, and, under appropriate triangulations **[St]**, the set of such points forms a subcomplex of M, which is denoted $S(f)$. For example, if $g\colon S^{n-1} \to S^{n+1}$ is a nontrivial knot and $f = Cg\colon D^n \to D^{n+2}$ is the cone on g, then the cone point 0 in D^n is a singular point for f. Note that PL embeddings $M^n \to W^{n+k}$ with $k > 2$ are always locally flat **[Hu]**. In what follows, all embeddings are understood to be PL and to have PL manifolds as domain and range.

The maximum dimension of the singular set of any embedding is easily seen to be $n - 2$, where n is the dimension of the domain manifold. Further, suppose f is a given embedding with singular dimension k (i.e., k is the dimension of the singular subcomplex of the domain of f). Then an embedding with singular dimension equal to any integer between k and $n - 2$, inclusive, can be obtained by "sewing" nontrivial knots into the original embedding. It can be arranged that the modified embeddings are pseudoisotopic to the given embedding **[CS3]**. Thus it is natural to ask, for a manifold M^n which is known to embed in another manifold W^{n+2}, what is the minimum of the singular dimensions of all possible embeddings of M in W?

Let $N \geqq 2$ be an arbitrary integer which is fixed for this discussion. A "classical lens space" will mean the orbit space of a linear action of the finite cyclic group $\boldsymbol{Z}/N$ on an odd dimensional sphere. If M^{2n-1} and W^{2n+1} have the homotopy type of classical lens spaces there is an embedding of M in W **[CS3]** but the obstruction to the existence of a locally flat embedding in most cases is nontrivial **[CS2]**. Hence

AMS (MOS) subject classifications (1970). Primary 57C35; Secondary 57E25, 57E30, 57D65.

homotopy lens spaces form a natural class of manifolds in which to study the above question of the minimum possible dimension of the singular set. The techniques employed require a modification of the question, in that attention is restricted to those embeddings which are induced from embeddings of the covering spheres (i.e., which induce isomorphisms on fundamental groups). In some cases this is no restriction, as can be seen by a characteristic class argument (e.g., see **[CS1]**). For such embeddings we have the following calculation **[H]**, which provides an answer in terms of computable invariants.

THEOREM. *Let M^{2n-1} and W^{2n+1} be two* PL *manifolds having the homotopy type of classical lens spaces, with $n \geqq 2$. Then there are reduced splitting invariants $\{r_{2k}: 1 \leqq k \leqq n-1\}$ and $\{t_{2k}: 1 \leqq k \leqq n-1\}$ associated to the pair of manifolds M, W, such that*

(i) *if $r_{2k} = t_{2k}$ for all k, then there exists an embedding $i: M \to W$ which is locally flat, and*

(ii) *otherwise,*

$$\min\{\dim S(i)\} = (2n-1) - \min\{2k: r_{2k} \neq t_{2k}\}$$

where i ranges over all embeddings of M in W which are induced by equivariant embeddings of the covering spheres.

The invariants are elements of $\mathbf{Z}/2$, $\mathbf{Z}/N$ or $\mathbf{Z}/2N$, depending on k. They are associated to the manifolds M and W as follows. Let $f: M \to L_1$ be a homotopy equivalence of M to a classical lens space. That there is a classical lens space L_2 in the homotopy class of W which contains L_1 as the $(2n-1)$-skeleton follows from the homotopy classification of classical lens spaces **[O]**, **[C]**. Let $g: W \to L_2$ be a homotopy equivalence.

Suppose that L_1 is in fact the standard lens space, which is the orbit space of $S^{2n-1} \subseteq \mathbf{C}^n$ under the action of $\mathbf{Z}/N$ given by multiplication of each coordinate by the Nth root of unity $\exp(2\pi i/N)$. In this case, observe that the action of $\mathbf{Z}/N$ is the restriction of the action of $S^1 \subseteq \mathbf{C}$ on S^{2n-1} given by multiplication, so that there is an S^1-fibration $\pi: L_1 \to CP(n-1)$, the complex projective space of (complex) dimension $n-1$. The homotopy equivalence f defines a normal invariant $\eta(f): L_1 \to G/\mathrm{PL}$, and a straightforward obstruction theory argument **[W]** shows that $\eta(f)$ factors through π, up to homotopy:

$$\begin{array}{ccc} L_1 & \xrightarrow{\eta(f)} & G/\mathrm{PL} \\ {\scriptstyle\pi}\downarrow & \nearrow_{\eta} & \\ CP(n-1) & & \end{array}$$

Let η be one such factorization. Then we may associate to η the usual splitting invariants: $s_{2k} = \theta(\eta \mid CP(k))$, the simply-connected surgery theory obstruction of the normal invariant obtained by restricting η to the $2k$-skeleton. Hence $s_{4k} \in \mathbf{Z}$ and $s_{4k+2} \in \mathbf{Z}/2$.

The invariants $\{s_{2k}\}$ depend upon the choice of η, which in general is not unique. This indeterminacy can be eliminated by the following reduction process. Put $r_{4k+2} = s_{4k+2}$ for all k. Write $N = 2^e M$ where $M = 2q + 1$.

(1) If $q \geqq 1$, define inductively the sets $\{a_k \in \boldsymbol{Z}: 0 \leqq a_k < M\}$ and $\{b_k \in \mathbf{Z}\}$ by the formula

$$a_k + b_k M = s_{4k} - \sum_{i=1}^{\min\{q, k-1\}} \binom{M}{2i+1} b_{k-i}.$$

Thus, at step k, the expression on the right involves only previously defined integers and b_k and a_k, $0 \leqq a_k < M$, are the quotient and remainder, respectively, upon division of this expression by M.

(2) If $q = 0$, define $\{a_k\}$ all to be zero.

(3) If $e \geqq 1$, define $r_4 \in \boldsymbol{Z}/2N$ to be the unique integer modulo $2N$ such that

$$r_4 \equiv s_4 \bmod 2^{e+1}, \qquad r_4 \equiv a_1 \bmod M.$$

Similarly, define $r_{4k} \in \boldsymbol{Z}/N$, for $2 \leqq k$ and for $k = 1$ if $e = 0$, such that

$$r_{4k} \equiv s_{4k} \bmod 2^e, \qquad r_{4k} \equiv a_k \bmod M.$$

The invariants $\{t_{2k}\}$ are defined in the same way using $g|(g^{-1}L_1)$ in place of f. Here $g^{-1}L_1$ is the transverse inverse image of L_1 under g, so the restriction of g to it is a normal map and has an associated normal invariant. Finally, if L_1 is not the standard lens space, then the above definition is applied to the images of $\eta(f)$ and $\eta(g|(g^{-1}L_1))$ under the bijection of [W] from $[L_1, G/\mathrm{PL}]$ to the normal invariants of the standard lens space.

The connection between these invariants and the original problem comes from a general result of [CS3]. If $\theta\colon M^{2n-1} \to W^{2n+1}$ is a Poincaré embedding, then there is a PL embedding which realizes θ and the minimum singular dimension among all such PL embeddings is $2n - 1$ minus the filtration degree of a normal invariant, denoted $\eta(\theta) \in [M, G/\mathrm{PL}]$, associated to the Poincaré embedding θ. Here the filtration on $[M, G/\mathrm{PL}]$ consists of the kernels of the maps induced by inclusions of the k-skeleta $M^{(k)} \to M$.

When M and W have the homotopy type of classical lens spaces and f and g are homotopy equivalences chosen as above, then $M \xrightarrow{f} L_1 \subseteq L_2 \xleftarrow{g} W$ determines a Poincaré embedding θ of M in W, which is realized by any embedding of M in W (induced by a map of spheres). The normal invariant of θ can be expressed in terms of those of f and g:

$$\eta(\theta) = f^*(\eta(g|g^{-1}L_1) - \eta(f)).$$

It remains to show, using results of [W], that the reduced splitting invariants of the theorem classify the normal invariants $\eta(f)$ and $\eta(g|(g^{-1}L_1))$, and that the filtration degree of their Whitney difference can be read off as the first subscript for which the two lists of reduced splitting invariants disagree. Details may be found in [H].

Note that there are other ways to reduce the splitting invariants $\{s_{2k}\}$ above so that the resulting list classifies the normal invariant $\eta(f)$ (or $\eta(g|g^{-1}L_1)$). The method given has the advantage of allowing the filtration degree to be read off in the simple manner described. It also reflects the difference in the structure of the group of normal invariants localized at and away from the prime 2. The $\{s_{2k}\}$ satisfy the relation $s_2 \equiv s_4 \bmod 2$, which the $\{r_{2k}\}$ and $\{t_{2k}\}$ inherit.

For a given homotopy equivalence $g\colon W \to L_2$, it is possible to find a homotopy equivalence $f: M \to L_1 = L_2^{(2n-1)}$ such that $\eta(f) = \eta(g|g^{-1}L_1)$ when (a) one of n, N is odd or (b) n, N are even and $t_{2n-2} = 0$ **[W]**. Combining this with part (i) of the theorem yields the results of **[CS2]** on locally flat embeddings of homotopy lens spaces. Combining the same fact with part (ii) of the theorem yields examples of embeddings with irreducibly high dimensional singular sets: Let W^{2n+1} be a given homotopy lens space and k be an integer satisfying $0 \leqq k < 2n - 1$ and $2n - 1 - k \equiv 0 \bmod 2$ (mod 4 if N is odd). Then there is a homotopy lens space M^{2n-1} for which there is a PL embedding i of M in W (induced by an embedding of spheres) with $\dim S(i) = k$ but no such embedding with strictly smaller dimensional singular set. The case $N = 2$ and $k = 2n - 3$ of this result was given in **[CS1]**.

Analogs of the theorem and of the corollaries just stated hold for real and complex projective spaces. The statements in these cases are simpler since the usual splitting invariants will serve for complex projective spaces, and the reduction process described above is quite simple for real projective spaces, where $N = 2$.

References

[C] M. Cohen, *A course in simple-homotopy theory*, Springer-Verlag, Berlin and New York, 1973. MR **50** #14762.

[CS1] S. Cappell and J. Shaneson, *Nonlocally flat embeddings, smoothings, and group actions*, Bull. Amer. Math. Soc. **79** (1973), 577–582. MR **47** #9635.

[CS2] ———, *The codimension two placement problem and homology equivalent manifolds*, Ann. of Math. (2) **99** (1974), 277–384. MR **49** #3978.

[CS3] ———, *Piecewise linear embeddings and their singularities*, Ann. of Math. (2) **103** (1976), 163–228.

[H] W. Homer, *Equivariant PL embeddings of spheres*, Ph.D. thesis, Princeton Univ., Princeton, N. J., 1975.

[Hu] J. F. P. Hudson, *Piecewise linear topology*, Benjamin, New York, 1969. MR **40** #2094.

[O] P. Olum, *Mappings of manifolds and the notion of degree*, Ann. of Math. (2) **58** (1953), 458–480. MR **15**, 338.

[St] D. Stone, *Stratified polyhedra*, Lecture Notes in Math., vol. 252, Springer-Verlag, Berlin and New York, 1972.

[W] C. T. C. Wall, *Surgery on compact manifolds*, Academic Press, New York, 1970.

Ohio State University

Proceedings of Symposia in Pure Mathematics
Volume 32, 1978

CHARACTERISTIC CLASSES AND KOSZUL COMPLEXES

FRANZ W. KAMBER[1,2] AND PHILIPPE TONDEUR[2]

1. Introduction. The characteristic classes of the title are the (real) generalized characteristic classes of foliated bundles as in [7], [8]. We recall that this characteristic class construction generalizes the Chern-Weil construction for ordinary bundles, which corresponds to the situation of a bundle trivially foliated by points. This construction provides characteristic invariants for a great variety of geometric settings, for which we refer to [8].

The universal characteristic classes appearing in this framework arise from the cohomology of the truncated relative Weil algebra $W(\mathfrak{g}, \mathfrak{h})_q$, defined for a pair $(\mathfrak{g}, \mathfrak{h})$ of Lie algebras and a nonnegative integer q, corresponding to the codimension of the underlying foliation. The present purely algebraic paper is concerned with the structure of the cohomology of differential graded algebras of similar type. These are the $\mathfrak{g}$-*DG*-algebras of §2, where the standard terminology on this subject is recapitulated. The structure theorems in §3 lead then, in particular, to the computation of $H(W(\mathfrak{g}, \mathfrak{h})_q)$, the main result of the earlier note [6]. See also Godbillon [5] for the case $(\mathfrak{gl}(q), \mathfrak{so}(q))$. This computation is presented in §4, together with other applications of these methods. As the reader will note, the algebraic context of the present paper is the same as in the early papers of Cartan [3] and Koszul [9].

2. $\mathfrak{g}$-*DG*-algebras. A *DG*-algebra $E^{\cdot}$ is a $\mathfrak{g}$-*DG*-algebra over a Lie algebra $\mathfrak{g}$ [8, p. 46], if it is equipped with derivations $i(x)\colon E^{\cdot} \to E^{\cdot -1}$ of degree -1 and derivations $\theta(x)\colon E^{\cdot} \to E^{\cdot}$ of degree 0 for every $x \in \mathfrak{g}$, such that the following conditions hold:

AMS (MOS) subject classifications (1970). Primary 57D20; Secondary 57D30.

Key words and phrases. Characteristic classes, foliations, Weil algebra, truncated Weil algebra, Koszul complex, cohomology.

[1]Talk given by the first author.

[2]This work was partially supported by a grant from the National Science Foundation.

$i(x)^2 = 0$ for all $x \in \mathfrak{g}$;
$\theta[x, y] = [\theta(x), \theta(y)]$ for all $x, y \in \mathfrak{g}$;
$[\theta(x), i(y)] = i[x, y]$ for all $x, y \in \mathfrak{g}$;
$\theta(x) = i(x)d + di(x)$ for all $x \in \mathfrak{g}$.

We consider pairs $(\mathfrak{g}, \mathfrak{h})$ of Lie algebras with $\mathfrak{h} \subset \mathfrak{g}$. The elements

$$E_{\mathfrak{h}} = \{a \in E \mid \theta(x)a = 0,\ i(x)a = 0 \text{ for } x \in \mathfrak{h}\}$$

form the $\mathfrak{h}$-basic elements of E. This defines a subcomplex of E, and we wish to describe the structure of the cohomology $H(E_{\mathfrak{h}})$. For the Weil algebra $E = W(\mathfrak{g})$, in particular, $W(\mathfrak{g})_{\mathfrak{h}} \equiv W(\mathfrak{g}, \mathfrak{h})$. More generally we have for $E = W(\mathfrak{g})_q$ the truncated Weil algebra $W(\mathfrak{g})_{\mathfrak{h}} = W(\mathfrak{g}, \mathfrak{h})_q$ [**8**, p. 67].

We make throughout the following technical assumptions. E is assumed to be semisimple as a $\mathfrak{g}$-module, with finite-dimensional simple components. The pair $(\mathfrak{g}, \mathfrak{h})$ is assumed to be a reductive pair of Lie algebras. We choose an $\mathfrak{h}$-equivariant retraction $\theta\colon \mathfrak{g} \to \mathfrak{h}$ of the inclusion $i\colon \mathfrak{h} \subset \mathfrak{g}$. This insures that each $\mathfrak{g}$-connection in E (i.e. homomorphism $W(\mathfrak{g}) \xrightarrow{k} E$, see [**8**, p. 97]) canonically gives rise to an $\mathfrak{h}$-connection in E, namely the composition $W(\mathfrak{h}) \xrightarrow{k(\theta)} W(\mathfrak{g}) \xrightarrow{k} E$ of k with the $\mathfrak{h}$-connection $k(\theta)$ determined by θ.

We make now use of the primitive subspace $P_{\mathfrak{g}} \subset (\bigwedge \mathfrak{g}^*)^{\mathfrak{g}} \cong H(\mathfrak{g})$ and a transgression $\tau_{\mathfrak{g}}\colon P_{\mathfrak{g}} \to I(\mathfrak{g})$ into the invariant polynomials on $\mathfrak{g}$. For a $\mathfrak{g}$-*DG*-algebra E with connection we define a graded algebra

$$A(E, \mathfrak{h}) = \bigwedge P_{\mathfrak{g}} \otimes I(\mathfrak{h}) \otimes E_{\mathfrak{g}}$$

and a differential d_A, a derivation of degree 1, which is 0 on $I(\mathfrak{h})$, equal to $d_E \mid E_{\mathfrak{g}}$ on $E_{\mathfrak{g}}$, and on $\bigwedge P_{\mathfrak{g}}$ uniquely characterized by the formula

$$d_A(y) = 1 \otimes 1 \otimes h(\tau_{\mathfrak{g}}(y)) - 1 \otimes i^*(\tau_{\mathfrak{g}}(y)) \otimes 1$$

for $y \in P_{\mathfrak{g}}$. Here $h\colon I(\mathfrak{g}) \to E_{\mathfrak{g}}$ denotes the restriction of the $\mathfrak{g}$-connection $W(\mathfrak{g}) \to E$ to the $\mathfrak{g}$-basic elements $W(\mathfrak{g}, \mathfrak{g}) = I(\mathfrak{g})$, and $i^*\colon I(\mathfrak{g}) \to I(\mathfrak{h})$ the restriction map on invariant polynomials. This construction of $A(E, \mathfrak{h})$ is functorial. Moreover there is a functorial *DG*-homomorphism $\varphi(E, \mathfrak{h})\colon A(E, \mathfrak{h}) \to E_{\mathfrak{h}}$ inducing an isomorphism

$$\varphi(E, \mathfrak{h})_*\colon H(A(E, \mathfrak{h})) \xrightarrow{\cong} H(E_{\mathfrak{h}}) \tag{2.1}$$

(see [**6**, Proposition 1] and [**8**, Theorem 5.82]).

In the course of this paper we shall further need the Samelson space $\hat{P} = \hat{P}(\mathfrak{g}, \mathfrak{h}) \subset P_{\mathfrak{g}}$ of the pair $(\mathfrak{g}, \mathfrak{h})$. For any $\mathfrak{g}$-*DG*-algebra E with connection we define then a graded algebra

$$\hat{A}(E) = \bigwedge \hat{P} \otimes E_{\mathfrak{g}} \tag{2.2}$$

with differential $d_{\hat{A}}$ equal to $d_E \mid E_{\mathfrak{g}}$ on $E_{\mathfrak{g}}$, and on $\bigwedge \hat{P}$ uniquely characterized by the formula

$$d_{\hat{A}}(y) = 1 \otimes h(\tau_{\mathfrak{g}}(y))$$

for $y \in \hat{P}$. The deficiency

$$d = \operatorname{def}(\mathfrak{g}, \mathfrak{h}) = \operatorname{rank} \mathfrak{g} - \operatorname{rank} \mathfrak{h} - \dim \hat{P} \tag{2.3}$$

is an integer $\geqq 0$. A (C)-pair (Cartan pair) is a pair $(\mathfrak{g}, \mathfrak{h})$ with zero deficiency. In [6] we introduced the stronger notion of a (CS)-pair $(\mathfrak{g}, \mathfrak{h})$ (Special Cartan pair), for which there exists a transgression $\tau_\mathfrak{g}$ such that

$$\ker(I(\mathfrak{g}) \xrightarrow{i^*} I(\mathfrak{h})) = \mathrm{Id}(\tau_\mathfrak{g}\hat{P}) \subset I(\mathfrak{g}). \tag{2.4}$$

We will further need the choice of a Samelson complement $\tilde{P}$ of the Samelson space $\hat{P}$ in $P_\mathfrak{g}$, i.e. $P_\mathfrak{g} = \hat{P} \oplus \tilde{P}$. Under a transgression $\tau_\mathfrak{g}$ the image in $I(\mathfrak{g})$ of $\hat{P}$ is denoted $\hat{V}$, and the image of $\tilde{P}$ is denoted $\tilde{V}$. For $V = \hat{V} \oplus \tilde{V}$ we have then $I(\mathfrak{g}) \cong S(V)$ (symmetric algebra over V).

3. Structure theorems for $H(E_\mathfrak{h})$. The results of this section provide a method for the computation of the cohomology of the $\mathfrak{h}$-basic elements $E_\mathfrak{h}$ in a $\mathfrak{g}$-DG-algebra E. In all these statements we assume that $(\mathfrak{g}, \mathfrak{h})$ is a reductive pair of Lie algebras, that E is equipped with a $\mathfrak{g}$-connection and is semisimple as a $\mathfrak{g}$-module, with finite-dimensional simple components.

THEOREM A. *There exists a multiplicative spectral sequence*

$$'E_2^{s,t} = \mathrm{Tor}_{I(\mathfrak{g})}^{s,t}(I^\cdot(\mathfrak{h}), H^\cdot(E_\mathfrak{g})) \Rightarrow H^n(E_\mathfrak{h}), \qquad n = s + t, \tag{3.1}$$

where $\mathrm{Tor}_{I(\mathfrak{g})}^{s,t}$ *is graded as in Baum* [1], *i.e.* $'E^{s,t} = 0$ *for* $s > 0$, *and* $t = n - s$ *is the complementary degree. Furthermore* $'E_2^{s,t} = 0$ *for* $-s > \operatorname{rank}\mathfrak{g} - \operatorname{rank}\mathfrak{h}$.

This spectral sequence is of Eilenberg-Moore type.

In order to obtain more specific information about $H(E_\mathfrak{h})$, we will have to impose some of the following conditions.

(a) There exists a transgression $\tau_\mathfrak{g}\colon P \to I(\mathfrak{g})$ such that $\tau_\mathfrak{g}\hat{P} \subset \ker(I(\mathfrak{g}) \to I(\mathfrak{h}))$.

(b) There exists a Samelson complement $\tilde{P} \subset P_\mathfrak{g}$ such that $\tau_\mathfrak{g}\tilde{P} \subset \ker(h_*\colon I(\mathfrak{g}) \to H(E_\mathfrak{g}))$, where h_* denotes the characteristic homomorphism of E ($h(\omega)$ is defined by the connection ω in E, but $h_* = (h(\omega))_*$ is independent of the particular choice of the connection).

(c) There exist a subalgebra $\mathscr{H} \subset E_\mathfrak{g}$ and a connection ω in E such that $h(\omega)\colon I(\mathfrak{g}) \to \mathscr{H}$, $d_{E_\mathfrak{g}}|\mathscr{H} = 0$ and the canonical map $\mathscr{H} \to H(E_\mathfrak{g})$ is an isomorphism.

3.2. COROLLARY. *If E satisfies condition* (c) *then the differentials in the spectral sequence* $'E$ *satisfy* $d_i = 0$ *for* $i \geqq 2$ *and*

$$H(E_\mathfrak{h}) \cong \mathrm{Tor}_{I(\mathfrak{g})}(I(\mathfrak{h}), H(E_\mathfrak{g})) \tag{3.3}$$

as a graded algebra.

Concerning the cohomology of the complex $\hat{A}(E)$ in (2.2) we have the following result.

THEOREM B. *There exists a multiplicative spectral sequence*

$$''E_2^{s,t} = \mathrm{Tor}_{I(\mathfrak{g})}^{s,t}(S(\tilde{V}), H(E_\mathfrak{g})) \Rightarrow H^n(\hat{A}(E)), \quad n = s + t, \tag{3.4}$$

where $''E_2^{s,t} = 0$ *except for* $0 \leqq -s \leqq \dim \hat{P}$. *If E satisfies condition* (c), *then the differentials in* $''E$ *satisfy* $d_i = 0$ *for* $i \geqq 2$ *and*

$$H(\hat{A}(E)) \cong \mathrm{Tor}_{I(\mathfrak{g})}(S(\tilde{V}), H(E_\mathfrak{g})) \tag{3.5}$$

as a graded algebra.

THEOREM C. *Suppose that either condition* (a), *or conditions* (b) *and* (c) *are satisfied. Then there exists a multiplicative spectral sequence*

$$(3.6)\qquad '''E_2^{s,t} = \mathrm{Tor}_{S(\bar V)}^{s,t}(I(\mathfrak{h}), H(\hat{A}(E))) \Rightarrow H^n(E_{\mathfrak{h}}), \qquad n = s + t,$$

whose nonzero terms are bounded by $0 \leqq -s \leqq d$. *Here* d *denotes the deficiency of* $(\mathfrak{g}, \mathfrak{h})$ *defined in* (2.3).

In order to obtain the spectral sequences in Theorems A, B, C, we associate a Koszul complex to the $\mathfrak{g}$-DG-algebra E and the subalgebra $\mathfrak{h} \subset \mathfrak{g}$. By (2.1) we may replace $E_{\mathfrak{h}}$ by the cohomologically equivalent complex $A(E, \mathfrak{h})$. Filtering $A(E, \mathfrak{h})$ by the decreasing sequence of differential ideals

$$(3.7)\qquad 'F^sA(E, \mathfrak{h})^n = \Big(\coprod_{p\leqq -s} \wedge_p(P) \otimes I(\mathfrak{h}) \otimes E_{\mathfrak{g}}\Big)^n,$$

we have then for the bigraded algebra associated to $'F^s$ clearly $'G^sA = 0$ for $s > 0$ and

$$'G^sA^{\cdot} \cong \wedge_{-s}(P) \otimes I(\mathfrak{h}) \otimes E_{\mathfrak{g}} \quad \text{for } s \leqq 0.$$

The algebra $'G^sA^{\cdot}$ can be identified with a Koszul complex as follows. Let $c_1, \cdots, c_r$ ($r = \mathrm{rank}\, \mathfrak{g}$) denote a set of homogeneous polynomial generators of $I(\mathfrak{g})$ forming a basis of the subspace $V = \tau_{\mathfrak{g}}P_{\mathfrak{g}}$ of $I(\mathfrak{g})$. The suspensions $x_j = \sigma(c_j)$ then form a basis of $P_{\mathfrak{g}}$ and $\tau_{\mathfrak{g}}x_j = c_j$ for $j = 1, \cdots, r$.

Let $R = I(\mathfrak{h}) \otimes I(\mathfrak{g})$ and consider the sequence of elements $a_j = 1 \otimes c_j - i^*c_j \otimes 1$ in R. As the c_j are polynomial generators, it follows that the sequence $\mathfrak{a} = (a_j)_{j=1,\cdots,r}$ is an R-sequence, i.e. a_1 is not a zero-divisor of R and a_{i+1} is not a zero-divisor in $R/R(a_1, \cdots, a_i)$ for $i = 1, \cdots, r - 1$. It follows that the Koszul complex ($s \leqq 0$)

$$(3.8)\qquad K_R^s(\mathfrak{a}) = \wedge_{-s}(P) \otimes R = \wedge_{-s}(x_1, \cdots, x_r) \otimes I(\mathfrak{h}) \otimes F[c_1, \cdots, c_r]$$

with differential $d_K\colon K_R^s(\mathfrak{a}) \to K_R^{s+1}(\mathfrak{a})$ given by

$$(3.9)\qquad \begin{aligned} &d_K(x_{i_1} \wedge \cdots \wedge x_{i_{-s}} \otimes f) \\ &\quad = \sum_{k=1}^{-s} (-1)^{k+1} x_{i_1} \wedge \cdots \wedge \widehat{x_{i_k}} \wedge \cdots \wedge x_{i_{-s}} \otimes a_{i_k} f \end{aligned}$$

is a resolution of the algebra $R/R(a_1, \cdots, a_r) \cong I(\mathfrak{h})$ (see [**4**, VIII, 4]):

$$(3.10)\qquad K_R^{\cdot}(\mathfrak{a}) \xrightarrow{\ \varepsilon\ } I(\mathfrak{h}) \to 0.$$

It then follows easily that

$$(3.11)\qquad 'G^sA(E, \mathfrak{h}) \cong K_R^s(\mathfrak{a}) \otimes_{I(\mathfrak{g})} E_{\mathfrak{g}},$$

where $E_{\mathfrak{g}}$ is considered as a DG-algebra with differential d_E. As $'d_0 = 'G(d_A)$ is induced by d_E, we have for the associated spectral sequence

$$(3.12)\qquad 'E_1^{s,t} = (K_R^s(\mathfrak{a}) \otimes_{I(\mathfrak{g})} H^{\cdot}(E_{\mathfrak{g}}))^{s+t=n}$$

and

$$(3.13) \qquad 'E_2^{s,t} = \mathrm{Tor}_{I(\mathfrak{g})}^{s,t}(I(\mathfrak{h}), H^{\cdot}(E_{\mathfrak{g}})),$$

since $'d_1 = d_K \otimes 1$. If n denotes the total degree in $A(E, \mathfrak{h})$, it follows that $'d_1$ preserves the complementary degree $t = n - s$ and thus (3.10) is a graded resolution of $I(\mathfrak{h})$ with respect to the complementary degree t. This establishes part of Theorem A.

To obtain the spectral sequence in Theorem C we consider on $A(E, \mathfrak{h})$ the filtration $'''F$ defined as follows:

$$(3.14) \qquad '''F^s A(E, \mathfrak{h})^n = \Big(\coprod_{p \leqq -s} \wedge_p(\tilde{P}) \otimes \wedge(\hat{P}) \otimes I(\mathfrak{h}) \otimes E_{\mathfrak{g}}\Big)^n,$$

which again satisfies $'''G^s(A) = 0$ for $s > 0$. In order to compute $'''G^s(A)$ for $s \leqq 0$, we define a Koszul complex as follows.

We choose the basis $c_1, \cdots, c_r$ of V compatible with the decomposition $V = \hat{V} \oplus \tilde{V}$, and we denote by $y_1, \cdots, y_{r'}$, and $z_1, \cdots, z_m$ the corresponding bases of the primitive spaces $\hat{P}$ and $\tilde{P}$, respectively, $r' = \dim \hat{P}$, $m = rk\mathfrak{g} - r'$. By definition, the Samelson space $\hat{P}$ satisfies

$$(3.15) \qquad i^*(\hat{V}) = i^*(V) \cap i^*(\tilde{V}) \cdot I(\mathfrak{h})^+,$$

and therefore we may choose fixed elements $\varphi_{j,k} \in I(\mathfrak{h})^+$, $j = 1, \cdots, r'$, $k = 1, \cdots, m$, such that

$$(3.16) \qquad i^*\tau_{\mathfrak{g}}(y_j) = \sum_{k=1}^{m} i^*\tau_{\mathfrak{g}}(z_k) \cdot \varphi_{j,k}, \qquad j = 1, \cdots, r'.$$

The sequence $\bar{\mathfrak{a}}$ of elements in R given by

$$(3.17) \qquad \bar{\mathfrak{a}}: \begin{cases} 1 \otimes \tau_{\mathfrak{g}}(z_k) - i^*\tau_{\mathfrak{g}}(z_k) \otimes 1, & k = 1, \cdots, m, \\ 1 \otimes \tau_{\mathfrak{g}}(y_j) - \sum_{k=1}^{m} \varphi_{j,k} \otimes \tau_{\mathfrak{g}}(z_k), & j = 1, \cdots, r', \end{cases}$$

is easily seen to be an R-sequence equivalent to the R-sequence $\mathfrak{a}$ considered before, i.e. $R(\mathfrak{a}) = R(\bar{\mathfrak{a}}) \subset R$. Consequently, the corresponding Koszul complexes are equivalent. An explicit multiplicative isomorphism of complexes

$$(3.18) \qquad \psi: K_R^{\cdot}(\bar{\mathfrak{a}}) \longrightarrow K_R^{\cdot}(\mathfrak{a})$$

is given by the identity on the factors $\wedge \tilde{P} \otimes I(\mathfrak{h}) \otimes I(\mathfrak{g})$ and by the mapping

$$(3.19) \qquad \psi(y_j \otimes 1 \otimes 1) = y_j \otimes 1 \otimes 1 - \sum_{k=1}^{m} z_k \otimes \varphi_{j,k} \otimes 1 \quad \text{for } y_j \in \hat{P}.$$

If $M^{\cdot}$ denotes the DG-algebra $E_{\mathfrak{g}}$ or $\mathscr{H}$ according to the cases (a) or (b), (c), we obtain

$$(3.20) \qquad K_R(\mathfrak{a}) \otimes_{I(\mathfrak{g})} M = K_{\tilde{R}}(\tilde{\mathfrak{a}}) \otimes_{S(\tilde{V})} (\wedge \hat{P} \otimes M),$$

where now $\tilde{R} = I(\mathfrak{h}) \otimes S(\tilde{V})$ and $\tilde{\mathfrak{a}}$ denotes the $\tilde{R}$-sequence in $\tilde{R}$ consisting of the first m elements in the sequence $\bar{\mathfrak{a}}$ (3.17). In either case ψ induces an isomorphism

$$(3.21) \qquad '''E_1^{s,t} = H^n('''G^s(A)) \cong (K_{\tilde{R}}^{s}(\tilde{\mathfrak{a}}) \otimes_{S(\tilde{V})} H(\hat{A}(E)))^n, \qquad n = s + t,$$

with $'''d_1$ corresponding to $d_K \otimes 1$. Thus the E_2-term in the spectral sequence associated to the filtration $'''F$ is given by

(3.22) $$'''E_2^{s,t} \cong \operatorname{Tor}^{s,t}_{S(\tilde{V})}(I(\mathfrak{h}), H(\hat{A}(E))).$$

The statements about the ranges of the spectral sequences in Theorems A, B and C follow essentially from known facts on the cohomology of commutative (graded) rings.

3.23. REMARK. Condition (a) implies that the restriction homomorphism i^* factorizes through $S(\tilde{V})$:

$$I(\mathfrak{g}) \twoheadrightarrow I(\mathfrak{g})/\mathrm{Id}(\hat{V}) \cong S(\tilde{V}) \longrightarrow I(\mathfrak{h}).$$

Under conditions (a) and (c) it follows (via Theorems A and B) that the spectral sequence in Theorem C can be naturally identified with the spectral sequence associated with the composition of functors

$$I(\mathfrak{h}) \otimes_{I(\mathfrak{g})} \quad = I(\mathfrak{h}) \otimes_{S(\tilde{V})} (S(\tilde{V}) \otimes_{I(\mathfrak{g})} \quad),$$

i.e. $'''E_2$ is of the form

$$'''E_2 = \operatorname{Tor}_{S(\tilde{V})}(I(\mathfrak{h}), \operatorname{Tor}_{I(\mathfrak{g})}(S(\tilde{V}), H(E_{\mathfrak{g}}))) \Rightarrow \operatorname{Tor}_{I(\mathfrak{g})}(I(\mathfrak{h}), H(E_{\mathfrak{g}})).$$

4. Applications.

4.1. *Theorem of Baum-Smith* [2]. We apply Corollary 3.2 to the De Rham complex $E = \Omega(P)$ of a principal G-bundle $P \to M$, with G compact connected and M compact symmetric. For a closed subgroup $H \subset G$ the De Rham complexes of M and $X = P/H$ are identified with $E_{\mathfrak{g}}$ and $E_{\mathfrak{h}}$. The isomorphism (3.3) is then the main result of Baum-Smith in [2].

4.2. *Cartan pairs.* These reductive pairs $(\mathfrak{g}, \mathfrak{h})$ are characterized by deficiency $d = 0$ [3]. If $(\mathfrak{g}, \mathfrak{h})$ also satisfies condition (a) of §3, it is a (CS)-pair as defined in (2.4). Theorem C then implies that the edge homomorphism $\bar{\beta}$ of the spectral sequence $'''E$ is an isomorphism

(4.3) $$H(E_{\mathfrak{h}}) \overset{\bar{\beta}}{\cong} H(\hat{A}(E)) \otimes_{I(\mathfrak{g})} I(\mathfrak{h}).$$

This is Theorem (5.107) in [8], where the map $\bar{\beta}$ is explicitly described.

If E satisfies condition (c) of §3, it follows from Theorem B that

(4.4) $$H(\hat{A}(E)) \cong \operatorname{Tor}_{I(\mathfrak{g})}(S(\tilde{V}), H(E_{\mathfrak{g}})),$$

and under conditions (b) and (c) it follows from Theorem C that

(4.5) $$H(E_{\mathfrak{h}}) \cong I(\mathfrak{h}) \otimes_{I(\mathfrak{g})} H(E_{\mathfrak{g}}) \oplus I(\mathfrak{h}) \otimes_{S(\tilde{V})} \left[\coprod_{-s=1}^{\dim \hat{P}} \operatorname{Tor}^{s,\cdot}_{I(\mathfrak{g})}(S(\tilde{V}), H(E_{\mathfrak{g}})) \right]$$

as graded algebras.

4.6. *Cohomology of reductive pairs of Lie algebras.* In this case we consider $E = \bigwedge \mathfrak{g}^*$. Since $E_{\mathfrak{g}} \cong F$ (the groundfield), conditions (b) and (c) of §3 are trivially satisfied. We have $\hat{A}(\bigwedge \mathfrak{g}^*) = \bigwedge \hat{P}$ with trivial differential. Theorem C therefore implies for the Chevalley-Eilenberg cohomology of the pair $(\mathfrak{g}, \mathfrak{h})$ the formula

(4.7) $$H(\mathfrak{g}, \mathfrak{h}) \cong \left[I(\mathfrak{h})/I(\mathfrak{g})^+ \cdot I(\mathfrak{h}) \oplus \coprod_{-s=1}^{d} \operatorname{Tor}^{s,\cdot}_{S(\tilde{V})}(I(\mathfrak{h}), F) \right] \otimes \bigwedge \hat{P}.$$

(See Baum [1], Cartan [3], and Koszul [9].)

4.8. *Cohomology of the relative truncated Weil algebra* $W(\mathfrak{g}, \mathfrak{h})_q$. In this case we consider $E = W(\mathfrak{g})_q$, $0 \leqq q \leqq \infty$, the truncated Weil algebra (see Definition 4.38 in [**8**]). We use Theorems B and C to compute the cohomology of the algebra $W(\mathfrak{g}\ , \mathfrak{h})_q$ of its $\mathfrak{h}$-basic elements. In this case we have

$$\hat{A}_q \equiv \hat{A}(W(\mathfrak{g})_q) = \wedge \hat{P} \otimes I(\mathfrak{g})_q. \tag{4.9}$$

As $I(\mathfrak{g})_q$ has zero differential, condition (c) is satisfied and we have from Theorem B an isomorphism of graded algebras

$$H(\hat{A}_q) \cong \mathrm{Tor}_{I(\mathfrak{g})}(S(\tilde{V}), I(\mathfrak{g})_q). \tag{4.10}$$

To compute this algebra, we proceed as follows. Choose the generators c_j ($j = 1, \cdots, r = \operatorname{rank} \mathfrak{g}$) of $I(\mathfrak{g})$ and a homogeneous basis y_i of $\hat{P}$ such that $\deg c_i \leqq \deg c_j$ for $i \leqq j$ and $\tau_{\mathfrak{g}} y_i = c_{\alpha_i}$ for $i = 1, \cdots, r' = \dim \hat{P}$ ($\alpha_1 \leqq \cdots \leqq \alpha_{r'}$). Filter the Koszul complex

$$\hat{A}_q = K_{I(\mathfrak{g})}(c_{\alpha_1}, \cdots, c_{\alpha_{r'}}) \otimes_{I(\mathfrak{g})} I(\mathfrak{g})_q$$

by

$$X_{(k)} = \wedge(y_1, \cdots, y_k) \otimes F[c_1, \cdots, c_r]_q. \tag{4.11}$$

We have $X_{(0)} = F[c_1, \cdots, c_r]_q$, $X_{(k)} \subset X_{(k+1)}$ and $X_{(r')} = \hat{A}_q$. Denote by

$$Z^s_{(k)} \subset X^s_{(k)} = \wedge_{-s}(y_1, \cdots, y_k) \otimes F[c_1, \cdots, c_t]_q \tag{4.12}$$

the subspace spanned by the standard monomial cocycles

$$z_{(i,j)} = y_{i_1} \wedge \cdots \wedge y_{i_{-s}} \otimes c_1^{j_1} \cdots c_r^{j_r}$$

satisfying the conditions (a), (b) and (c) of [**8**, 5.110]. Clearly $Z^s_{(k)} \neq 0$ only for $0 \leqq -s \leqq k$. There is a commutative diagram of exact sequences

$$\begin{array}{ccccccccc}
\cdots \to & H^s(X^{\cdot}_{(k-1)}) & \xrightarrow{\cdot c_{\alpha_k *}} & H^s(X^{\cdot}_{(k-1)}) & \xrightarrow{i_*} & H^s(X^{\cdot}_{(k)}) & \xrightarrow{(j_k)_*} & H^{s+1}(X_{(k-1)}) & \to \cdots \\
& \uparrow & & \uparrow & & \uparrow & & \uparrow & \\
\cdots \longrightarrow & Z^s_{(k-1)} & \xrightarrow{\cdot c_{\alpha_k}} & Z^s_{(k-1)} & \xrightarrow{i} & Z^s_{(k)} & \xrightarrow{j_k} & Z^{s+1}_{(k-1)} & \to \cdots
\end{array} \tag{4.13}$$

where $\cdot c_{\alpha_k}$ denotes multiplication with c_{α_k}. The map i_* and the vertical maps are induced by inclusions. $(j_k)_*$ is induced by the map defined by

$$j_k(y_{i_1} \wedge \cdots \wedge y_{i_{-(s+1)}} \wedge y_k \otimes \Phi) = y_{i_1} \wedge \cdots \wedge y_{i_{-(s+1)}} \otimes \Phi$$

on monomials containing a factor y_k, and zero elsewhere. Clearly

$$Z^0_{(k)} \cong H^0(X_{(k)}) \cong F[c_1, \cdots, c_r]_q/(c_{\alpha_1}, \cdots, c_{\alpha_k}).$$

By induction over a lexicographical ordering of the pairs (s, k), $0 \leqq -s \leqq k$, and the 5-lemma, we conclude that

$$Z^s_{(k)} \cong H^s(X^s_{(k)}) \quad \text{for all } 0 \leqq -s \leqq k \leqq r'.$$

This proves Theorem 5.110 in [8] and the results in [6], namely the algebra $Z_q = \coprod_{s \leqq 0} Z^s$ (with trivial multiplication for factors $z \in \wedge_{-s}(y_1, \cdots, y_{r'}) \otimes \mathrm{Id}(c_{\alpha_1}, \cdots, c_{\alpha_{r'}}) \subset Z^s$, $s < 0$) of monomial standard cocycles in $\hat{A}_q$ is isomorphic to $H(\hat{A}_q)$, i.e.

$$H(\hat{A}_q) \cong \mathrm{Tor}_{I(\mathfrak{g})}(S(\tilde{V}), I(\mathfrak{g})_q) \cong Z_q. \tag{4.14}$$

If the pair $(\mathfrak{g}, \mathfrak{h})$ now satisfies condition (a) of §3, we can invoke Theorem C. Together with the result that Z_q is a subalgebra of cocycles in $\hat{A}_q$ realizing the cohomology $H(\hat{A}_q)$, we obtain the following result.

4.15. THEOREM. *Let* $(\mathfrak{g}, \mathfrak{h})$ *be a reductive pair satisfying condition* (a). *Then there is an isomorphism of graded algebras*

$$H(W(\mathfrak{g}, \mathfrak{h})_q) \cong \mathrm{Tor}_{S(\tilde{V})}(I(\mathfrak{h}), Z_q).$$

In particular, if $(\mathfrak{g}, \mathfrak{h})$ *is a* (CS)*-pair* $(d = 0)$, *we have*

$$H(W(\mathfrak{g}, \mathfrak{h})_q) \cong I(\mathfrak{h}) \otimes_{I(\mathfrak{g})} Z_q.$$

The last equation is the result announced in [6]. For the pair $(\mathfrak{gl}(q), \mathfrak{so}(q))$ this gives the result of Vey proved in [5]. It would be interesting to find geometric realizations for the classes coming from $\mathrm{Tor}^{s,t}_{S(\tilde{V})}(I(\mathfrak{h}), Z_q)$, $s < 0$, in the case of a pair with positive deficiency.

REFERENCES

1. P. Baum, *On the cohomology of homogeneous spaces*, Topology **7** (1968), 15–38. MR **36** #2168.

2. P. Baum and L. Smith, *The real cohomology of differentiable fibre bundles*, Comment. Math. Helv. **42** (1967), 171–179. MR **36** #4574.

3. H. Cartan, *Cohomologie réelle d'un espace fibré principal différentiable*, Séminaire Cartan, Exposés 19 et 20 (1949–50).

4. H. Cartan and S. Eilenberg, *Homological algebra*, Princeton Univ. Press, Princeton, N.J., (1956). MR **17**, 1040.

5. C. Godbillon, *Cohomologies d'algèbres de Lie de champs de vecteurs formels*, Séminaire Bourbaki, Vol. 1972/73, 25e année, Exposé No. 421, Lecture Notes in Math., Vol. 383, Springer-Verlag, Berlin and New York, 1974, pp. 69–87. MR **50** #17.

6. F. Kamber and Ph. Tondeur, *Cohomologie des algèbres de Weil relatives tronquées*, C. R. Acad. Sci. Paris Ser. A-B **276** (1973), A459–A462. MR **47** #9660.

7. ———, *Characteristic invariants of foliated bundles*, Manuscripta Math. **11** (1974), 51–89.

8. ———, *Foliated bundles and characteristic* classes, Lecture Notes in Math., vol. 493, Springer-Verlag, Berlin and New York, 1975, pp. 1–208.

9. J. L. Koszul, *Sur un type d'algèbres différentielles en rapport avec la transgression*, Colloque de Topologie, Bruxelles, 1950, pp. 73–81. MR **13**, 109.

UNIVERSITY OF ILLINOIS AT URBANA-CHAMPAIGN

Proceedings of Symposia in Pure Mathematics
Volume 32, 1978

TRANSLATIONS ON $M \times R$

WENSOR LING*

0. Introduction. New advances are being made in the study of groups of automorphisms of manifolds; clearly the homotopy properties of these groups are especially interesting (cf. [1], [13], [32], [56], [57]). These lead to many novel and interesting problems that cut across some of the richest areas of topology and certainly guarantee much exciting research. Also interesting and more related to this paper is the algebraic simplicity of certain groups of automorphisms [2], [22], [24], [45], [46], [55]. Beyond these broad outlines, details are few.

Here we concern ourselves with the algebraic structure of $A(M \times R)$, the group of orientation preserving automorphisms of $M \times R$ for M an oriented topological, piecewise linear or smooth manifold. This group warrants study because it leads to an understanding of the automorphism groups of open manifolds that possess a boundary by utilizing the boundary's collar structure. The notion of translation on $M \times R$ is defined in §2. Write $M \times R = \cdots W_{-1} \cdot W_0 \cdot W_1 \cdots$, W_i an indexed copy of W, some h-cobordism with homeomorphic ends [54], a translation is briefly a homeomorphism that shifts each W_i to W_{i+1}. Translations are generators of $A(M \times R)$. The origin of translation was laid in the work of Schreier and Ulam [51] (cf. also [23], [49]). §3 shows $A(M \times R)$ is not much different from $A(M)$ except for a torsion invariant. A geometrical interpretation of $\pi_0 A(M \times R)$ in terms of inertial h-cobordisms is given in 3.9. Investigation of the relation of conjugation between translations is taken up in §4, where we proved the main theorem (4.1): Two translations conjugate if and only if their respective isotopy classes do so in $\pi_0 A(M \times R)$. This generalizes the case of M a point [51], [23]. For the remainder of this paper, we restrict our attention to translations on $S^n \times R$ and investigate how their conjugation is related to the stable homeomorphism theorem and the annulus theorem [36].

AMS (MOS) subject classifications (1970). Primary 57D50; Secondary 57E05.

*Research for this paper was done with the partial support of Massachusetts Institute of Technology's Undergraduate Research Opportunity Program and written while the author is a National Science Foundation Graduate Fellow.

In a sequel to this paper [**43**], we apply translations to reduce the algebraic problem of enumerating normal subgroups for the group of automorphisms of open manifolds that possess a boundary to the topological one of computing π_0 and π_1 of that group and certain related automorphism groups. Thus, the problem of normal subgroups is really of a topological nature.

Most of the research for this paper was done while I was a student at M.I.T. under the consecutive guidances of Professors John W. Morgan, William P. Thurston, Gerald A. Anderson, and written under the direction of Professor Thurston at Princeton. I wish to express my deep appreciation of their patient guidance and kind encouragement. It is also a pleasure to thank Professor Allen E. Hatcher who clarified some topological properties of automorphism groups for me. Lastly, results central to [**43**], obtained as an application of ideas presented here, were obtained independently without using translations by Fr. Paul A. Schweitzer, S.J., who has also guided me and his charity and warm fellowship, spiritual as well as mathematical, I gladly thank.

1. Notations and conventions. Let A represent one of the categories Top, PL, Diff^k, $1 \leqq k \leqq +\infty$, the last one being that of smooth manifolds and C^k-maps. The automorphism groups are equipped with the compact-open topology for the categories Top and PL, the C^k-topology for Diff^k. We recall the well-known fact that Diff$^j(V)$ is homotopy equivalent to Diff$^k(V)$ for $1 \leqq j, k \leqq +\infty$ and V any smooth manifold. We have the following useful subgroups of $A(M \times R)$: (i) $e_{M\times R}$, the identity; (ii) $A(M \times R, C) = \{h\colon h|_C = e\}$; (iii) $A_0(M \times R)$, the identity component of $A(M \times R)$; (iv) $A_K(M \times R) = \{h\colon \text{supp}(h) \text{ compact}\}$, where $\text{supp}(h) = \text{cl}\{x \in M \times R\colon hx \neq x\}$; (v) $A_i(M \times R) = \{h\colon h|M_r^i = e \text{ for some } r \in R\}$, where $i = +, -$, $M_r^+ = M \times [r, +\infty)$, $M_r^- = M \times (-\infty, r]$ and $M_r = M \times r$. We often drop the manifold $M \times R$ from our notation for convenience, e.g., A means $A(M \times R)$ and there should be no confusion with the category A from the context.

The automorphism $\boldsymbol{t}_r \in A(M \times R)$ is defined by $\boldsymbol{t}_r(x, s) = (x, s + r)$. Lastly, for $f, g \in A(R)$, we use $f > g$ to mean $f(s) > g(s)$ for all $s \in R$.

2. Translations as generators.

DEFINITION. Let $\mathcal{T} = \{T_i\colon i \in Z\}$ be a subcollection of $\{h(M_0^-)\colon h \in A\}$; then $\mathcal{T}$ is a *translation structure* on $M \times R$ provided

(a) $T_i \subset \text{int } T_{i+1}$ for each $i \in Z$,

(b) $\bigcap \mathcal{T} = \varnothing$,

(c) $\bigcup \mathcal{T} = M \times R$.

DEFINITION. A homeomorphism $t \in A$ is an *A-translation* (or simply *translation*) on $M \times R$ if there is a translation structure $\mathcal{T}$ on $M \times R$ satisfying $t(T_i) = T_{i+1}$ for each $i \in Z$, t is said to be *carried by* $\mathcal{T}$.

There are plenty of translation structures and translations.

EXAMPLE. The *canonical translation* on $M \times R$, $t = e_M \times \boldsymbol{t}_1$, is carried by the translation structure $\{M_n^-\colon n \in Z\}$. It is also carried by $\{M_{n+1/2}\colon n \in Z\}$.

Note that in our terminology, subtraction by 1 is not a translation on $* \times R$, but is rather the inverse of a translation. We proceed to show in 2.4 that translations generate A. For convenience, let $H(f) = e_M \times f$ for $f \in A(R)$, $F_+h(r) = \max\{\pi_R h(x)\colon x \in M_r\}$ and $F_-h(r) = \min\{\pi_R h(x)\colon x \in M_r\}$ for $h \in A$ and $r \in R$.

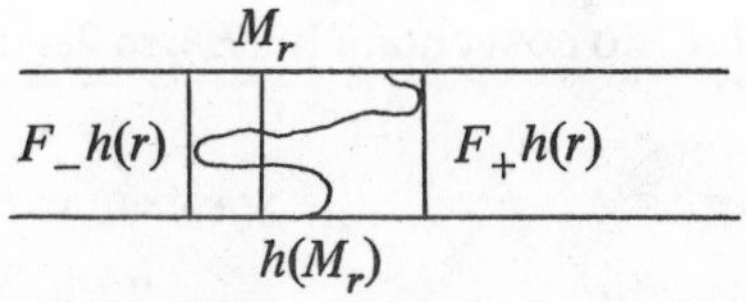

Note that H and F_i are well-defined maps. We prove that for F_i by noting

LEMMA 2.1. (i) $F_i \circ H = e_{\mathrm{Top}(R)}$,
(ii) $F_i(Hf \circ h) = f \circ F_ih$ *and* $F_i(h \circ Hf) = F_ih \circ f$ *for* $f \in \mathrm{Top}(R)$ *and* $h \in A$,
(iii) $F_+(h \circ g) \leqq F_+ h \circ F_+g$ *and* $F_-(h \circ g) \geqq F_-h \circ F_-g$ *for* $h, g \in A$,
(iv) $(F_+h)(F_-h^{-1}) = e_R$ *for* $h \in A$.

PROOF. The proof is rather straightforward, so we prove (iv) as an example. Given $r \in R$,

$$M^-_{(F_+h)(F_-h^{-1})(r)} \supseteq hM^-_{(F_-h^{-1})(r)} \subseteq (h^{-1}M^-_r) = M^-_r,$$

so $(F_+h)(F_-h^{-1})(r) \leqq r$. Conversely, by compactness of M_r, there is some $(x, r) \in M_r$ with $\pi_R h^{-1}(x, r) = F_-h^{-1}(r)$, so $h^{-1}(x, r) \in M_{F_-h^{-1}(r)}$, and

$$(F_+h)(F_-h^{-1})(r) = \max \pi_R h(M_{F_-h^{-1}(r)}) \geqq \pi_R h(h^{-1}(x, r)) = r.$$

COROLLARY 2.2. *For each* $h \in A$ *and* $i = +$ *or* $-$, F_ih *lies in* $\mathrm{Top}(R)$. *In other words,* F_i *are well defined.*

PROOF. Because M is compact, $F_ih(r)$ are well-defined real numbers. Next, the function F_+h is strictly increasing; by 2.1(iv), it is also surjective. Consequently, $F_+h \in \mathrm{Top}(R)$. Again by 2.1(iv), $F_+(h^{-1}) = (F_-h)^{-1}$; thus, $F_-h \in \mathrm{Top}(R)$ as well.

REMARK. It is not hard to show H and F_i are continuous. Therefore in the category Top, HF_i are retractions of Top onto $\mathrm{Im}(H)$ by 2.1(i).

Now, F_- furnishes a way of recognizing translations.

LEMMA 2.3. *Suppose* $h \in A$ *and* $F_-h > e_R$; *then* h *is an* A-*translation on* $M \times R$ *carried by* $\{h^i(M_0^-): i \in Z\}$.

PROOF. Let us check through properties (a) through (c) in the DEFINITION.

(a) $F_-h > e_R$ implies $\mathrm{int}\, h(M_0^-) \subset \mathrm{int}\, M^-_{F_-h(0)} \subset M_0^-$, so $\mathrm{int}\, h^{i+1}(M_0^-) \subset h^i(M_0^-)$.

(b) Observe that given $f \in \mathrm{Top}(R)$ with $f > e_R$ and $x \in R$, we must have $\lim_{n \to +\infty} f^n(x) = +\infty$, for otherwise $\lim_{n \to +\infty} f^n(x)$ is a fixed point of f, contradicting $f > e_R$. Now, 2.1(iii) yields $h^n(M_0^-) \supseteq M^-_{F_-h^n(0)} \supseteq M^-_{(F_-h)^n(0)}$; hence $\bigcup_{n \in Z} h^n(M_0^-) = M \times R$.

(c) is proved similarly.

Consider $S = \{h \in A: F_-h > e_R\}$; this collection is a set of translations which for the following reason is useful. (Compare [3].)

PROPOSITION 2.4. *If a subgroup* G *of* A *contains* $\mathrm{Im}(H)$, *then* $(G \cap S)^{-1}(G \cap S) = G$. *In particular,* $S^{-1}S = A$, *i.e., each orientation preserving homeomorphism can be factored as the product of a translation and the inverse of a translation.*

PROOF. Given $g \in G$, find $f \in A(R)$ with $f > e_R$ and $f > F_+g$, so $fF_-(g^{-1}) = f(F_+g)^{-1} > e_R$. But $F_-(Hf) = f > e_R$ and $F_-(Hf \circ g^{-1}) = fF_-(g^{-1}) > e_R$. Hence, $g = g(Hf)^{-1}(Hf) = (Hf \circ g^{-1})^{-1}(Hf) \in (G \cap S)^{-1}(G \cap S)$.

The subgroups A_+ and A_- do not contain Im(H), so 2.4 does not apply. However, analogues do hold: set

$$S_+ = \{h \in A_+ : F\_h|_{(-\infty, r)} > e_R, \text{ some } r \in R\}$$

and

$$S_- = \{h \in A_- : F\_h|_{(r, +\infty)} > e_R, \text{ some } r \in R\}.$$

The elements of S_i, though not translations, behave like it.

COROLLARY 2.5. $(S_i)^{-1}(S_i) = A_i$, $i = +$ *or* $-$.

3. Torsions on $A(M \times R)$**.** $A(M \times R)$ has $A(M)$ as subgroup via the map $h \mapsto h \times e_R$, the latter is not our full group $A(M \times R)$; this difference can be detected by a torsion invariant which we proceed to define now.

DEFINITION. An h-cobordism $(W; \partial_+ W, \partial_- W)$ is said to be *inertial* if $\partial_+ W$ is homeomorphic to $\partial_- W$. We set

$$I(M) = \{\tau \in \mathrm{Wh}(M) : \tau = \tau(W, M), \partial_+ W \cong \partial_- W = M\}.$$

EXAMPLE 1. The double of $(W^w; \partial_+ W, \partial_- W)$ obtained by gluing two copies of W along $\partial_- W$ is inertial; it has the torsion $\tau(W, \partial_- W) + (-1)^{w-1}\bar{\tau}(W, \partial_- W)$ [**47**, 11.4].

EXAMPLE 2. In §2, each T_i in a translation structure is an infinite product of inertial h-cobordisms. In fact, $T_i = T^*_{i-1} \cdot T^*_{i-2} \cdots$, where $T^*_j = T_{j+1} -$ int T_j deformation retracts onto each boundary component, $\partial_+ T^*_j \cong \partial_- T^*_j \cong M$. (As a deformation retraction of T^*_j onto $\partial_- T^*_j$, take for example the restriction to T^*_j of the deformation of $M \times R$ to ∂T_j, induced by the one from $M \times R$ to M_0, followed by composition with the retraction from $M \times R$ to T^*_j.) A translation t carried by $\mathscr{T}$ is thus a sequence of homeomorphisms $\{t|_{T^*_i} : i \in Z\}$ that shifts the sequence of h-cobordisms $\{T^*_i : i \in Z\}$ to the positive side by one.

EXAMPLE 3. For each $g \in A$, there is some $r \in R$ with $F\_(\boldsymbol{t}_r g)(0) > 0$. Then just as T^*_j above, $W_g = \boldsymbol{t}_r g(M_0^-) -$ int M_0^- is an inertial h-cobordism with torsion $\omega(g) = \tau(W_g, M_0) \in I(M)$. Since W_g is determined up to a homeomorphism preserving M_0, $\omega(g)$ is well defined. Of course, ω vanishes on the subgroup $A(M)$.

The map $\omega: A \to I(M)$ defined in Example 3 is the desired torsion invariant. We proceed to explore some of its properties; reminiscent of the formula for torsions of homotopy equivalences [**47**, 7.8], we have

PROPOSITION 3.1. *For* $f, g \in A$, $\omega(g \circ f) = \omega(g) + g_* \omega(f)$, *where* g_* *is the automorphism of* Wh(M) = Wh($\pi_1 M$) *induced by the automorphism of* $\pi_1 M \cong \pi_1(M \times R)$ *arising from* g.

PROOF. Write g' for $\boldsymbol{t}_r g$ and h' for $\boldsymbol{t}_s h$, r and s so chosen that $F_- g'h(0) > F_+ h'(0) > 0$. We claim that $W_{g'h}$ and $W_{g'h'}$ are homeomorphic by a homeomorphism that preserves M_0 pointwisely and so have the same torsion. For, choose a collar $c: M \times I \to M \times R$ on the negative side of $g'h(M_0)$ small enough not to intersect $h'(M_0)$; then since $W = h'(M_0^-) -$ int $h(M_0^-)$ is a product, $c(M \times 1)$ has both Im(c) and Im(c) $\cup$ $g'(W)$ as collars. Uniqueness of collars then furnishes a homeomorphism fixing $c(M \times 1)$ which upon extension by identity on the negative side of $c(M \times 1)$ yields the desired homeomorphism. But then

$$\omega(g \circ h) = \tau(W_{gh}, M_0) = \tau(W_{g'h'}, M_0) = \tau(W_{g'} \bigcup_{g'|M_0:\, \partial_- W_{h'} \to \partial_+ W_{g'}} W_{h'}, M_0)$$
$$= \tau(W_{g'}, M_1) + (dig|_{M_0})_* \tau(W_{h'}, M_0)$$
$$= \omega(g) + (dig|_{M_0})_* \omega(h) = \omega(g) + g_* \omega(h),$$

where $i\colon \partial_+ W_{g'} \to W_{g'}$ is the inclusion and $d\colon W_{g'} \to M_0$ a deformation retraction.

If W is an h-cobordism between $h_0(M_0)$ and $h_1(M_0)$, i.e., $W = h_0(M_0^-) - h_1(M_0^-)$, then we adopt the following convention in order that W has a unique torsion $\tau(W) \in \mathrm{Wh}(M_0)$, namely, set $\tau(W)$ to be $\omega(h_0) - \omega(h_1)$. Note that under this convention, the torsion of the product of two h-cobordisms is the sum of their respective torsions, which is what one desires. Note also that $\tau(W)$ does not necessarily lie in $I(M)$ because ω is neither a homeomorphism nor $I(M)$ necessarily a subgroup (cf. 3.3). Moreover, it is straightforward from 3.1 that $\tau(g(W)) = g_*\tau(W)$.

PROPOSITION 3.2. *$\omega\colon A \to I(M)$ is surjective.*

PROOF. Given $\tau \in I(M)$, we can find an h-cobordism $(W; \partial_- W, \partial_+ W)$ of torsion τ together with a homeomorphism h from $\partial_- W = M$ to $\partial_+ W$. Consider the infinite product $P = \cdots \cup_h W_{-1} \cup_h W_0 \cup_h W_1 \cup_h \cdots$, W_i a copy of W; then P is homeomorphic to $M \times R$ [**54**] say by a homeomorphism k that sends $\partial_- W_0 = M$ onto $M \times 0$ identically. Next, define f by sending W_i identically onto W_{i+1}, forgetting the indices, so the overlap $W_i \cap W_{i-1} = \partial_- W_i$ is sent to $\partial_+ W_i$ either way by h. Now, $W_{kfk^{-1}}$ is homeomorphic to W via k, so $\omega(kfk^{-1}) = \tau(W_{kfk^{-1}}, M_0) = \tau(W, M) = \tau$ as desired.

COROLLARY 3.3. *$I(M)$ is a subgroup of* $\mathrm{Wh}(M)$ *if and only if the action of A on* $\mathrm{Wh}(M)$ *leaves $I(M)$ invariant.*

PROOF. *If part.* Given $\sigma, \tau \in I(M)$, we first show $-\tau \in I(M)$. Find $r \in R$ such that $F_+(\boldsymbol{t}_r g)(0) > 0$; then $M_0^- - \operatorname{int} \boldsymbol{t}_r g(M_0^-)$ is the required h-cobordism. Next, we show $\tau + \sigma \in I(M)$. By 3.2, there is some $g \in A$ with $\tau(W_g, M) = \tau$. Let W be such that $\tau(W, M) = (g^{-1})_*\sigma$; this by assumption is inertial. Hence so is $W \cup_{g:\, \partial_- W \to \partial_+ W_g} W_g$ whose torsion is exactly $\tau + g_*(g^{-1})_* \sigma = \tau + \sigma$.

Only if part. Suppose W has torsion $g_*^{-1}\tau \notin I(M)$, some $\tau \in I(M)$; then $W \cup_{g:\, \partial_- W \to \partial_+ W_g} W_g$ is not inertial and has torsion $\tau + \omega(g)$, so $I(M)$ is not closed.

REMARK 3.4. It is interesting in the above context to note Hausmann's result [**31**, 2.3], who studied $I(M)$ in relation to torsions of homotopy equivalences. In general, $I(M)$ is quite complicated and warrants a separate study. Unlike $\mathrm{Wh}(M)$, $I(M)$ does not depend on $\pi_1(M)$ alone [**31**]. In some instances, $I(M)$ is actually the whole group $\mathrm{Wh}(M)$, e.g., when $\pi_1 M = Z_q$, q odd, dim M even and greater than 5 [**41**]. As another example, if the quotient map $Gl_n(Z\pi_1 M) \to \mathrm{Wh}(\pi_1 M)$ is onto, then for $2 \leqq p \leqq m-2$, $I(M \# n(S^p \times S^{m-p}))$ is the same as $\mathrm{Wh}(M \# n(S^p \times S^{m-p}))$ [**29**].

The map ω is an invariant of path components.

PROPOSITION 3.5. *ω induces $\omega'\colon \pi_0 A \to I(M)$, i.e., $\omega(gh) = \omega(hg) = \omega(g)$ for $g \in A$, $h \in A_0$.*

PROOF. Given $h \in A_0$, we claim there is a factorization $h = h_2 h_1$ with $h_i|_{M_{r_i}} = e$, some $r_i \in R$ and $h_i \in A$. Let $E \in A(M \times R \times I)$ be an isotopy from $e_{M \times R}$ to h; then there are reals r_1 and r_2 such that $E(M_{r_1} \times I) \cap (M_{r_2} \times I) = \varnothing$. Consider an isotopy $k_t\colon M_{r_1} \cup M_{r_2} \to M \times R$ defined by $k_t|_{M_{r_2}} = e$ and $k_t|_{M_{r_1}} = E|_{M_{r_1}\times t}$, the isotopy extension theorem ([21], [42] for Top; [32] for PL; [15], [50] for Diff^k) then furnishes K_t satisfying $K_0 = e$ and $K_t|_{(M_{r_1}\cup M_{r_2})\times I} = k_t$. By relative approximation we can make K_1 smooth to any degree necessary while satisfying the above. Setting $h_1 = K_1^{-1}h$ and $h_2 = K_1$ then gives the desired factorization.

Now since $h_i t_{r_i}|_{M_0} = e$ and $(h_i t_{r_i})_* = (e_{M\times R})_*$, so

$$\omega(h_i) = \omega(h_i t_{r_i}) + (h_i t_{r_i})_*\, \omega(t_{r_i}^{-1}) = 0 + 0 = 0$$

and $\omega(h) = \omega(h_2 h_1) = \omega(h_2) + (h_2)_*\omega(h_1) = 0$. Consequently, $\omega(gh) = \omega(g) + g_*\omega(h) = \omega(g) = \omega(hg)$.

The group $\pi_0 A$ inherits from A an action on $\mathrm{Wh}(M)$.

COROLLARY 3.6. *If $I(M)$ is nontrivial, then $H^1(\pi_0 A, \mathrm{Wh}(M))$ is nontrivial.*

PROOF. For $\omega'\colon \pi_0 A \to I(M)$ is a nontrivial cocycle by 3.1.

We proceed to study the group $\pi_0 A$. Consider the relation of concordance in $A(M)$; we denote the group of concordance classes by $\pi_0\tilde{A}(M)$ and the equivalence class of $h \in A(M)$ by $\langle h\rangle$.

PROPOSITION 3.7. *The map $i\colon \pi_0\tilde{A}(M) \to \pi_0 A$ defined by $i\langle h\rangle = (h \times e_R)A_0$ for $h \in A(M)$ is an injection.*

We need a lemma.

LEMMA 3.8. *If $f(M_r) = M_r$, then f is isotopic to $f|_{M_r} \times e_R$. Moreover, $A(M \times R, M_r) \simeq *$.*

PROOF. Consider the following isotopy $H_t\colon M \times R \to M \times R$, $t \in I$,

$$H_t(x, s) = \begin{cases} f(x, s-t') & \text{if } s \geqq r+t', \\ (\pi_M f(x, r), s) & \text{if } |s-r| \leqq t', \\ f(x, s+t') & \text{if } s \leqq r-t', \end{cases}$$

where $t' = t/(1-t)$. Evidently, $H_0 = f$ and $H_1 = f|_{M_r} \times e_R$. For $A = \mathrm{Diff}^k$, H_t may not be differentiable, but we can apply the uniqueness of C^k-collars to a bicollar of M first and then perform H_t which is now differentiable.

The next definition gives a useful geometrical interpretation of $\pi_0 A$; 3.7 will follow from this.

DEFINITION. Note that an inertial h-cobordism is characterized by a pair (W, h), where W is an h-cobordism and $h\colon \partial_- W = M \to \partial_+ W$ a homeomorphism. There is an obvious equivalence relation on these pairs, namely, $(W, h) \sim (W', h')$ whenever we have a homeomorphism $g\colon W \to W'$ preserving M and $gh = h'$. This is the same as requiring gh to be concordant to h' as one can verify. The collection of equivalence classes of these pairs is then defined to be $I_A(M)$, where $\tau(W, h)$ denotes the class of (W, h). $I_A(M)$ has a rule of composition: We take $\tau(W \cup_{h:\partial_- W' \to \partial_+ W} W', \hat{h}h')$ to be the product $\tau(W, h)\,\tau(W', h')$, where $\hat{h}$ is an extension of h to W'. This is well defined and inverses exist with $\tau(M \times I, e_M \times 1)$ as the identity, so $I_A(M)$ is actually a group.

Just as in Example 3, we can define a map $\nu\colon A \to I_A(M)$ by $\nu(g) = \tau(W_g, t_r g|_{M_0})$.

THEOREM 3.9. *ν induces an isomorphism $\nu'\colon \pi_0 A \to I_A(M)$.*

PROOF. That ν is a homomorphism follows from scrutinizing the proof of 3.1. Surjectivity and well-definedness of ν' are corollaries of the proofs of 3.2 and 3.5. To see injectivity, say $(W_f, f|_{M_0}) \sim (M \times I, e_M \times 1)$ via a homeomorphism g: $W_f \to M \times I$, then extend g to a full homeomorphism in A and apply 3.8, we see $f \in A_0$ as desired.

PROOF OF 3.7. Consider the map $\pi_0 \tilde{A}(M) \to I_A(M)$ defined by $\langle h \rangle \mapsto \tau(M \times I, h \times 1)$; this is equal to $\nu' i$ and clearly injective.

In dimensions where the s-cobordism theorem holds, $I_A(M)$ can be regarded as the collection of pairs (τ, h), $\tau \in I(M)$, with the rule of composition $(\tau, h)(\tau', h') = (\tau + h_* \tau', \hat{h}h')$. On the other hand, in these same dimensions, $I(M)$ can be treated as the collection of inertial h-cobordisms $(W; M, M')$ up to homeomorphism preserving M. If we adopt this definition in the low dimensions also, then this circumvents the need of s-cobordism theorem. To summarize this section, we have

THEOREM 3.10. $0 \to \pi_0 \tilde{A}(M) \to^i \pi_0 A \to^{\omega'} I(M) \to 0$ *is a short exact sequence in the sense that* $\omega'(fA_0) = \omega'(gA_0)$ *implies* $(f^{-1}g)A_0 \in \mathrm{Im}(i)$.

PROOF. Clearly $\omega' i = 0$. Conversely, denote by $\tau(W)$ the class of the inertial h-cobordism W; then $\omega'(fA_0) = \omega'(gA_0)$ implies $\tau(W_f) = \tau(W_g)$. Now, assume for convenience that $f, g \in S$, which we can do by 2.4 and 3.5, so by the proof of 3.1,

$$\begin{aligned}\tau(W_{f^{-1}g}) &= \tau\Big(W_{f^{-1}} \bigcup_{f^{-1}:\, \partial_- W_g \to \partial_+ W_{f^{-1}}} W_g\Big) \\ &= \tau\Big(W_{f^{-1}} \bigcup_{f^{-1}:\, \partial_- W_f \to \partial_+ W_{f^{-1}}} W_f\Big) \\ &= \tau(M \times I).\end{aligned}$$

This yields some $k \in A(M \times R, M_0)$ such that $kf^{-1}g(M_0) = M_1$. But then 3.8 implies $k \in A_0$ and $t_{-1}k(f^{-1}gA_0) = (t_{-1}kf^{-1}g|_{M_0} \times e_R)A_0 \in \mathrm{Im}(i)$, so 3.10 follows from the normality of A_0.

Note how the above proof circumvents the s-cobordism theorem; in fact, without the geometric definition of $I(M)$, we would have to show first $\omega(f^{-1}g) = 0$ using 3.1 and then apply the s-cobordism theorem to conclude $\tau(W_{f^{-1}g}) = \tau(M \times I)$.

COROLLARY 3.11. *If* $I(M)$ *vanishes, then* $\pi_0 A \cong \pi_0 \tilde{A}(M)$.

REMARK. This happens for example if M^m is a sphere except possibly the topological 4-sphere by the annulus theorem **[36]**; a surface of genus g **[8]**; is simply connected and $m \geqq 4$ **[6]**; has a finitely generated free abelian fundamental group and $m \geqq 5$ **[7]**; $\mathrm{Wh}(M) = 0$; is an odd dimensional lens space **[47]**; or is $S^1 \times S^3$ **[25]**, **[52]**.

For the computation of $\pi_0 A$ we have an aid in the following

REMARK 3.12. Let $C(M) = A(M \times I, M \times 0)$ and $A_e(M) = \{h \in A(M): \langle h \rangle = \langle e_M \rangle\}$. Using the fibrations $A(M \times I, \partial) \to C(M) \to^{\text{evaluation}} A_e(M)$, $A_e(M) \to A(M) \to \pi_0 \tilde{A}(M)$ and 3.10, together with the isomorphism $\pi_i A(M \times I, \partial) \cong \pi_{i+1} A(M \times R)$ **[43]**, we obtain a long exact sequence

$$\begin{aligned}\cdots &\to \pi_2 A \to \pi_1 C(M) \to \pi_1 A(M) \to \pi_1 A \to \pi_0 C(M) \\ &\to \pi_0 A(M) \to \pi_0 A \xrightarrow{\omega'} I(M) \to 0.\end{aligned}$$

The group $\pi_0 C(M)$ fits into a natural exact sequence [**28**], [**30**]

$$H_0(\pi_1 M; (\pi_2 M)[\pi_1 M]/(\pi_2 M)[1]) \to \pi_0 C(M) \\ \to \mathrm{Wh}_2(\pi_1 M) \oplus H_0(\pi_1 M; Z_2[\pi_1 M]/Z_2[1]) \to 0,$$

where M^m is a handlebody in case A = Top and $m = 5$ [**14**, p. 149]. If M has vanishing first Postnikov invariant $k_1 \in H^3(\pi_1 M; \pi_2 M)$, this is a short exact sequence that splits. If M is k-connected, we have more information [**14**, Chapter 3], namely, for $m \geqq 5$ or A = Top and $m \geqq 6$, $\pi_j C^{\mathrm{Diff}}(M) \cong \pi_j C^{\mathrm{Diff}}(D^m, \partial D^m)$ for $j \leqq 2k - 4$ and $\pi_j C^{\mathrm{Top}}(M) \cong \pi_j C^{\mathrm{PL}}(M)$ vanishes for $j \leqq \inf(2k - 3, k + 2)$.

The groups $\pi_0 A(S^n \times S^m \times R) \cong \pi_0 \tilde{A}(S^n \times S^m)$ has been fully computed for $(n, m) \neq (2, 2)$ [**33**], [**30**, Appendix].

4. Conjugating translations. We will first give a few propositions that elucidate the process of conjugation and then specialize our consideration to Top-translations on $S^n \times R$.

THEOREM 4.1. *Two translations t_1 and t_2 conjugate if and only if their respective isotopy classes do so in $\pi_0 A$.*

PROOF. Only if part is clear. Conversely, there is some u with $ut_1u^{-1} = t_0$ isotopic to t_2. We will first conjugate t_0 by f so that ft_0f^{-1} and t_2 are carried by the same translation structure. Assume for the moment that $t_j \in S$, $j = 0, 2$; then by 2.3, $\{t_j^i(M_0^-): i \in Z\}$ is a translation structure carrying t_j. Let $T^*_{j,i} = t_j^{i+1}(M_0^-) - \mathrm{int}\, t_j^i(M_0^-)$; then 3.5 implies $\tau(T^*_{0,0}) = \omega(t_0) = \omega(t_2) = \tau(T^*_{2,0})$, so the uniqueness of h-cobordism theorem [**47**, 11.3] or in low dimensions our geometrical definition of $I(M)$ (before 3.10) and the uniqueness of collars give some $f_0 \in A(M \times R, M_0)$ (in the differentiable case up to any degree of smoothness by relative approximation) with $f_0(T^*_{0,0}) = T^*_{2,0}$ and f_0 stationary (cf. [**4**, 3.1.3] for definition) with respect to the collars $t_j^0|_{M\times[0,\varepsilon]}$ and $t_j^1|_{M\times[-\varepsilon,0]}$. Thus, $f_0 \in A_0$ via 3.8. Suppose by induction we have defined for $i \geqq 0$, $f_i \in A_0$ with $f_i(T^*_{0,i}) = T^*_{2,i}$, stationary near $\partial T^*_{0,i}$ with respect to $t_j^i|_{M\times[0,\varepsilon_i]}$ and $t_j^{i+1}|_{M\times[\varepsilon_i,0]}$. Let $f'_{i+1} = t_2 f_i t_0^{-1}$, then there is some $v_{i+1} \in A$ such that $v_{i+1}(T^*_{2,i+1}) = T^*_{2,i+1}$, stationary near $\partial T^*_{2,i+1}$ with respect to $t_2^{i+1}|_{M\times[0,\varepsilon_{i+1}]}$ and $t_2^{i+2}|_{M\times[-\varepsilon_{i+1},0]}$, while $v_{i+1}|_{t_2^{i+1}(M_0)} = f_i(f'_{i+1})^{-1}$. Such a v_{i+1} exists since $f_i(f'_{i+1})^{-1} \in A_0$ implies $\tau(f_i(f'_{i+1})^{-1}T^*_{2,i+1}) = \tau(T^*_{2,i+1})$ so that the uniqueness of h-cobordisms or our geometrical definition of $I(M)$ and uniqueness of collars apply. Moreover, $v_{i+1} \in A_0$ by applying 3.8 to $v_{i+1}|_{t_2^{i+1}(M_0)}$. We can then define f_{i+1} to be $v_{i+1}f'_{i+1} \in A_0$. Similarly, we can define f_i for $i \geqq 0$. Clearly $\{f_i|_{T^*_{0,i}}: i \in Z\}$ patch to $f \in A_0$ (even if $A = \mathrm{Diff}^k$ because of the stationary properties) with

$$f(t_0^i(M_0^-)) = f\Big(\bigcup_{j=-\infty}^{i-1} T^*_{0,j}\Big) = \bigcup_{j=-\infty}^{i-1} T^*_{2,j} = t_2^i(M_0^-),$$

so that ft_0f^{-1} is carried by $\{t_2^i(M_0^-): i \in Z\}$.

Now we proceed to define a conjugating homeomorphism from $t_3 = ft_0f^{-1}$ to t_2. Note that t_3 is still isotopic to t_2. Moreover, $t_2^{-1}t_3(M_0) = M_0$, so $t_2^{-1}t_3|_{M_0}$ is concordant to e_{M_0} by 3.8 and 3.7. Write T_i^* for $T^*_{2,i}$ and let $g_0 \in A(T_0^*)$ be defined by a concordance from $t_2^{-1}t_3|_{M_0}$ to e_{M_0} on $M \times [0, \varepsilon]$ stationary near M_0, ε so small that $M_\varepsilon \subset \mathrm{int}\, T_0^*$, while $g_0 = e$ elsewhere. Also, define $g_i \in A(T_i^*)$ by $g_i = t_2^i g_0 t_3^{-i}$; then

$$g_i|_{\partial_- T_i^*} = (t_2)^i(t_2^{-1}t_3)(t_3)^{-i} = t_2^{i-1}\, t_3^{-(i-1)}$$

and

$$g_i|_{\partial_+ T_i^*} = (t_2)^i(e)(t_3)^{-i} = t_2^i\, t_3^{-i}.$$

Therefore, $\{g_i : i \in Z\}$ patch to $g \in A$ with

$$gt_3g^{-1}(x) = g_{i+1}t_3\, g_i^{-1}(x) = (t_2^{i+1}g_0\, t_3^{-(i+1)})(t_3)(t_3^i\, g_0^{-1}\, t_2^{-i})(x) = t_2(x)$$

for each $x \in T_i^*$, so that t_1 is indeed a conjugate of t_2.

Lastly, regarding the assumption $t_0, t_2 \in S$, suppose $\{T_i : i \in Z\}$ is a translation structure carrying t_0; we will show how to replace the given t_0 with an isotopic conjugate in S. Say $T_0 = h(M_0^-)$; then there is some "large" $a \in A(R)$ satisfying $a(F_-(t_0h)(0)) > F_+h(0)$ and $F_-(t_0) > a^{-1}$. Thus, $(Ha)t_0$ is a translation in S carried by $\{((Ha)t_0)^i h(M_0^-) : i \in Z\}$, isotopic to t_0 because $Ha \in A_0$ from letting M be a point in 3.10. Now we can construct a homeomorphism $f \in A_0$, just as before, to map $\{T_i : i \in Z\}$ to $\{((Ha)t_0)^i h(M_0^-) : i \in Z\}$ and then construct g that actually conjugates $ft_0\, f^{-1}$ to $(Ha)t_0$, thereby proving the theorem.

The process of conjugation in the light of the above proof is really in two steps and 3.10 shows where the obstruction for each step lies: The first conjugates up to translation structure, its obstruction comes from comparing $\omega(t_1)$ and $\omega(t_2)$ in $I(M)$; the second completes the conjugation and its obstruction $\langle t_2^{-1}t_3|_{M_0}\rangle \in \pi_0\tilde{A}(M)$ comes from patching $\{g_i : i \in Z\}$. As a corollary of the above proof, we have

COROLLARY 4.2. *If $t_1, t_2 \in S_j$, $j = +$ or $-$, then there is some $k \in A_j$ such that $(t_1)^{-1}(kt_2k^{-1}) \in A_K$.*

PROOF. Suppose $t_1, t_2 \in S_-$ (the case of S_+ is similar); then after conjugating t_1 by a translation we may assume $F_-(t_1)|_{[0,+\infty)} > e_R$. Just as in 4.1, consider the sets $T_i^* = t_2^{i+1}(M_0^-) - \text{int}\ t_2^i(M_0^-)$ for $i \geqq 0$. Modifying that proof slightly readily yields some $f_k \in A_-$ such that $(f_k\, t_k\, f_k^{-1})^i(M_0^-) = M_i^-$, $i \geqq 0$, $k = 1, 2$. Now, $t_k \in S_- \subset A_-$ implies $(f_1\, t_1\, f_1^{-1})^{-1}(f_2\, t_2\, f_2^{-1})|_{M_0}$ is concordant to e, so that modifying again furnishes $g \in A_-$ with $(f_1\, t_1 f_1^{-1})|_{M_0^+} = g(f_2\, t_2\, f_2^{-1})g^{-1}|_{M_0^+}$. Setting $k = f_1^{-1}\, gf_2$ then yields $(t_1)^{-1}\,(kt_2k^{-1}) \in A_- \cap A(M \times R, M_0^+) \subset A_K$.

Altering the above proof slightly shows t_1 and t_2 actually conjugate.

COROLLARY 4.3. *If $t_1, t_2 \in S_j$, $j = +$ or $-$, satisfy $F_-t_1|_{[\text{point}]_r^i} > e_R$ and $t_1|_{M_r^j} = t_2|_{M_r^j}$, where $\{i, j\} = \{+, -\}$, then there is some $k \in A$ such that $(t_1)^{-1}(kt_2k^{-1}) = e_{M\times R}$.*

PROOF. We prove for S_- alone. After conjugating t_i by a translation we may assume $r = 0$. Next, f_i in 4.2 can satisfy $f_1|_{T_0^*} = f_2|_{T_0^*}$ and $f_1|_{M_0^-} = f_2|_{M_0^-} = e$ since $T_0^* = t_1(M_0^-) - \text{int}\ M_0^- = t_2(M_0^-) - \text{int}\ M_0^-$. Consequently,

$$\begin{aligned}(f_1\, t_1 f_1^{-1})^{-1}(f_2\, t_2\, f_2^{-1})|_{M_0^-} &= f_1\, t_1^{-1}(f_1^{-1}f_2)\, t_2|_{M_0^-}\\ &= f_1 t_1^{-1}\, t_2|_{M_0^-} = f_1|_{M_0^-} = e.\end{aligned}$$

Therefore, the homeomorphism g in 4.2 can further possess the property $g|_{M_1^-} = e$. This implies $f_1t_1f_1^{-1}|_{M_0} = g(f_2t_2f_2^{-1})g^{-1}|_{M_0^-}$, aside from their equality on M_0^+ which we saw in 4.2. Hence, setting $k = f_1^{-1}\, gf_2$,

$$(t_1)^{-1}(kt_2\, k^{-1}) \in A(M \times R, M_0^-) \cap A(M \times R, M_0^+) = e_{M\times R}.$$

The rest of this section will be devoted to the special case of M an m-sphere, $m \in N$. Here, matters are especially simple because if $m \neq 3$, $\pi_0 A(S^m \times R) \cong \pi_0 \bar{A}(S^m)$ by 3.11. Let C_n denote the statement $\pi_0 \widetilde{\operatorname{Top}}(S^n) = 0$ and TC_n (translation conjugates) the statement: Given $t_j \in \operatorname{Top}(S^{n-1} \times R)$ translations carried by $\{T_{j \cdot i}: i \in Z\}$, $j = 1, 2$, then there is some $f \in \operatorname{Top}(S^{n-1} \times R)$ such that $ft_2 f^{-1} = t_1$ and $f(T_{2,i}) = T_{1,i}$ for each $i \in Z$. Consider the n-dimensional annulus theorem AT_n (conjecture if $n = 4$), it is simply $I(S^{n-1}) = 0$; also the n-dimensional stable homeomorphism theorem SHT_n (conjecture if $n = 4$), which by the torus lifting trick [**36**, Theorem 6] is equivalent to $\pi_0 \operatorname{Top}(R^n) = 0$ and which further reduces to $\pi_0 \operatorname{Top}(S^{n-1} \times R) = 0$. Theorem 3.10 for spheres is then exactly the statement: AT_n and $C_{n-1} \Leftrightarrow SHT_n$ (cf. [**11**, 9.3]). Here, the notion of translations fits in quite well. In fact, by 3.11 we actually showed in 4.1 the following

PROPOSITION 4.4. *For each $n \in N$, AT_n and $C_{n-1} \Rightarrow TC_n$.*

In a similar vein,

PROPOSITION 4.5. *For each $n \in N$, $TC_n \Leftrightarrow SHT_n$.*

To establish this, we need the notion of topological dilations on S^n. (Compare with [**48**] and [**49**].) Given $x \in R^n$ and $g: R^n - x \approx S^{n-1} \times R$, we can define g_*: $\operatorname{Top}(R^n, x) \cong \operatorname{Top}(S^{n-1} \times R)$ by

$$g_*(h)(y) = \begin{cases} x & \text{if } y = x, \\ ghg^{-1}(y) & \text{if } y \neq x. \end{cases}$$

Similarly, given $p, q \in S^n$ and $k: S^n - q \approx R^n$, $k(p) = x$, then we can define the obvious $k_*: \operatorname{Top}(S^n, p \cup q) \cong \operatorname{Top}(R^n, x)$.

DEFINITION. A map $d \in \operatorname{Top}(R^n, x)$ ($\operatorname{Top}(S^n, p \cup q)$) is said to be a *dilation* on R^n (resp. S^n) if $g_*(d)$ (resp. $g_* k_*(d)$) is a translation on $S^{n-1} \times R$. Correspondingly, a pair $(p, \mathscr{D})$, where $\mathscr{D} = \{D_i: i \in Z\}$ consists of n-cells in S^n, is a *dilation structure* on S^n if $\{gk(D_i - p \cup q): i \in Z\}$ is a translation structure.

Note that these definitions are independent of g and k. Thus, g, g_*, k, and k_* will be used without being explicitly described.

LEMMA 4.6. *TC_n implies if $d \in \operatorname{Top}(R^n)$ is a dilation, then $G(d) = \operatorname{Top}(R^n)$. Here $G(C)$ is the least normal subgroup of $\operatorname{Top}(R^n)$ containing C, a given subset of $\operatorname{Top}(R^n)$.*

PROOF. Given $h \in \operatorname{Top}(R^n)$, we can find $z_1, z_2 \in R^n - 0$ and $A \in \operatorname{Gl}_n(R)$ such that $h(z_1) = z_2 = A(z_1)$. Thus, $h = A(A^{-1}h)$ lies in $(\operatorname{Top}(R^n, 0))(t_{z_1} \operatorname{Top}(R^n, 0) t_{z_1}^{-1}) \subset G(\operatorname{Top}(R^n, 0))$, where t_{z_1} is the map of vector addition by z_1. But TC_n and 2.4 imply $G(g_*(d)) = \operatorname{Top}(S^{n-1} \times R)$, so we are done.

PROOF OF 4.5. *Only if part.* In view of 4.6 and 2.3, if $a \in \operatorname{Top}(R)$ satisfies $a > e_R$, then $G(g_*^{-1}(Ha)) = \operatorname{Top}(R^n)$, so it suffices to show a is stable. But certainly we can factor a into homeomorphisms each supported outside a nonempty open set.

If part. By scrutinizing the proof of [**49**, Theorem 1] (actually one needs to go back to [**48**]), we see that the corresponding result for dilations on R^n was proved. These are then translated via the maps g and g_* into what we want.

Let $S_+^n = \{(x_0, \cdots, x_n) \in S^n: x_n \geqq 0\}$ and $S_-^n = \operatorname{cl}(S^n - S_+^n)$. We will identify S^{n-1} with $S_+^n \cap S_-^n$.

PROPOSITION 4.7. *If f_1 and f_2 are two disjoint locally flat embeddings of S^{n-1} into S^n, then there exists dilation d on S^n such that $d(f_1) = f_2$.*

PROOF. We actually prove the somewhat stronger statement: For $i = 1, 2$, let $F_i \in \mathrm{Top}(S^n)$ be some extension of f_i guaranteed by the generalized Schoenflies extension theorem [9] and $F = F_2F_1^{-1}$; then there is a dilation d carried by $\{d^i(F_1S^n_+): i \in Z\}$ such that for some $U \supset f_1(S^{n-1})$, $d|_U = F$. (In particular, $d(f_1) = F(f_1) = F_2F_1^{-1}(f_1) = f_2$.) The dilation d is constructed by modifying F.

Consider $A = \bigcap_{i \in Z} F^i(F_1S^n_+)$, this is a cellular set, so by [20], for every neighborhood U_1 of A, there is some $h: S^n/A \to S^n$ such that

(i) $h \circ \pi|_{S^n - U_1} = e$, where $\pi: S^n \to S^n/A$ is the collapsing map. By the normality of S^n, we may choose U_1 so that cl U_1 lies in int $F_1(S^n_+)$. Thus,

(ii) $\mathrm{int}(S^n - U_1) \supset F_1(S^n_-)$.

Now, F induces $F' \in \mathrm{Top}\,(S^n)$ by

$$F'(x) = \begin{cases} (h\pi)F(h\pi)^{-1}(x) & \text{if } x \neq p, \\ p & \text{if } x = p, \end{cases}$$

where $p = h\pi(A)$. Applying the same procedure to map F' and the cellular set $A' = \bigcap_{i \in Z}(F')^i(F_1S^n_-)$, we obtain maps $S^n/A' \to^{h'} S^n \to^{\pi'} S^n/A'$ and a homeomorphism $d \in \mathrm{Top}(S^n)$ together with a neighborhood $U_2 \supset A'$ satisfying

(i′) $h'\pi'|_{S^n - U_2} = e$ and

(ii′) $\mathrm{int}(S^n - U_2) \supset F_2(S^n_+)$.

We proceed to check the analogues of (a) through (c) in the definition of translation structure for $\{d^i(F_1S^n_+): i \in Z\}$. Firstly, since f_1, f_2 disjoint and F_1, F_2 orientation preserving, the indices can be chosen so that $F_1(S^n_+) \subset \mathrm{int}\, F_2(S^n_+)$. Let $U = (S^n - \mathrm{cl}\, U_1) \cap (S^n - \mathrm{cl}\, U_2)$, then by properties (ii) and (ii′), $U \supset F_1(S^n_-) \cap F_2(S^n_+) \supset F_1(S^n_-) \cap F_1(S^n_+) = f(S^{n-1})$. Moreover, by (i) and (i′), $d|_U = F'|_{S^n - U} = F$. Consequently, int $d(F_1(S^n_+)) = \mathrm{int}\, F_2(S^n_+) \supset F_1(S^n_+)$, which implies condition (a). Next, by (i), (i′), (ii), and (ii′), we see

$$\bigcap_{i \in Z} d^i(F_1S^n_+) = \bigcap_{i \in Z}(F')^i(F_1S^n_+) = h\pi\left(\bigcap_{i \in Z} F^i(F_1\, S^n_+)\right) = h\pi(A) = p,$$

i.e., (b) is satisfied. Lastly, (ii) and (ii′) give

$$\bigcap_{i \in Z} d^i(F_1\, S^n_-) = h'\pi'\left(\bigcap_{i \in Z} (F')^i(F_1\, S^n_-)\right) = h'\pi'(A') = q \in S^n.$$

Thus, (c) is satisfied as well.

COROLLARY 4.8. *For each $n \in N$, $TC_n \Rightarrow AT_n$.*

PROOF. In the notation of the preceding proof, $d(F_1S^n_+) = F(F_1S^n_+) = F_2(S^n_+)$. Now, in TC_n take $t_1 = t_1$ and $t_2 = g_*k_*(d)$, we find $f \in \mathrm{Top}(S^{n-1} \times R)$ such that $f(gk(d^i(F_1S^n_+) - p \cup q)) = (S^{n-1})^-_i$. Therefore, we see that the region between f_1 and f_2 is

$$fgk(F_2(S^n_+) - \mathrm{int}\, F_1(S^n_+)) = (S^{n-1})^-_1 - \mathrm{int}(S^{n-1})^-_0 = S^{n-1} \times I.$$

Lastly, to complete the relationship between TC_n and AT_n,

PROPOSITION 4.9. *For each $n \in N$, AT_m for $m = 1, \cdots, n \Rightarrow TC_n$.*

PROOF. We prove this by induction. The case of $n = 1$ follows from 4.4 since C_0 is trivially true. By the induction hypothesis, AT_m for $m = 1, \cdots, n-1 \Rightarrow TC_{n-1}$, which in turn implies SHT_{n-1} by 4.5. Now, SHT_{n-1} if and only if each element of $\mathrm{Top}(S^{n-1})$ is stable, which is true if and only if $\mathrm{Top}(S^{n-1})$ is simple **[24]**, hence C_{n-1}. Consequently, AT_m for $m = 1, \cdots, n \Rightarrow AT_n$ and $C_{n-1} \Rightarrow TC_n$ by 4.4.

As a corollary, we obtain the well-known relationship between SHT_n and AT_n **[11]**, **[36]**.

COROLLARY 4.10. *For each* $n \in N$, (i) $SHT_n \Rightarrow AT_n$,
(ii) AT_m *for* $m = 1, \cdots, n \Rightarrow SHT_n$.

A translation is standard if regarded as a shift of $\cdots W_{-1} \cup_h W_0 \cup_h W_1 \cdots$, it takes W_i to W_{i+1} identically (cf. the proof of 3.2). By the proof of 3.2, for each class $\tau(W, h) \in I_A(M)$ there is a standard translation t such that $\nu(t) = \tau(W, \bar{h})$. This is in fact quite general:

PROPOSITION 4.11. *Each orientation preserving homeomorphism of* $M \times \mathbf{R}$ *can be translated to an isotopic conjugate of a standard translation.*

PROOF. This is clear by virtue of 2.4, 4.2 and 3.9.

REFERENCES

1. D. R. Anderson, Lecture given at 1976 Summer Research Institute.

2. ———, *The algebraic simplicity of certain groups of homeomorphisms*, Amer. J. Math. **80** (1958), 955–963. MR **20** #4607.

3. ———, *On homeomorphisms as products of a given homeomorphism and its inverse*, Topology of 3-Manifolds and Related Topics, M. K. Fort, ed., Prentice-Hall, Englewood Cliffs, N.J., 1962. MR **25** #3115.

4. P. L. Antonelli, D. Burghelea and P. J. Kahn, *The concordance-homotopy groups of geometric automorphism groups*, Lecture Notes in Math., vol. 251, Springer-Verlag, Berlin and New York, 1971, pp. 1–140.

5. R. Arens, *A topology for spaces of transformations*, Ann. of Math. (2) **47** (1946), 480–490. MR **8** #165.

6. D. Barden, *h-cobordisms between 4-manifolds*, Cambridge Univ. Press, London and New York, 1964.

7. H. Bass, A. Heller and R. Swan, *The Whitehead group of a polynomial extension*, Inst. Hautes Études Sci. Publ. Math. **22** (1964), 61–79. MR **30** #4806.

8. E. M. Brown, *Unknotting in* $M^2 \times I$, Trans. Amer. Math. Soc. **123** (1966), 480–505. MR **33** #6640.

9. M. Brown, *A proof of the generalized Schoenflies theorem*, Bull. Amer. Math. Soc. **66** (1960), 74–76. MR **22** #8470b.

10. ———, *Locally flat embeddings of topological manifolds*, Ann. of Math. (2) **75** (1962), 331–341. MR **24** #A3637.

11. M. Brown and H. Gluck, *Stable structures on manifolds*, Ann. of Math. (2) **79** (1964), 1–58.

12. D. Burghelea and R. K. Lashof, *The homotopy type of the space of diffeomorphisms*. I, II, Trans. Amer. Math. Soc. **196** (1975), 1–50.

13. D. Burghelea, *Automorphisms of manifolds*, these PROCEEDINGS, part 1, pp. 347–371.

14. D. Burghelea, R. K. Lashof and M. Rothenberg, *Groups of automorphisms of manifolds*, Lecture Notes in Math., vol. 473, Springer-Verlag, Berlin and New York, 1975.

15. J. Cerf, *Topologie de certains espaces de plongements*, Bull. Math. Soc. France **89** (1961), 227–380. MR **25** #3543.

16. ———, *Sur les difféomorphisms de la sphère de dimension trois* ($\Gamma_4 = 0$), Lecture Notes in Math., vol. 53, Springer-Verlag, Berlin and New York, 1968. MR **37** #4824.

17. ———, *The pseudo-isotopy theorem for simply connected differentiable manifolds*, Lecture

Notes in Math., vol. 197, Springer-Verlag, Berlin and New York, 1971, pp. 76–82. MR **44** #7585.

18. A. V. Cernavskii, *Local contractibility of the homeomorphism group of a manifold*, Math. USSR Sb. **8** (1969), 287–333.

19. M. M. Cohen, *A course in simple-homotopy theory*, Springer-Verlag, Berlin and New York, 1973.

20. A. Douady, *Plongements des sphères*, Seminaire Bourbaki, 13e année, 1960/61, Exposé 205, N. Bourbaki, Paris.

21. R. D. Edwards and R. C. Kirby, *Deformations of spaces of imbeddings*, Ann. of Math. (2) **93** (1971), 63–88. MR **44** #1032.

22. D. B. A. Epstein, *The simplicity of certain groups of homeomorphisms*, Compositio Math. **22** (1970), 165–173. MR **42** #2491.

23. N. J. Fine and G. E. Schweigert, *On the group of homeomorphisms of an arc*, Ann. of Math. (2) **62** (1955), 237–253. MR **17**, 288.

24. G. Fisher, *On the group of all homeomorphisms of a manifold*, Trans. Amer. Math. Soc. **97** (1960), 193–212. MR **22** #8487.

25. S. Fukuhara, *On the Hauptvermutung of 5-dimensional manifolds and s-cobordisms*, J. London Math. Soc. (2) **5** (1972), 549–555. MR **49** #9853.

26. M. E. Hamstrom, *Homotopy in homeomorphism spaces, Top and PL*, Bull. Amer. Math. Soc. **80** (1974), 207–230.

27. A. Hatcher, *Higher simple homotopy theory*, Ann. of Math. (2) **102** (1975), 101–137. MR **52** #4305.

28. ———, *Concordance spaces, higher simple-homotopy theory, and applications*, these PROCEEDINGS, part 1, pp. 3–21.

29. A. Hatcher and T. C. Lawson, *Stability theorem for "concordance implies isotopy" and "h-cobordism implies diffeomorphism"*, Duke Math. J. **43** (1976), 555–560.

30. A. Hatcher and J. Wagoner, *Pseudo-isotopies of compact manifolds*, Astèrique, No. 6, Société Mathématique de France, Paris, 1973. MR **50** #5821.

31. J. C. Hausmann, *h-cobordismes entres variétés homéomorphes*, Comment. Math. Helv. **50** (1975), 9–13. MR **51** #4279.

32. J. F. P. Hudson, *Piecewise linear topology*, Benjamin, New York, 1969. MR **40** #2094.

33. M. Kato, *A concordance classification of PL homeomorphisms of* $S^p \times S^q$, Topology **8** (1969), 371–383. MR **41** #1057.

34. M. A. Kervaire, *Le théorème de Barden-Mazur-Stallings*, Comment. Math. Helv. **40** (1965), 31–42. MR **32** #6475.

35. R. C. Kirby, *Stable homeomorphisms and the annulus conjecture*, Ann. of Math. (2) **89** (1969), 575–582. MR **39** #3499.

36. ———, *Lectures on triangulations of manifolds*, notes, Univ. of Calif, Los Angeles, 1969.

37. ———, *Some conjectures about four-manifolds*, Actes Congrèss Internat. Math. **2** (1970), 79–84.

38. R. C. Kirby and L. C. Siebenmann, *Some theorems on topological manifolds*, edited by N. H. Kuiper, Lecture Notes in Math., vol. 197, Springer-Verlag, Berlin and New York, 1971, pp. 1–7. MR **44** #1037.

39. ———, *Foundational essays on topological manifolds, smoothings, and triangulations*, Ann. of Math. Studies, no. 88, Princeton Univ. Press, Princeton, N.J.

40. J. M. Kister, *Isotopies in 3-manifolds*, Trans. Amer. Math. Soc. **97** (1960), 213–224. MR **22** #11378.

41. T. C. Lawson, *Inertial h-cobordism with finite cyclic fundamental group*, Proc. Amer. Math. Soc. **44** (1974), 492–496. MR **50** #11279.

42. J. Lees, *Immersions and surgeries of topological manifolds*, Bull. Amer. Math. Soc. **75** (1969), 529–534. MR **39** #959.

43. W. Ling, *Normal subgroups of the group of automorphisms of an open manifold that has boundary* (to appear).

44. ———, (in prep.).

45. J. N. Mather, *Commutators of diffeomorphisms*, Comment. Math. Helv. **49** (1974), 512–528. MR **50** #8600.

46. ———, *Commutators of diffeomorphisms*. II, Comment. Math. Helv. **50** (1975), 33–40. MR **51** #11576.

47. J. Milnor, *Whitehead torsion*, Bull. Amer. Math. Soc. **72** (1966), 358–426. MR **33** #4922.

48. E. Nunnally, *Dilations on invertible spaces*, Trans. Amer. Math. Soc. **123** (1966), 437–448. MR **34** #8384.

49. ———, *A factorization of stable homeomorphisms of* E^n, Proc. Amer. Math. Soc. **19** (1968), 387–389. MR **37** #2201.

50. R. Palais, *Local triviality of the restriction map for embeddings*, Comment. Math. Helv. **34** (1960), 305–312. MR **23** #A666.

51. J. Schreier and S. Ulam, *Bine bemerkung uber die gruppen der topologischen abbildungen der Kreislinie auf sich selbst*, Studia Math. **5** (1934), 155–159.

52. J. L. Shaneson, *Embeddings with codimension two of spheres in spheres and H-cobordism of* $S^1 \times S^3$, Bull. Amer. Math. Soc. **74** (1968), 972–974. MR **37** #5887.

53. L. C. Siebenmann, *Topological manifolds*, Actes Congrès Internat. Math. **2** (1970), 133–163.

54. J. R. Stallings, *On infinite processes leading to differentiability in the complement of a point*, Differential and Combinatorial Topology (Proc. Sympos. in Honor of Marston Morse), edited by S. S. Cairns, Princeton Univ. Press, Princeton, N.J., 1965, pp. 245–254. MR **31** #5213.

55. W. P. Thurston, *Foliations and groups of diffeomorphisms*, Bull. Amer. Math. Soc. **80** (1974), 304–307. MR **49** #4027.

56. J. B. Wagoner, *Diffeomorphisms*, K_2, *and analytic torsion*, these PROCEEDINGS, part 1, pp. 23–33.

57. F. Waldhausen, these PROCEEDINGS.

58. J. V. Whittaker, *Normal subgroups of some homeomorphism groups*, Pacific J. Math. **10** (1960), 1469–1478. MR **24** #A542.

59. E.C. Zeeman, *Isotopies and knots in manifolds*, Topology of 3-manifolds, edited by M. K. Fort, Jr., Prentice-Hall, Englewood Cliffs, N.J., 1962. MR **25** #3520.

PRINCETON UNIVERSITY

Proceedings of Symposia in Pure Mathematics
Volume 32, 1978

HOMOTOPY INVARIANCE OF ATIYAH INVARIANTS

WALTER D. NEUMANN

In the following, all manifolds are understood to be smooth, compact, and oriented. The invariants to be discussed are

(I). The α-invariants $\alpha(M^{2n-1}, g)$ of a smooth group action on a closed odd dimensional manifold M^{2n-1}. These were introduced by Atiyah and Singer **[2]** and have been extensively studied since then; see for instance Hirzebruch and Zagier **[6]** and the literature quoted there.

(II). Certain γ-invariants $\gamma(X^{2n-1}, \rho)$ associated to a representation $\rho: \pi_1(X) \to U(m)$ of the fundamental group of a closed odd dimensional manifold X^{2n-1}. These invariants arise for instance in Atiyah, Singer and Patodi **[1]** via the theory of spectral asymmetry. They came up in a different context in **[8]** and **[11]**.

(III). A "monodromy" $\mathscr{H}$ associated to a manifold X^{2n-1} and a map $X \to S^1$; $\mathscr{H} = (H, S, t)$ is an isometric structure, that is it consists of a vector space H plus a symmetric or skew-symmetric bilinear form S on H and an isometry $t: (H, S) \to (H, S)$. If $X \to S^1$ is a fibration with fiber F^{2n-2}, then $(H, S) = H^{n-1}(F; Q)$ with the cup product form $S(x, y) = \langle x \cup y, [F] \rangle$ and t is the monodromy of the fibration.

The latter invariant will be, by its very definition, a homotopy invariant. On the other hand, if the representation $\rho: \pi_1(X) \to U(m)$ in (II) factors over a free abelian group ($\rho = \pi_1(X) \to \boldsymbol{Z}^s \to U(m)$), then $\gamma(X, \rho)$ is calculable in terms of the monodromy invariants, and is hence also a homotopy invariant. Finally, in view of the intimate relationship between α- and γ-invariants, we obtain also homotopy invariant calculations of α-invariants in certain situations.

There seems to be a certain analogy of the homotopy invariance proved here with the homotopy invariance of higher Novikov signatures proved by Farrel and Hsiang **[7]** and Lusztig **[9]**. This analogy deepens a feeling the author has often had,

---

AMS (MOS) subject classifications (1970). Primary 57A65, 57D20, 57E15, 58G10.

that so-called "peripheral invariants", such as the α- and γ-invariants of odd dimensional manifolds, are connected in some deeper way than has yet been discovered to characteristic class and surgery type invariants.

The present paper is a fairly exact version of the talk given at Stanford; in particular no proofs are included. Some proofs appeared in the preliminary manuscript **[11]** (see also **[13]**). Complete proofs and further applications and examples will appear in the final version of **[11]**.

1. α-invariants of group actions. Let N^{2n} be a closed manifold with a smooth G-action and $g \in G$. Then the Atiyah-Singer fixed point theorem **[2]** calculates the equivariant signature sign(N, g) as a polynomial in the characteristic classes of the fixed point set N^g and its normal bundle $\nu(N^g)$:

$$\operatorname{sign}(N, g) = f(N^g, \nu(N^g)).$$

If N has boundary M^{2n-1}, this equation is no longer valid, but a standard argument shows that if g has no fixed points in M, then the "error"

$$\alpha(M, g) = \operatorname{sign}(N, g) - f(N^g, \nu(N^g))$$

is an invariant of (M, g). More generally, given (M, G) and $g \in G$ acting without fixed points on M, some disjoint multiple $q(M, g)$ bounds, and one may define $\alpha(M, g) = \alpha(q(M, g))/q$. Denote by $\alpha(M, G)$ the (partially defined) map $G \to \boldsymbol{C}$ whose value for $g \in G$ is $\alpha(M, g)$.

DEFINITION. Let G act effectively on a manifold M. We say an element $g \in G$ of finite order k is *S^1-induced* if there exists an equivariant map $(M, g) \to (S^1, e^{2\pi iq/k})$ for some q prime to k. In particular g acts freely.

We say G acts *h-injectively* on M if a dense set of elements of G are S^1-induced.

EXAMPLES. (i) One can show that a homologically injective action in the sense of Conner and Raymond [3] is h-injective.

(ii) If G is connected and acts on M then the following conditions are equivalent:

(a) (M, G) is h-injective;

(b) some nontrivial $g \in G$ is S^1-induced;

(c) some finite covering of G has the form $H \times S^1$ and the induced S^1-action on M is homologically injective.

(iii) If G is finite, any free orientation preserving action on a surface is h-injective. Very many free actions on Seifert spaces are (in a sense that can be made precise).

THEOREM 1. *If G acts h-injectively on M^{2n-1} then $\alpha(M, G)$ is a homotopy invariant of (M, G).*

Our results in fact give a reasonably calculable intrinsic description of $\alpha(M, G)$ for h-injective actions.

EXAMPLE. Given a cyclic action $(N^{2n-2}, \boldsymbol{Z}/k)$ we can form the S^1-manifold $M = N \times_{\boldsymbol{Z}/k} S^1$. The S^1-action on M is h-injective; in fact by Conner and Raymond [3], this construction gives all homologically injective S^1-actions. For $t \in S^1 = \{t \in \boldsymbol{C} \mid \|t\| = 1\}$ our calculation gives the formula (n even)

$$\alpha(M, t) = -2 \sum_{q=1}^{[(k-1)/2]} a_q \frac{t^{q'} + t^{q'+1} + \cdots + t^{k'-q'-1}}{1 + t + \cdots + t^{k'-1}},$$

where q'/k' is q/k in lowest terms and a_q is the integer (!)

$$a_q = \frac{1}{k} \sum_{r=0}^{k-1} e^{-2\pi i q r/k} \operatorname{sign}(N, g^r),$$

where g is a generator of the $\boldsymbol{Z}/k$-action on N. If one compares this with Ossa's calculation **[14]** of the α-invariant of S^1-actions on 3-manifolds, one obtains some interesting identities between rational functions.

A similar result holds if n is odd.

REMARK. For non-h-injective actions the α-invariant is in general not homotopy invariant, as can be seen for instance by the standard free cyclic actions on the 3-sphere; see the remark at the end of §2.

2. γ-invariants. Given a compact connected manifold Y^{2n} and a unitary representation $\rho\colon \pi_1(Y) \to U(m)$ of its fundamental group, there is an induced local coefficient system (= locally trivial sheaf) $\Gamma \to Y$ with fiber $(\boldsymbol{C}^m, h)$, where h is the standard hermitian metric on $\boldsymbol{C}^m$. These metrics on the fibers of Γ fit together to give a bilinear map of sheaves $b\colon \Gamma \times \Gamma \to \boldsymbol{C}$, where $\boldsymbol{C}$ also denotes the trivial sheaf over Y with fiber $\boldsymbol{C}$. Define a cup product form on $H^n(Y, \partial Y; \Gamma)$ by

$$S_{Y,\Gamma}\colon H^n(Y, \partial Y; \Gamma) \otimes H^n(Y, \partial Y; \Gamma) \to H^{2n}(Y, \partial Y; \Gamma \otimes \Gamma) \to H^{2n}(Y, \partial Y; \boldsymbol{C}) = \boldsymbol{C},$$

where the first map is cup product and the second is the coefficient map induced by b. This $S_{Y,\Gamma}$ is a (in general not nondegenerate) hermitian or skew-hermitian form, according as n is even or odd. Define $\operatorname{sign}(Y, \rho) = \operatorname{sign}(S_{Y,\Gamma})$, where, if $S_{Y,\Gamma}$ is skew-hermitian we mean signature of the hermitian form $+iS_{Y,\Gamma}$.

Let $X^{2n-1} = \partial Y$ and denote the composed representation $\pi_1(X) \to \pi_1(Y) \to U(m)$ also by ρ. Then

$$\gamma(X, \rho) = \operatorname{sign}(Y, \rho) - n \cdot \operatorname{sign}(Y)$$

is an invariant of (X, ρ). If (X, ρ) does not bound, but some disjoint multiple $q(X, \rho)$ does, we can define $\gamma(X, \rho) = \gamma(q(X, \rho))/q$.

This invariant also arises analytically via Atiyah, Patodi and Singer theory in **[1]**. In particular it can be defined even if no multiple of (X, ρ) bounds. If $\operatorname{Im}(\rho\colon \pi_1(X) \to U(m))$ is abelian I can give a purely topological description of the invariant also in this case, but in general this appears to be still an open problem.

One can also carry through the above definition permitting indefinite unitary representations $\rho\colon \pi_1(X) \to U(p, q)$, where $U(p, q)$ is the group of isometries of $\boldsymbol{C}^{p+q}$ with the indefinite hermitian form of type (p, q). $\gamma(X, \rho)$ is then only well defined under suitable additional assumptions, for instance if one restricts $\operatorname{Im}(\rho)$ to be abelian (or more generally to be a central extension of a finite group), see **[13]**. The results to be described hold also for this more general definition.

THEOREM 2. *If the representation $\rho\colon \pi_1(X) \to U(m)$ factors over a free abelian group, $\pi_1(X) \to \boldsymbol{Z}^s \to U(m)$, then $\gamma(X, \rho)$ is a homotopy invariant.*

In view of the following relationship between α- and γ-invariants, Theorem 1 is in fact an easy consequence of this Theorem 2.

THEOREM 3. *Let G be a finite group acting freely on M^{2n-1}. Then the covering*

$M^{2n-1} \to M/G = X^{2n-1}$ is classified by a homomorphism $f\colon \pi_1(X) \to G$. Let $\rho_i\colon G \to U(n_i)$, $i = 1, \cdots, r$, be all irreducible representations of G. Then

$$\alpha(M, g) = \sum_{j=1}^{r} \mathit{trace}(\rho_j(g))\gamma(X, \rho_j f),$$
$$\gamma(X, \rho_i f) = \frac{1}{|G|} \sum_{g \neq 1} (\mathit{trace}(\rho_i(g^{-1})) - n_i)\alpha(M, g).$$

Thus to calculate $\alpha(M^{2n-1}, G)$ for general G one can look at the $g \in G$ of finite order which act freely and compute $\alpha(M, g)$ for these via the γ-invariant. Then try to extend to arbitrary $g \in G$ using the continuity properties of $\alpha(M, G)$. If G acts freely or if G is connected this works and gives a complete calculation of $\alpha(M, G)$ in terms of γ-invariants.

Theorem 3 is a quite easy character computation and was proved in **[1]** and **[11]**.

REMARK. Some condition is necessary in Theorems 1 and 2 to conclude homotopy invariance. For example the lens spaces $L(7, 1)$ and $L(7, 2)$ are homotopy equivalent. The γ-invariants of $L(7, 1)$ with respect to the six nontrivial irreducible representations $\boldsymbol{Z}/7 \to U(1)$ are respectively: $-3/7$, $-13/7$, $-17/7$, $-17/7$, $-13/7$, $-3/7$, while for $L(7, 2)$ they are: $1/7$, $-3/7$, $-5/7$, $-5/7$, $-3/7$, $1/7$. In fact, 3-dimensional lens spaces are classified up to diffeomorphism by their γ-invariants; equivalently free linear cyclic actions on S^3 are classified up to equivariant diffeomorphism by their α-invariants.

γ-invariants of lens spaces in any dimension were completely calculated in **[11]**. They are generalized Dedekind sums (see also **[15]**).

3. Monodromy. Suppose we are given a closed manifold X^{2n-1} and a homomorphism $f_\sharp\colon \pi_1(X) \to \boldsymbol{Z}$. Since S^1 is a $K(\boldsymbol{Z}, 1)$, we can represent $f_\sharp$ by a unique map $f\colon X \to S^1$ up to homotopy. If this map f can be chosen as a fibration with fiber F^{2n-2} say, then one has the monodromy transformation $H^{n-1}(F) \to H^{n-1}(F)$ which preserves the cup product form. It is this monodromy that we wish to generalize to the case that $f\colon X \to S^1$ is not a fibration.

Let $\bar{X} \to X$ be the infinite cyclic covering classified by the homomorphism $f_\sharp\colon \pi_1(X) \to \boldsymbol{Z}$. Equivalently $\bar{X}$ is the pullback

$$\begin{array}{ccc} \bar{X} & \xrightarrow{\bar{f}} & \boldsymbol{R} \\ \downarrow & & \downarrow \\ X & \xrightarrow{f} & S^1 \end{array}$$

If $t \in S^1$ is a regular value of f and $N = f^{-1}(t)$, then $\bar{X}$ can be constructed by cutting X open along N and pasting infinitely many copies of the resulting manifold with boundary together end to end.

Let $\hat{f} \in H_{2n-2}(\bar{X})$ be the homology class represented by one copy of N in $\bar{X}$. Equivalently, $\hat{f}$ is the image of $1 \in \boldsymbol{Z}$ in the composition $\boldsymbol{Z} = H^1_c(\boldsymbol{R}) \to H^1_c(\bar{X}) \cong H_{2n-2}(\bar{X})$ induced by the proper map $\bar{f}\colon \bar{X} \to \boldsymbol{R}$ and Poincaré duality, so $\hat{f}$ only depends on the homotopy class of f.

Define a bilinear form

$$S_0\colon H^{n-1}(\bar{X}; \boldsymbol{Q}) \otimes H^{n-1}(\bar{X}; \boldsymbol{Q}) \to \boldsymbol{Q}, \qquad S_0(x, y) = \langle x \cup y, \hat{f}\rangle.$$

This form is degenerate in general, but it induces a nondegenerate form S on

$H = H^{n-1}(X; \boldsymbol{Q})/\mathrm{Rad}\, S_0$, where $\mathrm{Rad}\, S_0 = \{x \in H^{n-1}(\bar{X}; \boldsymbol{Q}) \mid S_0(x, y) = 0$ for all $y\}$.

LEMMA 4. *(H, S) is a finite dimensional vector space with nondegenerate $(-1)^{n-1}$-symmetric form. The covering transformation $\bar{X} \to \bar{X}$ induces an isometry $t\colon H \to H$.*

DEFINITION. The $(-1)^{n-1}$-symmetric isometric structure $\mathscr{H}(X, f) = (H, S, t)$ will be called the (middle-dimensional) *monodromy* of (X, f).

We have defined the monodromy over $\boldsymbol{Q}$. We could equally well have used other coefficients. If K is a field of characteristic 0 then using universal coefficient theorems it is easy to see $\mathscr{H}^K(X, f) \cong \mathscr{H}^{\boldsymbol{Q}}(X, f) \otimes K$, where the superscript indicates coefficients, so $\mathscr{H}^{\boldsymbol{Q}}(X, f)$ contains the most information (it is however false that $\mathscr{H}^{\boldsymbol{Q}}(X, f)$ equals $\mathscr{H}^{\boldsymbol{Z}}(X, f) \otimes \boldsymbol{Q}$; in fact the precise relation between the $\mathscr{H}^R(X, f)$ for different coefficient rings R remains unclear in general).

$\mathscr{H}(X, f) = \mathscr{H}^{\boldsymbol{Q}}(X, f)$ is a very rich invariant. Not only is the set of isometric structures over $\boldsymbol{Q}$ extremely abundant, but every isometric structure occurs as monodromy, at least in the skew-symmetric case. In fact:

THEOREM 5. *For any skew-symmetric isometric structure $\mathscr{H} = (H, S, t)$ over $\boldsymbol{Q}$ there exists a 3-manifold M^3 and a map $f\colon M^3 \to S^1$ such that $\mathscr{H}(M^3, f) \cong \mathscr{H}$.*

The relation between the monodromy and our previous invariants is given in the simplest case by the following theorem.

THEOREM 6. *Given $\rho\colon \pi_1(X) \to U(m)$ such that $\rho = \tau f_\#$ for some $f_\#\colon \pi_1(X) \to \boldsymbol{Z}$ and $\tau\colon \boldsymbol{Z} \to U(m)$, then $\gamma(X, \rho)$ only depends on $\mathscr{H}^{\boldsymbol{R}}(X, f)$.*

Here is a precise description of the dependence in the antisymmetric case (that is n even); the result in the symmetric case is similar. Let S_q be the $(-1)^{q-1}$-symmetric bilinear form given by the $q \times q$ matrix

$$\begin{pmatrix} & & & & & 1 \\ & & & & -1 & \\ & 0 & & 1 & & \\ & & -1 & & & \\ & \cdot^{\cdot^{\cdot}} & & & 0 & \\ (-1)^q & & & & & \end{pmatrix}$$

and let t_q be the isometry of S_q having matrix of the form

$$\begin{pmatrix} 1 & \frac{1}{2} & * & \cdots & * \\ & 1 & \frac{1}{2} & & \vdots \\ & & \ddots & & \\ & 0 & & 1 & \frac{1}{2} \\ & & & & 1 \end{pmatrix},$$

(t_q is uniquely determined by this). Define

$$\begin{aligned} \mathscr{H}_{\pm 1}^{(q)} &= (\boldsymbol{R}^q, S_q, \pm t_q), && q \text{ even}, \\ &= \left(\boldsymbol{R}^{2q}, \begin{pmatrix} 0 & S_q \\ -S_q & 0 \end{pmatrix}, \pm(t_q \oplus t_q)\right), && q \text{ odd}, \\ \mathscr{H}_\lambda^{(q)} &= \left(\boldsymbol{R}^{2q}, \begin{pmatrix} 0 & -1 \\ 1 & 0 \end{pmatrix}^q \otimes S_q, \begin{pmatrix} \cos\theta & -\sin\theta \\ \sin\theta & \cos\theta \end{pmatrix} \otimes t_q\right), && \lambda = e^{i\theta},\ 0 < \theta < \pi. \end{aligned}$$

For any isometric structure $\mathscr{H} = (H, S, t)$ define $-\mathscr{H}$ to mean $(H, -S, t)$.

By Milnor [**10**], any skew-symmetric isometric structure $\mathscr{H}$ over $\boldsymbol{R}$ is an orthogonal sum of an $\mathscr{H}_0 = (H_0, S_0, t_0)$ such that t_0 has no eigenvalue of unit length and a sum of isometric structures of the form $\pm \mathscr{H}_\lambda^{(q)}$, $\lambda = e^{i\theta}$, $0 \leqq \theta \leqq \pi$, defined as above.

For $A \in U(m)$ define an invariant $\gamma(\mathscr{H}, A)$ by requiring $\gamma(-\mathscr{H}, A) = -\gamma(\mathscr{H}, A)$ and $\gamma(\mathscr{H} \oplus \mathscr{H}', A) = \gamma(\mathscr{H}, A) + \gamma(\mathscr{H}' A)$, and putting

$$\begin{aligned}
\gamma(\mathscr{H}_0, A) &= 0, && \mathscr{H}_0 \text{ as above;}\\
\gamma(\mathscr{H}_1^{(q)}, A) &= -(-1)^{q/2}\cdot\operatorname{rank}(A - I), && q \text{ even},\\
&= 0, && q \text{ odd};\\
\gamma(\mathscr{H}_{-1}^{(q)}, A) &= (-1)^{q/2}\cdot\operatorname{corank}(A + I), && q \text{ even},\\
&= 0, && q \text{ odd};\\
\gamma(\mathscr{H}_\lambda^{(q)}, A) &= \operatorname{corank}(A - \lambda I) + \operatorname{corank}(\lambda A - I), && q \text{ even},\\
&= 2 \sum_{-\theta \leqq \psi < 2\pi - \theta} d(\psi, \theta)\cdot\operatorname{corank}(A - e^{i\psi} I), && q \text{ odd};
\end{aligned}$$

where

$$\begin{aligned}
d(\psi, \theta) &= 0 \quad \text{if } -\theta < \psi < \theta,\\
&= 1 \quad \text{if } \psi = \pm\theta,\\
&= 2 \quad \text{if } \theta < \psi < 2\pi - \theta.
\end{aligned}$$

THEOREM $6\frac{1}{2}$. *In Theorem* 6 $\gamma(X, \rho) = \gamma(\mathscr{H}^{\boldsymbol{R}}(X, f), \rho(1))$.

THEOREM 7. *If* $\rho: \pi_1(X) \to U(m)$ *factors as* $\rho = \tau g_\#$, *where* $g_\#: \pi_1(X) \to \boldsymbol{Z}^s$ *and* $\tau: \boldsymbol{Z}^s \to U(m)$, *then* $\gamma(X, \rho)$ *can be calculated via a limiting process from the monodromies* $\mathscr{H}(X, hg)$ *where h runs through all maps* $h: \boldsymbol{Z}^s \to \boldsymbol{Z}$. *Alternatively one can give an explicit calculation in terms of finitely many monodromies of X calculated with suitable local coefficients on X.*

4. Application to signature defect. The α- and γ-invariants arise naturally as correction terms to multiplicativity of signature of branched coverings and coverings of bounded manifolds; see for instance Hirzebruch [**5**].

EXAMPLE. Let $N^{4k} \to Y^{4k}$ be a d-fold covering of oriented manifolds with boundaries $M^{4k-1} = \partial N \to X^{4k-1} = \partial Y$. Then the error to multiplicativity of signature, namely $\operatorname{sign}(N) - d\cdot\operatorname{sign}(Y)$, is an invariant of $M \to X$ which is denoted "signature defect":

$$\operatorname{def}(M \to X) = \operatorname{sign}(N) - d\cdot\operatorname{sign}(Y).$$

In fact if $\pi_1(M) \subset \pi_1(X)$ is the induced inclusion of fundamental groups and $H \subset \pi_1(M)$ is a normal subgroup of $\pi_1(X)$ of finite index (H exists), and if $\rho_i: \pi_1(X)/H \to U(n_i)$, $i = 1, \cdots, r$, are all the irreducible representations of $\pi_1(X)/H$ and m_i is the dimension of the trivial component of $\rho_i|(\pi_1(M)/H)$ for each i, then

$$\operatorname{def}(M \to X) = \sum_{i=1}^{r} m_i \gamma(X, \rho_i).$$

For a proof see [**11**, Chapter III].

One obtains similar results for branched coverings by cutting out the branch locus and considering the resulting unbranched covering of manifolds with boundary.

Our calculations lead to the following periodicity result for the "signature defect" def($M \to X$).

THEOREM 8. *Let M_r^{4k-1}, $r = 1, 2, \cdots$, be a family of closed manifolds, and whenever r divides s let an (s/r)-fold cyclic covering $M_s \to M_r$ be given such that all the obvious diagrams commute. Then* def($M_s \to M_1$) *is a linear plus an almost periodic function of s.*

By an *almost periodic function* is meant the restriction of a linear combination of periodic functions $\boldsymbol{R} \to \boldsymbol{R}$. If the periods in question are rationally dependent then this linear combination is itself periodic. In fact in the above theorem the coverings $M_s \to M_1$ can all be pulled back from the standard coverings of S^1 via a suitable map $M_1 \to S^1$ and the periods in question are the numbers $1/q$, where $e^{2\pi i q}$, $0 < q \leqq 1$, runs through the eigenvalues of unit length of the monodromy of this map.

As a corollary one obtains the periodicity statement for signature of cyclic suspension of knots announced in **[12]**. This result has also been shown subsequently (for fibered knots, but the proof works for arbitrary knots) by Durfee and Kauffman **[4]** and by Cappell and Shaneson (unpublished) using an alternative method.

REMARK. def($M \to X$) also arises as the error to multiplicativity for coverings of the Atiyah-Patodi ν-invariant of a riemmanian manifold and of the Atiyah-Kreck δ-invariant of a framed manifold, so in particular we get similar "linear plus almost periodic" statements for these invariants, see **[11]**.

REFERENCES

1. M. F. Atiyah, V. K. Patodi and I. M. Singer, *Spectral assymetry and riemannian geometry*. II, Math. Proc. Cambridge Philos. Soc. **78** (1975), 405–432. MR **53** #1655B.

2. M. F. Atiyah and I. M. Singer, *The index of elliptic operators*. III, Ann. of Math. (2) **87** (1968), 546–604. MR **38** #5245.

3. P. E. Conner and F. Raymond, *Injective operations of toral groups*, Topology **10** (1971), 283–296. MR **43** #6937.

4. A. Durfee and L. Kauffman, *Periodicity of branched cyclic covers*, Math. Ann. **218** (1975), 157–174. MR **52** #6731.

5. F. Hirzebruch, *The signature of ramified coverings*, Collected Math. Papers in Honor of Kodaira, Tokyo Univ. Press, Tokyo, 1969, pp. 253–265. MR **41** #2707.

6. F. Hirzebruch and D. Zagier, *The Atiyah-Singer theorem and elementary number theory*, Math. Lecture Series 3, Publish or Perish, Boston and Berkeley, 1970.

7. W. C. Hsiang, *A splitting theorem and the Künneth formula in algebraic K-theory*, Algebraic *K*-Theory and its Geometric Applications, Lecture Notes in Math., vol. 108, Springer-Verlag, Berlin and New York, 1969, pp. 72–77. MR **40** #6560.

8. K. H. Knapp, *Signaturdefekte modulo Z für freie G-Aktionen*, Dissertation, Bonn 1975.

9. G. Lusztig, *Novilov's higher signature and families of elliptic operators*, J. Differential Geometry **7** (1972), 229–256. MR **48** #1250.

10. J. Milnor, *On isometries of inner product spaces*, Invent. Math. **8** (1969), 83–97. MR **40** #2764.

11. W. D. Neumann, *Signature related invariants of manifolds* (mimeographed notes, Chapters 1-3, Bonn 1974; revised version in preparation).

12. ———, *Cyclic suspension of knots and periodicity of signature for singularities*, Bull. Amer. Math. Soc. **80** (1974), 977–981. MR **50** #11256.

13. ———, *Multiplicativity of signature*, J. Pure Appl. Algebra (to appear).

14. E. Ossa, *Fixpunktfreie S^1-aktionen*, Math. Ann. **186** (1970), 45–52. MR **41** #7698.
15. D. Zagier, *Higher-dimensional Dedekind sums*, Math. Ann. **202** (1973), 149–172. MR **50** #9801.

UNIVERSITY OF MARYLAND

Proceedings of Symposia in Pure Mathematics
Volume 32, 1978

COVERING DIMENSION AND CHARACTERISTIC CLASSES FOR FOLIATIONS

HERBERT SHULMAN*

A flat G-bundle has characteristic classes arising from the group cohomology of G. These classes vanish in dimensions greater than the dimension of the nerve of an open cover which trivializes the bundle (e.g., any cover consisting of simply-connected open sets).

The main result of this paper is that for characteristic classes of foliations there is a similar phenomenon, but with a shift in dimension. The proof will depend on results of [**2**].

First recall the following:

DEFINITION. If an open cover $\{U_\alpha\}$ of a space has a nonempty $(r+1)$-fold intersection but no $(r+2)$-fold intersections, then $\dim(\mathrm{Nerve}\{U_\alpha\}) = r$.

Also, given a smooth codimension k foliation (actually only a Γ_k-structure is needed) on M, there is a natural map $H^*(WO_k) \to H^*(M; R)$ where WO_k is the well-known finite dimensional DGA defined in [**1**]. Elements of $H^*(WO_k)$ are referred to as the characteristic or exotic classes of foliations and also as continuous cohomology classes of $B\Gamma_k$.

Our main theorem is

THEOREM. *Let $\{U_\alpha; \varphi_\alpha: U_\alpha \to R^k\}$ be a set of charts for a codimension k foliation on a manifold M, and $r = \dim(\mathrm{Nerve}\{U_\alpha\})$. Then the characteristic class map $H^s(WO_k) \to H^s(M; R)$ is zero for $s > k + r$.*

PROOF. In the notation of [**2**], $\Gamma = \Gamma_k$ is the topological category having obj $= R^k$ and morph $=$ germs of diffeos of R^k, $(N\Gamma)_*$ is the associated simplicial space, and Ω^* denotes the de Rham functor.

AMS (MOS) subject classifications (1970). Primary 55F40, 57D20, 57D30; Secondary 55B05, 55B45, 58A10.

*Partially supported by NSF Grant MCS 76 06763.

$(N\Gamma)_*$ is a smooth simplicial manifold of dimension k, i.e., $\forall\, p$, $(N\Gamma)_p$ is a smooth k-dimensional manifold (though non-Hausdorff).

Let M_U be the usual category associated to $\{U_\alpha\}$, having objects $= \{(x, V): x \in V, V$ a finite intersection of U_α's$\}$ (see [4]). The associated simplicial space $(NM_U)_*$ is also a smooth simplicial manifold (of the same dimension as M).

By applying Ω^* to $(N\Gamma)_*$ and $(NM_U)_*$ we get double complexes using the de Rham and simplicial differentials.

We then obtain a factoring:

$$\begin{array}{ccc} H^*_{\text{total}}(\Omega^*(N\Gamma)_*) & \xrightarrow{\Phi} & H^*_{\text{total}}(\Omega^*(NM_U)_*) \\ {\scriptstyle P}\uparrow & & \uparrow{\scriptstyle W} \\ H^*(WO_k) & \xrightarrow{\Psi} & H(M;R) \end{array}$$

Here W is induced by embedding closed differential forms on M into the first "column" $\Omega^*((NM_U)_0)$.

Φ is induced by the transition functions $\varphi_{\alpha\beta}: U_\alpha \cap U_\beta \to \Gamma$ associated to the charts φ_α of the foliation.

P is explained on p. 54 of [2]. (We are using $H(WO_k) \approx H(\Omega^*NGL_k/F_k\,\Omega^*NGL_k)$.) See also [5]. Although we do not make use of it here, the recent deep theorem of J. Petro [3] states that P is an isomorphism.

Given all the above, the proof is not hard:

If $\alpha \in H^s(WO_k)$, then $P\alpha \in H^s_{\text{total}}(\Omega^*(N\Gamma)_*)$ has a cochain representative

$$\overline{P\alpha} \in \bigoplus_{p+q=s;\, q\leqq k} \Omega^q((N\Gamma)_p)$$

($q \leqq k$ since $(N\Gamma)_p$ is k-dimensional). Then $\Phi\overline{P\alpha} \in \bigoplus_{p+q=s;\, q\leqq k} \Omega^q((NM_U)_p)$ represents $\Phi P\alpha$.

If $s > k + r$, then $\Phi\overline{P\alpha} \in \bigoplus_{p>r} \Omega^q((NM_U)_p)$. But since $\dim(\text{Nerve}\{U_k\}) = r$, $(NM_U)_p$ has only degenerate elements for $p > r$.

Thus $\Phi\overline{P\alpha}$ maps into the degenerate subcomplex and hence $\Phi P\alpha = 0$.

An application of this result is the following: A codimension one foliation of S^3 with nonzero Godbillon-Vey invariant cannot be trivialized by just 2 open sets.

More generally, if a codimension k foliation can be defined by a system of charts with no $(k + 2)$-fold nonempty intersections, then all the exotic classes vanish.

Hopefully, by further use of this method, one can get a bound on the exotic numbers of a foliation in terms of the number of simplices in the nerve of a covering by charts.

REFERENCES

1. R. Bott, *Lectures on characteristic classes and foliations*, Lecture Notes in Math., vol. 279, Springer-Verlag, Berlin and New York, 1972, pp. 1–94.

2. R. Bott, H. Shulman and J. Stasheff, *On the de Rham theory of certain classifying spaces*, Advances in Math. **20** (1976), 43–56. MR **53** #6583.

3. J. Petro, *The continuous cohomology of BΓ*, Ph.D. thesis, Temple University, 1976.

4. G. Segal, *Classifying spaces and spectral sequences*, Inst. Hautes Études Sci. Publ. Math. **34** (1968), 105–112. MR **38** #718.

5. H. Shulman and J. Stasheff, *De Rham theory for BΓ* (Proc. of Rio Conf. on Foliations, Jan. 1976), Lecture Notes in Math., Springer-Verlag, Berlin and New York (to appear).

YESHIVA UNIVERSITY

Proceedings of Symposia in Pure Mathematics
Volume 32, 1978

A CONNECTED SUM DECOMPOSITION FOR COMPLETE INTERSECTIONS

JOHN W. WOOD*

The general problem is, given an even dimensional manifold X, to find a decomposition $X = M \# k(S^n \times S^n)$ where k is as large as possible. The main result here is such a decomposition when X is the transversal intersection of nonsingular hypersurfaces in complex projective space.

THEOREM 1. *If X_n is a complete intersection of even complex dimension $n > 2$ and degree $d > 1$, then there is a smooth manifold M and a differentiable connected sum decomposition $X = M \# k(S^n \times S^n)$ where rank $H_n M =$ Sign M for $n \equiv 0$ mod 4 or rank $H_n M - 2 = -$ Sign M for $n \equiv 2$ mod 4.*

This result extends, and relies on, the case of a nonsingular hypersurface treated in a joint paper with Ravi Kulkarni **[4]**.

There are some obvious necessary conditions on a general manifold X for such a splitting. Adding a few technical hypotheses we can state

THEOREM 2. *Let X be a closed manifold of dimension $2n > 4$, n even, with $\pi_1 X = 0$. Suppose $H_n(X; Z) = A \oplus B$. Then the following two sets of conditions are equivalent:*

(1) $X = M \# N$ where $H_n M = A$, $H_n N = B$, and $N = S^n \times S^n \# \cdots \# S^n \times S^n$.

(2) $B \subset \operatorname{im}\{\pi_n X \to H_n X\}$, further if $\beta \in B$ there is a map $f\colon S^n \to X$ such that $f_*[S^n] = \beta$ and $f^*\tau X$ is stably trivial, and there is a basis for B with respect to which the intersection pairing has matrix $\left(\begin{smallmatrix} 0 & I \\ I & 0 \end{smallmatrix}\right)$.

PROOF. The splitting (1) easily implies the conditions (2), for example τN is actually trivial on the n-skeleton. Conversely the symplectic basis for B in (2) can be represented by maps $S^n \to X$ such that the pull-backs of τX are stably trivial.

AMS (MOS) subject classifications (1970). Primary 57D65, 14M10.
*Research partially supported by the National Science Foundation.

Since X is simply connected and of dimension > 4, by results of Whitney and Haefliger [2] these maps are homotopic to embeddings. Further by the Whitney process we obtain disjoint embeddings of $S^n \vee S^n$ (each with a point of transversal intersection). The normal bundles are stably trivial since $\tau X \mid S^n$ and τS^n are. But since n is even and the self-intersections are 0, the normal bundles are trivial [1, p. 88]. This means that each $S^n \vee S^n$ sits in X exactly as $S^n \times * \cup * \times S^n \subset S^n \times S^n$, so $N - D^{2n} \subset X$ representing the summand B. The splitting (1) follows.

A complete intersection X_n (of complex dimension n) is the transversal intersection of $m - n$ hypersurfaces in CP_m. Thus there is a sequence $X_n \subset X_{n+1} \subset \cdots \subset CP_m$ where X_j is the transversal intersection of X_{j+1} with a hypersurface. Let $i : X_n \subsetneq CP_m$. To prove Theorem 1 we will verify that complete intersections satisfy conditions (2) of Theorem 2.

LEMMA 1. *If $f\colon S^n \to X_n, f^* \tau X_n$ is stably trivial.*

PROOF. Let $\nu_j = \nu(X_j \subset X_{j+1})$ so that $f^*\tau X_n \oplus f^*\nu_n \oplus f^*\nu_{n+1} \mid X_n \oplus \cdots = f^*i^*\tau CP_m$. Since $n > 2$, $f \circ i;\ S^n \to CP_m$ is null-homotopic, so $f^*i^*\tau CP_m$ is trivial. Also each ν_j is an oriented 2-plane bundle, so $f * \nu_j \mid X_n$ is trivial (because $\pi_{n-1} SO(2) = 0$). Therefore $f^*\tau X_n$ is stably trivial.

The image of the Hurewicz homomorphism can be described in terms of the intersection pairing on $H_n X_n$.

LEMMA 2. *There is a class $h \in H_n X$ with $h \cdot h = d$, the degree of X_n, and* $\operatorname{im}\{\pi_n X \to H_n X\} = h^\perp = \{u \in H_n X \colon u \cdot h = 0\}$.

PROOF. The inclusion $i\colon X_n \subsetneq CP_m$ is an n-equivalence by the Lefschetz theorem so for n even there is a diagram:

$$\begin{array}{ccccccccc} 0 & \longrightarrow & H_{n+1}(CP_m, X_n) & \longrightarrow & H_n X & \xrightarrow{i_*} & H_n CP_m & \longrightarrow & 0 \\ & & \uparrow{=} & & \uparrow & & & & \\ & & \pi_{n+1}(CP_m, X_n) & \xrightarrow{=} & \pi_n X & & & & \end{array}$$

Therefore $\operatorname{im}\{\pi_n X \to H_n X\} = \ker\{i_*\colon H_n X \to H_n CP_m\}$. Now let $x \in H^2(CP_m)$ be the generator satisfying $x \cap [CP_m] = [CP_{m-1}]$ and let $h = i^* x^{n/2} \cap [X] \in H_n X_n$. Then $h \cdot u = (x^{n/2} \cap i_* u)\,[CP_m]$ so $h \cdot u = 0$ if and only if $i_* u = 0$. Thus $h^\perp = \ker i_*$. If the hypersurfaces whose intersection is X_n have degrees $d_1, \cdots, d_{m-n}$ and d is their product then $i_*[X] = dx^{m-n} \cap [CP_m]$ and $h \cdot h = d > 0$.

This situation is the same as in the case of even dimensional hypersurfaces studied in [4]. We have a free abelian group $H = H_n X$ with a unimodular bilinear form given by the intersection pairing (so H is an inner product space) and a subspace $h^\perp$. To split $h^\perp$ and H we quote an algebraic result from [4, Theorem 10.1]:

THEOREM 3. *Let H be an inner product space over Z with $|\operatorname{Sign} H| \geqq 5$ and let $h \in H$ satisfy $h \cdot h > 0$. Then $h^\perp$ and H are isomorphic to orthogonal direct sums $h^\perp = A \oplus B$, $H = A' \oplus B$, where A is a definite bilinear form space, $A' = \operatorname{span}\{A, h\}$, and B has intersection matrix $\left(\begin{smallmatrix} 0 & I \\ I & 0 \end{smallmatrix}\right)$ or $\left(\begin{smallmatrix} I & 0 \\ 0 & -I \end{smallmatrix}\right)$ with respect to a suitable basis.*

For the proof see [4, §10], the difficulty is caused by the fact that the intersection form on $h^\perp$ is not unimodular. Note A' is definite if and only if $\operatorname{Sign} H > 0$ (since $h \cdot h > 0$).

By Lemma 1 the self-intersection of any spherical cycle is even (since $w_n \nu = 0$) so B has intersection matrix $\left(\begin{smallmatrix} 0 & I \\ I & 0 \end{smallmatrix}\right)$. Hence Theorem 3 will provide the summand B satisfying the conditions (2) of Theorem 2 once we have verified the condition on the signature.

LEMMA 3. *If X_{2m} is a complete intersection of degree d and complex dimension $2m$, then $(-1)^m$ Sign $X_{2m} \geqq 5$ for $m \geqq 2$ and $d \geqq 3$.*

PROOF. Let $X_n^{\underline{d}}$ denote the transversal intersection of hypersurfaces of degrees $\underline{d} = (d_1, \cdots, d_r)$ in CP_{n+r}, so $d = d_1 \cdots d_r$. The signatures of these complete intersections in various dimensions are related to the signature of the hypersurfaces by the equation

$$(*) \qquad \sum_{m=0}^{\infty} (-1)^m \operatorname{Sign} X_{2m}^{\underline{d}}\, t^{2m} = (1+t^2)^{r-1} \prod_{j=1}^{r} \left\{ \sum_{m=0}^{\infty} (-1)^m \operatorname{Sign} X_{2m}^{d_j}\, t^{2m} \right\}.$$

This is a consequence of formula (2) of [**3**, §22]. We may assume that no $d_j = 1$. Then the coefficients $(-1)^m$ Sign $X_{2m}^{d_j}$ on the right are all $\geqq 0$. In fact, for $d_j \geqq 3$, $(-1)^m$ Sign $X_{2m}^{d_j} \geqq 5$ by [**4**, Lemma 8.4] and in case some $d_j = 2$, $\sum_{m=0}^{\infty} (-1)^m \operatorname{Sign} X_{2m}^2 t^{2m} = 2(1 - t^4)^{-1}$. Thus if any $d_j \geqq 3$, the conclusion of the lemma is clear from $(*)$ and if all $d_j = 2$ and $m \geqq 2$ it is easily checked.

Lemma 3 shows that the algebraic splitting Theorem 3 and hence the conditions (2) of Theorem 2 hold for complete intersections of degree $\geqq 3$. In case degree $X_n = 2$, $M = X$ satisfies the conclusion of Theorem 1. This completes the proof of Theorem 1.

REFERENCES

1. W. Browder, *Surgery on simply-connected manifolds*, Springer-Verlag, Berlin and New York 1971.

2. A. Haefliger, *Differentiable imbeddings*, Bull. Amer. Math. Soc. **67** (1961), 109–112. MR **23** #A665.

3. F. Hirzebruch, *Topological methods in algebraic geometry*, 3rd ed., Springer, Berlin and New York, 1966. MR **34** #2573.

4. R. S. Kulkarni and J. W. Wood, *Topology of nonsingular hypersurfaces* (to appear).

UNIVERSITY OF ILLNOIS AT CHICAGO CIRCLE

Proceedings of Symposia in Pure Mathematics
Volume 32, 1978

NORMAL SINGULARITIES OF SURFACES

STEPHEN SHING-TOUNG YAU

Let p be the unique singularity of a normal two dimensional Stein space V. Let $\pi: M \to V$ be a resolution of V. It is known that the geometric genus $=$ $\dim H^1(M, \mathcal{O})$ is independent of resolution. In [1], M. Artin developed a theory for those singularities with $\dim H^1(M, \mathcal{O}) = 0$ (rational singularities). Recently Laufer [5] has developed a theory for those Gorenstein singularities with $\dim H^1(M, \mathcal{O}) = 1$ (minimally elliptic singularities). In [9], we developed a theory for those Gorenstein singularities with $\dim H^1(M, \mathcal{O}) = 2$. If the geometric genus of p is equal to one, then p is called a strongly elliptic singularity. Suppose π is the minimal good resolution. Let Γ denote the associated weighted dual graph, including the genera of the irreducible components. In [6], Laufer developed a deformation theory preserving Γ. This theory allows him to introduce the notion of a property of the associated singularity holding generically for Γ. Let Z be the fundamental cycle. Let $\mathcal{O}(-Z)$ be the sheaf of germs of holomorphic functions on M whose divisors are at least Z. Let $\mathcal{O}_Z = \mathcal{O}/\mathcal{O}(-Z)$. Then $\chi(Z) = \dim H^0(M, \mathcal{O}_z) - \dim H^1(M, \mathcal{O}_Z)$ may be computed from Γ via Riemann-Roch Theorem. Weak ellipticity is $\chi(Z) = 0$. If $\chi(Z) = 0$, Laufer proved that generically $\dim H^1(M, \mathcal{O}) = 1$. Therefore, it is interesting to build up a theory for those non-Gorenstein singularities such that $\dim H^1(M, \mathcal{O}) = 1$. In this paper, we complete the theory for strongly elliptic singularities.

Let m be the maximal ideal in ${}_V\mathcal{O}_p$, the local ring of germs of holomorphic functions at p. One important question in the theory of normal two-dimensional singularities is "the identification of m." We first define the maximal ideal cycle which serves to identify the maximal ideal. It is known that the maximal ideal cycle can have lower estimate in terms of fundamental cycle [1]. We give a nonlower estimate for maximal ideal cycle in terms of the canonical divisor which is computable via the topological information, i.e., the weighted dual graph of the singular-

AMS (MOS) subject classifications (1970). Primary 13C40, 13C45, 14B05, 14E15, 32C40; Secondary 14J15, 14C20.

ity. In this paper, we also develop a theory for a general class of weakly elliptic singularities which satisfy a maximality condition. Maximally elliptic singularities may have the geometric genus arbitrarily large. Also minimally elliptic singularities in the sense of Laufer are maximally elliptic singularities. We prove that maximally elliptic singularities are Gorenstein singularities. We are able to identify the maximal ideal. Therefore, the important invariants of the singularities are extracted from the topological information. The topological classification of hypersurface maximally elliptic singularities can be obtained. But the list is too long to be included. For weakly elliptic singularities, we introduced a new concept called elliptic sequence. This is defined purely topologically, i.e., it can be computed explicitly via the intersection matrix. It turns out that weakly elliptic singularities can be effectively studied by the method of elliptic sequences. We prove that $-K$, where K is the canonical divisor, is equal to the summation of the elliptic sequence. Moreover, the analytic data dim $H^1(M, \mathcal{O})$ is bounded by the topological data, the length of the elliptic sequence. It is a pleasure to acknowledge the help and encouragement of Professor Henry Laufer in this research. We would also like to thank Professor Hironaka and Professor Mumford for their interest in our work.

1. Maximal ideal cycle.

DEFINITION 1.1. Let $A = \pi^{-1}(p)$ be the exceptional set in a resolution $\pi: M \to V$. Suppose that $\{A_i\}$, $1 \leqq i \leqq n$, are the irreducible components of A. If $f \in m$, then the divisor of f, $(f) = [f] + D$ where $[f] = \sum n_i A_i$ and D does not involve any of A_i. Let Y be the positive cycle such that $Y = \inf_{f \in m} [f]$. Then Y is called the *maximal ideal cycle*.

DEFINITION 1.2. Use the notation of Definition 1.1. Let K be the canonical divisor on M. We define the *negative cycle* $K' = \sum k_i A_i$ on A where $k_i \in \mathbf{Z}$, to be a cycle such that $A_i \cdot K' = A_i \cdot K$ for all $A_i \subseteq A$ (K' does not always exist).

THEOREM 1.3. *Let $\pi: M \to V$ be the minimal resolution of a normal two-dimensional Stein space with p as its only singular point. Suppose K' exists and* dim $H^1(M, \mathcal{O}) \geqq 2$; *then the maximal ideal cycle relative to π cannot be greater than or equal to* $-K'$.

If ${}_V\mathcal{O}_p$ is a Gorenstein ring, then K' exists. We remark that all hypersurface singularities and complete intersections are Gorenstein.

THEOREM 1.4. *If we assume ${}_V\mathcal{O}_p$ is Gorenstein in Theorem* 1.3, *then the same result holds even if π is not necessarily the minimal resolution.*

2. Elliptic sequence. Let Z be the fundamental cycle [**1**, p. 132] of A. A singularity is weakly elliptic if $\chi(Z) = 0$ [**2**], [**4**]. In [**2**], Laufer defines a minimally elliptic cycle E.

THEOREM 2.1. *Let $\pi: M \to V$ be a resolution of the normal two-dimensional Stein space with p as its only singularity. Suppose ${}_V\mathcal{O}_p$ is Gorenstein and $H^1(M, \mathcal{O}) = \mathbf{C}^2$. Then p is a weakly elliptic singularity.*

DEFINITION 2.2. Let D be a positive cycle on A. Then Z_D is the fundamental cycle on the support of D $(= |D|)$.

DEFINITION 2.3. Let A be the exceptional set of the minimal good resolution

$\pi: M \to V$ where V is a normal two-dimensional Stein space with p as its only weakly elliptic singularity. If $E \cdot Z < 0$, we say that the *elliptic sequence* is $\{Z\}$ and the *length of the elliptic sequence* is equal to one. Suppose $E \cdot Z = 0$. Let B_1 be the maximal connected subvariety of A such that $B_1 \supseteq |E|$ and $A_i \cdot Z = 0$ for all $A_i \subseteq B_1$. Since A is an exceptional set, $Z \cdot Z < 0$. So B_1 is properly contained in A. Suppose $Z_{B_1} \cdot E = 0$. Let B_2 be the maximal connected subvariety of B_1 such that $B_2 \supseteq |E|$ and $A_i \cdot Z_{B_1} = 0$ for all $A_i \subseteq B_2$. By the same argument as above, B_2 is properly contained in B_1. Continuing this process, we finally obtained B_m with $Z_{B_m} \cdot E < 0$. We call $Z_{B_0} = Z, Z_{B_1}, \cdots, Z_{B_m}$ the *elliptic sequence* and the *length of elliptic sequence is* $m + 1$.

THEOREM 2.4. *Use the notation of Definition* 2.3. *Suppose p is not a minimally elliptic singularity and K′ exists. Then the elliptic sequence is of the following form*:

$$Z_{B_0} = Z,\ Z_{B_1}, \cdots, Z_{B_l}, Z_{B_{l+1}} = Z_E, \quad l \geqq 0.$$

Moreover, $-K' = \sum_{i=0}^{l} Z_{B_i} + E$ *and* $\dim H^1(M, \mathcal{O}) \leqq l + 2 =$ *the length of the elliptic sequence.*

3. Maximally elliptic singularities. Let V be a normal 2-dimensional Stein space with p as its only weakly elliptic singularity. Let $\pi: M \to V$ be the minimal good resolution. Suppose K' exists. If $\dim H^1(M, \mathcal{O}) =$ length of the elliptic sequence, then p is called a *maximally elliptic singularity*.

THEOREM 3.1. *If p is a maximally elliptic singularity, then* ${}_V\mathcal{O}_p$ *is Gorenstein.*

THEOREM 3.2. *Use the above notation. Suppose p is a maximally elliptic singularity. If* $Z_E \cdot Z_E \leqq -2$, *then* $m\mathcal{O} = \mathcal{O}(-Z)$. *If* $Z_E \cdot Z_E \leqq -3$, *then*

$$H^0(M, \mathcal{O}(-Z)) \underset{C}{\otimes} H^0(M, \mathcal{O}(-nZ)) \longrightarrow H^0(M, \mathcal{O}(-(n+1)Z))$$

is surjective for all $n < 1$. *If we assume further that the length of the elliptic sequence is equal to two, then the above map is surjective for all* $n \geqq 1$. *In this case* $m^n \approx H^0(A, \mathcal{O}(-nZ))$ *for all* $n \geqq 1$ *where* $A = \pi^{-1}(p)$.

4. Laufer sequence.

DEFINITION 4.1. Suppose π is the minimal good resolution and p is a weakly elliptic singularity. Let E be the minimally elliptic cycle. If $E \cdot Z < 0$, we say that *the Laufer sequence* is $\{Z\}$ and *the length of Laufer sequence* is equal to 1. Suppose $E \cdot Z = 0$. Let L_1 be the maximal connected subvariety of A such that $L_1 \supseteq |E|$ and $A_i \cdot Z = 0$ for all $A_i \subseteq L_1$. Since A is an exceptional set, L_1 is properly contained in A. Let Z_{L_1} be the fundamental cycle on L_1. Suppose $Z_{L_1} \cdot E = 0$. Let L_2 be the maximal connected subvariety of A such that $L_2 \supseteq |E|$ and $A_i \cdot (Z + Z_{L_1}) = 0$ for all $A_i \subseteq L_2$. Having defined L_{i-1}, let L_i be the maximal connected subvariety containing $|E|$ such that for all $A_j \subseteq L_i$, $A_j \cdot (Z_{L_0} + Z_{L_1} + \cdots + Z_{L_{i-1}}) \leqq 0$ where $Z_{L_0} = Z$. Continuing this process, we finally obtain L_m with $(\sum_{i=0}^{m} Z_{L_i}) \cdot E = Z_{L_m} \cdot E < 0$ (this will be justified in Proposition 1.2). We call $\{Z_{L_0}, Z_{L_1}, \cdots, Z_{L_m}\}$ *the Laufer sequence* and *the length of the Laufer sequence* is $m + 1$.

The definition of Laufer sequence is purely topological, i.e., it can be computed via intersection matrix. The reader should observe the difference between the Laufer sequence and elliptic sequence which we defined in 2.3.

PROPOSITION 4.2. *The Laufer sequence is well defined in the sense that the process above is stopped after a finite number of steps. Let* $\{Z_{L_0}, Z_{L_1}, \cdots, Z_{L_m}\}$ *be the Laufer sequence. Then* $\chi(\sum_{i=0}^{h} Z_{L_i}) = 0$, $0 \leqq h \leqq m$ *and* $A_j \cdot (\sum_{i=0}^{h} Z_{L_i}) \leqq 0$ *for all* $A_j \subseteq A$. *Moreover*, $\dim H^1(M, \mathcal{O}) \leqq m + 1$.

5. Calculation of the multiplicities and Hilbert functions. Suppose $\dim H^1(M, \mathcal{O}) = 1$. We identify the maximal ideal m of ${}_V\mathcal{O}_p$. In particular, we get a formula for the multiplicity of the singularity. We also calculate the Hilbert function of ${}_V\mathcal{O}_p$. In particular, the dimension of the Zariski tangent space is computed. Hence we know the lowest possible embedding dimension of the singularity. We remark that all the results here are sharp.

THEOREM 5.1. *Let* $\pi: M \to V$ *be a minimal good resolution of a normal two-dimensional Stein space with p as its only strongly elliptic singularity. Let* $\{Z_{L_0}, Z_{L_1}, \cdots, Z_{L_n}\}$ *be the Laufer sequence. Then* $m\mathcal{O} \subseteq \mathcal{O}(-\sum_{i=0}^{n} Z_{L_i})$. *If* $Z_{L_n} \cdot Z_E \leqq -2$, *then* $m\mathcal{O} = \mathcal{O}(-\sum_{i=0}^{n} Z_{L_i})$ *provided that either one of the following holds*: (1) $Z_E = E$, *i.e.,* π *is the minimal resolution*; (2) $Z_{L_n}/|E| = Z_E$.

If $Z_{L_n} \cdot Z_E \leqq -3$, then $\dim m^j/m^{j+1} = -j(\sum_{i=0}^{n} Z_{L_i}^2)$ provided that one of the above (1) or (2) holds.

REFERENCES

1. M. Artin, *On isolated rational singularities of surfaces*, Amer. J. Math. **88** (1966), 129–136. MR **33** #7340.

2. R. Gunning, *Lectures on Riemann surfaces*, Princeton Univ. Press, Princeton, N.J., 1966. MR **34** #7789.

3. H. Laufer, *Normal two-dimensional singularities*, Ann. of Math. Studies, No. 71, Princeton Univ. Press, Princeton, N.J., 1971. MR **47** #8904.

4. ———, *On rational singularities*, Amer. J. Math. **94** (1972), 597–608. MR **48** #8837.

5. ———, *On minimally elliptic singularities*, Amer. J. Math. (to appear).

6. ———, *Deformations of resolutions of two-dimensional singularities*, Rice Univ. Studies **59** (1973), 53–96.

7. P. Wagreich, *Elliptic singularities of surfaces*, Amer. J. Math. **92** (1970), 419–454. MR **45** #264.

8. Stephen S.-T. Yau, *On maximally elliptic singularities*, Trans. Amer. Math. Soc. (to appear).

9. ———, *On almost minimally elliptic singularities*, Bull. Amer. Math. Soc. **83** (1977), 362–364.

THE INSTITUTE FOR ADVANCED STUDY

Current Address: HARVARD UNIVERSITY

H-SPACES, LOOP SPACES, AND *CW* COMPLEXES

Proceedings of Symposia in Pure Mathematics
Volume 32, 1978

ON THE TOPOLOGY OF FINITE H-SPACES

JAMES P. LIN*

1. Introduction. During the past four years I have been fascinated with homological and homotopical properties of finite H-spaces. This lecture will attempt to survey some of the more recent developments in this field. All proofs and technicalities will be pushed aside; instead, I would like to motivate and propagandize for a technique that seems to explain a lot of the work done on the topology of Lie groups (resp. finite H-spaces) in the last twenty-five years. This technique was first described in Alex Zabrodsky's thesis **[21]**, and has now been refined to the point where it admits many applications.

Here is the basic idea. If X is an H-space with finitely many cells in each dimension, then the cohomology $H^*(X; \boldsymbol{Z}_p)$ and the homology $H_*(X; \boldsymbol{Z}_p)$ are dual Hopf algebras. One can ask, given an element t in $PH_*(X; \boldsymbol{Z}_p)$, when is its pth power t^p nonzero? The dual question in cohomology is, given an element x in $H^*(X; \boldsymbol{Z}_p)$ with x dual to t (written $\langle x, t\rangle \neq 0$), when is there an x_1 in $H^*(X; \boldsymbol{Z}_p)$ with $\langle x_1, t^p\rangle \neq 0$?

The duality means

$$\begin{aligned}\langle x_1, t^p\rangle &= \langle \bar{\Delta} x_1, t^{p-1} \otimes t\rangle \\ &= \langle(\bar{\Delta} \otimes 1)\bar{\Delta} x_1, t^{p-2} \otimes t \otimes t\rangle \\ &= \langle \bar{\Delta}^{p-1} x_1, t \otimes t \otimes \cdots \otimes t\rangle\end{aligned}$$

where $\bar{\Delta}^{p-1} = (\bar{\Delta} \otimes \cdots \otimes 1)(\bar{\Delta} \otimes \cdots \otimes 1) \cdots (\bar{\Delta} \otimes 1)\bar{\Delta}$. One sees that a sufficient condition for x_1 to be dual to t^p is if $\bar{\Delta}^{p-1} x_1 = x \otimes x \otimes \cdots \otimes x$. Then

$$\begin{aligned}\langle x_1, t^p\rangle &= \langle \bar{\Delta}^{p-1} x_1, t \otimes \cdots \otimes t\rangle \\ &= \langle x \otimes \cdots \otimes x, t \otimes \cdots \otimes t\rangle \\ &= \langle x, t\rangle^p \neq 0.\end{aligned}$$

AMS (MOS) subject classifications (1970). Primary 57F10, 57F25; Secondary 57F05, 55G20, 55D45.

*The author was partially supported by the National Science Foundation and the Sloan Foundation during the preparation of this paper.

An example of such an x_1 is a pth divided power. It turns out that there exist "universal examples" of spaces that contain elements dual to pth powers. These examples have been worked out by many people (e.g. [22]), so we do not have to work them out ourselves. These people have shown that there are two-stage Postnikov systems associated to factorizations

$$\beta_1\mathscr{P}^n = \sum a_i b_i \quad \text{for } p \text{ odd,}$$
$$Sq^{n+1} = \sum a_i b_i \quad \text{for } p = 2.$$

Given such a two-stage system E, there are elements $v \in H^{2np}(E; \boldsymbol{Z}_p)$, $u \in H^{2n}(E; \boldsymbol{Z}_p)$ with $\bar{\Delta}^{p-1}(v) = u \otimes u \otimes \cdots \otimes u$ for p odd, $v \in H^{2n}(E; \boldsymbol{Z}_2)$, $u \in H^n(E; \boldsymbol{Z}_2)$ with $\bar{\Delta} v = u \otimes u$ for $p = 2$.

We can use these spaces to detect elements in the cohomology of H-spaces. That is

(1) Find a map $f: X \to E$ with $f^*(u) = x$.

(2) Measure the deviation D_f of f from being an H-map.

Then

$$\bar{\Delta}^{p-1} f^*(v) = x \otimes \cdots \otimes x + \text{terms involving } D_f.$$

The key to applying this technique is to measure D_f. Often, one can make D_f involve primary cohomology operations. In this case, either x is in the image of primary operations, or there is a new generator x_1 in the mod p cohomology.

If x_1 is nonzero, then applying an operation to x_1, one detects an x_2 that is dual to a primative pth power. Often, this process detects an infinite sequence of nonzero cohomology classes. (See Lin [16] for all the details.) This contradicts the finite dimensionality of $H^*(X; \boldsymbol{Z}_p)$. Hence one concludes that x must be in the image of primary Steenrod operations. In this manner, we obtain information about the action of the Steenrod algebra on $H^*(X; \boldsymbol{Z}_p)$.

2. Applications. A. *The loop space conjecture.* Historically, a key result in the study of Lie groups is the following theorem due to R. Bott [6]:

BOTT'S THEOREM. *Let G be a simply connected Lie group and let ΩG be the base-pointed loops on G. Then the integral homology $H_*(\Omega G; \boldsymbol{Z})$ is torsion free.*

Bott's proof of this theorem involves Morse theory and relies heavily on the differentiable manifold structure of a Lie group. Bott and Samelson [7] use this result to explore homotopy theoretic properties of Lie groups. Later, Araki [3] observed that Bott's theorem restricts the structure of the Postnikov towers of the compact exceptional groups E_7 and E_8 and he uses this fact to compute their mod p cohomology.

After Bott proved this result, it was conjectured that the integral homology of the loops on a finite simply connected associative H-space should be torsion free. This conjecture has been the topic of exciting research that brought together the theory of fibre bundles, secondary operations and Hopf algebras. It has been known for years [9] that if the mod p cohomology of an H-space X has no even degree algebra generators, then the homology of ΩX has no p-torsion. The loop space conjecture therefore is reduced to the case where the mod p cohomology contains even degree algebra generators. In this case, work of many authors **[11]** shows that if the module of even degree indecomposables has the form

$$QH^{\text{even}}(X; Z_p) = \sum_{l=1}^{\infty} \beta_1 \mathscr{P}^l QH^{2l+1}(X; Z_p)$$

then $H_*(\Omega X; Z)$ has no p-torsion. An application of the secondary operation proves this fact for odd primes p. Hence

THEOREM A. *Let X be a simply connected finite H-space. Then $H_*(\Omega X; Z)$ has no odd torsion.*

B. *Torsion coefficients in $H^*(X; Z)$.* In a series of papers, Araki and Borel **[3]**, **[5]** use Cartan's classification theorem to make case-by-case computations of the p-torsion coefficients for the integral cohomology of all simply connected Lie groups. Their work has been enriched, recently, by the discovery of finite, simply connected H-spaces that have p-torsion in their integral cohomology and are not homotopy equivalent to any Lie group. These examples are due to Harper **[12]** who has constructed for each odd prime p a finite simply connected H-space whose integral cohomology has nonzero p-torsion. In Harper's examples and in computations of Araki and Borel, it was discovered that the p-torsion coefficients are of order at most p. We can now generalize these theorems:

THEOREM B. *For p odd, X finite simply connected H-space, the p torsion coefficients of $H^*(X; Z)$ are of order at most p.*

Note that this follows because $QH^{\text{even}}(X; Z_p)$ is in the image of β_1.

C. *The K-theory of a finite H-space.* The K-theory of simply connected Lie groups has a simpler structure than the ordinary cohomology groups. This was first observed by Hodgkin **[13]**, who discovered that the K-theory of a simply connected Lie group is torsion free and has the structure of an exterior algebra. The key step in the proof of this fact is his use of the classifications theorem for Lie groups to show every even generator of the mod p cohomology of a simply connected Lie group is in the image of Milnor's primary operation Q_1. Then a simple application of the Atiyah-Hirzebruch spectral sequence for mod p K-theory proves the theorem. Araki and Atiyah **[4]** asked whether this theorem depends on the classification theorem for simply connected Lie groups. They were able to prove Hodgkin's theorem "without classification", by appealing to an argument that involved the maximal torus of a Lie group.

By applying the secondary operation described, one can show that $QH^{\text{even}}(X; Z_p)$ is in the image of Q_1. This is enough to show

THEOREM C. *If X is a simply connected finite H-space, the K-theory $K^*(X; Z)$ has no odd torsion and the* mod p *K-theory is an exterior algebra. Thus, for odd primes, Hodgkin's theorem is an invariant of the H-structure alone.*

D. *The Hurewicz map.* The loop space theorem implies many facts about the Hurewicz map. An application of the suspension map shows that *the kernel of the Hurewicz map*

$$\pi_n(X) \otimes \mathbf{Z}_{(p)} \xrightarrow{h_n \otimes \mathbf{Z}_{(p)}} PH_n(X; \mathbf{Z}_{(p)})$$

is precisely the p-torsion of $\pi_n(X)$. Therefore, the Hurewicz theorem implies *the first nonvanishing homotopy group of X has no odd torsion.* These theorems completely codify the Hurewicz map for odd primes. They generalize theorems of Weingram [19] and Browder [10]. For finite H-spaces, Weingram shows h_{2n} is always trivial and Browder proves the mod p Hurewicz map

$$\pi_n(X) \otimes \mathbf{Z}_p \xrightarrow{h_n \otimes \mathbf{Z}_p} PH_n(X; \mathbf{Z}_p)$$

is trivial in even degrees.

E. *The even degree cohomology generators at an odd prime.* One of the ultimate goals in the theory of H-spaces is the classification of their mod p cohomology rings over the Steenrod algebra. It is known from the Borel structure theorem for mod p Hopf algebras that the cohomology $H^*(X; \mathbf{Z}_p)$ splits as algebras into the tensor product of an exterior algebra on generators of odd degree with a polynomial algebra on generators of even degree. Borel shows that each generator of the polynomial algebra either has infinite height or is truncated at height p^f for some positive integer f. On the other hand, Adams [1] has shown that every odd sphere S^{2n+1} localized at an odd prime p is an H-space. Therefore by considering a product of odd dimensional spheres localized at p, one can construct H-spaces whose mod p cohomology is any exterior algebra on generators of odd degree. Hence, any classification of mod p cohomology rings cannot place any restrictions on the degrees of the odd generators.

However, one can say a great deal about even generators of the mod p cohomology ring. Careful analysis shows

THEOREM E. *Let p be an odd prime. The even degree algebra generators of* $H^*(X; \mathbf{Z}_p)$ *are concentrated in degrees of the form* $2p^j - 2$ *and* $(2p^n - 2) + 2(p^{j+1} + p^{j+2} + \cdots + p^t)$. *Moreover, every generator of the second form is in the image of a Steenrod reduced power.*

Thus, for example the mod 3 cohomology of the compact exceptional group E_8 contains two even generators, one in degree 20 and one in degree 8. The 20 degree generator is connected to the 8 degree generator by the Steenrod reduced power $\mathcal{P}^3$. This type of result was motivated by a theorem due to Kane [15]. We generalize his result to H-spaces whose mod p homology ring is not associative.

F. *The height theorem for p odd.* In the case of a finite H-space, every even mod p cohomology generator must be truncated, because the mod p cohomology is a finite dimensional vector space. It is interesting to note that all the known examples of finite, simply connected H-spaces have their even generators truncated at height p. A Hopf algebra theorem of Milnor and Moore [17] then implies the mod p homology is primitively generated. This leads one to ask if every finite simply connected H-space has a primitively generated mod p homology ring.

Our tight hold on the even generators implies

THEOREM F. *Every even mod p cohomology generator is truncated at height p or* p^2.

3. Concluding remarks. After proving these results, I began to wonder if much of the previous work on finite H-spaces could be interpreted in the context of this secondary operation. I discovered that the following theorems can be proven by essentially detecting pth powers in the homology of an H-space.

THEOREM 1 (ADAMS [2]). *The spheres that are H-spaces are* S^1, S^3 *and* S^7.

For details, see Zabrodsky's article [23].

THEOREM 2 (BROWDER [10]). *The second homotopy group of a finite H-space is trivial.*

This can be proven by looking at the two-stage Postnikov system associated to $a_1 = \beta_1, \mathscr{P}^n = b_1, \beta_1\mathscr{P}^n = a_1b_1$.

THEOREM 3 (THOMAS [18]). *If X is a finite H-space with primitively generated* mod 2 *cohomology, then for all r, k* > 0

$$PH^{2^r+2^{r+1}k-1}(X; Z_2) = Sq^{2^r}PH^{2^{r+1}k-1}(X; Z_2).$$

Thomas works in the projective plane. But if one builds a Postnikov system based on the relations he uses, it is easy to see that he is really detecting the dual of a square.

THEOREM 4 (HUBBUCK [14]). *Every finite homotopy commutative, homotopy associative H-space is a torus.*

This is implied by Zabrodsky [20] and Browder [8].

REFERENCES

1. J. F. Adams, *H-spaces with few cells*, Topology **1** (1962), 67–72. MR **26** #5574.

2. ———, *On the nonexistence of elements of Hopf invariant one*, Ann. of Math. (2) **72** (1960), 20–104. MR **25** #4530.

3. S. Araki, *Differential Hopf algebras and the cohomology* mod 3 *of the compact exceptional groups* E_7 *and* E_8, Ann. of Math. (2) **73** (1961), 404–436. MR **23** #A1372.

4. ———, *Hopf structures attached to K-theory: Hodgkin's theorem*, Ann. of Math. (2) **93** (1970), 508–525.

5. A Borel, *Topology of Lie groups and characteristic classes*, Bull. Amer. Math. Soc. **61** (1955), 397–432. MR **17**, 282.

6. R. Bott, *On torsion in Lie groups*, Proc. Nat. Acad. Sci. U.S.A. **40** (1954), 586–588. MR **16**, 12.

7. R. Bott and H. Samelson, *The cohomology ring of G/T*, Proc. Nat. Acad. Sci. U.S.A. **41** (1955), 490–493. MR **17**, 182.

8. W. Browder, *Homotopy commutative H-spaces*. Ann. of Math. (2) **75** (1962), 283–311. MR **27** #765.

9. ———, *On differential Hopf algebras*, Trans. Amer. Math. Soc. **107** (1963), 153–176. MR **26** #3061.

10. ———, *Torsion in H-spaces*, Ann. of Math. (2) **74** (1961), 24–51. MR **23** #A2201.

11. A. Clark, *Homotopy commutativity and the Moore spectral sequence*, Pacific J. Math. **15** (1965), 65–74. MR **31** #1679.

12. J. Harper, *H-spaces with torsion* (to appear).

13. L. Hodgkin, *On the K-theory of Lie groups*, Topology **6** (1967), 1–36. MR **35** #4950.

14. J. Hubbuck, *Automorphisms of polynomial algebras and homotopy commutativity in H-spaces*, Osaka J. Math. **6** (1969), 197–209. MR **41** #6212.

15. R. Kane, *The module of indecomposables for finite H-spaces*, Trans. Amer. Math. Soc. (to appear).

16. J. Lin, *Torsion in H-spaces*. I, Ann. of Math. (2) **103** (1976), 457–487.

17. J. Milnor and J. C. Moore, *On the structure of Hopf algebras*, Ann. of Math. (2) **81** (1965), 211–264. MR **30** #4259.

18. E. Thomas, *Steenrod squares and H-spaces*, Ann. of Math. (2) **77** (1963), 306–317. MR **26** #3057.

19. S. Weingram, *On the incompressibility of certain maps*, Ann. of Math. (2) **93** (1971), 476–485. MR **46** #890.

20. A. Zabrodsky, *Cohomology operations and homotopy commutative H-spaces*, Steenrod Algebra and its Applications, Lecture Notes in Math., vol. 168, Springer-Verlag, Berlin and New York, 1970, pp. 308–317.

21. ———, *Implications in the cohomology of H-spaces*, Illinois J. Math. **14** (1970), 363–375. MR **41** #6217.

22. ———, *Secondary operations in the cohomology of H-spaces*, Illinois J. Math. **15** (1971), 648–655. MR **48** #1234.

23. ———, *On sphere extensions of classical Lie groups*, Proc. Sympos. Pure Math., vol. 22, Amer. Math. Soc., Providence, R. I., 1971. MR **47** #5869.

UNIVERSITY OF CALIFORNIA AT SAN DIEGO

PRINCETON UNIVERSITY

Proceedings of Symposia in Pure Mathematics
Volume 32, 1978

ON THE CONSTRUCTION OF MOD p H-SPACES

JOHN R. HARPER[1]

The use of localization to construct new finite H-spaces depends on having mod p ingredients different from Lie groups. The methods outlined in this note form a systematic framework in which to make constructions. We apply these methods to the construction of mod p H-spaces with p-torsion in their homology (c.f. **[2]**, **[8]**).

The new mod p H-space is described as follows. Let T_p denote the algebra over the mod p Steenrod algebra given by

$$T_p = \bigwedge(x_3, x_{2p+1}) \otimes Z_p[x_{2p+2}]/(x_{2p+2}^p)$$

where $\mathscr{P}^1 x_3 = x_{2p+1}$ and $\beta x_{2p+1} = x_{2p+2}$.

THEOREM A. *There exists a finite simply connected CW complex $K(p)$ whose localization at p is an H-space and whose* mod p *cohomology is T_p.*

The mod p H-space $K(p)$ has the rational type of a rank 2 H-space with type $(3, 2p^2 + 2p - 1)$. A routine application of Zabrodsky's mixing process and Theorem A yields

THEOREM B. *There exists a simply connected finite H-space $X(p)$ rationally equivalent to* Sp(n) *where $n = \frac{1}{2}p(p+1)$ and having $K(p)$ as a* mod p *factor. In particular, $H_{2p+1}(X(p); Z)$ contains a cyclic subgroup of order p.*

Our work is based on ideas developed by Massey and Peterson in **[9]**. Results of Zabrodsky **[14]** apply to show that $K(p)$ admits no mod p homotopy associative H-structure. Further work of Kane **[7]** clarifies necessary conditions on the mod p cohomology Hopf algebra of a finite H-space having p-torsion and a homotopy associative H-structure. Our work leaves open the question of whether or not p-torsion for $p \geqq 7$ can appear in the homology of simply connected finite H-spaces admitting classifying spaces or even just homotopy associative multiplications.

AMS (MOS) subject classifications (1970). Primary 57F25, 55D45, 55G20; Secondary 57F05, 57F10, 55J20.

[1]Research supported by NSF Grant MCS 76-07157.

This note represents a condensation of a paper to appear [5]. We have attempted to treat just the ideas leaving the elaboration of details to the other paper.

During the symposium Haynes Miller made a remark which appears here as Proposition 1.1. This result serves as a basis for replacing rather elaborate calculations on which this work was originally set with well-known and accessible calculations in the cohomology of the Steenrod algebra. I am most grateful to Miller for his interest, insight and discussions. I am also grateful to A. Liulevicius and A. Zabrodsky whose influences, both implicit and explicit, pervade this work.

1. First we recall some features of unstable modules over the Steenrod algebra. References are [3], [9], [10], [11].

Let $\mathfrak{A}(p)$ denote the mod p Steenrod algebra. Let $\mathfrak{UM}$ denote the category of unstable modules over the Steenrod algebra. We explicitly deal with odd primes, leaving the case $p = 2$ to the reader or [9].

The category $\mathfrak{UM}$ has a pair of adjoint functors Σ and Ω. If M is an object, then ΣM is the usual suspension. To define Ω we first recall the action $\lambda : M \to M$ given by the formula

$$\lambda x = \begin{cases} p^k x & \text{if } \deg x = 2k, \\ \beta p^k x & \text{if } \deg x = 2k+1. \end{cases}$$

Then ΩM is defined degree-wise by $(\Omega M)^{k-1} = (M/\lambda M)^k$. From the definitions it follows that in $\mathfrak{UM}$ $\operatorname{Hom}(\Omega M, N) = \operatorname{Hom}(M, \Sigma N)$. The exactness properties of Ω are studied in [9]. If we let Ω_r denote the left derived functors of Ω then $\Omega_r = 0$ for $r \geqq 2$ and $\Omega_1 M$ is isomorphic to $\ker \lambda$ as Z_p-modules but with regrading

$$(\Omega_1 M)^{2np-1} = (\ker \lambda)^{2n}, \qquad (\Omega_1 M)^{2np+1} = (\ker \lambda)^{2n+1}.$$

Next we consider the derived functors $\operatorname{Ext}^s(M, -)$ of $\operatorname{Hom}(M, -)$ in $\mathfrak{UM}$. Since we deal with degree preserving maps, we treat the usual second grading by writing $\operatorname{Ext}^s(M, \Sigma^t N)$. The calculation of certain Ext's is a major part of our work.

PROPOSITION 1.1. *Let M, N be in $\mathfrak{UM}$. There exists a long exact "EHP" sequence*

$$\cdots \to \operatorname{Ext}^s(\Omega M, N) \xrightarrow{E} \operatorname{Ext}^s(M, \Sigma N) \xrightarrow{H} \operatorname{Ext}^{s-1}(\Omega_1 M, N) \xrightarrow{P} \operatorname{Ext}^{s+1}(\Omega M, N) \to \cdots.$$

PROOF. Apply the composition of functors spectral sequence of [4] to the equation $\operatorname{Hom}(\Omega M, N) = \operatorname{Hom}(M, \Sigma N)$. One obtains a spectral sequence with

$$E_2^{s,t} = \operatorname{Ext}^s(\Omega_t M, N)$$

converging to $\operatorname{Ext}^{s+t}(M, \Sigma N)$. Since $\Omega_t M = 0$ for $t \geqq 2$ we have a two-row spectral sequence from which the long exact sequence is extracted. The map p is the differential d_2.

Let $B(n) \subset \mathfrak{A}(p)$ denote the left ideal of operations annihilating cohomology classes of dimension $\leqq n$. Set $F(n) = \Sigma^n \mathfrak{A}(p)/B(n)$. Then $F(n)$ is projective in $\mathfrak{UM}$.

One can show that if P is projective in $\mathfrak{UM}$ then P can be expressed as a sum of modules of type $F(n)$. Hence there is a G.E.M. denoted by $K(P)$ realizing $U(P)$ in the sense

$$H^*K(P) = U(P).$$

An important property of projectives P is that they are free λ-modules in the sense that if $x \in P$,

$$\lambda x = 0 \Leftrightarrow x = 0.$$

Let $F(n)\cdot\beta$ denote the left $\mathfrak{A}(p)$-submodule of $F(n)$ generated by the Bockstein β. Set

$$F'(n) = F(n)/F(n)\cdot\beta.$$

While $F'(n)$ is not projective, it is a free λ-module and is realized by $K(Z, n)$. Let Q be a module which is a direct sum of modules of type $F(n)$ or $F'(n)$. Then ΩQ is again of the same type and if Q is realized by the G.E.M. $K(Q)$ then ΩQ is realized by $\Omega K(Q)$, i.e.

$$H^*(\Omega K(Q)) = U(\Omega Q).$$

2. This section is concerned with cohomology properties of certain fibre spaces. We state a result of Massey and Peterson **[9]** proved for p odd by Barcus **[1]**. We also obtain the formula for principal action in coholomogy.

Let $f\colon B \to B_0$ be a map. Consider the principal fibration over B induced by f from the path-loop fibration over B_0, $\Omega B_0 \to^i E \to^p B \to^f B$. We call this fibration a *Massey-Peterson fibration* if the following conditions are satisfied **[9**, §II**]**.

(1) B_0 is simply connected G.E.M. of finite type. We write $B_0 = K(Q)$ in the notation of §1.

(2) B is a space of finite type such that $H^*(B) = U(Z)$ as algebras over $\mathfrak{A}(p)$ and $Z = Z_0 \oplus Z_1$ with Z_1 a free λ-module.

(3) $f\colon B \to B_0$ satisfies $f^*|Q\colon Q \to Z_1$.

THEOREM 2.1 **[1]**, **[9]**. (a) *Suppose $f\colon B \to B_0$ induces a Massey-Peterson fibration; then as algebras over Z_p, $H^*(E) \cong U(Z') \otimes U(\Omega Q')$ where $Z' = \operatorname{coker} f^*|Q = Z_0 \oplus Z_1/\operatorname{inf}^*$ and $Q' = \ker f^*|Q$.*

(b) *The structure of $H^*(E)$ as an $\mathfrak{A}(p)$-module is given by the fundamental sequence of $\mathfrak{A}(p)$-modules*

$$0 \to U(Z') \xrightarrow{p^*} N \xrightarrow{i^*} \Omega Q' \to 0$$

where $N \subset H^*(E)$ is the submodule given by the extension. This sequence is natural for maps of Massey-Peterson fibrations.

Note that if the sequence splits over $\mathfrak{A}(p)$ then $H^*(E) \cong U(Z' \oplus \Omega Q')$ as algebras over $\mathfrak{A}(p)$.

Let $\eta\colon \Omega B_0 \times E \to E$ denote the principal action.

PROPOSITION 2.2. *The principal action in a Massey-Peterson fibration is primitive in the sense that if $e \in N$ then $\eta^*(e) = 1 \otimes e + i^*(e) \otimes 1$.*

PROOF. We apply the naturality of the fundamental sequence to the following diagram embedding η as a map of Massey-Peterson fibrations:

$$\begin{array}{ccccccc}
\Omega B_0 \times \Omega B_0 & \xrightarrow{1\times i} & \Omega B_0 \times E & \xrightarrow{p\pi E} & B & \xrightarrow{c\times f} & B_0 \times B_0 \\
\downarrow{\scriptstyle \Omega m_0} & & \downarrow{\scriptstyle \eta} & & \| & & \downarrow{\scriptstyle m} \\
\Omega B_0 & \xrightarrow{i} & E & \xrightarrow{p} & B & \xrightarrow{f} & B_0
\end{array}$$

where c is the constant map and m the canonical multiplication. Since $H^*(B_0 \times B_0)$

$= U(Q \oplus Q)$ we obtain the following diagram of fundamental sequences from which the proposition follows:

$$\begin{array}{ccccccccc}
0 \to & U(Z') & \to & \Omega Q \oplus N & \xrightarrow{1\oplus i^*} & \Omega Q \oplus \Omega Q' & \to 0 \\
& \Vert & & \uparrow{\scriptstyle \eta^*} & & \uparrow{\scriptstyle \Omega m_0^*} & \\
0 \to & U(Z') & \to & N & \xrightarrow{\quad i^* \quad} & \Omega Q' & \to 0
\end{array}$$

3. Here we give a reformulation of material in [9] concerning the notion of a geometric realization of a resolution. Let $M \in \mathfrak{UM}$ and suppose $\mathscr{R}$ is a projective resolution $P^* \to^{\varepsilon} M$. We write $d_s: P_{s+1} \to P_s$, $s \geqq 0$, for the maps in $\mathscr{R}$, and assume each P_s is s-connected,

$$P_s^k = 0 \quad \text{for } k \leqq s.$$

If M is connected, a minimal resolution of M will have this connectivity property.

Using the free λ-structure of projectives one can prove

LEMMA 3.1. *Let $\mathscr{R}: P_* \to^{\varepsilon} M$ be a projective resolution of M. For $s \geqq 0$ the following sequence is exact*:

$$\Omega^s P_s \xleftarrow{\Omega^s d_s} \Omega^s P_{s+1} \leftarrow \cdots$$

and $\operatorname{im} \Omega^s d_s = \Omega \ker \Omega^{s-1} d_{s-1}$.

DEFINITION. *A geometric realization* of $\mathscr{R}$ is a tower of principal fibrations

$$\begin{array}{ccccccc}
& K(\Omega^s P_{s+1}) & & & K(\Omega P_2) & & K(P_1) \\
& \uparrow{\scriptstyle f_s} & & & \uparrow{\scriptstyle f_1} & & \uparrow{\scriptstyle K(d_0)} \\
E_{s+1} \xrightarrow{p_s} & E_s & \longrightarrow \cdots \longrightarrow & & E_1 & \xrightarrow{\quad p_0 \quad} & K(P_0) \\
& \uparrow{\scriptstyle i_s} & & & \uparrow{\scriptstyle i_1} & & \\
& K(\Omega^s P_s) & & & K(\Omega P_1) & &
\end{array}$$

where the fibrations are induced from path-loop fibrations by maps $f_s: E_s \to K(\Omega_s P_{s+1})$ determined as follows. For each $s \geqq 1$ an $\mathfrak{A}(p)$-monomorphism k_s is postulated:

$$k_s: \Omega \ker \Omega^{s-1} d_{s-1} \to H^*(E_s)$$

such that $f_s^* \,|\, \Omega^s P_{s+1} = k_s \circ \Omega^s d_s$ and $i_s^* \circ k_s: \Omega \ker \Omega^{s-1} d_{s-1} \to \Omega^s P_s$ is the inclusion.

Given an abstract resolution $\mathscr{R}$, a realization need not exist, since there is nothing in $\mathscr{R}$ to determine the maps k_s.

The results of §III of [9] can be stated as the following.

SPLITTING THEOREM 3.2 [THEOREM 25.1 OF [9]]. *The fibrations in a geometric realization are Massey-Peterson fibrations and*

$$H^*(E_s) \cong U(M) \otimes U(\Omega \ker \Omega^{s-1} d_{s-1})$$

as algebras over $\mathfrak{A}(p)$.

REALIZATION THEOREM 3.3 [PROPOSITION 26.1 OF [9]]. *If X is a simply connected complex of finite type with $H^*(X) = U(M)$ and $\mathscr{R}$ is a resolution of M, then $\mathscr{R}$ has a realization and there are maps $q_s: X \to E_s$, $s \geqq 1$, $q_0: X \to K(P_0)$ such that*

$$q_0^* \,|\, P_0 = \varepsilon: P_0 \to M$$

and q_s lifts q_{s-1}.

CONVERGENCE THEOREM 3.4 [§27 OF [9]]. *Let Y be a simply connected finite complex. If for each $s \geqq 1$ there exists a map $g_s\colon Y \to E_s$, $g_0\colon Y \to K(P_0)$ such that g_s lifts g_{s-1}, then there exists a map of localizations $g\colon Y_{(p)} \to X_{(p)}$ covering all the g_s. In particular, g^* is determined by g_0.*

4. Here we apply the earlier material to obtain results concerning the realizations of spaces and maps from algebraic data. These results are analogous to methods of Toda [**12**] in stable homotopy. We assume spaces are simply connected.

PROPOSITION 4.1. *Let X be a space with $H^*(X) = U(M)$ and Y be a finite complex. Let $g_0\colon H^*(X) \to H^*(Y)$ be a map of algebras over the Steenrod algebra. If* $\operatorname{Ext}^{s+1}(M, \Sigma^s H^*(Y)) = 0$ *for $s \geqq 1$ then there exists a map of localizations $g\colon Y_{(p)} \to X_{(p)}$ such that $g^* = g_0$.*

PROOF. Let $\mathscr{R}$ be a resolution $P_* \to M$ of M with P_s s-connected. By 3.3 $\mathscr{R}$ has a realization. We construct g by lifting through the tower provided by the realization and then applying 3.4. To get started we have $g_0\colon Y \to K(P_0)$ inducing the abstractly given map g_0. Since $g_0^* d_0 = 0$ we have a lifting $g_1\colon Y \to E_1$. Assume inductively we have constructed a lifting $g_s\colon Y \to E_s$. The obstructions to lifting into E_{s+1} are represented by the map f_s in the diagram

$$\begin{array}{ccccc} & & E_{s+1} & & \\ & & \Big\downarrow{\scriptstyle p_s} & & \\ Y & \xrightarrow{g_s} & E_s & \xrightarrow{f_s} & K(\Omega^s P_{s+1}) \end{array}$$

and $g_s^* f_s^* \in \operatorname{Hom}(\Omega^s P_{s+1}, H^*Y) = \operatorname{Hom}(P_{s+1}, \Sigma^s H^*Y)$, which is the module of cochains for $\operatorname{Ext}^{s+1}(M, \Sigma^s H^*Y)$. Next we apply the splitting of the fundamental sequence of E_s by the map k_s to see that $g_s^* f_s^*$ is a cocycle. We have $f_s^* \,|\, \Omega^s P_{s+1} = k_s \circ \Omega^s d_s$. Thus

$$g_s^* f_s^* \circ \Omega^s d_{s+1} = k_s \circ \Omega^s d_s \circ \Omega^s d_{s+1} = 0.$$

By hypothesis $[g_s^* f_s^*] = 0$ in Ext^{s+1}. Hence there is a map $w\colon \Omega^s P_s \to H^*(Y)$ such that $w \circ \Omega^s d_s = -\, g_s^* f_s^*$. Use the principal action η of E_s to alter g_s to another lifting g_s'. Since η^* is primitive we have

$$(g_s')^* f_s^* = g_s^* f_s^* + w^* i_s^* f_s^* = g_s^* f_s^* + w^* \Omega^s d_s = 0.$$

Hence g_s' lifts to E_{s+1}, concluding the inductive step.

There is a similar result concerning the existence of a realization of an abstractly given resolution.

PROPOSITION 4.2. *Let $\mathscr{R}$ be a resolution $P_* \to^{\varepsilon} M$ with P_s s-connected. If* $\operatorname{Ext}^{s+2}(M, \Sigma^s U(M)) = 0$ *for $s \geqq 1$ then $\mathscr{R}$ has a realization.*

PROOF. The proof is similar to 4.1.

Next we place some results of Zabrodsky [**13**] in our setting. We state a modification of 4.1 which takes advantage of some extra structure.

DEFINITION. A self-map $\varphi\colon X \to X$ is called a *power map* if the induced map φ^* on the quotient of indecomposables $QH^*(X)$ acts as multiplication by a fixed $\mu \in Z_p$. We call the pair (X, φ) a *power space*. We denote by $E\mu H^*(X)$ the μ-characteristic vectors of φ^* acting on $H^*(X)$,

$$E\mu H^*(X) = \{X \in H^*(X) \mid \varphi^* X = \mu X\}.$$

Note that $E\mu H^*(X)$ is an $\mathfrak{A}(p)$-submodule.

DEFINITION. If (X, φ), (Y, ψ) are power spaces and $f: X \to Y$ is a map, we call f a *P-map* if for some integer $r \geqq 0$ the following diagram commutes up to homotopy:

$$\begin{array}{ccc} X & \xrightarrow{f} & Y \\ {\scriptstyle \varphi^{p^r}} \downarrow & & \downarrow {\scriptstyle \psi^{p^r}} \\ X & \xrightarrow{f} & Y \end{array}$$

Note that $(\varphi^{p^r})^* = \varphi^*$.

PROPOSITION 4.3. *Let (X, φ), (Y, ψ) be power spaces with $H^*(X) = U(M)$ and Y a finite complex. Suppose φ^* and ψ^* induce multiplication by μ. Let g_0: $H^*(X) \to H^*(Y)$ be a map of algebras over the Steenrod algebra such that $g_0\varphi^* = \psi^* g_0$. If* $\operatorname{Ext}^{s+1}(M, \Sigma^s E\mu H^*Y) = 0$ *for $s \geqq 1$ then there exists a map of localizations g: $Y_{(p)} \to X_{(p)}$ such that $g^* = g_0$.*

The proof uses the lifting theorem of [**13**].

5. In this section we outline the application of the previous results to the construction of the new mod p H-space $K(p)$.

Consider the unstable $\mathfrak{A}(p)$-module

$$M = \langle x_3, \mathscr{P}^1 x_3, \beta\mathscr{P}^1 x_3 \rangle.$$

Then $U(M)$ is the algebra T_p given in the introduction. As an algebra, T_p has dimension $2p^2 + 2p + 2$.

PROPOSITION 5.1. *There exists an H-space (E, m) satisfying the following, $m: E \times E \to E$:*

(a) *The* mod p *cohomology $H^*(E) = U(M)$ in dimensions $\leqq 2p^2 + 4p - 4$ with $U(M)$ embedded as a primitively generated sub-Hopf algebra.*

(b) *The homology approximation of E through dimensions $2p^2 + 2p + 2$ realizes $U(M)$ and is the space $K(p)$.*

(c) *Given $\mu \in Z_p$, let φ be the power map on E determined by m and inducing multiplication by μ. Then m is a P-map of the power spaces $(E \times E, \varphi \times \varphi)$ and (E, φ).*

We obtain E as a 3-stage Postnikov system over $K(Z, 3)$. First construct the 2-stage space E_1 over $K(Z, 3)$ with k-invariant $\mathscr{P}^p\mathscr{P}^1\iota_3$. We can use Theorem 2.1 to calculate its cohomology, obtaining

$$H^*(E_1) = U(M) \otimes U(X_1)$$

as algebras over Z_p. The $\mathfrak{A}(p)$-generators of X_1 in dimensions $\leqq 2p^2 + 2p + 2$ are classes e_1 of dimension $2p^2 + 2p - 2$ and e_2 of dimension $2p^2 + 2p$. We find for purely dimensional reasons that the fundumental sequence of E_1 splits and that e_2 is represented by an integral class. We then take E to be the space induced by killing e_1 and e_2 in the diagram

$$\begin{array}{l} E \\ \downarrow \\ E_1 \to K(Z_p, 2p^2 + 2p - 2) \times K(Z, 2p^2 + 2p). \end{array}$$

We can again use Theorem 2.1 to compute cohomology. To obtain the H-structure we can use the results of [6]. We find that e_1 is primitive but e_2 is not. The relevant information for this calculation is given by the defining relations for e_1 and e_2:

$$e_1\colon \mathscr{P}^1(\mathscr{P}^p\mathscr{P}^1\iota_3) = \mathscr{P}^{p+1}\mathscr{P}^1\iota_3 = 0,$$
$$e_2\colon \beta\mathscr{P}^1\beta(\mathscr{P}^p\mathscr{P}^1\iota_3) = \beta\mathscr{P}^{p+1}\iota_3 = 0.$$

Using the coproduct information we find that the loop multiplication on E_1 can be altered by a map $E_1 \wedge E_1 \to K(Z_p, 2p^2)$ so as to make e_2 primitive and keep e_1 primitive. These are the main steps in the proof of 5.1.

The power map on E induces one on the subcomplex $K(p)$ and we have the diagram

$$\begin{array}{c} \qquad\qquad\qquad\qquad\qquad\quad K(p) \\ \qquad\qquad\qquad\qquad\qquad\quad \downarrow \\ K(p) \times K(p) \to E \times E \to E \end{array}$$

with the horizontal composition a P-map. The main part of our argument is the analysis of the obstruction to compressing the horizontal composition into $K(p)$. Since the inclusion of $K(p)$ in E induces a mod p isomorphism in dimensions $\leqq 2p^2 + 2p + 2$, the mod p obstruction to compression lies in dimension $> \dim K(p)$; hence any compression will automatically cover the folding map.

We use the realization theory of §3 to interpose a tower of fibrations between $K(p)$ and E. The space E has been built as a realization of a partial resolution of M:

$$\mathscr{R}'\colon 0 \leftarrow M \xleftarrow{\varepsilon} F'(3) \xleftarrow{d_0} F(2p^2+1) \xleftarrow{d_1} Q$$

where $Q = F(2p^2 + 2p - 1) \oplus F'(2p^2 + 2p + 1)$ and d_1 is given by the equations defining the classes e_1 and e_2, d_0 given by the first k-invariant. Here we have $\ker \varepsilon = \operatorname{im} d_0$ but only $\ker d_0 \supset \operatorname{im} d_1$. We can expand $\mathscr{R}'$ to a resolution

$$\mathscr{R}\colon 0 \leftarrow M \leftarrow F'(3) \leftarrow F(2p^2+1) \leftarrow Q \oplus P_2 \leftarrow P_3 \leftarrow \cdots$$

with P_i projective and connectivity of $\Omega^i P_i > 2p^2 + 2p + 2$. By the Realization Theorem 3.3, $\mathscr{R}$ has a geometric realization, and our compression problem is described by the following diagram:

$$\begin{array}{ccl} & & K(p) \\ & & \downarrow \\ & & \vdots \\ & & \downarrow \\ & & E_2 \to K(\Omega^2 P_3) \\ & & \downarrow \\ K(p) \times K(p) \to E \times E & \to & E \to K(\Omega P_2) \\ & & \downarrow \nearrow \\ & & E_1 \to K(\Omega Q) \\ & & \downarrow \\ & & K(Z,3) \to K(Z_p, 2p^2+1). \end{array}$$

The dotted arrow indicates we are killing the classes in $H^*(E)$ coming from $H^*(E_1)$, having divided the realization of this portion of $\mathscr{R}$ into 2 parts.

We can now use Proposition 4.3. Let

$$L = E\mu H^*(K(p) \times K(p)).$$

One finds dim $L = 4p^2 + 2p - 3$. To obtain Theorem A we have to show

PROPOSITION 5.2. $\mathrm{Ext}^{s+1}(M, \Sigma^s L) = 0$ *for* $s \geqq 1$.

Using Proposition 1.1 we have the exact sequence

$$\mathrm{Ext}^{s+1}(\Omega M, \Sigma^{s-1}L) \xrightarrow{E} \mathrm{Ext}^{s+1}(M, \Sigma^s L) \xrightarrow{H} \mathrm{Ext}^s(\Omega_1 M, \Sigma^{s-1}L).$$

Now $\Omega M = \langle x_2, \mathcal{P}^1 x_2 \rangle$ happens to be the generating module for complex projective space $\mathrm{CP}(p^2 - 1)$. If we let $S(m)$ be the generating module for the sphere S^{2m-1},

$$S(m) = \begin{cases} Z_p & \text{in dim } 2m - 1, \\ 0 & \text{elsewhere,} \end{cases}$$

then we have

$$\mathrm{Ext}^{s+1}(\Omega M, \Sigma^{s-1}L) = \mathrm{Ext}^s(S(p^2), \Sigma^{s-2}L) \quad \text{for } s \geqq 1.$$

L is sufficiently connected so that $\Sigma^{-1}L$ still makes sense as an unstable $\mathfrak{A}(p)$-module. We also have $\Omega_1 M = \langle x_{2p^2+1}, \mathcal{P}^1 x_{2p^2+1} \rangle$ so $\Omega_1 M$ is an extension

$$0 \to S(p^2 + p) \to \Omega_1 M \to S(p^2 + 1) \to 0.$$

Thus we are reduced to calculating $\mathrm{Ext}^s(S(m), B)$ for $m = p^2, p^2 + 1, p^2 + p$ and $B = \Sigma^{s-\varepsilon}L$ where $\varepsilon = 1, 2$. For this calculation we can use known results on the cohomology of the Steenrod algebra, and standard algebraic methods to obtain 5.2.

REFERENCES

1. W. D. Barcus, *On a theorem of Massey and Peterson*, Quart. J. Math. Oxford Ser. (2) **19** (1968), 33–41. MR **37** #5877.

2. A. Borel, *Sous-groupes commutatifs et torsion des groupes de Lie compacts connexes*, Tôhoku Math. J. (2) **13** (1961), 216–240. MR **24** #5094.

3. A. K. Bousfield and D. M. Kan, *The homotopy spectral sequence of a space with coefficients in a ring*, Topology **11** (1972), 79–106.

4. H. Cartan and S. Eilenberg, *Homological algebra*, Princeton Univ. Press, Princeton, N. J., 1956. MR **17**, 1040.

5. J. R. Harper, *H-spaces with torsion* (to appear).

6. J. R. Harper and C. Schochet, *Coalgebra extensions in two-stage Postnikov systems*, Math. Scand. **29** (1971), 232–236. MR **46** #9985.

7. R. Kane, *Torsion in homogopy associative H-spaces*, Illinois J. Math. **20** (1976), 476–485.

8. J. Lin, *Torsion in H-spaces*. II, Ann. of Math. (to appear).

9. W. S. Massey and F. P. Peterson, *On the* mod 2 *cohomology structure of certain fibre spaces*, Mem. Amer. Math. Soc. No. 74 (1967). MR **37** #2226.

10. L. Smith, *Hopf fibration towers and the unstable Adams spectral sequence*, Proc. Sympos. Pure Math., vol. 17, Amer. Math. Soc., Providence, R. I., 1970, pp. 129–160. MR **41** #4544.

11. N. E. Steenrod and D. B. A. Epstein, *Cohomology operations*, Ann. of Math. Studies, no. 50, Princeton Univ. Press, Princeton, N.J., 1962. MR **24** #3056.

12. H. Toda, *On spectra realizing exterior parts of the Steenrod algebra*, Topology **10** (1971), 53–75. MR **42** #6814.

13. A. Zabrodsky, *Power spaces*, Inst. Adv. Study (mimeograph).

14. ———, *Implications in the cohomology of H-spaces*, Illinois J. Math. **14** (1970), 363–375. MR **41** #6217.

UNIVERSITY OF ROCHESTER

Proceedings of Symposia in Pure Mathematics
Volume 32, 1978

SMOOTHINGS OF *H*-SPACES

ERIK KJAER PEDERSEN

Consider a partition of all primes $\Pi = P_1 \cup \cdots \cup P_k$, and let G_i be quasi-finite complexes such that $(G_i)_{p_i}$, G_i localized at P_i, are H-spaces, that are rationally equivalent as H-spaces. The homotopy pull-back of the diagram

$$\begin{array}{ccc} (G_1)_{p_1} & \longrightarrow & (G_1)_0 \\ & & \wr\| \\ (G_2)_{p_2} & \longrightarrow & (G_2)_0 \\ & & \wr\| \\ & & \vdots \\ & & \wr\| \\ (G_k)_{p_k} & \longrightarrow & (G_k)_0 \end{array}$$

is well known to be a quasi-finite H-space [**1**]. We will say that X is obtained by Zabrodsky mixing of $G_1, \cdots, G_k$ at $P_1, \cdots, P_k$. Note that X depends on the rational equivalences chosen.

We consider H-spaces obtained by mixing the following types of spaces:

At the prime 2: products of Lie groups and S^7 and $\boldsymbol{RP}^7$.

At odd primes: products of

(a) Lie groups, S^7, $\boldsymbol{RP}^7$ and

(b) any stably reducible quasi-finite P.D. space with no 3-dimensional generator of the rational exterior algebra cohomology and finitely presented fundamental group, and

(c) principal S^3-bundles with basespace a stably reducible quasi-finite P.D. space.

We prove the following theorems:

THEOREM. *Let X be as above,* 1-*connected; then X is of the homotopy type of a parallelizable differentiable manifold.*

AMS (*MOS*) *subject classifications* (1970). Primary 55D45, 57D10, 57D65.

We notice that by choosing all G_i equal, but choosing different rational equivalences we get the following:

COROLLARY. *Let Y be in the genus of a simply connected Lie-group G_1 (i.e., $Y_p \cong G_p$ for all primes p_1); then Y is homotopy equivalent to a parallelizable manifold.*

In the nonsimply connected context we prove the following:

THEOREM. *Let X be as above; then X is of the homotopy type of a finite complex, and in the genus of X is a parallelizable differentiable manifold.*

INDICATION OF PROOF. The case where $X_{(2)} \cong (RP^3)^k \times (S^7)^l \times (RP^7)^m$ requires special arguments. In all other cases we proceed by constructing a fibration $S^1 \to X \to Y$, where Y is a stably reducible P.D. space. The map p induces isomorphisms on fundamental groups, and the fibration is orientable. We then use [2] to compute the Wall finiteness obstruction for X. The formula in this case says $\sigma(X) = X(S') \cdot \sigma(Y)$, X the euler characteristic; hence $\sigma(X) = 0$, so X is homotopy equivalent to a finite CW complex.

We then consider X as a P.D. boundary of the corresponding D^2-fibration $D^2 \to E \to Y$, and notice that the classifying map $E \to BG$ reduces to BO since Y is stably reducible and S^1-fibrations are equivalent to $O(2)$-bundles. This allows us to set up a surgery problem

$$(M, \delta M) \longrightarrow (E, Y)$$

and we proceed to show the surgery problem $\delta M \to Y$ has obstruction 0. The reduction of E is trivial when restricted to Y; hence Y is of the homotopy type of a parallelizable differentiable manifold.

REFERENCES

1. P. Hilton, G. Mislin and J. Roitberg, *Localization of nilpotent groups and spaces*, Math. Studies, No. 15, North-Holland, Amsterdam, 1975.

2. E. K. Pedersen and L. R. Taylor, *The Wall finiteness obstruction for fibrations*, Amer. J. Math. (to appear).

ODENSE UNIVERSITY

Proceedings of Symposia in Pure Mathematics
Volume 32, 1978

THE HOMOTOPY-DIMENSION OF NILPOTENT SPACES

PETER J. KAHN

The following describes joint work with K. S. Brown. Full details will appear in [1].

The homotopy-dimension of a space (= CW complex) X, written ho dim X, is the smallest dimension of all spaces in its homotopy type (allowing ∞ as a possible value). Results of Wall [8], Stallings [6] and Swan [7] imply that ho dim X is given by the cohomological dimension of X,

(1) $$\operatorname{cd} X = \sup\{i \mid H^i(X, M) \neq 0, \text{ for some } M\},$$

provided cd $X \neq 2$. Here, M is allowed to range over all $\pi_1 X$-modules, and cohomology is taken with twisted coefficients.

Restricting to untwisted coefficients in (1), we obtain $\operatorname{cd}_s X$, the simple cohomological dimension of X. It is well known that, for any 1-connected X,

(2) $$\text{ho dim } X = \operatorname{cd}_s X.$$

Of course this follows from the previous paragraph when ho dim $X \neq 2$, but it can be seen directly by using a homology decomposition of X.

Our aim is to obtain (2) for nilpotent spaces. On the one hand, we view this as part of the general program of extending the homotopy theory of 1-connected spaces to nilpotent spaces (e.g., see [3]). From this perspective, one immediate difficulty is that nilpotent spaces are not known to admit homology decompositions. On the other hand, we also view our goal as an attempt to obtain an easy method for computing and developing properties of ho dim X for a class of X significantly larger than the class of 1-connected spaces. From this perspective, cd$X$ is regarded as hard to deal with and $\operatorname{cd}_s X$ as easy.

Our main result is:

AMS (MOS) subject classifications (1970). Primary 55D99; Secondary 20J05.

THEOREM 1. *If X is connected and nilpotent and $\pi_1 X$ is finitely-generated, then* ho dim $X = \mathrm{cd}_s X$.

We apply Theorem 1 to study the effect of localization on homotopy-dimension. If P is a proper subset of the primes and X is nilpotent, let X_P denote its P-localization. It follows from (2), together with basic facts about localization, that when X is 1-connected,

$$\text{ho dim } X_P \leqq \text{ho dim } X + 1. \tag{3}$$

In fact, it is not hard to see that (3) is essentially the only relation between ho dim X and ho dim X_P that is valid for all 1-connected X. Our extension to nilpotent spaces is given by

THEOREM 2. *If X is nilpotent and either* (a) *$\pi_1 X$ is finite or* (b) *$H_* X$ is countable, then* ho dim $X_P \leqq$ ho dim $X + 1$.

Under hypothesis (a), the conclusion follows from Theorem 1 in the same way that (3) is obtained from (2) for 1-connected X. With hypothesis (b), however, additional techniques are required because $\pi_1 X$ and $\pi_1 X_P$ need not be finitely-generated.

We do not know whether the cardinality restrictions on $\pi_1 X$ or $H_* X$ in Theorems 1 and 2 are essential.

We now briefly describe the methods used for Theorem 1.

We first prove that ho dim $X = \mathrm{cd}\, X$ for all nilpotent X. That is, we show that the possible dimension-two exceptions to the Wall-Stallings-Swan result (cf. second paragraph) cannot be nilpotent. This involves classifying all nilpotent X satisfying $\mathrm{cd}\, X \leqq 2$.

By the above, the equality of Theorem 1 is equivalent to $\mathrm{cd}\, X = \mathrm{cd}_s X$. We restrict to nilpotent X with $\pi_1 X$ finite. Clearly, it suffices to show that neither cd nor cd_s changes upon passing to finite, regular covers of X.

Suppose that X is finite-dimensional. The fact that cd does not change is essentially the well-known Shapiro's Lemma [**2**, pp. 116–119] of homological algebra. No nilpotence is used. The fact that cd_s does not change, however, does require nilpotence, together with a transfer argument.

The case of infinite-dimensional X is more troublesome and requires techniques like those of the Nakayama-Rim theory of cohomologically trivial modules [**5**].

To deal with the case of finitely-generated $\pi = \pi_1 X$, we prove the following result simultaneously with Theorem 1, by induction on the rank (or Hirsch number) of π (see [**4**, p. 149]):

THEOREM 3. *If X is nilpotent with universal cover $\tilde{X}$ and if $\pi = \pi_1 X$ is finitely-generated, then* ho dim $X =$ ho dim $\tilde{X}$ + *rank* π, *and* $\mathrm{cd}_s X = \mathrm{cd}_s \tilde{X}$ + *rank* π.

We can use Theorem 3 to get some interesting lower bounds on the dimension of nilpotent spaces with a fixed fundamental group.

THEOREM 4. *Suppose X is nilpotent with fundamental group π. Then*: (a) ho dim $X \geqq$ *rank* π. (b) *If $X \neq K(\pi, 1)$, then* ho dim $X \geqq$ *rank* $\pi + 2$. (c) *If π has nonzero torsion, then* ho dim $X \geqq$ *rank* $\pi + 3$.

This theorem raises the following question: Given a nilpotent group π, what is the minimum dimension of all nilpotent spaces X satisfying $\pi_1 X = \pi$?

REFERENCES

1. K. S. Brown and P. J. Kahn, *Homotopy dimension and simple cohomological dimension of spaces*, Comment. Math. Helv. **52** (1977), 111–127.

2. H. Cartan and S. Eilenberg, *Homological algebra*, Princeton Univ. Press, Princeton, N. J., 1956. MR **17,** 1040.

3. E. Dror, *A generalization of the Whitehead theorem*, Sympos. on Algebraic Topology, Lecture Notes in Math., vol. 249, Springer-Verlag, Berlin and New York, 1971, pp. 13–23.

4. K. Gruenberg, *Cohomological topics in group theory*, Lecture Notes in Math., vol. 143, Springer-Verlag, Berlin and New York, 1970. MR **43** #4923.

5. J.-P. Serre. *Corps locaux*, Hermann, Paris, 1968.

6. John Stallings, *Group theory and three-dimensional manifolds*, Yale Math. Monographs 4, Yale Univ. Press, New Haven, Conn., 1971.

7. Richard Swan, *Groups of cohomological dimension one*, J. Algebra **12** (1969), 585–610. MR **39** #1531.

8. C. T. C. Wall, *Finiteness conditions for CW-complexes*. II, Proc. Roy. Soc. Ser. A **295** (1966), 129–139. MR **35** #2283.

CORNELL UNIVERSITY

Proceedings of Symposia in Pure Mathematics
Volume 32, 1978

COMPLEXES OF COHOMOLOGICAL DIMENSION TWO

JOEL M. COHEN*

We say that a complex (i.e., CW complex) X has cohomological dimension (c-dim) $\leqq n$ if $H^2(X; A) = 0$ for $i > n$ and for all $\pi_1 X$-modules A. If X is an n-complex, then it has c-dim $\leqq n$.

Problem. If c-dim $X \leqq n$, is X of the homotopy type of ($\simeq$) an n-complex?

In **[6]** Wall showed that c-dim $X \leqq n$ is the same as saying that there is an n-complex K dominating X—i.e., there are maps $f: X \to K$, $g: K \to X$ with $gf \sim 1_X$. He further showed that the answer is yes for $n > 2$. Stallings and Swan **[5]** showed it for $n = 1$. This is a report of progress on the case $n = 2$.

We shall outline a proof of the following

THEOREM 1. *If* c-dim $X \leqq 2$, *then there is a wedge W of 2-spheres such that* $X \vee W \simeq$ *a 2-complex. If X is finite, W is finite.*

Furthermore, we shall construct a 3-complex X and a 2-complex K such that $X \vee S^2 \simeq K \vee S^2$ but $X \not\simeq K$. I have not yet proved that $X \not\simeq$ a 2-complex but I will indicate reasons for considering it unlikely.

There are some related results—one, a condition for realizing homomorphism $\pi_* K \to \pi_* K$; the second, a relation between stably equivalent modules and stably equivalent complexes.

The details of the arguments appear in **[3]**.

In **[1]** it is proved that if X is a nilpotent space and c-dim $\leqq 2$, then $X \simeq$ a 2-complex.

This paper owes a great deal to the late George Cooke. In particular the lemma in §1 is due essentially to him.

AMS (MOS) subject classifications (1970). Primary 55A20; Secondary 55A05.

*Research partially supported by the National Science Foundation.

1. Proof of Theorem 1.

LEMMA. *Let Y be an n-complex,* $n > 1$, $\pi = \pi_1 X$, $\Lambda = Z\pi$. *Assume that* $\pi_n Y \cong M \oplus F$ *where F is a free Λ-module with a basis* $\{x_\alpha\}$. *Assume that* $\{h(x_\alpha)\}$ *is the basis for a summand of* $H_n Y$, *where h is the Hurewicz map. Let X be Y with* $(n + 1)$-*cells attached by the* x_α. *Then* $\pi_n X \cong M$ *and* $X \vee \bigvee_\alpha S^n \simeq Y$.

We use this as follows: Let X be an $(n + 1)$-complex of c-dim $\leqq n$. Then X is dominated by an n-complex L. Let $f: X \to L$, $g: L \to X$ be such that $g \circ f \sim 1_X$. g may be assumed to be cellular. Thus $g(L) \subset X^n$, the n-skeleton. Let $h: X \to X^n$ be given by $g \circ f$. Then $i \circ h \sim 1_X$ where $i: X^n \subsetneq X$ is the inclusion. But then

$$\pi_n X^n \cong \pi_n X \oplus \pi_{n+1}(X, X^n) \quad \text{and} \quad H_n X^n \cong H_n X \oplus H_{n+1}(X, X^n)$$

where these splittings are natural with respect to the Hurewicz homomorphism since they are induced by maps. Now $\pi_{n+1}(X, X^n)$ (resp. $H_{n+1}(X, X^n)$) is a free Λ- (resp. Z-) module on the cells of $X - X^n$. h maps one basis to the other so the lemma shows that $X^n \simeq X \vee \bigvee_\alpha S^n$, α running through the cells of $X - X^n$.

If X has c-dim $\leqq 2$, then it has a priori c-dim $\leqq 3$ so by Wall's result **[5]** X may be assumed to be a 3-complex. By the above argument $X \vee \bigvee_\alpha S^2 \simeq X^2$ and Theorem 1 is proved.

2. The example. We say that Λ-modules M and N are stably equivalent, $M \doteq N$, if for some finitely generated free Λ-module F, $M \oplus F \cong N \oplus F$. We say that two complexes X and Y are stably equivalent, $X \doteq Y$, if $X \vee \bigvee_{i=1}^n S^2 \simeq Y \vee \bigvee_{i=1}^n S^2$ for some integer n.

The relation between the two notions is the following

THEOREM 2. *Let K be a 2-complex with* $\pi_1 K$ *finite. Let N be a* $Z[\pi_1 K]$-*module such that* $N \doteq \pi_2 K$. *Then there is a 3-complex X such that* $X \doteq K$ *and* $\pi_2 X \cong N$.

The construction of X is obtained by looking at $\pi_2(K \vee \bigvee_1^n S^2) \cong \pi_2 K \oplus F \cong N \oplus F$ and then attaching 3-cells to kill F. The proof that $X \doteq K$ (which is not true for $\pi_1 K$ infinite) used the following case of a more general result in **[3]** about realizing maps from a 2-complex to a complex.

LEMMA. *Let K be a 2-complex and* $\tilde{K}$ *its universal cover. Let* $\varphi: C_2\tilde{K} \to C_2\tilde{K}$ *be a* $Z\pi_1 K$-*module map with* $\partial_2 \varphi = \partial_2$. Let $\hat{\varphi}$ *be the induced endomorphism of* $\pi_2 K$ $(= \ker \partial_2)$. *Then there is a map* $f: K \to K$ *with* $\pi_1 f$ *the identity and* $\pi_2 f = \hat{\varphi}$.

Let π be the generalized quaternionic group of order 32. It may be presented as $\{x, y \mid yxy^{-1} x, y^2 x^8\}$. Let K be the 2-complex of this presentation. Let $\Lambda = Z\pi$. $\varphi = \sum_{\alpha \in \pi} \alpha$ generates a copy of $Z \subset \Lambda$.

PROPOSITION. $\pi_2 K \cong \Lambda/Z$.

Modifying a result of Swan **[4]** we get

PROPOSITION. *There is a noncyclic Λ-module M with* $M \oplus \Lambda \cong \Lambda/Z \oplus \Lambda$.

Thus using Theorem 2, we get a 3-complex X such that $X \vee S^2 \simeq K \vee S^2$ and $\pi_2 X \cong M$. I suspect that X cannot be $\simeq$ a 2-complex because M has the following "bad" property: It is noncyclic although $Z_p \otimes M$ is cyclic for every prime P.

π_2(a 2-complex) is a very restricted sort of module, very closely related to a presentation of a group, and I suspect that this "bad" property should not be possible. For example, if $\pi = F/R$ where F is finitely generated free and π is finite, then the abelianization R^a (which is closely related to π_2(2-complex) has the property that its number of generators is the maximum number of generators required for each $Z_p \otimes R^a$ (cf. **[2]**, for example).

I further conjecture that this is the only way to get an exception to a complex of c-dim $\leqq 2$ not being $\simeq$ a 2-complex: I suspect that any counterexample must require a finite group π which has a free periodic resolution of period 4 of Z. Equivalently there is a 2-complex K with $\pi_1 K = \pi$ and $\pi_2 K \cong Z\pi/Z$; again equivalently π acts freely on a 3-complex $\simeq S^3$. It would further require the existence of a noncyclic module $M \doteq Z\pi/Z$.

Bibliography

1. K. S. Brown and P. J. Kahn, *Homotopy dimension and simple cohomological dimension of spaces*, Comment. Math. Helv. **52** (1977), 111–127.

2. J. M. Cohen, *On the number of generators of a module*, J. Pure Appl. Algebra (to appear).

3. ———, *Complexes dominated by a 2-complex*, Topology (to appear).

4. R. G. Swan, *Projective modules over group rings and maximal orders*, Ann. of Math. (2) **76** (1962), 55–61. MR **25** #3066.

5. ———, *Groups of cohomological dimension one*, J. Algebra **12** (1969), 585–601. MR **39** #1531.

6. C. T. C. Wall, *Finiteness conditions for CW complexes*. I, Ann. of Math (2) **81** (1965), 55–69. MR **30** #1515.

University of Maryland

Proceedings of Symposia in Pure Mathematics
Volume 32, 1978

THE STABLE DECOMPOSITION FOR THE DOUBLE LOOP SPACE OF A SPHERE

F. R. COHEN, M. E. MAHOWALD AND R. J. MILGRAM*

Let $\mathscr{C}_n(j)$ denote the space of Boardman and Vogt's "j little n-cubes" [6]. $\mathscr{C}_n(j)$ supports a free Σ_j-action where Σ_j denotes the symmetric group on j letters. Consider any based space X and let $X^{[j]}$ denote the smash product of X with itself j times. Giving $X^{[j]}$ the natural Σ_j-action induced by permutation of coordinates, we form the "equivariant half-smash product", $D_{n,j}(X)$, where

$$D_{n,j}(X) = \mathscr{C}_n(j) \times_{\Sigma_j} X^{[j]}/\mathscr{C}_n(j) \times_{\Sigma_j} *.$$

$D_{n,j}(X)$ is a generalization of the quadratic construction extensively used in homotopy theory. In fact, using J. P. May's approximation to $\Omega^n\Sigma^n X$ [6], V. Snaith's generalization [8] of a result of D. S. Kahn [5] shows that the space $\Omega^n\Sigma^n X$ splits stably into a wedge $\bigvee_{j\geq 0} D_{n,j}(X)$ if X is a connected, compactly generated Hausdorff space with nondegenerate base point.

In the special case that X is a sphere, work in [7, §§1, 2] shows that $D_{n,j}(S^m)$ is the Thom space of the m-fold Whitney sum $m(P_{n,j})$, where $P_{n,j}$ is the j-dimensional vector bundle over $\mathscr{C}_n(j) \times_{\Sigma_j} *$,

$$\mathscr{C}_n(j) \times_{\Sigma_j} \mathbf{R}^j \longrightarrow \mathscr{C}_n(j) \times_{\Sigma_j} *$$

induced from the action of Σ_j on $\mathbf{R}^j$ which permutes coordinates. $\mathscr{C}_n(j)$ has the homotopy type of a finite complex [6] and from [7] it follows that $\varphi(j,n)P_{n,j}$ is trivial for some finite integer $\varphi(j,n)$. Hence

$$D_{n,j}(S^{m+\varphi(j,n)}) = \Sigma^{j\varphi(j,n)} D_{n,j}(S^m)$$

and more generally

$$D_{n,j}(\Sigma^{\varphi(j,n)}X) = \Sigma^{j\varphi(j,n)} D_{n,j}(X).$$

AMS (MOS) subject classifications (1970). Primary 55B20, 55D35, 55E45; Secondary 55F50.
*The authors were in part supported by National Science Foundation grants.

We do not know much about $\varphi(j, n)$ for general j and n, however when $n = 2$, we have

THEOREM 1. *$2P_{2,j}$ is trivial over $\mathscr{C}_2(j) \times_{\Sigma_j} *$. Hence*

$$D_{2,j}(S^q) = \begin{cases} \Sigma^{jq}\mathscr{C}_2(j) \times_{\Sigma_j} *, & q \text{ even}, \\ \Sigma^{j(q-1)} D_{2,j}(S^1), & q \text{ odd}. \end{cases}$$

REMARKS. (a) $\mathscr{C}_2(j) \times_{\Sigma_j} *$ has the homotopy type of the classical configuration space of all unordered sets of j distinct points in $\mathbf{R}^2$. Its fundamental group is the Artin braid group B_j on j strings and thus $\mathscr{C}_2(j) \times_{\Sigma_j} *$ is the Eilenberg-Mac Lane space $K(B_j, 1) = B(B_j)$. (See for example **[2**, Chapter 1] for details.)

(b) Recently M. Mahowald observed that the spaces $D_{2,2j}(S^1)$ are representatives for the Brown-Gitler spectra at the prime 2. That is to say

$$\tilde{H}^*(D_{2,2j}(S^1); \mathbf{Z}_2) \cong \mathscr{A}(2)/[\chi(Sq^{j+1}), \chi(Sq^{j+2}), \cdots, \chi(Sq^{j+n}), \cdots] \otimes e_{2j}.$$

(See for example **[3]** for details.) (Here e_j is a copy of $\mathbf{Z}_2$ concentrated in degree j.)

(c) Any stable cohomology theory or homology theory depends only on the stable homotopy type of a space, thus

$$\mathscr{H}\Omega^2\Sigma^{2+m}X = \coprod_j \mathscr{H}(D_{2,j}(\Sigma^m X))$$

and Theorem 1 gives us much tighter control of these groups.

(d) Related to (c) is Mahowald's recent proof that the classes h_1h_i in the Adams spectral sequence are infinite cycles. While not strictly dependent on Theorem 1, Mahowald's proof is considerably shortened by it.

PROOF OF THEOREM 1. We prove that $2P_{2,j}$ is trivial over $\mathscr{C}_2(j) \times_{\Sigma_j} *$. Since $2P_{2,j} = P_{2,2j}$, we prove this last assertion by giving a map

$$\nu\colon \mathscr{C}_2(j) \times_{\Sigma_j} (\mathbf{R}^2)^j \longrightarrow (\mathbf{R}^2)^j$$

which, for each point $c \in \mathscr{C}_2(j)$, ν restricts to a nonsingular linear map of $(\mathbf{R}^2)^j$. It clearly suffices to exhibit such a map ν with $\mathscr{C}_2(j)$ replaced by $F(\mathbf{R}^2, j)$, the space of j ordered distinct points in $\mathbf{R}^2$ **[6]**. Regard $\mathbf{R}^2$ as the complex numbers and define $\nu'\colon F(\mathbf{R}^2, j) \times_{\Sigma_j} (\mathbf{R}^2)^j \to (\mathbf{R}^2)^j$ by the formula

$$\nu'((w_1, \cdots, w_j), (\xi_1, \cdots, \xi_j)) = \left(\sum_i \xi_i, \sum_i w_i\xi_i, \sum_i w_i^2\xi_i, \cdots, \sum_i w_i^{j-1}\xi_i\right).$$

This suffices.

We remark that the map ν' is given by a linear transformation of each fibre which is described by the Vandermonde matrix. This description has essentially been discussed by V. I. Arnol'd **[1]**.

Note. Our original proof of Theorem 1 was rather complicated but recently E. H. Brown, Jr. found this elementary proof which we include with his kind permission.

THEOREM 2. *There is a cofibration at the prime 2 given by*

$$\Sigma^{2j(r-2)+1}D_{2,2j-1}(S^1) \longrightarrow \Sigma^{2j(r-2)}D_{2,2j}(S^1) \longrightarrow \Sigma^{2j(r-1)}D_{2,j}(S^1)$$

for r even and much larger than j.

REMARK. The cofibre of the analogous map to that given in Theorem 2, $\Sigma D_{2,2j}(S^1) \to D_{2,2j+1}(S^1)$, is trivial at the prime 2.

PROOF. Let $J_r\colon \Omega S^{r+1} \to \Omega S^{2r+1}$ be the James-Hopf invariant map, and $\Omega J_r\colon \Omega^2 S^{r+1} \to \Omega^2 S^{2r+1}$ be its loop map. Notice that the composite $S^{r-1} \to^{i} \Omega^2 S^{r+1} \to^{\Omega J_r} \Omega^2 S^{2r+1}$ is homotopically trivial (where i is the standard inclusion). If X is a connected CW complex, then May's approximation, $C_n X$, of $\Omega^n \Sigma^n X$ and $\Omega^n \Sigma^n X$ are of the same homotopy type [6]. Regard ΩJ_r as a map from $C_2 S^{r-1}$ to $C_2 S^{2r-1}$ which, by the cellular approximation theorem, we may assume to be cellular. Recall that $C_n X$ is filtered with ith filtration denoted by $F_i C_n X$. Assume that r is even and much larger than j. Then for dimensional reasons we see that ΩJ_r restricts to

$$\Omega J_r| \colon F_i C_2 S^{r-1} \longrightarrow F_{[i/2]} C_2 S^{2r-1}, \qquad i \leqq 2j.$$

Since $D_{n,j}(X) = F_j C_n X / F_{j-1} C_n X$ [6], ΩJ_r yields a map $\overline{\Omega J_r}\colon D_{2,2j}(S^{r-1}) \to D_{2,j}(S^{2r-1})$ by passage to quotients. Let μ denote the first coordinate loop multiplication in $C_2 X$; observe that ΩJ_r is a loop map and that $F_1 C_2 S^{r-1}$ has the homotopy type of S^{r-1}. Clearly the composite

$$F_1 C_2 S^{r-1} \times F_{2j-1}\, C_2 S^{r-1} \xrightarrow{\ \mu\ } F_{2j} C_2 S^{r-1} \xrightarrow{\ \Omega J_r|\ } F_j C_2 S^{2r-1}$$

is deformable into $F_{j-1} C_2 S^{2r-1}$. Thus, the composite

$$(*) \qquad \Sigma^{r-1} D_{2,2j-1}(S^{r-1}) \xrightarrow{\ \bar{\mu}\ } D_{2,2j}(S^{r-1}) \xrightarrow{\ \overline{\Omega J_r}\ } D_{2,j}(S^{2r-1})$$

is homotopically trivial where $\bar{\mu}$ is induced by μ.

To finish the proof of Theorem 2, it suffices to show that $(*)$ is short exact in mod 2 homology. Here, let e_k denote the class in degree k of $H_*(\Omega S^{k+1}; Z_2)$. Recall [4] that

$$J_{r*}(e_r^{2S+\varepsilon}) = \begin{cases} e_{2r}^S & \text{if } \varepsilon = 0, \\ 0 & \text{if } \varepsilon = 1. \end{cases}$$

Let Q_1 denote the first nontrivial Dyer-Lashof operation (which is defined for second fold loop spaces). Note that $\sigma_* Q_1 x = (\sigma_* x)^2$ where σ_* is the homology suspension and $(\Omega J_r)_*$ is a map of Hopf algebras. Moreover, Q_1 of a primitive is primitive, so it follows that

$$(\Omega J_r)_* \underbrace{Q_1 \cdots Q_1}_{i \text{ times}} (e_{r-1}) = \underbrace{Q_1 \cdots Q_1}_{i-1 \text{ times}} (e_{2r-1}),$$

and $(\Omega J_r)_*(e_{r-1}) = 0$. Hence, in $(*)$ the homology sequence is seen to be exact. The result follows.

REMARK. Theorem 2 is due to Mahowald and with a somewhat different proof given in his work on $h_1 h_i$. It is possible to use Theorem 2 and a simple induction to prove the assertion in remark (b) following Theorem 1. The key observation is that $(Q_1 e_{r-1})^j$ is dual to χSq^j by the Cartan formula, and that $(\Omega J_r|)_*(Q_1 e_{r-1})^j = (e_{2r-1})^j$. The first few examples are given below.

$$\begin{aligned}
(\chi Sq^1)_*(Q_1 e_{r-1}) &= Sq^1_*(Q_1 e_{r-1}) = (e_{r-1})^2, \\
(\chi Sq^2)_*(Q_1 e_{r-1})^2 &= Sq^2_*(Q_1 e_{r-1})^2 = (e_{r-1})^4, \\
(\chi Sq^3)_*(Q_1 e_{r-1})^3 &= (Sq^2 Sq^1)_*(Q_1 e_{r-1})^3 = (e_{r-1})^6, \\
(\chi Sq^4)_*(Q_1 e_{r-1})^4 &= (Sq^4_* + (Sq^3 Sq^1)_*)\,(Q_1 e_{r-1})^4 = Sq^4_*\,(Q_1 e_{r-1})^4 = (e_{r-1})^8.
\end{aligned}$$

REFERENCES

1. V. I. Arnol'd, *Topological invariants of algebraic functions*. II, Funkcional. Anal. i Priložen. **4** (1970), 1–9 = Functional Anal. Appl. **4** (1970), 91–98. MR **43** #1991.

2. J. Birman, *Braids, links, and mapping class groups*, Ann. of Math. Studies, No. 82, Princeton Univ. Press, Princeton, N. J., 1974. MR **51** #11477.

3. E. Brown and S. Gitler, *A spectrum whose cohomology is a certain cyclic module over the Steenrod algebra*, Topology **12** (1973), 283–295. MR **52** #11893.

4. D. Husemoller, *Fibre bundles*, McGraw-Hill, New York, 1966. MR **37** #4821.

5. D. S. Kahn and S. Priddy, *Applications of the transfer to stable homotopy theory*, Bull. Amer. Math. Soc. **78** (1972), 981–987. MR **46** #8220.

6. J. P. May, *The geometry of iterated loop spaces*, Lecture Notes in Math., vol. 271, Springer-Verlag, Berlin and New York, 1972.

7. R. J. Milgram, *Group representations and the Adams spectral sequence*, Pacific J. Math. **41** (1972), 157–182. MR **46** #3598.

8. V. Snaith, *A stable decomposition for $\Omega^n S^n X$*, J. London Math. Soc. (2) **7** (1974), 577–583. MR **49** #3918.

NORTHERN ILLINOIS UNIVERSITY

THE INSTITUTE FOR ADVANCED STUDY

NORTHWESTERN UNIVERSITY

STANFORD UNIVERSITY

Proceedings of Symposia in Pure Mathematics
Volume 32, 1978

H_∞ RING SPECTRA AND THEIR APPLICATIONS

J. P. MAY

A few years ago, Frank Quinn, Nigel Ray, and I introduced the notion of an "E_∞ ring spectrum". This concept was geared towards applications in infinite loop space theory, in particular to the analysis on the infinite loop level of such classifying spaces in geometric topology as BS Top and BSF. A summary of the results obtained in this direction may be found in [**12**], the details of the theory being given in [**11**] and the details of the applications of the theory to the study of characteristic classes being given in [**4**].

The theory of E_∞ ring spectra made clear that the familiar cohomology theories—classical theories, cobordism theories, topological and algebraic K-theories—were represented by spectra with enormously rich internal structure, and it seemed natural to try to exploit this structure to make calculations in stable homotopy theory. Since the Thom spectra occur in nature as E_∞ ring spectra and are so well understood, a good starting point seemed to be the analysis of MO and MU and, in particular, the study of the relationship between such new structure as the multiplicative homology operations on their zeroth spaces and the known stable information. Exploratory calculations by Stewart Priddy and Mark Steinberger made clear that the existing theory was inadequate for such computations. Again, the homotopy groups of E_∞ ring spectra carry operations, known in the 2-primary case as $\bigcup_i$-products, and it seemed natural to study the relationship between these operations and the Adams spectral sequence. Bob Bruner began work on this and found that the existing theory was also inadequate for this purpose.

That was the state of affairs before the Stanford conference. The difficulty was that E_∞ ring spectra were defined in terms of very precise point-set level algebraic structure on spectra. While this precision was vital to the applications in infinite loop space theory, it was ill adapted to work in stable homotopy theory. Shortly after the conference, I discovered how to express a great deal of the structure of

AMS (*MOS*) *subject classifications* (1970). Primary 55B20, 55E10, 55E45, 55G25, 55H15; Secondary 55D35.

E_∞ ring spectra solely in terms of maps in the stable category. The resulting theory of H_∞ ring spectra is the subject of this paper. I should emphasize from the start that the new theory is virtually independent of the old. In particular, its understanding requires no knowledge of infinite loop space theory.

While the foundations of the new theory are all in place, its calculational exploitation has barely begun. It already seems apparent, however, that it provides the unifying principle behind a variety of seemingly disparate stable phenomena. The present paper is mainly a preliminary announcement, and will contain few proofs. H_∞ ring spectra are defined in terms of maps $D_jE \to E$, where D_jE is the jth extended power of the spectrum E. We discuss extended powers in §1 and define and display examples of H_∞ ring spectra in §2.

We discuss the existing and potential applications in the last two sections. One rather immediate application will be an explanation in the new context of Nishida's theorem **[15]** which asserts the nilpotency of the ring π_*^s of stable homotopy groups of spheres. The other applications, some of which are work in progress, are not my work but that of my students. The mod p homology of an H_∞ ring spectrum has operations analogous to the by now familiar opreations on the homology of infinite loop spaces. These new operations have been analyzed by Mark Steinberger **[17]**, who has obtained simple conceptual proofs of the splittings of such Thom spectra as MO, MSO (at 2), and MSF. The homotopy operations on H_∞ ring spectra and their relationship to the Steenrod operations in the E_2 term and to the differentials of the Adams spectral sequence have been analyzed by Bob Bruner **[3]**, who has used this theory to simplify Milgram's proof **[14]** of the Hopf invariant one differentials and to give a new proof of Toda's key odd primary differentials **[18]**. There are also Steenrod type operations in the cohomology theories on spaces represented by H_∞ ring spectra. These include the classical Steenrod operations, the Adams operations in complex K-theory, and the cobordism Steenrod operations of tom Dieck **[5]**. Such cohomology theories also admit a new multiplicative transfer. These cohomology operations will be studied by others of a truly exceptional group of students presently at Chicago.

A more comprehensive summary has been circulated in preprint form. Full details will appear in due course.

1. Extended powers of spectra. Much of homotopy theory in the last few years has centered around exploitation of extended powers of spaces and variants thereof, and it has become manifest that such constructions are absolutely basic to the subject. Let Σ_j denote the symmetric group on j letters and let $E\Sigma_j$ be a fixed contractible space with a free action by Σ_j, say for definiteness the standard bar construction CW-resolution of Σ_j. For a based space X with basepoint $*$, define

$$D_jX = E\Sigma_j \ltimes_{\Sigma_j} X^{(j)} = E\Sigma_j \times_{\Sigma_j} X^{(j)}/E\Sigma_j \times_{\Sigma_j} *,$$

where $X^{(j)}$ denotes the j-fold smash product of X with itself. D_jX is called the jth extended power of X. By convention, $D_0X = S^0$ and $D_1X = X$; D_2X is also known as the (infinite) quadratic construction on X.

The theory of H_∞ ring spectra begins with the construction and analysis of analogous extended powers of spectra. Such generalizations of the smash product of spectra exist in any good stable category and were first obtained by Tsuchiya **[19]**, who worked in Boardman's category. For our theory, it is vital to work in the stable category $H\mathscr{S}$ defined in [**11**, II], the detailed treatment of which will appear in [**13**]. For the present note, it will suffice to define a spectrum E to be a sequence of spaces E_i with E_i homeomorphic to ΩE_{i+1}. A map $f: E \to F$ is a sequence of maps $f_i: E_i \to F_i$ such that f_i agrees under the given homeomorphisms with Ωf_{i+1}. There is a cylinder functor $E \wedge I^+$ on spectra, hence a resulting homotopy category $h\mathscr{S}$. The stable category $H\mathscr{S}$ is obtained from $h\mathscr{S}$ by formally inverting its weak homotopy equivalences. We shall also need the stabilization, or suspension spectrum, functor Q_∞ from spaces to spectra. It is specified by $Q_\infty X = \{Q\Sigma^i X\}$, where $QX = \operatorname{inj\,lim} \Omega^n\Sigma^n X$, and it preserves wedges, homotopy colimits and cofiberings, and smash products. We write S for the sphere spectrum $Q_\infty S^0$.

THEOREM 1.1. *There are functors $D_j: H\mathscr{S} \to H\mathscr{S}$, with $D_0E = S$ and $D_1E = E$ for spectra E, and there are natural isomorphisms $Q_\infty D_j X \cong D_j Q_\infty X$ of spectra for spaces X.*

REMARKS 1.2. (i) Just as we can define $D^n_G X = (EG)^n \ltimes_G X^{(j)}$ for any subgroup G of Σ_j, where $(EG)^n$ denotes the n-skeleton of EG, so we can define DnE for spectra E. These more general extended powers have all the properties one would expect. In particular, just as for spaces,

$$D^n_G E / D^{n-1}_G E = (BG)^n/(BG)^{n-1} \wedge E^{(j)}.$$

(ii) The filtration of $D_G E$ given by the $D^n_G E$ gives rise to a spectral sequence which converges from $H_*(G; k_* E^{(j)})$ to $k_* D_G E$ for any homology theory k. When k is ordinary homology with coefficients in a field, the spectral sequence collapses.

(iii) Let T be any set of primes and let E be a connective spectrum. If $\lambda: E \to E_T$ is the localization of E at T, then, by the homological characterization of T-local spectra and of localization, $D_G E_T$ is T-local and $D_G\lambda: D_G E \to D_G E_T$ is the localization of $D_G E$ at T. If $\gamma: E \to \hat{E}_T$ is the completion of E at T, then, while $D_G \hat{E}_T$ need not be T-complete, the homological characterization of completions implies that the completion at T of $D_G\gamma: D_G E \to D_G \hat{E}_T$ is an equivalence. (See [**13**] for localizations and completions of spectra.)

The spectrum level analogs of the various elementary properties of extended powers of spaces scattered through the literature are all true, but rather more difficult to prove. These properties center around transformations which correspond to the inclusion of the trivial group in Σ_j, the diagonal of Σ_j, the sum $\Sigma_j \times \Sigma_k \to \Sigma_{j+k}$, and the wreath sum $\Sigma_j \int \Sigma_k \to \Sigma_{jk}$. Let $E^{(j)}$ denote the j-fold smash product of E with itself.

PROPOSITION 1.3. *There are natural transformations $\iota_j: E^{(j)} \to D_j E$, $\tau_j: D_j(E \wedge F) \to D_j E \wedge D_j F$, $\pi_{jk}: D_j E \wedge D_k E \to D_{j+k} E$, and $\rho_{jk}: D_j D_k E \to D_{jk} E$ which are compatible with their evident space level analogs.*

REMARKS 1.4. These transformations specialize in obvious ways when j or k is zero, with the important exception of $\rho_{j0}: D_j S \to S$, which we rename ξ_j. $D_j S^0$ is just $B\Sigma_j^+$, the union of $B\Sigma_j$ and a disjoint basepoint 0. Define $\delta: B\Sigma_j^+ \to S^0$ by $\delta(0) = 0$ and $\delta(x) = 1$ for $x \in B\Sigma_j$. Then ξ_j is the composite

$$D_j S = D_j Q_\infty S^0 \cong Q_\infty D_j S^0 = Q_\infty B\Sigma_j^+ \xrightarrow{Q_\infty \delta} Q_\infty S^0 = S.$$

REMARKS 1.5. The transformations ι_1, τ_1, ρ_{j1}, and ρ_{1k} are given by identity maps, and $\pi_{1,1} = \iota: E \wedge E \to D_2E$.

The preprint version gave a systematic list of commutative diagrams involving these transformations, the simplest examples asserting that each τ_j is commutative and associative, that $\{\pi_{jk}\}$ is a commutative and associative system, and that $\{\rho_{jk}\}$ is an associative system. Such diagrammatic results are vital to the rigorous development but will not be elaborated here. Similarly, Nishida's important observations [15] about the extended powers of wedges of spaces remain true for spectra. These yield a decomposition of $D_j(E_1 \vee \cdots \vee E_k)$ as a wedge, and this in turn yields a deomposition of $D_jk: D_jE \to D_jE$ as a sum of maps. Specialization yields the following basic facts.

LEMMA 1.6. *If $k = p^iq$ with p prime, $i \geqq 1$, and q prime to p, then $D_pk: D_pE \to D_pE$ can be written in the form $p^ig + (p, k - p)\iota_ph_p$ for a certain map $h_p: D_pE \to E^{(p)}$.*

2. H_∞ ring spectra. By a (commutative) ring spectrum we understand a spectrum E together with maps $e: S \to E$ and $\phi: E \wedge E \to E$ such that the following diagrams commute in $H\mathscr{S}$, where γ is the transposition:

$$\begin{array}{ccccc} E\wedge S & \xrightarrow{1\wedge e} & E\wedge E & \xleftarrow{e\wedge 1} & S\wedge E \\ & \searrow & \downarrow{\scriptstyle\phi} & \swarrow & \\ & & E & & \end{array} \qquad \begin{array}{ccc} E\wedge E\wedge E & \xrightarrow{1\wedge\phi} & E\wedge E, \\ {\scriptstyle\phi\wedge 1}\downarrow & & \downarrow{\scriptstyle\phi} \\ E\wedge E & \xrightarrow{\phi} & E \end{array}$$

and

$$\begin{array}{ccc} E\wedge E & & \\ & \searrow{\scriptstyle\phi} & \\ {\scriptstyle\gamma}\downarrow & & E \\ & \nearrow{\scriptstyle\phi} & \\ E\wedge E & & \end{array}$$

In fact, this familiar notion incorporates only a small fraction of the structure generally available.

DEFINITION 2.1. An H_∞ ring spectrum is a spectrum E together with maps ξ_j: $D_jE \to E$ for $j \geqq 0$ such that $\xi_1 = 1$ and the following diagrams commute in $H\mathscr{S}$ for $j, k \geqq 0$:

$$\begin{array}{ccccc} D_jE\wedge D_kE & & \xrightarrow{\pi_{jk}} & & D_{j+k}E \\ {\scriptstyle\xi_j\wedge\xi_k}\downarrow & & & & \downarrow{\scriptstyle\xi_{j+k}} \\ E\wedge E & \xrightarrow{\iota_2} & D_2E & \xrightarrow{\xi_2} & E \end{array} \quad\text{and}\quad \begin{array}{ccc} D_jD_kE & \xrightarrow{\rho_{jk}} & D_{jk}E \\ {\scriptstyle D_j\xi_k}\downarrow & & \downarrow{\scriptstyle\xi_{jk}} \\ D_jE & \xrightarrow{\xi_j} & E \end{array}$$

A map $f: E \to F$ between H_∞ ring spectra is an H_∞ ring map if $\xi_j \circ D_jf = f \circ \xi_j$ for all $j \geqq 0$.

LEMMA 2.2. *With $e = \xi_0: S \to E$ and $\phi = \xi_2\iota_2: E \wedge E \to E$, an H_∞ ring spectrum is a ring spectrum and an H_∞ ring map is a ring map.*

LEMMA 2.3. *With ξ_j as in Remarks 1.4, S is an H_∞ ring spectrum and $e\colon S \to E$ is an H_∞ ring map for any H_∞ ring spectrum E.*

LEMMA 2.4. *If E and F are H_∞ ring spectra, then $E \wedge F$ is an H_∞ ring spectrum with respect to the composites*

$$\xi_j\colon D_j(E \wedge F) \xrightarrow{\tau_j} D_jE \wedge D_jF \xrightarrow{\xi_j \wedge \xi_j} E \wedge F.$$

LEMMA 2.5. *If E is an H_∞ ring spectrum, then $\xi_j \iota_j\colon E^{(j)} \to E$ is the j-fold iterated product on E and is itself an H_∞ ring map.*

Remarks 1.2(iii) imply the following important closure property of the category of H_∞ ring spectra.

LEMMA 2.6. *If E is a connective H_∞ ring spectrum, then its localization E_T and completion $\hat{E}_T$ at any set of primes T admit unique H_∞ ring structures such that λ: $E \to E_T$ and $\gamma\colon E \to \hat{E}_T$ are H_∞ ring maps.*

A much less structured notion of "Γ-spectrum" was exploited by Nishida **[15]**, and we shall discuss an analog of his definition at the end of the section. Essentially the present dofinition of H_∞ ring spectrum was given by Tsuchiya **[19]** but, because he was working in Boardman's stable category and therefore did not have very explicit control on the D_jE, the only example he was able to give was S itself. Our construction of the D_jE dovetails with the definition of E_∞ ring spectra **[11**, IV. 1.1 and 1.2] to give the following result.

THEOREM 2.7. *E_∞ ring spectra admit nautral H_∞ ring structures.*

Therefore, as we shall shortly make clear, H_∞ ring spectra abound. Basically, the theorem means that the notion of E_∞ ring spectrum was appropriate not only for infinite loop space theory but also for stable homotopy theory. It is the zeroth space E_0 which features in the earlier applications, and there is a useful consistency statement which ties together the space and spectrum level points of view. For any spectrum E, there is a natural map $\tau\colon Q_\infty E_0 \to E$ which induces the suspension $k_*E_0 \to k_*E$ for any homology theory k. While the zeroth space of D_jE is not well understood, we have the composite natural map

$$\omega\colon D_jE_0 \xrightarrow{\subset} QD_jE_0 = (Q_\infty D_jE_0)_0 \cong (D_jQ_\infty E_0)_0 \xrightarrow{(D_j\tau)_0} (D_jE)_0.$$

An E_∞ ring spectrum E comes equipped with maps $\xi_j\colon D_jE_0 \to E_0$ as a key part of its structure, and these are determined by the spectrum level maps ξ_j via the commutative diagram

$$\begin{array}{ccc} D_jE_0 & \xrightarrow{\xi_j} & \\ {\scriptstyle\omega}\downarrow & & E_0 \\ (D_jE)_0 & \xrightarrow{(\xi_j)_0} & \end{array}$$

There are three well-understood classes of examples.

EXAMPLES 2.8. If X is an E_∞ space with zero, with structural maps $\xi_j\colon D_jX \to X$, then $Q_\infty X$ is an E_∞ ring spectrum by **[11**, IV, §1] and the associated H_∞ ring structure is given by the maps $Q_\infty\xi_j\colon D_jQ_\infty X \cong Q_\infty D_jX \to Q_\infty X$. When $X = S^0$, we recover the H_∞ ring structure on $S = Q_\infty S^0$.

EXAMPLES 2.9. Let G be a bundle or fibration theory, such as O, U, Sp, Spin, Top, F and their oriented versions SO, SU, etc. (Technically, let G be a monoid-valued $\mathscr{I}_*$ functor [**11**, I] which maps to F.) As observed in [**11**, IV, §2], the Thom spectrum MG is an E_∞ ring spectrum. We can also construct Thom spectra $M(\Sigma_j \int G)$, and it turns out that D_jMG is equivalent to $M(\Sigma_j\int G)$ and that $\xi_j\colon M(\Sigma_j\int G) \to MG$ is the map of Thom spectra obtained by passage to Thom spaces and then to spectra from the maps

$$E\Sigma_j \times_{\Sigma_j} BG(n)^j = B(\Sigma_j \int G(n)) \to BG(nj)$$

of classifying spaces induced by the evident wreath product homomorphism $\Sigma_j\int G(n) \to G(nj)$. (The precise geometry needed to make sense of and prove these assertions in general is more elaborate than indicated, but the sketch is correct as given when G is O, U, or Sp.)

EXAMPLES 2.10. Let R be a commutative ring and consider the Eilenberg-Mac Lane spectrum $HR = K(R, 0)$. As pointed out in [**11**, VIII, §1], HR is an E_∞ ring spectrum. Here the associated H_∞ ring structure is trivial to describe. Remarks 1.2(ii) imply that $\iota_j\colon HR^{(j)} \to D_jHR$ induces an isomorphism in R-cohomology in degree 0. The j-fold tensor product of the fundamental class with itself is thus an element of $H^0(D_jHR; R)$, and this element provides (or rather, is) the map ξ_j: $D_jHR \to HR$.

The following remarks should be regarded as in some sense providing counits for H_∞ ring spectra.

REMARKS 2.11. Let E be a connective H_∞ ring spectrum. Then $H^0(D_jE; R) \cong H^0(E^{(j)}; R)$ is the j-fold tensor product of $H^0(E; R)$ with itself. Let $i : E \to HR$ be a map such that $ie\colon S \to HR$ is the unit of HR. In practice, $e^*\colon H^0(E; R) \to H^0(S; R)$ will be an isomorphism; hence i will be unique. Clearly $\xi_jD_ji = i\xi_j$: $D_jE \to HR$ for $j \geqq 1$ since the fundamental class of $H^0(HR; R)$ pulls back to the j-fold tensor product of its image under i^* under both composites. Therefore i is an H_∞ ring map.

The deepest class of examples comes from the general machinery developed in [**11**, VI and VII]. This allows one to construct E_∞ ring spectra from spaces with suitable internal structure, namely E_∞ ring spaces. The most important examples are the classifying spaces of categories with suitable internal structure, namely bipermutative categories. In particular, as explained in [**11**, VIII, §1], the cited machinery applies to give the connective spectra kO and kU of topological K-theory and the connective spectra of the algebraic K-theory of commutative rings E_∞ ring structures and therefore H_∞ ring structures.

Thus most common cohomology theories are already known to be represented by H_∞ ring spectra. Note too that smash products and localizations and completions yield a great many H_∞ ring spectra not known to be E_∞ ring spectra. Indeed, it would not be reasonable to expect such intrinsically homotopical constructions to preserve the precise algebraic structure of E_∞ ring spectra. Similarly, I have no reason to believe the E_∞ ring level version of the following conjectures, but I am fairly confident that they are valid as stated.

Conjecture 2.12. The p-local Brown-Peterson spectrum BP is an H_∞ ring spectrum and the natural map $MU_p \to BP$ is an H_∞ ring map.

Conjecture 2.13. The periodic spectra KO and KU are H_∞ ring spectra and the natural maps $kO \to KO$ and $kU \to KU$ are H_∞ ring maps.

Conjecture 2.14. The classical orientation maps $MU \to kU$ and M Spin $\to kO$ are H_∞ ring maps.

Jim McClure is working on the first and has proven the others [added in proof].

For some applications, the following ad hoc generalization of Nishida's notion of a Γ-spectrum [**15**, 1.5] is more appropriate than the notion of an H_∞ ring spectrum.

DEFINITION 2.16. A spectrum E is a pseudo H_∞ ring spectrum if

(i) E is the telescope of a sequence of connective spectra E_n,

(ii) E is a (weak) ring spectrum with unit induced from a map $S \to E_0$ and product induced from a unital, associative, and commutative system of compatible maps $E_m \wedge E_n \to E_{m+n}$, and

(iii) for each $j \geqq 0$ and $n \geqq 0$ and each integer s divisible by n, there exists a map $\xi_j\colon D_j\Sigma^s E_n \to \Sigma^{js}E_{jn}$ whose composite with $\iota_j\colon \Sigma^{js}E_n^{(j)} \cong (\Sigma^s E_n)^{(j)} \to D_j\Sigma^s E_n$ is the Σ^{js}-fold suspension of the iterated product.

The point is that no compatibility between the ξ_j for different n and s is required, the crux of the distinction between H_∞ ring spectra and pseudo H_∞ ring spectra being the lack of any commutation relation between extended powers and suspension. The weaker structure is just sufficient to allow certain key homological calculations to go through, but is too weak to be of general significance. Its utility comes from the following definition and lemma.

DEFINITION 2.17. Let (F, f) be a spectrum together with a map $f\colon S^r \to F$ for some integer r. Define $D(F, f) = \operatorname{Tel} \Sigma^{-rn}D_nF$, where the telescope is defined with respect to the maps

$$\Sigma^{-rn}D_nF = \Sigma^{-r(n+1)}(D_nF \wedge S^r) \xrightarrow{\Sigma^{-r(n+1)}(\pi_{n1}\cdot 1\wedge f)} \Sigma^{-r(n+1)}D_{n+1}F.$$

Define $D^rE = D(\Sigma^rE, \Sigma^re)$ if E is a spectrum with unit $e\colon S \to E$.

LEMMA 2.18. *$D(F, f)$ is a pseudo H_∞ ring spectrum with product given by suspensions of the maps $\pi_{mn}\colon D_mF \wedge D_nF \to D_{m+n}F$ and with $\xi_j = \rho_{jn}$ on $D_j\Sigma^{rn}(\Sigma^{-rn}D_nF)$ $= D_jD_nF$.*

3. Homology operations and the nilpotency of π_*^s. Let E be an H_∞ ring spectrum. We have the maps $\xi_j\colon D_jE \to E$, and our aim is to explain what they are good for. Let us first consider mod p homology for a fixed prime p. Remarks 1.2 imply that for $G \subset \Sigma_p$ we have $H_*D_GE = H_*(G; (H_*E)^p)$. When $G = \pi$ is cyclic of order p, an element $x \in H_*X$ gives rise to an element $e_i \otimes x^p \in H_*D_\pi E$. Its image under ξ_{p*} is written Q_ix. With a suitable reindexing (just as in [**4**, p.7] for the homology of E_∞ spaces), we obtain natural homomorphisms $Q^s\colon H_*E \to H_*E$ of degree s if $p = 2$ and of degree $2s(p-1)$ if $p > 2$. The following theorem is due to Steinberger [**17**]. The key point is the Nishida relations, which are subtly different from their E_∞ space level analogs.

THEOREM 3.1. *The operations Q^s have the following properties.*

(1) $Q^sx = 0$ *if* $2s <$ *degree* x [*if* $s <$ deg x *when* $p = 2$].

(2) $Q^sx = x^p$ *if* $2s =$ *degree* x [*if* $s =$ deg x *when* $p = 2$].

(3) $Q^s(1) = 0$ *if* $s \neq 0$, *where* $1 \in H_0E$ *is the identity element.*

(4) *The external and internal Cartan formulas hold:*

$$Q^s(x \otimes y) = \sum_{i+j=s} Q^i x \otimes Q^j y \quad \text{and} \quad Q^s(xy) = \sum_{i+j=s} Q^i x Q^j y.$$

(5) $\tau_*\colon \tilde{H}_*E_0 \to H_*E$ *commutes with the operations, where* $(\xi_p)_0\omega\colon D_pE_0 \to E_0$ *induces the operations on* $\tilde{H}_*E_0$.

(6) *The Adem relations hold: If* $p \geqq 2$ *and* $r > ps$, *then*

$$Q^rQ^s = \sum_i (-1)^{r+i}(pi - r, r - (p-1)s - i - 1)Q^{r+s-i}Q^i;$$

if $p > 2$, $r \geqq ps$, *and* β *denotes the* mod p *Bockstein, then*

$$Q^r\beta Q^s = \sum_i (-1)^{r+i}(pi - r, r - (p-1)s - i)\beta Q^{r+s-i}Q^i$$
$$- \sum_i (-1)^{r+i}(pi - r - 1, r - (p-1)s - i)Q^{r+s-i}\beta Q^i.$$

(7) *The Nishida relations hold: Let* $P^r_*\colon H_*E \to H_*E$ *denote the dual of the Steenrod operation* P^r, *with* $P^r = Sq^r$ *when* $p = 2$. *Then, with n chosen much larger than r and s,*

$$P^r_*Q^s = \sum_i (-1)^{r+i}(r - pi, p^n + s(p-1) - pr + pi)Q^{s-r+i}P^i_*;$$

hence $\beta Q^s = (s-1)Q^{s-1}$ *if* $p = 2$, *and, if* $p > 2$,

$$P^r_*\beta Q^s = \sum_i (-1)^{r+i}(r - pi, p^n + s(p-1) - pr + pi)\beta Q^{s-r+i}P^i_*$$
$$+ \sum_i (-1)^{r+i}(r - pi - 1, p^n + s(p-1) - pr + pi)Q^{s-r+i}P^i_*\beta.$$

(*In* (6) *and* (7), (i, j) *is* $(i+j)!/i!\,j!$ *if* $i \geqq 0$ *and* $j \geqq 0$ *and zero otherwise.*)

This should be compared with [4, I, 1.1]. When E is connective, $Q^s = 0$ for $s < 0$ by (1). While we have very few examples of nonconnective H_∞ ring spectra, they do have nontrivial operations for $s < 0$. Part (5) is particularly pertinent for E_∞ ring spectra; the relevant space level operations were denoted $\tilde{Q}^s$ in [4, II] to distinguish them from the additive operations Q^s present in any infinite loop space.

EXAMPLE 3.2. Consider the Thom spectra MG, $G = O$, U, and Sp. By Examples 2.9, the Q^s on H_*MG are determined by commutation with the Thom isomorphism from the known (additive!) homology operations Q^s on H_*BG. In turn the Q^s on H_*MG determine the $\tilde{Q}^s$ on $H_*(MG)_0$ modulo the kernel of τ_*. The same idea applies to other G, such as SF.

EXAMPLES 3.3. $H_*Q_\infty X = \tilde{H}_*X$ for any space X (as a matter of definition). By Examples 2.8, the operations Q^s on $H_*Q_\infty X$ coincide with the operations $\tilde{Q}^s$ on $\tilde{H}_*X$ if X is an E_∞ space with zero.

The next example, again due to Steinberger [17], is more surprising. Recall that the homology operations on Eilenberg-Mac Lane spaces are trivial for trivial reasons [4, I. 6.1]. The situation for spectra is rather different.

EXAMPLES 3.4. $H_*(HZ_p) = A_*$ is generated as an algebra over the Dyer-Lashof algebra by ξ_1 if $p = 2$ and by τ_0 if $p > 2$. In fact, if χ denotes the conjugation in the dual of the Steenrod algebra, then

$$Q^{2^n-2}(\xi_1) = \chi\xi_n \quad \text{if } p = 2 \text{ and } n \geqq 2$$

and $Q^{1+p+\cdots+p^{n-1}}\tau_0 = (-1)^n\chi\tau_n$ and $\beta\chi\tau_n = \chi\xi_n$ if $p > 2$ and $n \geqq 1$.

Steinberger has also computed the operations Q^s for HZ, HZ_{p^i}, kO, and kU and has obtained a simple conceptual proof of the following version of a result due to Nishida [**15**, §3].

PROPOSITION 3.5. *If E is a connective pseudo H_∞ ring spectrum with $\pi_0 E = Z_p$, then E is equivalent to a wedge of suspensions of HZ_p.*

For actual H_∞ ring spectra, the proof exploits the H_∞ ring map $i\colon E \to HZ_p$ of Remarks 2.11. Since $\pi_0 E = Z_p$, i_* hits ξ_1 if $p = 2$ or τ_0 if $p > 2$; hence i_* is an epimorphism by Examples 3.4. Therefore H^*E is a free A-module. Since E is p-complete (because $p\pi_*E = 0$ by the ring structure), the required splitting follows. Of course, the splitting of MO is an obvious consequence. Other familiar splittings of Thom spectra, such as those of MSO (at 2) and of MSF, follow by the same method and the fact that $H_*(HZ)$ is generated over the Dyer-Lashof algebra by only a few low dimensional elements. The proof of Proposition 3.5 for general pseudo H_∞ ring spectra proceeds along the same lines, but with HZ_p replaced by a suitable wedge of Eilenberg-Mac Lane spectra. Nishida exploits this result by combining it with the following observation [**15**, 2.5], which is immediate from the calculations in [**4**, I, §§4–5].

PROPOSITION 3.6. *Let X be an $(r-1)$ connected space with $H_r X = Z_p$, where $r > 0$ and r is even if p is odd. Let $f\colon S^r \to X$ be nontrivial on H_r. Then the homomorphism $H_i \Sigma^r D_k X \to H_i D_{k+1} X$ induced by the composite*

$$D_k X \wedge S^r \xrightarrow{1\wedge f} D_k X \wedge X \xrightarrow{\pi_{k,1}} D_{k+1} X$$

is a monomorphism and is an isomorphism if $i < (k+1)(r + (2p-3)/p)$.

The main theorem [**15**, 4.1] of the first part of Nishida's paper translates as follows in our context (except that the dimensional range that Nishida states is not quite as sharp as his arguments allow).

THEOREM 3.7. *Let E be an H_∞ ring spectrum. If $x \in \pi_r E$ is of order p, where $r > 0$ and r is even if p is odd, and if $y \in \pi_t S$ with $0 < t < (2p-3)(n+1)/p - 1$, then $x^n y = 0$; in particular, with $E = S$, $y^{n+1} = 0$ if $py = 0$.*

Indeed, x extends to $\tilde{x}\colon \Sigma^r M \to E$, where M is the Moore spectrum $S \cup_p CS$. The pseudo H_∞ ring spectrum $D^r M$ of Definition 2.17 splits as a wedge of Eilenberg-Mac Lane spectra, and the natural map $D_n \Sigma^r M \to \Sigma^{rn} D^r M$ is an equivalence in the specified range. Since $x^n\colon S^{rn} \to E$ factors as the composite

$$S^{rn} \xrightarrow{\iota_n} D_n S^r \longrightarrow D_n \Sigma^r M \xrightarrow{D_n \tilde{x}} D_n E \xrightarrow{\xi_n} E$$

by a trivial diagram chase, the conclusion follows.

While this argument fails for x of order p^i when $i > 1$, there is an analog of Proposition 3.5 which can be used to prove a result of the form $x^n y = 0$ if $p^i x = 0$ and x satisfies certain other conditions involving compositions with elements in low stems.

The framework for the full nilpotency result is the following consequence of

Lemma 1.6 applied to $D_p(p^i)$: $D_pS^{rn} \to D_pS^{rn}$ and a diagram chase based on the factorization of x^{pn} for $x \in \pi_r E$ in the form

$$S^{prn} \xrightarrow{\iota_p} D_pS^{rn} \xrightarrow{D_p(x^n)} D_pE \xrightarrow{\xi_p} E.$$

THEOREM 3.8. *Let E be an H_∞ ring spectrum. If $x \in \pi_r E$ is of order p^i and if $y \in \pi_t S$ factors as a composite*

$$S^{prn+t} \xrightarrow{\bar{y}} D_pS^{rn} \xrightarrow{h_p} S^{prn},$$

then $p^{i-1}(pz + x^{pn}y) = 0$ for some $z\colon S^{prn+t} \to E$, hence $p^{i-1}x^{pn+1}y = 0$; in particular, with $E = S$, $p^{i-1}y^{pn+2} = 0$ if $p^iy = 0$.

By iteration (and $2y^2 = 0$ for y in an odd stem), the nilpotency of π_*^s will follow once it is shown that for every even $t > 0$ there exists $n > 0$ such that h_{p*}: $\pi_{ptn+t}D_pS^{tn} \to \pi_{ptn+t}S^{ptn}$ is an epimorphism. The known method of proof (explained below) gives a very poor estimate on n, and one wonders if a better result might be obtainable by different techniques. Note in particular the tantalizing fact that if $y \in \pi_t S$ is of order 2, then $y^4 = 0$ provided only that y factors as $S^{3t} \xrightarrow{\bar{y}} D_2S^t \xrightarrow{h_2} S^{2t}$.

We require information about the homotopy types of the spectra $D_\pi^k S^n$, π cyclic of order p. (Here S^n is a sphere spectrum, but by Theorem 1.1 it amounts to the same thing to study the stable homotopy type of the space with the same name.) Let L^k be the k-skeleton of $L = S^\infty/\pi$ with its standard cell structure and let $L_j^k = L^k/L_{j-1}$ with $L_0^k = (L^k)^+$. The L_j^k are called stunted lens spaces or, if $p = 2$, stunted projective spaces and are written P_j^k in the latter case. The following results are well known [6], [14], [3].

THEOREM 3.9. *Let $p = 2$ and let $\phi(k)$ be the number of $j \equiv 0, 1, 2$ or $4 \bmod 8$, such that $0 < j \leq k$.*

(i) *$D_\pi^k S^n$ and $\Sigma^n Q_\infty P_n^{n+k}$ are equivalent spectra.*
(ii) *P_m^{m+k} and P_n^{n+k} are stably equivalent if $m \equiv n \bmod 2^{\phi(k)}$.*
(iii) *P_n^{n+k} is stably reducible if and only if $n + k + 1 \equiv 0 \bmod 2^{\phi(k)}$.*

By (i) and (ii), $D_\pi^k S^n$ and $\Sigma^{2n} Q_\infty P_0^k$ are equivalent if $n \equiv 0 \bmod 2^{\phi(k)}$.

THEOREM 3.10. *Let $p > 2$ and let $\psi(k) = [k/2(p-1)]$.*

(i) *$D_\pi^k S^n$ and $\Sigma^n L_{n(p-1)}^{n(p-1)+k}$ are equivalent spectra.*
(ii) *L_{2m}^{2m+k} and L_{2n}^{2n+k} are stably equivalent if $m \equiv n \bmod p^{\psi(k)}$.*
(iii) *L_{2n}^{2n+k} is stably reducible if and only if k is odd and $2n + k + 1 \equiv 0 \bmod p^{\psi(k)}$.*

By (i) and (ii), $D_\pi^k S^n$ is equivalent to $\Sigma^{pn} Q_\infty L_0^k$ if $n \equiv 0 \bmod p^{\psi(k)}$.

As Adams has explained [1], the Kahn-Priddy theorem [7] can be stated as follows. (Adams' L is not $B\pi$ but a p-local version of $B\Sigma_p$; however, mimicry or consideration of the transfer $Q_\infty B\Sigma_p \to Q_\infty B\pi$ shows the theorem to be valid as stated.)

THEOREM 3.11. *Localize all spectra at p. For $1 < k$ or $k = \infty$, any map $Q_\infty L^k \to S$ which induces an isomorphism on π_{2p-3} induces an epimorphism on π_t for $0 < t < k$.*

Let $r > 0$ with r even if $p > 2$, let $t > 0$, and let $n > 0$ be minimal such that there exists $k > t$ with $rn \equiv 0 \bmod 2^{\phi(k)}$ if $p = 2$ or mod $p^{\psi(k)}$ if $p > 2$. By

Theorems 3.9 and 3.10 and the fact that $L_0^k = (L^k)^+$, $D_\pi^k S^{rn}$ is equivalent to $S^{prn} \vee \Sigma^{prn} Q_\infty L^k$. By Theorem 3.11, the following result will complete the proof that π_*^s is nilpotent since it will imply that Theorem 3.8 applies to all y under the specified restrictions on r, t, and n.

THEOREM 3.12. *Let $q \geqq 0$ be even and divisible by p and let $j = pq + 2p - 3$. Then h_{p*} maps $\pi_j D_p S^q = Z_p \oplus Z_p$ surjectively to $\pi_j S^{pq} = Z_p$ and annihilates the image Z_p of $\pi_j S^{pq}$ under ι_{p*}.*

Of course, since $\pi_j D_\pi^k S^q \to \pi_j D_p S^q$ is an isomorphism for $k > 2p - 3$, the result remains true upon restriction to $D_\pi^k S^q$. This is the heart of the nilpotency theorem. After killing bottom homotopy groups and localizing at p, so that the relevant dimension j is the Hurewicz dimension, one can give a quite simple homological proof. Details are in the preprint version and will appear elsewhere.

4. Homotopy operations and cohomology operations. We next consider the homotopy operations of an H_∞ ring spectrum E. Let $f \in \pi_n E$ and consider the composite $\xi_j \circ D_j f \colon D_j S^n \to E$. Any element $g \in \pi_q D_j S^n$ determines $\bar{g}(f) = g \circ \xi_j \circ D_j f \in \pi_q E$. For each fixed g, we may consider $\bar{g}$ as an operation $\pi_n E \to \pi_q E$. When $j = p$, we may restrict $\xi_j \circ D_j f$ to $D_\pi^k S^n$ and choose g in $\pi_q D_\pi^k S^n$, the advantage being the known homotopy type of $D_\pi^k S^n$.

In practice, the operations themselves are only a small part of the story, the main interest lying in the relationship between the geometry of H_∞ ring spectra and the computation of differentials and extensions in the Adams spectral sequence. The relevant theory has been worked out by Bruner [3]. We give a very partial sketch of his results, going only far enough to explain the computation of certain key differentials in the classical case $E = S$. Results along these lines for $p = 2$ and $E = S$ were originally obtained by Kahn [6], his work having been carried further by Milgram [14] and Makinen [8]. Even in this case, I find the present conceptualization a significant simplification.

Working in $H\mathscr{S}$, we first choose an Adams resolution

$$\begin{array}{ccccccc} E = E_0 & \longleftarrow & E_1 & \longleftarrow & E_2 & \longleftarrow & \cdots \\ \searrow & \nearrow & \searrow & \nearrow & & & \\ & W_0 & & W_1 & & & \end{array}$$

with each $E_{i+1} \to E_i$ a cofibration. We then define

$$F_n = \bigcup_{i_1 + \cdots + i_p = n} E_{i_1} \wedge \cdots \wedge E_{i_p} \quad \text{and} \quad Z_n = \bigvee_{i_1 + \cdots + i_p = n} W_{i_1} \wedge \cdots \wedge W_{i_p}.$$

There results an Adams resolution

$$\begin{array}{ccccccc} E^{(p)} = F_0 & \longleftarrow & F_1 & \longleftarrow & F_2 & \longleftarrow & \cdots \\ \searrow & \nearrow & \searrow & \nearrow & & & \\ & Z_0 & & Z_1 & & & \end{array}$$

The extended power construction of §1 generalizes to give inclusions $D_\pi^k F_n \subset D_\pi^k E$ which extend $F_n \subset E^{(p)}$. For $n \geqq k$, the restrictions of $\xi_p \colon D_p E \to E$ to $D_\pi^k F_n$ lift to

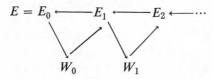

E_{n-k}, compatibly as n and k vary. In particular, we obtain a comparison of cofibrations

$$\begin{array}{ccccc} D_\pi^k F_{n+1} \cup D_\pi^{k-1} F_n & \longrightarrow & D_\pi^k F_n & \longrightarrow & (B\pi)^k/(B\pi)^{k-1} \wedge Z_n \\ \downarrow{\scriptstyle \xi_p} & & \downarrow{\scriptstyle \xi_p} & & \downarrow{\scriptstyle \theta_p} \\ E_{n-k+1} & \longrightarrow & E_{n-k} & \longrightarrow & W_{n-k} \end{array}$$

The maps in mod p homology induced by the θ_p are suitable for the construction of Steenrod operations in $E_2 = \mathrm{Ext}_{A_*}(Z_p, H_*E)$ according to the general prescription of **[10]**. The properties of the resulting operations are given in detail in **[10**, 11.8].

Now suppose given a map $f\colon S^n \to E$ of filtration i, so that f lifts to E_i but not to E_{i+1}. Let $\bar{f}$ denote the projection of f to W_i (so that $\bar{f}$ is nonzero). If $D_\pi^k S^n$ is reducible (see Theorems 3.9 and 3.10), we choose a splitting map $g\colon S^{pn+k} \to D_\pi^k S^n$ and define $D_k f = \bar{g}(f) \in \pi_{pn+k} E$. For any k and n, we also write D_k for the precursors $E_1^{i,i+n} \to E_1^{pi-k,pi+pn}$ of the Steenrod operations in E_2 (compare **[10**, p. 227]). When $p = 2$, D_k is usually called cup-k. We have $E_1^{s,t} = \pi_{t-s} W_s$, and a diagram chase shows that if $D_k f$ is defined, then $D_k \bar{f}$ is a permanent cycle which survives to $D_k f$. It can happen that $D_k \bar{f}$ is zero in E_∞ but that $D_k f$ is nonzero by virtue of having a filtration greater than $pi - k$.

When $D_k f$ is not defined, the nontriviality of the attaching map $S^{pn+k-1} \to D_\pi^{k-1} S^n$ of the top cell of $D_\pi^k S^n$ produces nontrivial differentials on the $D_k \bar{f}$. To illustrate, let $p > 2$ and let $a_0 \in E_2^{1,1}$ and $h_0 \in E_2^{1,2p-2}$ survive to p and to the element α_1 of mod p Hopf invariant one. A cursory examination of the top level of the attaching map shows that $d_2 D_k \bar{f} = a_0 D_{k-1} \bar{f}$ if k is even and $d_{2p-1} D_k \bar{f} = h_0 D_{k-2(p-1)} \bar{f}$ if k is odd and $k \not\equiv n - 1 \bmod p$. (Here and below, we work up to nonzero constants.) However, this ignores deeper levels of the attaching map and can be misleading because of filtration changes. Indeed, consider the case $k = 2(p - 1)$ and n even. Theorem 3.10 implies that the top cell of $D_\pi^{2p-2} S^n$ is attached by

$$nm\alpha_1 + p : S^{pn+2p-3} \to S^{pn} \vee S^{pn+2p-3} \subset D_\pi^{2p-3} S^n, \qquad m = \tfrac{1}{2}(p - 1).$$

Here both $D_0 f = f^p$ and $D_{2p-3} f$ are defined, and the factorization $S^{pn} \vee S^{pn+2p-3} \subset D_\pi^{2p-3} S^n \subset D_\pi S^n \xrightarrow{D_\pi f} D_\pi E \xrightarrow{\xi_p} E$ implies the following new relation. To repeat, this result and the rest of the material above is due to Bruner [3].

THEOREM 4.1. *For $f \in \pi_n E$ with n even, $nm\alpha_1 f^p + pD_{2p-3} f = 0$.*

It follows that the first possible nontrivial differential d_r on $D_{2p-2} \bar{f}$ kills an element of E_r which survives to $\alpha_1 f^p + pD_{2p-3} f$. How this element looks in E_r depends on the behavior of the specified geometric operations with respect to filtration. As an important example, take $E = S$ and consider the element $b_1^0 \in E_2^{2,2p(p-1)}$ which survives to β_1 **[9]**. We have $D_{2(p-1)} b_1^0 = b_1^1 \in E_2^{2,2p^2(p-1)}$. It is immediate from the calculation of E_2 in **[9]** that the $2p^2(p - 1) - 3$ stem has p-torsion Z_p or 0. Therefore $pD_{2p-3}\beta_1 = 0$. Thus the theorem has the following immediate consequence.

COROLLARY 4.2. *$\alpha_1 \beta_1^p = 0$ and $d_{2p-1} b_1^1 = h_0 (b_1^0)^p$.*

This is the main result of Toda's paper **[18]**.

For simplicity, we have restricted attention to the case when $\bar{f}$ is a permanent cycle in our discussion of differentials on the $D_k\bar{f}$. More subtle arguments apply in general; in particular the Hopf invariant one differentials $d_2h_i = h_0h_{i-1}^2$ if $p = 2$ and $d_2h_i = a_0D_{p-2}h_{i-1}$ if $p > 2$ can easily be obtained in this way (compare Milgram [14] and Makinen [8]). Indeed, all known differentials in the classical Adams spectral sequence can be derived conceptually by these techniques together with use of Massey products and Toda brackets.

Our final application, the construction and analysis of generalized Steenrod operations, is in more rudimentary form. The following sharpening of the notion of H_∞ ring spectrum seems to provide the correct framework.

DEFINITION 4.3. An H_∞^* ring spectrum is a spectrum E together with maps ξ_j: $D_j\Sigma^qE \to \Sigma^{jq}E$ for $j \geqq 0$ and all integers q such that $\xi_1 = 1$ and the following diagrams commute in $H\mathscr{S}$ for $j, k \geqq 0$ and all q and r:

$$\begin{array}{ccc} D_j\Sigma^qE \wedge D_k\Sigma^qE & \xrightarrow{\pi_{jk}} & D_{j+k}\Sigma^qE \\ \downarrow{\scriptstyle \xi_j\wedge\xi_k} & & \downarrow{\scriptstyle \xi_{j+k}} \\ \Sigma^{jq}E \wedge \Sigma^{kq}E & \xrightarrow{\phi} & \Sigma^{(j+k)q}E \end{array} \qquad \begin{array}{ccc} D_jD_k\Sigma^qE & \xrightarrow{\rho_{jk}} & D_{jk}\Sigma^qE \\ \downarrow{\scriptstyle D_j\xi_k} & & \downarrow{\scriptstyle \xi_{jk}} \\ D_j\Sigma^{kq}E & \xrightarrow{\xi_j} & \Sigma^{jkq}E \end{array}$$

and, up to the sign $(-1)^{j(j-1)qr/2}$,

$$\begin{array}{ccccc} D_j(\Sigma^qE \wedge \Sigma^rE) & \xrightarrow{\tau_j} & D_j\Sigma^qE \wedge D_j\Sigma^rE & \xrightarrow{\xi_j\wedge\xi_j} & \Sigma^{jq}E \wedge \Sigma^{jr}E, \\ \downarrow{\scriptstyle D_j\phi} & & & & \downarrow{\scriptstyle \phi} \\ D_j\Sigma^{q+r}E & & \xrightarrow{\xi_j} & & \Sigma^{jq+jr}E \end{array}$$

where the maps ϕ: $\Sigma^qE \wedge \Sigma^rE \cong \Sigma^{q+r}(E \wedge E) \to \Sigma^{q+r}E$ are induced by the product $\xi_2\iota_2$: $E \wedge E \to E$. We have a similar notion of H_∞^{d*} ring spectrum, for which q is restricted to be congruent to zero modulo d.

DEFINITION 4.4. Let E be an H_∞^* ring spectrum and let Y be any spectrum. Recall that $E^qY = [Y, \Sigma^qE]$ and define an external power operation $\mathscr{P}_j$: $E^qY \to E^{jq}D_jY$ by letting $\mathscr{P}_j(f)$ be the composite $D_jY \xrightarrow{D_jf} D_j\Sigma^qE \xrightarrow{\xi_j} \Sigma^{jq}E$. For a based space X, recall that $\tilde{E}^qX = E^qQ_\infty X$ and define an internal power operation P_j: $\tilde{E}^qX \to \tilde{E}^{jq}(B\Sigma_j^+ \wedge X)$ by $P_j(f) = d^*\mathscr{P}_j(f)$, where d: $B\Sigma_j^+ \wedge X \to D_jX$ is induced by the reduced diagonal Δ: $X \to X^{(j)}$.

We denote the external product $E^qY \wedge E^rZ \to E^{q+r}(Y \wedge Z)$ by juxtaposition; it is induced by ϕ: $\Sigma^qE \wedge \Sigma^rE \to \Sigma^{q+r}E$. The following result is immediate from the definitions and easy diagram chases.

LEMMA 4.5. *The external powers $\mathscr{P}_j$ satisfy the following properties, where $f \in E^qY$, $g \in E^rZ$, and $\sigma = j(j-1)qr/2$:*

(1) $\mathscr{P}_1(f) = f$ *and* $\iota_j^*\mathscr{P}_j(f) = f^j$, $\quad \iota_j$: $Y^{(j)} \to D_jY$.
(2) $\mathscr{P}_j(fg) = (-1)^\sigma\tau_j^*(\mathscr{P}_j(f)\mathscr{P}_j(g))$, $\quad \tau_j$: $D_j(Y \wedge Z) \to D_jY \wedge D_jZ$.
(3) $\mathscr{P}_j(\Sigma^rf) = (-1)^\sigma\tau_j^*(\mathscr{P}_j(f)\mathscr{P}_j(\Sigma^re))$, $\quad \tau_j$: $D_j(Y \wedge S^r) \to D_jY \wedge D_jS^r$.
(4) $\pi_{jk}^*\mathscr{P}_{j+k}(f) = \mathscr{P}_j(f)\mathscr{P}_k(f)$, $\quad \pi_{jk}$: $D_jY \wedge D_kY \to D_{j+k}Y$.
(5) $\rho_{jk}^*\mathscr{P}_{jk}(f) = \mathscr{P}_j(\mathscr{P}_k(f))$, $\quad \rho_{jk}$: $D_jD_kY \to D_{jk}Y$.

With $j = p$, tom Dieck [5] has studied the formal consequences of (1) and (2), which he took as axioms for Steenrod external power operations.

DEFINITION 4.6. Let E be an H_∞^* ring spectrum. If the Kunneth map $E^*B\Sigma_j \otimes_{\pi_* E} E^*X \to E^*(B\Sigma_j^+ \wedge X)$ is an isomorphism and $E^*B\Sigma_j$ is π_*E-free on the basis $\{w_i\}$, then each w_i gives an operation $\bar{w}_i$ by the formula $P_j(f) = \sum_i w_i \otimes \bar{w}_i(f)$. Alternatively, without a Kunneth assumption, each element $e \in E_k(B\Sigma_j;\ Z_n)$ gives an operation $\bar{e}$ by $\bar{e}(f) = P_j(f)/e$. Here mod n coefficients have been used since $\tilde{E}^*B\Sigma_j$ is a torsion group and thus finite coefficient rings will be of greatest interest.

EXAMPLE 4.7. For a commutative ring R, HR is an H_∞^{2*} ring spectrum, the requisite maps ξ_j: $D_j\Sigma^q HR \to \Sigma^{jq}HR$ being obtained precisely as in Examples 2.10. With $j = p$ and $R = Z_p$, we recover the classical Steenrod operations. Indeed, the slant product approach gives precisely Steenrod's original definition [**16**] carried out purely homotopically.

While some of what is true is clear enough, the following examples are conjectural at this writing since no details have yet been written down. They are being studied by Jim McClure. [ADDED IN PROOF: Details are now in place.]

EXAMPLES 4.8. It is not clear to me whether or not kU, KU, kO, and KO can reasonably be expected to be H_∞^* ring spectra. More likely they are H_∞^{2*} and H_∞^{8*} ring spectra. However, taking E to be the H_∞ ring spectrum kU and restricting attention to elements of degree zero, the Kunneth approach above certainly gives a new way of looking at Atiyah's method [**2**] of defining the Adams operations in classical complex K-theory.

EXAMPLES 4.9. The geometry of Examples 2.9 can surely be generalized to give that MO, MU, and M Sp are H_∞^*, H_∞^{2*}, and H_∞^{4*} ring spectra and thus to recover tom Dieck's Steenrod operations in the classical cobordism theories [**5**].

I am confident that the ideas explained above will in due course yield a useful coherent theory of Steenrod operations in generalized cohomology theories.

As a last observation, we define a curious new multiplicative transfer. Recall from [**7**] that a connected cover ν: $X \to B$ of degree j canonically gives rise to a map $\bar{\nu}$: $B \to E\Sigma_j \times_{\Sigma_j} X^j$. For $x \in E^qX$, the transfer $\tau(x) \in E^qB$ is defined to be the composite

$$B^+ \xrightarrow{\bar{\nu}^+} (E\Sigma_j \times_{\Sigma_j} X^j)^+ = D_j(X^+) \xrightarrow{D_j(x)} D_jE_q \xrightarrow{\theta_j} E_q,$$

where x is regarded as a based map $X^+ \to E_q$, with $E_q = \Omega^{-q}E_0$ for $q < 0$, and where θ_j is the Dyer-Lashof map for the infinite loop space E_q. The following analog may prove to be of interest.

DEFINITION 4.11. Let E be an H_∞^* ring spectrum and let ν: $X \to B$ be a connected cover of degree j. Define the multiplicative transfer $\tau_\otimes$: $E^qX \to E^{jq}B$ by $\tau_\otimes(x) = \bar{\nu}^*\mathscr{P}_j(x)$.

Brian Sanderson has noted that the internal power operation can be recovered as a special case of this transfer.

BIBLIOGRAPHY

1. J. F. Adams, *The Kahn -Priddy theorem*, Proc. Cambridge Philos. Soc. **73** (1973), 45–55.

2. M. F. Atiyah, *On power operations in K-theory*, Quart. J. Math. Oxford (2) **17** (1966), 165–193.

3. R. Bruner, *The Adams spectral sequence of H_∞ ring spectra*, Thesis, Univ. of Chicago, 1977.

4. F. Cohen, T. Lada and J. P. May, *The homology of iterated loop spaces*, Lecture Notes in Mathematics, vol. 533, Springer, 1976.

5. T. tom Dieck, *Steenrod-Operationen in Kobordismen-Theorie*, Math. Z. **107** (1968), 380–401.

6. D. S. Kahn, *Cup-i products and the Adams spectral sequence*, Topology **9** (1970), 1–9.

7. D. S. Kahn and S. B. Priddy, *Applications of the transfer to stable homotopy theory*, Bull. Amer. Math. Soc. **76** (1972), 981–987.

8. J. Makinen, *Boundary formulae for reduced powers in the Adams spectral sequence*, Ann. Acad. Sci. Fenn. Ser A. I Math. 562 (1973).

9. J. P. May, *The cohomology of restricted Lie algebras and of Hopf algebras; applications to the Steenrod algebra*, Thesis, Princeton Univ., 1964.

10. ———, *A general algebraic approach to Steenrod operations*, Lecture Notes in Math., vol. 168, Springer, 1970, pp. 153–231.

11. J. P. May (with contributions by F. Quinn, N. Ray, and J. Tornehave), *E_∞ ring spaces and E ring spectra*, Lecture Notes in Maths., vol. 533, Springer, 1977.

12. J. P. May, *Infinite loop space theory*, Bull. Amer. Math. Soc. **83** (1977), 456–494.

13. ———, *The homotopical foundations of algebraic topology*, Academic Press (In preparation).

14. R. J. Milgram, *Group representations and the Adams spectral sequence*, Pacific J. Math. **41** (1972), 157–182.

15. G. Nishida, *The nilpotency of lements of the stable homotopy groups of spheres*, J. Math. Soc. Japan **25** (1973), 707–732.

16. N. E. Steenrod, *Homology groups of symmetric groups and reduced power operations. Cyclic reduced powers of cohomology classes*, Proc. Nat. Acad. Sci. U.S.A. **39** (1953), 213–223.

17. M. Steinberger, *The homology operations of H_∞ ring spectra*, Thesis, Univ. of Chicago, 1977.

18. H. Toda, *An important relation in homotopy groups of spheres*, Proc. Japan Acad. **43** (1967), 839–842.

19. A. Tsuchiya, *Homology operations on ring spectrum of H^∞ type and their applications*, J. Math. Soc. Japan **25** (1973), 277–316.

UNIVERSITY OF CHICAGO

Proceedings of Symposia in Pure Mathematics
Volume 32, 1978

LITTLE CUBES AND THE CLASSIFYING SPACE FOR n-SPHERE FIBRATIONS

F. R. COHEN*

We give an unstable version of a theorem due to J. Tørnehave [12] relating BSF to the symmetric groups. Although our generalization is an immediate consequence of known results, it is both interesting and has useful applications. (See [4] for a more detailed discussion of the homological implications of our theorem.) Furthermore, this generalization describes the relationship between the Artin braid groups and the classifying space for integrally oriented two sphere fibrations with a canonical section.

Let Σ_j be the symmetric group on j letters, let $\mathscr{C}_n(j)$ be Boardman and Vogt's space of j "little n-cubes" [7], and let γ be the structure map for J. P. May's little cubes operad [7]. $\mathscr{C}_n(j)$ has the equivariant homotopy type of the classical configuration space of j ordered distinct points in $\mathbf{R}^n$. In [4], we study a multiplication $c_n\colon \mathscr{C}_n(j) / \Sigma_j \times \mathscr{C}_n(k) / \Sigma_k \to \mathscr{C}_n(jk) / \Sigma_{kj}$ which is given on points by the formula $c_n(a, b) = \gamma(b, a^k)$. Let p be a fixed prime and consider $(\coprod_{r\geqq 0} \mathscr{C}_n(p^r)/\Sigma_{p^r})_\otimes$ which as a space is the disjoint union of the $\mathscr{C}_n(p^r) / \Sigma_{p^r}$ and has the multiplication denoted by $\otimes$ which is induced by c_n. It is shown in [4] that this space together with $\otimes$ is an associative H-space with an identity and hence has a classifying space.

Let $SF(n)$ be the monoid of degree one based homotopy equivalences of S^n with multiplication given by composition of maps. Let $X[1/p]$ denote the localization of X away from p (when this localization exists).

THEOREM 1. *The universal cover of* $B(\coprod_{r\geqq 0} \mathscr{C}_n(p^r) / \Sigma_{p^r})_\otimes$ *is* $BSF(n)[1/p]$ *up to homotopy type.*

AMS (MOS) subject classifications (1970). Primary 55D35, 55F25; Secondary 55G30, 55H20.
*Author partially supported by an NSF grant.

REMARK 2. The cases $n = \infty$ and $n = 2$ are exceptional and interesting; they are discussed below.

In case $n = \infty$, $\mathscr{C}_\infty(j)/\Sigma_j$ is $K(\Sigma_j, 1)$ [7] and it is easy to see that the multiplication $c_\infty: K(\Sigma_j, 1) \times K(\Sigma_k, 1) \to K(\Sigma_{jk}, 1)$ is homotopic to the classifying map of the tensor product representation $\otimes: \Sigma_j \times \Sigma_k \to \Sigma_{jk}$. Here, Theorem 1 specializes to Tørnehave's result [12].

In case $n = 2$, $\mathscr{C}_2(j)/\Sigma_j$ is $K(B_j, 1)$ where B_j is the Artin braid group on j strings [7]. For a detailed discussion of the braid groups, see [2]. We now define the "tensor product of braids" homomorphism, $\otimes: B_j \times B_k \to B_{jk}$, intuitively with the following two pictures and leave the rigorous description (in terms of presentations) to the reader:

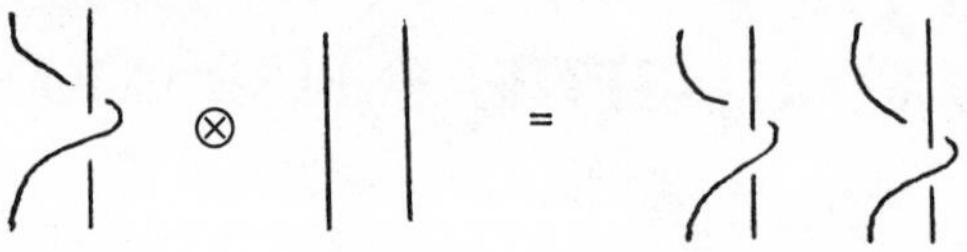

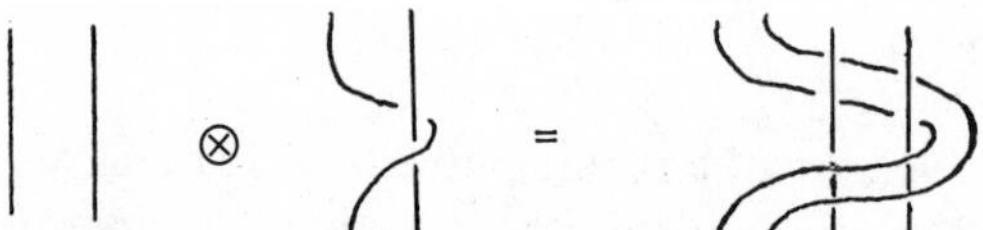

It is an easy exercise to show that $c_2: K(B_j, 1) \times K(B_k, 1) \to K(B_{jk}, 1)$ is homotopic to the classifying map of the "tensor product of braids" homomorphism. Furthermore, if $\sigma_j: B_j \to \Sigma_j$ is the standard epimorphism [2], then the following diagram commutes.

$$\begin{array}{ccc} B_j \times B_k & \xrightarrow{\otimes} & B_{jk} \\ \Big\downarrow{\scriptstyle \sigma_j \times \sigma_k} & & \Big\downarrow{\scriptstyle \sigma_{jk}} \\ \Sigma_j \times \Sigma_k & \xrightarrow{\otimes} & \Sigma_{jk} \end{array}$$

REMARK 3. The singular chains of $(\coprod_{r\geq 0} \mathscr{C}_2(p^r)/\Sigma_{p^r})_\otimes$ can be replaced by $\sum_{r\geq 0} B[B_{p^r}]$ where $B[G]$ is the classical (algebraic) bar construction for the group G [6]. The multiplication induced by $\otimes$ is of course given on chains by

$$B[B_j] \otimes B[B_k] \xrightarrow{\text{shuff}} B[B_j \times B_k] \xrightarrow{B[\otimes]} B[B_{jk}].$$

In addition, $H_*((\coprod_{r\geq 0} K(B_{p^r}, 1))_\otimes; Z_q)$ for q prime is completely known as an algebra by the methods in [3] and [4]. Furthermore, all elements in homology have representatives in the bar construction which are given in terms of two braids α and β where

$\alpha =$ 

and

$\beta =$ 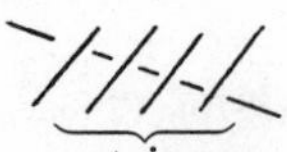

q strings

These observations together with Theorem 1 give means to compute Massey products and consequently differentials in the Eilenberg-Moore spectral sequence converging to $H_*(BSF(2); \mathbf{Z}_q)$ where q is prime. This is interesting if $q > 2$ [**4**]. Results and details of these computations will presumably appear elsewhere.

PROOF OF THEOREM 1. Let $\Omega^n_j S^n$ denote the jth component of $\Omega^n S^n$. Let $(\coprod_{r\geq 0} \Omega^n_{p^r} S^n)_\otimes$ be the disjoint union of the $\Omega^n_{p^r} S^n$ as a space with multiplication denoted by $\otimes$ and which is given by composition of maps. Consider the map α_n: $C_n S^0 \to \Omega^n S^n$ as defined by J. P. May in [**7**]. It is shown in [**4**] that α_n restricts to a homomorphism with respect to

$$\bar{\alpha}_n : \Bigl(\coprod_{r\geq 0} \mathscr{C}_n(p^r)/\Sigma_{p^r}\Bigr)_\otimes \longrightarrow \Bigl(\coprod_{r\geq 0} \Omega^n_{p^r} S^n\Bigr)_\otimes .$$

We prove Theorem 1 in two steps. The first step is to show that $B\bar{\alpha}_n$ is a homotopy equivalence and the second step is to show that the universal cover of $B(\coprod_{r\geq 0} \Omega^n_{p^r} S^n)_\otimes$ is homotopy equivalent to $BSF(n)[1/p]$.

Let $M_1 = (\coprod_{r\geq 0} \mathscr{C}_n(p^r)/\Sigma_{p^r})_\otimes$ and $M_2 = (\coprod_{r\geq 0} \Omega^n_{p^r} S^n)_\otimes$. Consider the commutative diagram

$$\begin{array}{ccc} M_1 & \xrightarrow{\bar{\alpha}_n} & M_2 \\ \downarrow{\scriptstyle \iota_1} & & \downarrow{\scriptstyle \iota_2} \\ \Omega BM_1 & \xrightarrow{\Omega B\bar{\alpha}_n} & \Omega BM_2 \end{array}$$

where ι_j: $M_i \to \Omega BM_i$, $i = 1, 2$, is the natural inclusion. Let $\pi = \pi_0 M_1$ and observe that $\pi_0\bar{\alpha}_n : \pi \to \pi_0 M_2$ is an isomorphism. By abuse of notation we identify $\pi_0 M_2$ as π. By [**9**], the vertical maps ι_j are group completions and we have an additional commutative diagram

$$\begin{array}{ccc} H_*M_1\,[\pi^{-1}] & \xrightarrow{\bar{\alpha}_{n*}[\pi^{-1}]} & H_*M_2\,[\pi^{-1}] \\ \downarrow{\scriptstyle \iota_{1*}[\pi^{-1}]} & & \downarrow{\scriptstyle \iota_{2*}[\pi^{-1}]} \\ H_*\Omega BM_1 & \xrightarrow{\Omega B\bar{\alpha}_{n*}} & H_*\Omega BM_2 \end{array}$$

where $H_*M_i[\pi^{-1}]$ is the ring obtained by inverting the elements of $\pi = \pi_0 M_i$ by right fractions. Furthermore, by [**3**], the map α_n: $C_n S^0 \to \Omega^n S^n$ is a group completion in homology. It follows directly from a check of definitions that $\bar{\alpha}_{n*}[\pi^{-1}]$ is an isomorphism. Since the vertical maps $\iota_{j*}[\pi^{-1}]$ are isomorphisms by [**9**], it follows that $\Omega B\bar{\alpha}_{n*}$ is an isomorphism. Consequently the map $\Omega B\bar{\alpha}_n$ restricted to connected components induces an isomorphism on homotopy groups. Since BM_i has the homotopy type of a CW-complex [**10**], [**11**], the first part of the proof follows from the Whitehead theorem.

The second part is apparently well known [**1**], [**8**], [**9**]; we include some details for completeness. The proof given here parallels, in part, those given in [**8**, § VII.5.3] and [**10**, Example (iii)], but we require a bit of extra care because the composition pairing is not generally homotopy commutative. Let p and -1 denote fixed based self-maps of S^n of degree p and -1 respectively. Let p^r be the r-fold composition of p with itself and let $-p^r$ be the composition $p^r \circ (-1)$. Consider the following diagram

$$
\begin{array}{lccccccc}
(1) & \Omega_1^n S^n & \xrightarrow{p} & \Omega_p^n S^n & \xrightarrow{p} & \Omega_{p^2}^n S^n & \longrightarrow & \cdots \\
& \downarrow {\scriptstyle *(-1)} & & \downarrow {\scriptstyle *(-p)} & & \downarrow {\scriptstyle *(-p^2)} & & \\
(2) & \Omega_0^n S^n & \xrightarrow{p} & \Omega_0^n S^n & \xrightarrow{p} & \Omega_0^n S^n & \longrightarrow & \cdots
\end{array}
$$

where the horizontal arrows are given by $f \mapsto p \circ f$ and the vertical arrows are given by $f \mapsto f * (-p^r)$ where $*$ is a fixed choice for the loop sum in $\Omega^n S^n$. (This choice is given by the formula $a * b = \theta_n(c_2; a, b)$ where c_2 is a fixed element in $\mathscr{C}_n(2)$ and θ_n is the action map of the little cubes operad [7].) By the distributivity diagram in [4, § 2], $p \circ (f * (-p^r)) = (p \circ f) * (p \circ (-p^r))$. By definition, $-p^r = p^r \circ (-1)$ and so the above diagram commutes on the nose. Let T_i be the mapping telescope of the sequence of maps given as (i) in the above diagram, $i = 1, 2$. Using the above remarks, it follows that T_1 and T_2 are homotopy equivalent. Furthermore, the natural map of $\Omega_0^n(S^n[1/p])$ to T_2 is clearly a homotopy equivalence. (Compare this to Example (iii) in [9].) By [5], $\Omega_0^n(S^n[1/p])$ and $(\Omega_0^n S^n)[1/p]$ are homotopy equivalent. Consequently T_1 is $SF(n)[1/p]$ up to homotopy type.

Theorem 1 follows directly from the above remarks together with the group completion theorem of [9].

REFERENCES

1. J. F. Adams, *On the groups $J(X)$*. I, Topology **2** (1963), 181–195. MR **28** #2553.

2. J. Birman, *Braids, links, and mapping class groups*, Ann. of Math. Studies, No. 82, Princeton Univ. Press, Princeton, N.J., 1974. MR **51** #11477.

3. F. Cohen, *The homology of $\mathscr{C}_{n+1}$-spaces, $n \geqq 0$*, Lecture Notes in Math., vol. 533, Springer-Verlag, Berlin and New York.

4. ———, *The homology of* SF(n + 1), Lecture Notes in Math., vol. 533, Springer-Verlag, Berlin and New York.

5. P. Hilton, G. Mislin and J. Roitberg, *Homotopical localization*, Proc. London Math. Soc. **26** (1973), 693–706. MR **48** #5063.

6. S. Mac Lane, *Homology*, Springer-Verlag, New York, 1967.

7. J. P. May, *The geometry of iterated loop spaces*, Lecture Notes in Math., vol. 271, Springer-Verlag, Berlin and New York, 1971.

8. ———, *E_∞ ring spaces and E_∞ ring spectra*, Lecture Notes in Math., vol. 577, Springer-Verlag, Berlin and New York, 1977.

9. D. McDuff and G. Segal, *Homology fibrations and the "group completion" theorem*, Invent. Math. **31** (1976), 279–284.

10. R. J. Milgram, *The bar construction and abelian H-spaces*, Illinois J. Math. **11** (1967), 234–241.

11. J. Milnor, *On spaces having the homotopy type of a* CW-*complex*, Trans. Amer. Math. Soc. **90** (1959), 272–280.

12. J. Tørnehave, *On BSG and the symmetric groups* (preprint).

NORTHERN ILLINOIS UNIVERSITY

Current Address: Institute for Advanced Study

PROBLEMS

Proceedings of Symposia in Pure Mathematics
Volume 32, 1978

SOME PROBLEMS ON HOMOTOPY THEORY, MANIFOLDS AND TRANSFORMATION GROUPS

W. BROWDER AND W. C. HSIANG

There were three sections of problems posed at the Stanford Conference, 1976; one for low-dimensional topology edited by R. Kirby, one for algebraic K-theory, L-theory and surgery theory edited by C. T. C. Wall, and the rest of them are here. We compile them together with some of the interesting problems posed at Indiana University in 1974. There are some discussions and references after each problem (if we have any) to help the reader.

We thank Dan Gottlieb and Reinhard Schultz for their help in the preparation of this manuscript.

A. Homotopy theory and homology theory. Here are some problems in homotopy theory and homology theory. We also include some questions posed at Indiana University, 1974.

1. Let p be an (odd) prime, and let $V(n)$ be the spaces introduced by L. Smith **[1]** and H. Toda **[2]** which satisfy

$$\tilde{H}^*(V(n); Z_p) \cong E[Q_1, \cdots, Q_n]$$

as a module over the mod p Steenrod algebra $\mathscr{A}^*(p)$, where $Q_i \in \mathscr{A}^{2p^i-1}(p)$ are the Milnor primitives.

CONJECTURE. (a) $V(\infty)$ does not exist.
More strongly,
(b) $V(0), \cdots, V(p-2)$ exist, but $V(p-1)$ does not.
The first open case is whether
(c) $V(4)$ exists for some primes.

REFERENCES. **[1]** L. Smith, *On realizing complex bordism modules*, Amer. J. Math.

AMS (*MOS*) *subject classifications* (1970). Primary 55Bxx, 55B20, 55Dxx, 55D15, 55D35, 22Dxx, 22C05, 57D85, 57Exx, 57E15.

92 (1970), 793–856; *Applications to the stable homotopy of spheres.* II, Amer. J. Math. **93** (1971), 226–263.

[2] H. Toda, *On spectra realizing exterior parts of the Steenrod algebra*, Topology **10** (1971), 53–65.

2 (M. Barratt). Nishida has shown that each element of π_*^S is nilpotent. Is it possible that the p-primary component $\pi_*^S(p)$ is nilpotent? That is, does there exist an integer $d(p)$ such that $x^{d(p)} = 0, \forall x \in \pi_n^S(p), n > 0$?

It is unknown if $x^4 = 0, \forall x \in \pi_n^S(2), n > 0$, or if $x^{2p^2} = 0, \forall x \in \pi_n^S(p), n > 0$, and p odd.

REFERENCE. [1] G. Nishida, *The nilpotency of elements of the stable homotopy groups of spheres*, J. Math. Soc. Japan **25** (1973), 707–732.

3 (J. C. Moore). For a complex X, let

$$P(X; z) = \sum_{i=1}^{\infty} \dim H_i(X; Q) z^i.$$

CONJECTURE. For X finite and simply connected, $P(\Omega X; z)$ is a rational function of z with pole at $z = -1$ or order equal to the order of the zero of $P(X; z)$ at $z = 1$.

DISCUSSION. One easily checks this for X of Cat $\leqq 2$, or H-spaces. A. Clark and L. Smith have shown that if the conjecture holds for A and B, then it also does for $A \vee B$.

4 (D. Sullivan). For a simply connected finite complex X, let

$$Q(X; z) = \sum_{i=0}^{\infty} \dim(\pi_i(X) \otimes Q) z^i.$$

CONJECTURE. $Q(X, 1) < \infty$ implies $Q(X, -1) \leqq 0$.

5 (D. Sullivan). CONJECTURE. There are no essential maps $BG \to X =$ finite complex for G a compact Lie group. In particular, there is no essential map $\boldsymbol{RP}(\infty) \to X =$ finite complex.

DISCUSSION. It is perhaps worthwhile noting in this connection that a conjecture of Segal seems related.

CONJECTURE (SEGAL). Let π be a finite group and $A(\pi)$ its Burnside ring. Then

$$A(\pi)^\wedge \cong \pi_S^0(B\pi)$$

where $A(\pi)^\wedge$ is the completion of $A(\pi)$. tom Dieck extends the definition of the Burnside ring to all compact Lie groups, and it follows easily from his definition that $A(S^1) = 0$. Hence, if Segal's conjecture is true for compact connected groups, $\pi_S^0(CP(\infty)) = 0$, so that there would be no *stable* essential maps $BS^1 \to X$, for X a finite complex.

It is also worth noting that Lin has proved that there is no stable essential map $\boldsymbol{K}(Z_2) \to \boldsymbol{X}$ where $\boldsymbol{K}(Z_2)$ is the mod 2 Eilenberg-Mac Lane spectrum and $\boldsymbol{X}$ is the suspension spectrum of a finite complex X.

REFERENCES. [1] T. Y. Lin, *Homological dimensions of stable homotopy modules and their geometric characterizations*, Trans. Amer. Math. Soc. **172** (1972), 473–490.

[2] ———, *Homological algebra of stable homotopy ring π_* of spheres*, Pacific J. Math. **38** (1971), 117–143.

6 (D. H. Gottlieb). A compact connected CW complex is said to be prime if

there is no Hurewicz fibration $F \to E \to B$ such that both F and B are compact, noncontractible CW complexes.

CONJECTURE. A suspension ΣX is prime if $\tilde{H}_*(\Sigma X; Z)$ is not a finite group and $\chi(\Sigma X) \neq 0$.

DISCUSSION. If the torsion of $H_*(\Sigma X; Z)$ is relatively prime to $\chi(\Sigma X)$, then ΣX is prime. On the other hand, the two hypotheses are necessary. For example, $\chi(S^3) = 0$ and S^3 is not a prime. Also, the Moore space $M(Z_p \oplus Z_q, n)$ is a suspension and it is homotopy equivalent to $M(Z_p, n) \times M(Z_q, n)$, so it is not prime, but its reduced homology is finite. For more information, see [**1**].

REFERENCE. [**1**] D. H. Gottlieb, *Fibering suspensions* (preprint).

7 (D. H. Gottlieb). Let $F \to E \to^p S^n$ be a fibration where F is a finite CW complex, $\chi(F) \neq 0$ and $n > 1$.

CONJECTURE. $p_*: H_*(E; Z) \to H_*(S^n; Z)$ is onto.

DISCUSSION. This conjecture is equivalent to the statement that

$$\pi_n(F^F; 1_F) \xrightarrow{\omega_*} \pi_n(F) \xrightarrow{h} H_n(F; Z)$$

is the trivial homomorphism where $\omega: F^F \to F$ is the evaluation map (at the base point), and h is the Hurewicz homomorphism. The following special cases of the conjecture are known:

(a) True for n odd [**1**, §2].

(b) True for fiber bundle with Lie group as the structure group [**2**, Theorem 5].

(c) True if $H_{n-1}(F; Z)$ has square-free torsion [**3**, Theorem 4.1].

(d) True if $n = 2$ [**4**, Theorem IV-1].

REFERENCES. [**1**] D. H. Gottlieb, *Witnesses, transgression and the evaluation map*, Indiana Univ. Math. J. **24** (1975), 825–836.

[**2**] J. C. Becker and D. H. Gottlieb, *Applications of the evaluation map and transfer map theorems*, Math. Ann. **211** (1974), 277–288.

[**3**] D. H. Gottlieb, *Evaluation subgroups of homotopy groups*, Amer. J. Math. **91** (1969), 729–755.

[**4**] ———, *A certain subgroup of the fundamental group*, Amer. J. Math. **87** (1965), 840–856.

8 (tom Dieck, Quillen and Oliver). Let X be a G-space and let H be a closed subgroup of G. Does there exist a transfer map

$$t: H^*(X/H; \ ) \to H^*(X/G; \ )$$

such that the composite with the natural map $H^*(X/G; \ ) \to H^*(X/H; \ )$ is the multiplication by $\chi(G/H)$?

DISCUSSION. It is probably false. If it is true, one would have another proof of Conner's conjecture [**1**].

REFERENCE. [**1**] R. Oliver, *A proof of the Conner conjecture*, Ann. of Math. **103** (1976), 637–644.

9 (R. Douglas and R. Body). Does unique factorization under direct product hold for 0-local spaces?

REFERENCES. [**1**] R. Douglas and R. Body, *Formal rational homotopy types satisfy unique factorization* (preprint).

[**2**] ———, *Positive weight rational homotopy types satisfy unique factorization* (in preparation).

10 (P. Kahn). For a group π, what is the minimum dimension $d(\pi)$ of all connected CW complexes X such that

(a) $\pi_1 X = \pi$,

(b) $\pi_1 X$ acts nilpotently on the homology of the universal cover of X?

DISCUSSION. For a nilpotent group π, the following is known:

(a) $d(\pi) \geqq \operatorname{rank} \pi$ (= Hirsch number),

(b) if π has nontrivial torsion, then $d(\pi) \geqq (\operatorname{rank} \pi) + 3$.

11 (D. S. Kahn and V. P. Snaith). The Hopf construction of spectrum from homotopy associative commutative H-spaces gives the KU-theory when applied to $\boldsymbol{CP}^\infty$. Specifically, the spectrum $\boldsymbol{X} = \{X_{2n} = S^2\boldsymbol{CP}^\infty\}$ maps by a split epimorphism onto KU-theory. So, $\boldsymbol{X} = \boldsymbol{KU} \times \boldsymbol{F}$ (as spectra). In addition, $\boldsymbol{F}$ has finite homotopy groups.

Question. What is $\boldsymbol{F}$?

DISCUSSION. $\boldsymbol{F}$ is nontrivial [1], [2], [3].

REFERENCES. [1] V. P. Snaith, *Toward algebraic cobordism*, Bull. Amer. Math. Soc. **83** (1977), 384–385.

[2] ———, *Algebraic cobordism and K-theory*, Mem. Amer. Math. Soc. (to appear).

[3] ———, *Cobordism invariants of varieties.*

12 (D. S. Kahn). The following problem was posed by Norman Steenrod many years ago and was settled by R. Swan [4]: Let A be a finitely generated abelian group on which a finite group π acts. Does there exist a Moore space X for A with a π action that realizes the given action of π on A? If one insists that π acts cellularly on a finite complex, Swan gave an example of a π-module which is not realizable [3], [4].

Question. What happens if one allows X to be a complex of finite type?

REFERENCES. [1] J. E. Arnold, *On Steenrod's problem of cyclic p-groups* (to appear).

[2] ———, *A solution of a problem of Steenrod for cyclic groups of prime order*, Proc. Amer. Math. Soc. **62** (1977), 177–182.

[3] H. Pittie, *Group actions and algebraic K-theory*, Thesis, Princeton University, Princeton, N. J., 1970.

[4] R. Swan, *Invariant rational functions and a problem of Steenrod*, Invent. Math. **7** (1969), 148–158.

Here is a set of problems on H-spaces by J. Lin. Let X be a 1-connected finite complex which is an H-space.

13. Is $H_*(\Omega X)$ torsion free?

Lin [4] showed there is no odd torsion.

14. Is the 2-torsion of $H^*(X)$ of exponent 2?

For odd primes see [4].

15. Is $H_*(X; Z_p)$ (p odd) primitively generated?

This is equivalent to the statement that $x^p = 0$ for any $x \in H^*(X; Z_p)$. It is known that $x^{p^2} = 0$ [4, part II].

16. Is $Q^{\text{even}} H^*(X; Z_p)$ (p odd) concentrated in degrees $2p + 2$, $2p^2 + 2$?

See [4, part II] for approximations to this result.

17. Is the first nonvanishing homotopy group of X in degree 3 or 7?

Clark [2] showed if X is a group, then it is in degree 3. Thomas [6] showed this degree is 3, 7 or 15 if $H^*(X; Z_2)$ is primitively generated, and if further $H^*(X)$ is 2-torsion free then the degree is 3, 7.

18. Is there a choice of generators g for $H^*(X; Z_2)$ so that their reduced coproducts $\mu^*g - p_1^*g - p_2^*g \in PH^*(X; Z_2) \otimes PH^*(X; Z_2)$? ($\mu: X \times X \to X$ is the product, p_i the projection on the ith factor.) A reference for this is Moore and Smith [5].

19. Is $H^*(X; Q)$ isomorphic as an algebra to $H^*(Y; Q)$ where Y is a product of Lie groups and seven spheres?

An example of Harper [3] indicates the corresponding statement for Z_p coefficients is false, $p > 5$; Wilkerson [7] shows this is true if X has a "maximal torus" in an appropriate sense.

20. Is $H^*(X; Z_2)$ isomorphic to $H^*(Y; Z_2)$ (as algebra) where Y is a product of a Lie group and S^7's?

DISCUSSION. Is it true for Z_p coefficients, $p \leqq 5$?

REFERENCES. [1] W. Browder, *Torsion in H-spaces*, Ann. of Math. (2) **74** (1961), 24–51.

[2] A. Clark, *On π_3 of finite-dimensional H-spaces*, Ann. of Math. (2) **78** (1963), 193–196.

[3] J. Harper, *On the construction of* mod *p H-spaces*, these PROCEEDINGS, part 2, pp. 207–214.

[4] J. Lin, *Torsion in H-spaces*. I, II, Ann. of Math. (2) **103** (1976), 457–487; (to appear).

[5] J. C. Moore and L. Smith, *Hopf algebras and multiplicative fibrations*. I, II, Amer. J. Math. **90** (1967), 752–780; 1113–1150.

[6] E. Thomas, *Steenrod squares and H-spaces*. I, II, Ann. of Math. (2) **77** (1963), 306-317; **81** (1965), 473–495.

[7] C. Wilkerson, *Classifying spaces, Steenrod operations and algebraic closure*, Topology **6** (1977), 227–237.

B. Manifolds. Again, we include several problems posed at Indiana University, 1974.

1 (M. Mahowald). *Question.* Let M^n be a closed spin manifold which has k linearly independent vector fields. Does the L class of M^n satisfy certain divisibility conditions? Specifically, is the class $\prod_{p \text{ primes}} p^{[j/(p-1)]} L_j(M^n) \in H^{4j}(M^n; Z)$ divisible by $2^{((4j-n+k+\varepsilon)/2)+\delta}$ where $\varepsilon = 0, 1, 2, 3$ is chosen such that $4j - n + k + \varepsilon \equiv 0\,(4)$ and δ depends on (the parity of) $(4j - n + k + \varepsilon)/2$?

2 (M. Mahowald). Let M^{2n+1} be a closed manifold, $n \neq 1, 3$, τ its tangent bundle with Thom space $T(\tau)$, and let $i \in \pi_{2n+1}(T(\tau))$ be the canonical class.

CONJECTURE. The Whitehead product $[i, i] \neq 0$ if and only if $\chi_2(M) \neq 0$ where

$$\chi_2(M) = \sum_{i=0}^{n} \dim H_i(M; Z_2)$$

is the mod 2 semicharacteristic.

3 (M. Hirsch). CONJECTURE. Every π-manifold M^n embeds in $\boldsymbol{R}^m$ where $m \geqq 3(n+1)/2$.

DISCUSSION. If M^n is $([n/4]+1)$-connected, then the embedding exists [1]. It is very closely related to a hard homotopy problem, viz., if $S^{n+k} \to S^n$ is stably trivial, then $\Sigma^{[(k-n)/2]} f$ is already trivial.

REFERENCE. [1] R. de Sapio, *Embedding π-manifolds*, Ann. of Math. (2) **82** (1965), 213–224.

4 (W. Thurston). *Question.* Let M^n be a closed n-dim manifold ($n > 3$). Does there exist a closed differentiable aspherical N^n (i.e., of type $K(\pi, 1)$) such that there is a map $f: N \to M$ inducing a homology isomorphism? In particular, can $M^n = S^n$ be homology equivalent to a closed differentiable aspherical manifold?

DISCUSSION. There exist many 3-dim aspherical manifolds which are homology spheres. It seems unlikely that these higher dimensional aspherical manifolds always exist, but no counterexample is known.

REFERENCE. **[1]** D. M. Kan and W. Thurston, *Every connected space has the homology of a* $K(\pi, 1)$, Topology **15** (1976), 253–258.

5 (F. Raymond). Suppose that M^n is a closed $K(\pi, 1)$ manifold such that π contains a normal subgroup Γ which is the fundamental group of N^k, a closed $K(\Gamma, 1)$ manifold.

Question. Does $\pi/\Gamma = Q$ act properly discontinuously on $\boldsymbol{R}^{n-k}$ with compact quotient? This is especially interesting when $\Gamma = Z^k$.

REFERENCE. **[1]** P. E. Conner and F. Raymond, *Deforming homotopy equivalences to homeomorphisms in aspherical manifolds*, Bull. Amer. Math. Soc. **83** (1977), 36–85.

6 (Bruce Williams). Let $f_0: M^m \to \boldsymbol{R}^{n+k}$ be an embedding with k linearly independent normal vectors.

Question. Is f_0 concordant to an embedding $f_1: M^m \to \boldsymbol{R}^{n+k}$ such that $M^m \to^{f_1} \boldsymbol{R}^{n+k} \to^{p} \boldsymbol{R}^n$ is an immersion where p is the obvious projection?

DISCUSSION. According to Brian Sanderson, if one is willing to take cobordism instead of concordance it follows from **[1]** that the answer to this question is affirmative.

REFERENCE. **[1]** U. Koschorke and B. Sanderson, *Self-intersections and higher Hopf invariants* (to appear).

7 (Bruce Williams). Let $c: S^{n+k} \to T(\eta^k)$ be a degree one map such that η^k ($k > 2$) is a PL block bundle over a closed PL manifold M^n, stably equivalent to the stable normal bundle and c is stably equivalent to a normal invariant of M^n. Develop an obstruction theory outside of the so-called metastable range to finding a homotopy of c to a map c_1 such that $c_1^{-1}(M^n)$ is PL homeomorphic to M.

DISCUSSION. W. Browder **[1]** proved that $\sum c$ is homotopic to a map d such that $d^{-1}(M^n) \subset S^{n+k+1}$ is PL homeomorphic to M^n.

REFERENCES. **[1]** W. Browder, *Embedding* 1-*connected manifolds*, Bull. Amer. Math. Soc. **72** (1966), 225–231.

[2] ———, *Embedding smooth manifolds*, Proc. Internat. Congr. Math. (Moscow, 1966), Mir, Moscow, 1968, pp. 712–719.

8 (B. Sanderson). Let us define a codim k submersion by

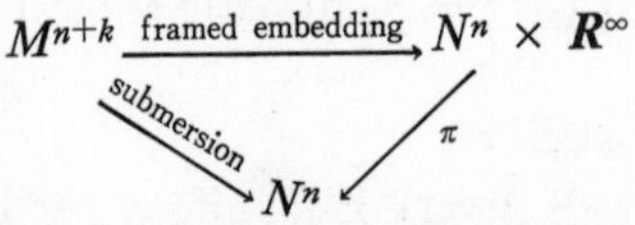

where π is the obvious projection. Let $Q = \Omega^\infty S^\infty$. Then, $Q(S^k)$ classifies codim k framed immersions for $k > 0$, and $Q(S^0)$ is the group completion of 'finite covering spaces'.

Question. Is $Q(S^k)$ related to 'codim k framed submersion' for $k < 0$?

REFERENCE. **[1]** U. Koschorke and B. Sanderson, *Self-intersections and higher Hopf invariants* (to appear).

9 (E. Miller). Consider the category $\mathcal{M}$ of n-dim compact manifolds (with boundary, in general) and embeddings (boundary may go to interior). For each manifold N, an assignment of a cohomology class $C(N) \in H^*(N; Z_2)$ is called 'strictly characteristic' if $\forall$ embedding $i: N \to N'$ (dim N = dim N') $i^*C(N') = C(N)$. In particular, for each diffeomorphism $i: N \to N$, $i^*C(N) = C(N)$.

Question. In dimension less than dim N, are the Stiefel-Whitney classes the only 'strictly characteristic' classes?

DISCUSSION. In the full subcategory of $\mathcal{M}$ of orientable $(4k - 1)$-dim compact manifolds and embeddings, there is a new strictly characteristic class $\alpha(M) \in H^{2k}(M; Z_2)$ where $\alpha(M)$ is defined by

$$(\alpha(M) \cup x)[M, \partial M] = (x \cup Sq^1 x)[M, \partial M].$$

This uses the fact that

$$\begin{aligned} H^{2k-1}(M, \partial M; Z_2) &\longrightarrow Z_2 \\ x &\longrightarrow (x \cup Sq^1 x)[M, \partial M] \end{aligned}$$

is a homomorphism.

Here are two classical problems recalled again by E. Thomas.

10. Given an almost complex manifold N^{2n} $(n \geqq 3)$, does it carry an actual complex structure?

DISCUSSION. For $n = 2$, there are examples due to van der Ven [1] and Yau [2].

REFERENCES. [1] A. van der Ven, *On the Chern number of certain complex and almost complex manifolds*, Proc. Nat. Acad. Sci. U.S.A. **55** (1966), 1624–1627.

[2] S. T. Yau, *Parallelizable manifolds without complex structure*, Topology **15** (1976), 51–53.

11. Given a complex manifold M and a smooth complex bundle W over M, does W come from a complex holomorphic structure? Moreover, if M is algebraic, is W algebraic?

DISCUSSION. If M is Stein, then this is just the Oka principle.

Here is a set of problems on the Kervaire invariant given by W. Browder. The outstanding problem is as ever.

12. Is the Kervaire manifold K^{2n} smoothable?

We know the answer is *no* for $n \neq 2^i - 1$ and for $n = 2^i - 1$, K^{2n} is smoothable if and only if h_i^2 in the Adams spectral sequence represents a homotopy element (called θ_i) [1]. The answer is known to be *yes* for $n = 1, 3, 7, 15$ and recently Barratt and Mahowald have shown it for $n = 31$. While $n = 1, 3, 7$ come from the existence of elements of Hopf invariant 1 in those dimensions and products, higher dimensions must be represented by geometric constructions, such as J. Jones's construction in dimension 30, which is an $S^7 \times S^7 \times S^7 \times S^7$ bundle over a surface of genus 5.

Subsidiary problems, which if solved affirmatively would lead to an inductive construction of θ_i [4] are:

13. Suppose θ_i exists.

(a) Is $2\theta_i = 0$?

(b) Is $(\theta_i)^2 = 0$?

A perhaps easier problem is based on the following fact:

The Kervaire invariant is defined for manifolds M^{2n} with the structure of

(1) an immersion of codim 1 if $n \neq 2^i - 2$ (hence all odd n),
(2) an oriented immersion in codim 2 if $n \neq 2^i - 3$,
(3) a spin immersion in codim 4 if $n \neq 2^i - 5$.

The cobordism of such immersions are represented by $\pi^S_{2n+1}(\boldsymbol{RP}^\infty)$, $\pi^S_{2n+2}(\boldsymbol{CP}^\infty)$ or $\pi^S_{2n+4}(M(\mathrm{Spin}(4)))$ respectively (where π^S denotes the stable homotopy group). Hence, if a framed M^{2n} exists, it represents nonzero elements in these cobordism groups for appropriate n, but the problem may be easier there.

14. Do $\pi^S_{2n+1}(\boldsymbol{RP}^\infty)$, $\pi^S_{2n+2}(\boldsymbol{CP}^\infty)$, $\pi^S_{2n+4}(M(\mathrm{Spin}(4)))$ contain elements with Kervaire invariant 1, in particular when $n = 2^i - 1$?

From the point of view of characterizing the homotopy of manifolds, it is important to develop some more calculational techniques for the Kervaire invariant. Sullivan's formula, for example, requires a normal map where the target is already equipped with a manifold structure. The thesis of Krevitt indicates that in the cobordism theory of manifold M^{2n} with all Stiefel-Whitney classes zero, the Kervaire invariant is a Z_2 cohomology characteristic number for $n \neq 2^i - 1$.

15. As one allows more and more Stiefel-Whitney classes to vanish, when does the Kervaire invariant become a cohomology number?

16. Calculate this characteristic number, in various contexts. For example, is it zero for smooth manifolds?

These questions are pointed toward problems such as:

17. If X is a 1-connected Poincaré space of dim $4k + 2$, give computable conditions that X be of the homotopy type of a closed smooth manifold.

The analogous problem in all dimensions when $\pi_1 X \neq 0$ is interesting and will involve more invariants.

In [4] it was shown that if there is an element $f \in \pi^S_{2n-1}(S^0; Z_2)$ of Hopf invariant 1, then θ_i exists (where $n = 2^i$), i.e., $f\colon S^{2n-2} \cup_2 e^{2n-1} \to S^0$ and $Sq^{2n}_f \neq 0$. Consider the converse:

18. Suppose $f \in \pi^S_{2n-1}(\tau(\xi); Z_2)$, $f | S^{2n-2} \in \pi^S_{2n-1}(\tau(\xi))$ represents a manifold of Kervaire invariant 1, where ξ is a bundle such that this makes sense and $h(f) = 0$ in $H_{2n-1}(\tau(\xi); Z_2)$. Under what circumstances does this imply $Sq^{2n}_f(U) \neq 0$ (where U denotes the Thom class)?

REFERENCES. [1] W. Browder, *The Kervaire invariant of framed manifolds and its generalization*, Ann. of Math. (2) **90** (1969), 157–186.

[2] J. Jones Ph.D. Thesis, Oxford, 1976.

[3] J. Krevitt, *The Kervaire invariant of manifolds with trivial total Stiefel-Whitney class*, Thesis, Massachusetts Institute of Technology, 1977.

[4] M. Mahowald, *Some remarks on the Kervaire invariant from the homotopy point of view*, Proc. Sympos. Pure Math., vol. 22, Amer. Math. Soc., Providence, R.I., 1971, pp. 165–170.

C. Compact transformation groups. Compact transformation groups deal with the compact subgroups of the automorphism group of the structure of a manifold or a space, i.e., the compact subgroups of Diff(M), PL(M), Top(M) for M a manifold or Top(X) for X a topological space. If G is such a subgroup, we shall say that we have a smooth, PL or topological action of G on M or X. The main problem is to classify these actions or to find some invariants to distinguish various actions. During the past two decades, the development of differential topology has led to excellent applications in the theory of compact transformation groups. Here are some of the working problems.

1 (R. Schultz). What is the classification of linear representations of a Lie group G up to equivariant topological equivalence?

DISCUSSION. For G noncompact (e.g., $G = Z$), Kuiper and Robbin [3] have shown that equivariant topological equivalence of linear representations does not imply linear equivalence. If we insist that the representations are *orthogonal*, then the problem is open. For G compact, it is likely that the answer is affirmative. See R. Schultz for some partial answer to the problem [4]. This problem is related to the equivariant smoothing theory of Lashof and Rothenberg [1]. Perhaps, it is also related to Lee and Wasserman [2]. We may try out some well-understood classes of representations of discrete groups and Lie groups as test cases. Also, the corresponding problem for representations into infinite dimensional linear spaces (e.g., representations into the Hilbert spaces) might deserve attention.

REFERENCES. [1] R. Lashof and M. Rothenberg, *G-smoothing theory* (preprint).

[2] C. N. Lee and A. Wasserman, *On the group JO(G)*, Mem. Amer. Math. Soc. **159** (1975).

[3] N. H. Kuiper and J. W. Robbin, *Topological classification of linear endomorphisms*, Invent. Math. **19** (1973), 83–106.

[4] R. Schultz, *On the topological classification of linear representations*, Topology **16** (1977), 263–269.

2 (R. Schultz). Suppose that a compact Lie group G acts smoothly on $\boldsymbol{R}^n$ such that

(a) the fixed point set of every subgroup (even G) is diffeomorphic to a Euclidean space;

(d) if $H \supsetneqq K$, the fixed point set of H is smoothly unknotted in the fixed point set of K.

Question. Is the action (smoothly or topologically) equivalent to a linear representation?

DISCUSSION. It might be profitable to put some further restrictions on the problem. For example, some dimensional restrictions to avoid the usual low dimensional problems are reasonable. Also, in (b) if the fixed point set of H is of codim $\geqq 3$ in the fixed point set of K, then it is automatically unknotted. Special cases are known: Connell, Montgomery and Yang proved the result for most actions with two orbit types [1], and it is straightforward to extend their argument in some cases (e.g., linearly ordered isotropy subgroups). Perhaps $G = Z_p \times Z_p$ is a good test case for this problem.

REFERENCE. [1] E. Connell, D. Montgomery and C. T. Yang, *Compact groups in E^n*, Ann. of Math. (2) **80** (1964), 94–103.

3 (S. Weintraub). Let Z_p act freely and orthogonally on S^{2k-1} so that the quotient is a classical lens space. It is known that, for $k \leqq p - 1$, p copies of this action bounds as a Z_p-manifold. Find an explicit coboundary.

DISCUSSION. Solving this problem for $k \leqq p - 1$ would solve the analogue in general. If the quotient of the action is the classical lens space $L^{2k-1}(p; a_1, \cdots, a_k)$ it is easy to write down the required coboundary when the a_i's are all distinct. In the case when the a_i's are the same, a lower bound for the rank of the homology of the coboundary was given in [1]. The exact bound is complicated, but it is greater than $(p/\pi)^k$. It would be interesting to know if this lower bound can be realized.

REFERENCE. [1] S. Weintraub, *Z_p-actions and the rank of $H_n(N^{2n})$*, J. London Math. Soc. (to appear).

4 (R. Schultz). *Question.* Given a smooth G-manifold M, which elements in $\mathcal{N}_*^G$ are representable by smooth G-manifolds topologically G-equivalent to M? One can ask a similar question for the oriented case.

DISCUSSION. If $G = Z_2$, it was solved by Conner and Stong. If $G = Z_{2^r}$ or Z_{p^r} (p odd), then calculations of Ewing and Schultz yield infinite families for M an orthogonal sphere.

REFERENCE. [1] R. E. Stong, *Notes on cobordism theory*, Math. Notes, Princeton Univ. Press, Princeton, N. J., 1968.

5 (R. Schultz). Suppose that we are given an orthogonal representation of G on the unit sphere S.

Question. Which classes in $\widetilde{KO}_G(S)$ can be represented by the tangent bundles of smooth G-manifolds, G-homeomorphic to S?

DISCUSSION. Perhaps it is easier to study $\widetilde{KO}_G(S) \otimes \boldsymbol{Q}$ instead. Much is known if G is finite cyclic.

6 (R. Schultz). *Question.* Does every smooth homotopy sphere of dim $\geqq 5$ admit an effective smooth S^1 action? A similar question can be stated for a fixed cyclic group.

DISCUSSION. It is true for $5 \leqq \dim \leqq 17$.

REFERENCES. [1] R. Schultz, *Circle actions on homotopy spheres bounding plumbing manifolds*, Proc. Amer. Math. Soc. **36** (1972), 297–300.

[2] ———, *Circle actions on homotopy spheres bounding generalized plumbing manifolds*, Math. Ann. **205** (1973), 201–210.

[3] ———, *Circle actions on homotopy spheres not bounding spin manifolds*, Trans. Amer. Math. Soc. **213** (1975), 89–98.

7 (F. Raymond and R. Schultz). It is generally felt that a manifold 'chosen at random' will have very little symmetry. Can this intuitive notion be made more precise? In connection with this intuitive feeling, we have the following specific question.

Question. Does there exist a closed simply connected manifold on which no finite group acts effectively? (A weaker question, no involution?)

DISCUSSION. There exist examples of such manifolds if $\pi_1 \neq 1$ in dim 3 [4], dim 4 [1], dim 7, 11, 19, 29, 37 [2], [3]. All are $K(\pi, 1)$'s except in dim 4. In dim 7, the manifolds can be chosen to be solvmanifolds.

REFERENCES. [1] E. M. Bloomberg, *Manifolds with no periodic homeomorphisms*, Trans. Amer. Math. Soc. **202** (1975), 67–78.

[2] P. E. Conner and F. Raymond, *Manifolds with few periodic homeomorphisms*, Proc. Second Conference on Transformation Groups, Part II, Lecture Notes in Math., vol. 299, Springer-Verlag, New York, 1972, pp. 1–75.

[3] ———, *Deforming homotopy equivalence to homeomorphisms in aspherical manifolds*, Bull. Amer. Math. Soc. **83** (1977), 36–85.

[4] F. Raymond and J. L. Tollefson, *Closed 3-manifolds with no periodic map*, Trans. Amer. Math. Soc. **221** (1976), 403–418.

8 (R. Schultz). Given a spin manifold N^n, let $\delta(N^n) \in \pi_n(BO) \simeq KO_n(\text{pt})$ be the image of the KO-theoretic fundamental class $[N^n]_{KO} \in KO_n(N^n)$ under the Kronecker index map, $\langle 1_{KO^*(N)}, [N^n]_{KO}\rangle \in KO_n(\text{pt})$.

Suppose that M is a closed spin manifold admitting a smooth S^1-action with the fixed point set $F^{8k+1} = M^{S^1}$ also a spin manifold.

Question. Is $\delta(F^{8k+1}) \in Z_2$ the trivial element?

DISCUSSION. The answer is 'yes' if M is a $Z_{(2)}$-homology sphere. See **[1]** that the corresponding result for dim $F = 8k + 2$ is false.

REFERENCE. **[1]** R. Schultz, *Homotopy sphere pairs admitting semifree differentiable actions*, Amer. J. Math. **96** (1974), 308–323.

The next three problems come from studying differentiable actions of simple Lie groups on homotopy spheres. See M. Davis, W. C. Hsiang and W. Y. Hsiang, *Differentiable actions of compact simple Lie groups on homotopy spheres and Euclidean spaces*, these PROCEEDINGS, part 1, pp. 313–323, for motivations and references.

9 (M. Davis, W. C. Hsiang and W. Y. Hsiang). Analyze differentiable actions of Coxeter groups (e.g., of the permutation group on n letters, Σ_n) on homotopy spheres and acyclic manifolds.

DISCUSSION. A special case one might try first is Σ_3 action on a homotopy sphere such that the fundamental generators have fixed point sets of codim 2. Such actions of these groups are interesting because they are basically the Weyl groups of simple Lie groups. A differentiable action of a simple Lie group on a homotopy sphere or on a Euclidean space induces an action of the Weyl group on the fixed point set of the maximal torus, which carries important information on the original action. It should also be of independent interest.

REFERENCE. **[1]** D. Montgomery and C. T. Yang, *A generalization of Milnor's theorem and differentiable dihedral group actions* (preprint).

10 (M. Davis, W. C. Hsiang and W. Y. Hsiang). Let M^n be a differentiable manifold and let $N(M^n)$ be the degree of symmetry of M^n **[1]**.

Question. Is it possible to construct a homotopy sphere $\Sigma^n \notin bP_{n+1}$ whose degree of symmetry $N(\Sigma^n)$ is $\geqq n/2$?

DISCUSSION. If $\Sigma^n \notin bP_{n+1}$, we can show that $N(\Sigma^n) \leqq 2n$ **[2]**. One probably can find some homotopy spheres $\Sigma^n \notin bP_{n+1}$ such that $n \geqq N(\Sigma^n) \geqq n/2$, but it is unlikely that there exists such a sphere with $N(\Sigma^n) > n$.

REFERENCES. **[1]** W. Y. Hsiang, *On the bound of the dimension of isometry group of all possible riemannian metric on an exotic sphere*, Ann. of Math. (2) **85** (1967), 351–358.

[2] M. Davis, W. C. Hsiang and W. Y. Hsiang, *Differentiable actions of compact simple Lie groups on homotopy spheres and Euclidean spaces*, these PROCEEDINGS, part 1, pp. 313–323.

11 (M. Davis, W. C. Hsiang and W. Y. Hsiang). Let ρ_n be the standard representation of $G_n = O_n$, U_n or Sp_n. Study differentiable actions modeled on $k\rho_n$ for $k > n$. Same question for $\mathrm{Ad}_{SO_n} \oplus \rho_n$, $\mathrm{Ad}_{SU_n} \oplus \rho_n$, or $\mathrm{Ad}_{Sp_n} \oplus \rho_n$.

DISCUSSION. For actions modeled on $k\rho_n$ ($k \leqq n$) and actions modeled on the adjoint representations, we have rather good understanding **[1]**, **[2]**, **[3]**, **[4]**. So this is the natural question of the next level.

REFERENCES. **[1]** M. Davis, *Regular O_n, U_n, Sp_n manifolds* (to appear).

[2] M. Davis and W. C. Hsiang, *Concordance classes of regular U_n and Sp_n action on homotopy spheres*, Ann. of Math. (2) **105** (1977), 325–341.

[3] M. Davis, W. C. Hsiang and J. Morgan, *Concordance classes of regular O_n actions on homotopy spheres* (in preparation).

[4] M. Davis, W. C. Hsiang and W. Y. Hsiang, *Differentiable actions of adjoint type and near adjoint type* (in preparation).

12 (R. Lashof and M. Rothenberg). Let ψ be a PL action of a finite group G on $(M \times \boldsymbol{R})$ such that it is equivariantly PL homeomorphic to a product. Let $(M \times \boldsymbol{R})_\gamma$ be a G-smoothing of $M \times \boldsymbol{R}$.

Prove a G equivariant Cairns-Hirsch theorem, i.e., show that there is a G-smoothing M_α of M and a G-isotopy of $\mathrm{id}_{M\times\boldsymbol{R}}$ to a G-diffeomorphism of $M_\alpha \times \boldsymbol{R}$ with $(M \times \boldsymbol{R})_\gamma$.

Editors' note. The equivariant Cairns-Hirsch theorem follows from a mild modification of the original proof of the Cairns-Hirsch theorem **[1]** as also observed by R. Miller. Let us sketch the proof here.

Since the PL action is a product, the orbit space $(M \times \boldsymbol{R})/G$ can be written as a PL product $(M/G) \times \boldsymbol{R}$ where M/G is the *simplicial* orbit space of M after a suitable triangulation. We have a simplicial projection (after appropriate subdivision if necessary)

(1) $$q\colon (M/G) \times \boldsymbol{R} \to \boldsymbol{R}.$$

Choose a fixed one simplex $\Delta^1 \subset \boldsymbol{R}$. Let

(2) $$\pi\colon M \times \boldsymbol{R} \to (M \times \boldsymbol{R})/G$$

be the projection to the orbit space. Then, the composite map

(3) $$\begin{array}{ccc} M \times \boldsymbol{R} & & \\ \downarrow \pi & \searrow^{q\circ\pi} & \\ (M \times \boldsymbol{R})/G & \xrightarrow{q} & \boldsymbol{R} \end{array}$$

induces an *equivariant continuous* vector field ξ on the open set $U = (q \circ \pi)^{-1}(\mathrm{Int}\Delta^1)$ by pulling back the standard vector field on Int Δ^1. Clearly, ξ is transverse to the codim 1 submanifold $M = (q \circ \pi)^{-1}(a)$ where $a \in \mathrm{Int}\,\Delta^1$ denotes the barycenter of Δ^1. M is an invariant submanifold under G and it has a G-smoothing M_α. Perturbing the vector field ξ and averaging it over G, we have an equivariant smooth vector field η which is close to the original vector field ξ.

Following the argument of [1], we see that the vector field η produces an equivariant smoothing $M_\alpha \times \boldsymbol{R}$ and a G-isotopy of $\mathrm{id}_{M\times\boldsymbol{R}}$ to $(M \times \boldsymbol{R})_\gamma$.

REFERENCE. [1] M. Hirsch and B. Mazur, *Smoothings of piecewise linear manifolds*, Ann. of Math. Studies, no. 80, Princeton Univ. Press, Princeton, N.J., 1974.

13 (R. Lashof and M. Rothenberg). *Question.* Let W be a smooth G-manifold G-homeomorphic to $M \times I$ (where $M = \partial_0 W$ is one component of ∂W). Is W G-diffeomorphic to $M \times I$? If not, what is the obstruction to G-diffeomorphism?

Editors' note. The answer is *no*, in general.

If M is a G-manifold, (H) the conjugacy class of an isotropy subgroup, let $M^{(H)}$ denote the submanifold of M consisting of orbits of type (H) (i.e., G/H' where $H' \in (H)$). Then M is *stratified* by $M^{(H)}$, which induces a stratification of M/G by $M^{(H)}/G$. If M is a smooth G-manifold, then $M^{(H)}$ has a smooth G-tubular neighborhood $N(M^{(H)})$ in M, which is a G-linear vector bundle.

By choosing these tubular neighborhoods $N(M^{(H)})$ inductively sufficiently small, one gets

$$\bar{M}^{(H)} = \text{Closure}\Big(M^{(H)} - \bigcup_{H \subsetneq H'} N(M^{(H')})\Big)$$

are smooth G-manifolds with 'faces'; that is $\partial \bar{M}^{(H)} = \bigcup_{H \subsetneq H'} \bar{M}^{(H)} \cap N(M^{(H')})$.

Now suppose W is a smooth G-cobordism of M, which is *isovariant*; that is, W has the same orbit types as M. Then each stratum $W^{(H)}$ of W is a cobordism of the corresponding stratum $M^{(H)}$ of M, and if we choose our tubular neighborhoods in W with care, we get $\bar{W}^{(H)}$ is a cobordism of $\bar{M}^{(H)}$, with faces being cobordisms of faces. Such a cobordism W is called an isovariant h-cobordism if $M \to W$ and $M' \to W$ (where $\partial W = M \cup M'$) are homotopy equivalences in the category of isovariant maps, or equivalently, $\bar{M}^{(H)} \to \bar{W}^{(H)}$, $\bar{M}'^{(H)} \to \bar{W}^{(H)}$ are homotopy equivalences for each isotropy subgroup H. In that case, a torsion can be defined for each stratum:

$$\tau(\bar{W}^{(H)}/G, \bar{M}^{(H)}/G) \in \text{Wh}(\pi_1 \bar{M}^{(H)}/G))$$

and we have the isovariant s-cobordism theorem of [**1**]:

THEOREM 1. *Let W be a smooth isovariant G-h-cobordism of M, and suppose* $\dim M^{(H)}/G \neq 2, 3, 4$ *for every isotropy subgroup H. Then W is G-diffeomorphic to* $M \times [0, 1]$ *if and only if all the torsions $\tau(\bar{W}^{(H)}/G, \bar{M}^{(H)}/G)$ vanish.*

This is proved inductively using the s-cobordism theorem on a 'stratum' $\bar{M}^{(H)}/G \to \bar{W}^{(H)}/G$, where on the boundary faces it is already a product.

Now suppose W is a smooth G-cobordism of M, and that there is a *G-homeomorphism h*: $W \to M \times [0, 1]$. Then it follows easily that W is a smooth isovariant G-h-cobordism (since a G-homeomorphism is clearly an isovariant homotopy equivalence).

To construct an example with nonzero torsion, let us consider the *semifree* case, i.e., a finite group G acting on M with orbits G or points.

THEOREM 2. *Let ψ be a smooth semifree action of a finite group G on $W = M \times$* $[0, 1]$ *with fixed set $W^G \subset W$. Suppose further that there is G-homeomorphism h: $W \to W$ from ψ to*

(i) $\rho \times I$ *where* $\rho = \psi | M \times 0$, $I =$ *trivial action on* $[0, 1]$,

(ii) $\dim W \geqq \dim W^G + 3$,

(iii) $\pi_1 W = \pi_1 W^G = 0$.

Then one can define an element $O(\psi) \in \text{Wh}(G)$ *such that if* $\dim W^G \geqq 6$, *then ψ is G-diffeomorphic to $\rho \times I$ if and only if $O(\psi) = 0$.*

THEOREM 3. *Given a smooth semifree action ρ of G on M with* $\dim M \geqq \dim M^G + 3 \geqq 7$ *and* $\pi_1 M = \pi_1 M^G = 0$, *and an element* $\tau \in \text{Wh}(G)$, *then there is a smooth semifree action ψ of G on $W = M \times [0, 1]$, G-homeomorphic to $\rho \times I$, such that $\psi | M \times 0 = \rho$ and $O(\psi) = \tau$.*

For Theorem 2 we note that under the given hypothesis, Theorem 1 yields torsion elements $\tau(\bar{W}^{(H)}/G, \bar{M}^{(H)}/G)$ for each isotropy subgroup H. Since ψ is semifree, $H = 1$ or G. Since $W^{(G)}/G = W^{(G)} = W^G$ is 1-connected $\tau(\bar{W}^{(G)}/G, \bar{M}^{(G)}/G) \in$

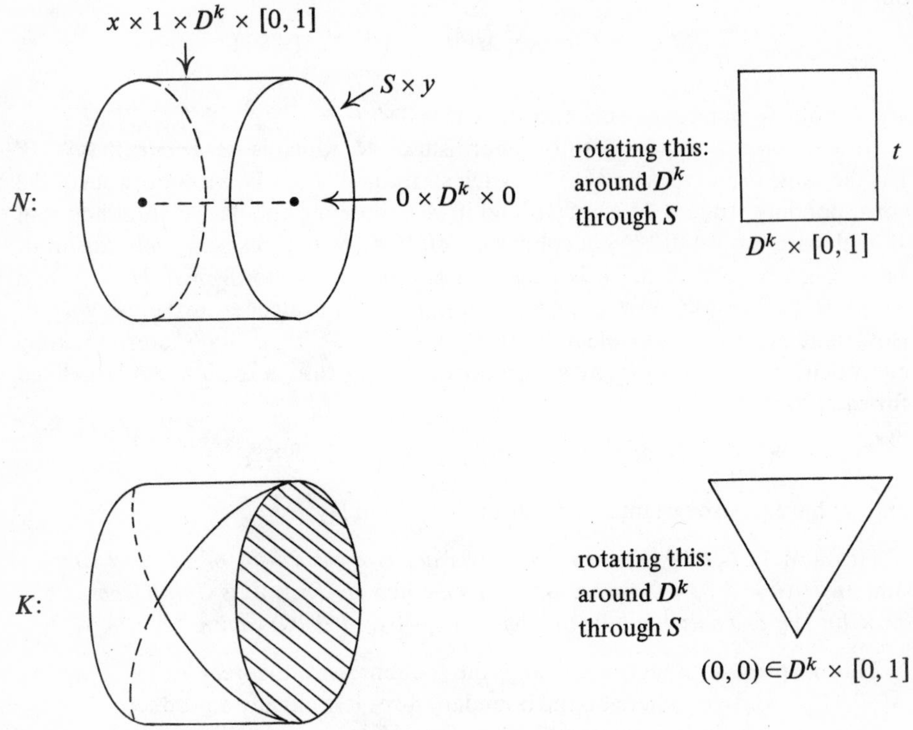

FIGURE 1

$\mathrm{Wh}(1) = 0$, so define $O(\psi) = \tau(\bar{W}^{(1)}/G,\ \bar{M}^{(1)}/G) \in \mathrm{Wh}(\pi_1(W^{(1)}/G))$. Since W and W^G are 1-connected and codim $W^G \geqq 3$, it follows that $W^{(1)} = W - W^G$ is 1-connected, so $\pi_1(W^{(1)}/G) = G$.

To prove Theorem 3, let us first consider a disc $D^k \times [0, 1] \subset M^G \times [0, 1]$ ($k = \dim M^G$), and $N =$ the G-tubular neighborhood of $M^G \times [0, 1]$ restricted to $D^k \times [0, 1]$, $N = V \times D^k \times [0, 1]$, where V is the unit ball in the linear representation space for G, the normal representation of G at a point of $D^k \subset M^G$. Let $S = \partial V$, and use coordinates $v = (x, t)$ in V where $x \in S$, $t = |v|$, and let $y \in D^k$. Then these coordinates represent N as the mapping cylinder of the projection $S \times D^k \times [0, 1] \to D^k \times [0, 1]$, where t is the cylinder coordinate.

Define $K \subset N$ as the mapping cylinder of $S \times D^k \times [0, 1] \to (0, 0) \in D^k \times [0, 1]$ by using t coordinate to shrink the coordinate in D^k. Explicitly

$$k(x, t, y, s) = (x, t, ty, ts) \in N,$$

define a map $k\colon N \to N$ such that $k(0 \times D^k \times [0, 1]) = (0, 0, 0, 0)$. (See Figure 1.)

Now define $K_n \subset K$,

$$K_n = K \cap (S \times [1/(n-1), 1/n] \times D^k \times [0, 1]).$$

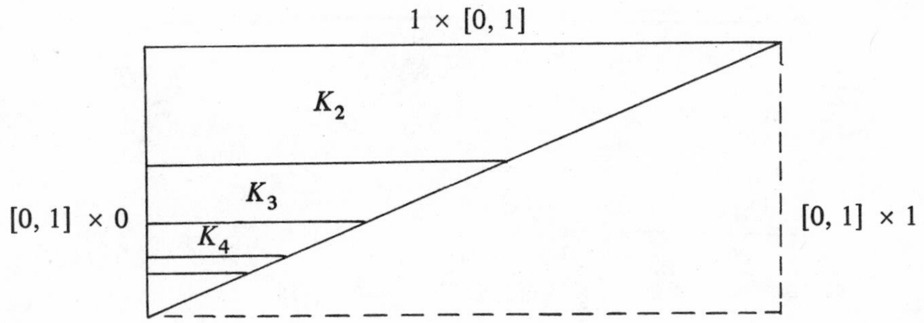

FIGURE 2

(See Figure 2.) Clearly

$$\partial K_n = (S \times [1/(n-1), 1/n] \times D^k \times 0) \cup (S \times \{1/(n-1), 1/n\} \times D^k \times [0, 1]) \cup K_n^0$$

(where K_n^0 is the slanted part of the boundary in Figure 2) and K_n^0 is diffeomorphic to $S \times [1/(n-1), 1/n] \times D^k \times 0$, and K_n is diffeomorphic to $K_n^0 \times [0, 1]$.

Note that all the manifolds defined above are G-submanifolds (with corners) of $M \times [0, 1]$, so that the orbit spaces embed in $(M/G) \times [0, 1]$.

Given $\tau \in \mathrm{Wh}(G)$, since $\pi_1(S/G) = G$, and $\pi_1(K_n^0/G) \cong \pi_1(S/G)$, we can find an h-cobordism U_2 of K_2^0/G which is a product on $\partial K_2^0/G$ with torsion $\tau(U_2, K_2^0/G) = \tau \in \mathrm{Wh}(G)$. But

$$\partial K_n = K_n^0 \cup (\partial K_n^0 \times [0, 1]) \cup (S \times [1/(n-1), 1/n] \times D^k \times 0),$$

so let us define a manifold W by

$$W = ((M \times [0, 1]) - K_2) \cup \tilde{U}_2$$

where $\tilde{U}_2 \to U_2$ is the universal cover of U_2, and the union is along $K_2^0 \cup (\partial K_2^0 \times [0, 1])$. (See Figure 3.) It follows easily that W is a G-isovariant h-cobordism of M and ρ and $\tau = \tau(\bar{W}^{(1)}/G, \bar{M}^{(1)}/G)$. It remains to construct a G-homeomorphism of $M \times [0, 1]$ to W.

To construct such a homeomorphism it suffices to construct a G-homeomorphism of $\bar{K} = \mathrm{Closure}(\tilde{U}_2 \cup K_3 \cup \cdots)$ to $K' = \mathrm{Closure}(K_2 \cup K_3 \cup \cdots)$ which is the identity on the $(\partial \bar{K}) \cap (\text{Interior } W)$. Following **[2]**, we note that K_i is G-diffeomorphic to K_j, all i and j, so define U_i to be the cobordism of K_i^0/G, $i > 2$, by U_{2j} is diffeomorphic to U, so that $\tau(U_{2j}, K_{2j}^0/G) = \tau$, and let U_{2j+1} be diffeomorphic to each other, with $\tau(U_{2j+1}, K_{2j+1}^0/G) = -\tau$. Then $U_{2j-1} \cup U_{2j}$ is a cobordism of $(K_{2j-1}^0/G) \cup (K_{2j}^0/G)$ with torsion $\tau - \tau$, so by the s-cobordism theorem, there is a G-diffeomorphism:

$$f_{2j}\colon \tilde{U}_{2j-1} \cup \tilde{U}_{2j} \to K_{2j-1} \cup K_{2j}$$

which is the identity on $(K_{2j-1}^0 \cup K_{2j}^0) \cup ((\partial(K_{2j-1}^0 \cup K_{2j}^0)) \times [0, 1])$. Then

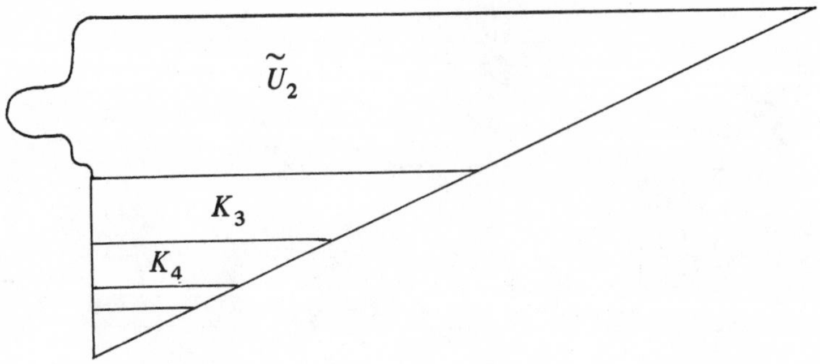

FIGURE 3

$$\bigcup_j f_{2j}: (\tilde{U}_3 \cup \tilde{U}_4) \cup (\tilde{U}_5 \cup \tilde{U}_6) \cup \cdots \\ \to (K_3 \cup K_4) \cup (K_5 \cup K_6) \cup \cdots$$

is a G-diffeomorphism which is identity on the boundary, and hence extends to a homeomorphism of the 1-point compactification $\bar{U}/G$ of $\tilde{U}_3/G \cup \tilde{U}_4/G \cup \cdots$ to that of $K_3/G \cup K_4/G \cup \cdots$. Thus we can define a homeomorphism

$$f': \tilde{U}_2 \cup \bar{U} \to \tilde{U}_2 \cup (K' - K_2) = \bar{K}.$$

Similarly we have a diffeomorphism g_{2j+1}: $\tilde{U}_{2j} \cup \tilde{U}_{2j+1} \to K_{2j} \cup K_{2j+1}$ which is 1 on the appropriate boundary pieces, so we get a G-homeomorphism $\bar{g}$: $\tilde{U}_2 \cup \bar{U} \to K'$. Hence $\bar{g} \circ f'^{-1}$: $\bar{K} \to K'$ is the G-homeomorphism desired.

REFERENCES. [1] W. Browder and F. Quinn, *A surgery theory for G-manifolds and stratified sets*, Manifold-Tokyo, Univ. of Tokyo Press, Tokyo, 1973, pp. 27–36.

[2] J. Stallings, *On infinite processes leading to differentiability in the complement of a point*, Differential and Combinatorial Topology, Princeton Univ. Press, Princeton, N.J., 1965, pp. 245–254.

14 (R. Schultz). Given a spin manifold M^n, one can define the *higher $\hat{A}$ genus* in $H^*(\pi_1(M^n); Q)$ in analogy with the Novikov higher signature.

Question. How is the higher $\hat{A}$-genus of a spin manifold M^n related to smooth actions of compact *connected* Lie group actions on M^n?

Editors' note. In a forthcoming note by W. Browder and W. C. Hsiang, this problem and its applications will be discussed in full detail.

15 (J. Cohen). Let π be a finite group of order m and let p be a prime number such that $(p, m) = 1$. Then, for any odd $n > 1$, there is a finite n-dim simply connected complex X which is a Z_p-homology sphere and admits a free π action.

Question. Can X be chosen to be a manifold (smooth, PL or Top) with a free π action?

DISCUSSION. Since $(p, m) = 1$, $m =$ order π, it follows that q_*: $H_*(X;Z_p) \to H_*(X/\pi; Z_p)$ is an isomorphism, where q: $X \to X/\pi$ is the projection. Since X is a 1-connected Z_p-homology sphere, it follows from Serre's $\mathscr{C}$-theory that there is a map $f\colon S^n \to X$ such that $f_*\colon H_*(S^n; Z_p) \to H_*(X; Z_p)$ is an isomorphism. Then, the composite

$$g = q \circ f\colon S^n \to X/\pi$$

is a normal map when we take a bundle map of ν_{S^n} (the normal bundle of S^n) into the trivial bundle over X/π, and a Z_p-homology equivalence. Hence we get a smooth surgery problem with the obstruction $\sigma(g) \in L_n(Z_p[\pi])$ (or $L_n(Z_{(p)}[\pi])$) and such that image $\sigma(g) = 0$ in $L_n(Z_p)$.

If $\sigma(g) \neq 0$, one could take a map in $[X/\pi, G/H]$ ($H = 0$, PL or Top) to get another normal map to try. The group $L_n(Z_p[\pi])$ is an elementary 2-group, but is not zero in general [**1**], [**3**]. The set of all these obstructions $\{\sigma(g)\} \subset L_n(Z_p[\pi])$ depends on the homotopy type of X/π which we may also vary, in particular, as we vary X. The question then becomes whether there is a choice of X, action of π on X, element in $[X/\pi, G/H]$ which gives us $\sigma(g) = 0$.

If m is odd and $p = 2$, then $L_n(Z_2[\pi]) = 0$ for $n = \pm 1$ by [**1**], [**2**], and if we take rational coefficients Q, $L_{-1}(Q[\pi]) = 0$ by [**1**]. Hence†

THEOREM. *If π is a finite group, then π acts freely and smoothly on a simply connected rational homology n-sphere, $n = 4k - 1$, $k > 1$. If π is of odd order, π acts freely and smoothly on a simply connected Z_2 homology n-sphere, for odd $n \geqq 5$.*

REFERENCES. [**1**] W. Pardon, *An invariant determining the Witt class of a unitary transformation over a semisimple ring*, J. Algebra **44** (1977), 396–410.

[**2**] D. S. Passman and T. Petrie, *Surgery with coefficients in a field*, Ann. of Math. (2) **95** (1972), 385–405.

[**3**] C. T. C. Wall, *Classification of Hermitian forms*, Invent. Math. **19** (1973), 59–71.

PRINCETON UNIVERSITY

†*Added in proof.* We are informed by W. Pardon that he can prove the following:

THEOREM. *For $n = 4k + 3$, $L_n^h(Z_{(p)}[\pi]) = 0$ if π is finite and $(p, |\pi|) = 1$.*

So, this problem is solved for $n = 4k + 3$.

Proceedings of Symposia in Pure Mathematics
Volume 32, 1978

FOLIATION PROBLEM SESSION (R. BOTT)

COMPILED BY

MARK MOSTOW AND PAUL SCHWEITZER

1. (R. Bott). Find a codimension 3 foliation with $h_3c_2 \neq 0$. (This is the first rigid class.) More generally, prove that $H^*(WO_k) \to H^*(B\Gamma_k)$ is injective, i.e., that all of the potential continuous characteristic classes for foliations are in fact nontrivial. (Partial results in this direction have been obtained by Bott, Haefliger, Heitsch, Kamber, Tondeur, Morita, Thurston, et al.)

Counterconjectures. a. (Jack Petro) Prove that $H^*(WO_k) \to H^*(B\Gamma_k)$ is *not* injective.

b. (Herb Shulman) Prove that $H^*(F\Gamma_{k+1}) \to H^*(F\Gamma_k)$ is zero, at least with real coefficients. (This would explain the vanishing of the rigid classes observed to date.)

2. (Herb Shulman). If M is a non-Hausdorff smooth manifold (not necessarily second countable), is the natural morphism from de Rham to singular cohomology an injection? What about the special case Γ_q, and the simplicial manifold $N\Gamma_q$?

Suggestions. Try using hypercovers to compute singular cohomology (P. Trauber).

Look at the Čech-singular and Čech-de Rham double complexes (M. Mostow). For the Hausdorff case see

A. Weil, *Sur les Théorèmes de de Rham*, Comment. Math. Helv. **26** (1952), 119–145.

3. (Paul Schweitzer and folklore). If the holonomy of a foliation is 0, is the Godbillion-Vey class $gv = 0$?

Evidence (Michel Herman, *Foliations of T^3 by planes*, preprint). Any smooth foliation of T^3 by planes has $gv = 0$.

AMS (MOS) subject classifications (1970). Primary 55 F40, 57 D20, 57D30; Secondary 57E30.

4. (Herb Shulman). Let F be a codimension k foliation on a $(2k + 1)$-dimensional orientable manifold M. An open cover of M is *trivializing* for F if the foliation restricted to any open set of the cover can be defined by a submersion. For each such cover consider the number of unordered $(k+2)$-tuples of distinct open sets with nonempty intersection. Define the covering dimension $\mathrm{cd}(F)$ to be the minimum (over all trivializing covers) of these numbers. Let $gv(F)$ be the Godbillon-Vey class of F evaluated on the fundamental class of M.

Conjecture. $|gv(F)| \leqq \mathrm{cd}(F)$.

Background. If $\mathrm{cd}(F) = 0$, then all the exotic classes vanish (Shulman, preprint).

5. (Philip Trauber). Find an index theorem for differential operators on a foliated manifold which are elliptic in the leaf directions. (Luke Hodgkin has proved that such an index can be defined.)

6. (P. Trauber). Let Emb^δ be the monoid of self-embeddings of $\boldsymbol{R}^n$, and let Diff^δ be the subgroup of diffeomorphisms, both with the discrete topology.

THEOREM (G. SEGAL, P. TRAUBER). *$B(\mathrm{Emb}^\delta)$ and $B\Gamma_n$ have the same weak homotopy type.*

Conjecture 1. The inclusion $B\,\mathrm{Diff}^\delta \to B\,\mathrm{Emb}^\delta$ is a homology equivalence (i.e., induces isomorphisms in homology).

Conjecture 2. It also induces an isomorphism in continuous cohomology (using two topologies on Diff and on Emb).

Background. Every space is homology equivalent to a $K(\pi, 1)$ (Thurston and Kan, preprint).

$\Omega^n B\bar{\Gamma}_n$ is $\boldsymbol{Z}$-homology equivalent to $B\,\overline{\mathrm{Diff}}$, where $\overline{\mathrm{Diff}}$ is the homotopy-theoretic fiber of $\mathrm{Diff}_K^\delta \to \mathrm{Diff}_K$, K denoting compact support (Mather, Thurston).

Also compare Quillen's plus construction.

7. (Mark Mostow).

Background. M. Mostow, *Continuous cohomology of spaces with two topologies*, Mem. Amer. Math. Soc., no. 175, Amer. Math. Soc., Providence, R. I. (1976), and *Variations, characteristic classes, and the obstruction to mapping smooth to continuous cohomology*, Trans. Amer. Math. Soc. **240** (1978), 163–182.

R. Bott, *Some remarks on continuous cohomology*, Proc. of Internat. Conf. on Manifolds and Related Topics in Topology, (Tokyo, 1973), pp. 161–170.

For any smooth space $(X' \to X)$ with two topologies there are natural maps

$$T_{DR}(X' \longrightarrow X) \xrightarrow{A} T_{011}(X' \longrightarrow X) \xleftarrow{B} T_c(X' \longrightarrow X),$$

where T_{DR} is de Rham cohomology (of spaces with two topologies), T_c is continuous singular cohomology (for 2 topologies), and T_{011} is continuous singular cohomology based on cochains C^0 with respect to C^1 variations of C^1 simplices. An example (Mostow, *Variations* ...) shows that B need not be an isomorphism, and hence that it is not always possible to map T_{DR} to T_c. Also, A is not always an isomorphism. Nonetheless, A and B *are* isomorphisms on $(B(G_{\text{discrete}}) \to BG)$ (by the van Est Theorem).

Conjecture. A and B are isomorphisms for $(B\Gamma_q \to BJ_q)$.

REMARK. $T_{DR}(B\Gamma_q \to BJ_q) = H^*(WO_q) = H_{G.F.}(\mathfrak{A}_q, O_q)$ (Bott, Haefliger).

8. (Jim Perchik). Cohomology classes of the Hamiltonian formal vector fields on $\boldsymbol{R}^2$ are detected by coefficients of x^0t^r in the expansion of

$$\left(\frac{1}{2}\right)\prod(1 - t^r x^q) = t^{-2} + 2 - t^8 - t^{14} \cdots + 3t^{62} - 8t^{64} + 9t^{66} + 10t^{70} + \cdots$$

(+ terms in x and x^{-1}). The above product is taken over $r = -1, 0, 1, 2, \cdots; q = -(r+2), -r, \cdots, r, r+2$; except for the case $r = q = 0$.

Question. Are infinitely many of these coefficients nonzero?

Reference

J. Perchik, *Cohomology of Hamiltonian and related formal vector field Lie algebras*, Topology **15** (1976), 395–404.

North Carolina State University

Pontificia Universidade Catholica

Proceedings of Symposia in Pure Mathematics
Volume 32, 1978

PROBLEMS IN LOW DIMENSIONAL MANIFOLD THEORY

ROB KIRBY

AMS (MOS) subject classifications (1970). Primary 14J99, 20F05, 53A35, 55A05, 55A10, 55A20, 55A25, 55A35, 55A40, 55A99, 55C35, 57A35, 57A45, 57A50, 57C15, 57C25, 57D25, 57D30, 57D50, 57D60, 57D65, 57D80, 57E25, 57E30.

Introduction. This problem list began at the Stanford Conference and grew, tripling in size, through January 1977. I have had vital help from many topologists, particularly Andrew Casson, Bob Edwards and Cameron Gordon. No attempt was made to discover the original authors of these problems; the name associated with a problem is almost always the person who suggested it and helped write the version here; it does not follow that unauthored problems are mine—usually the authorship is divided or the problem is folklore.

Low dimensional manifold theory is intrinsically beautiful, interesting, and intricate, and could survive in isolation. Still, its appreciation will be wider and its vigor greater if its relations with other fields are encouraged. To this end, I have sought problems involving algebraic geometry (mainly complex surfaces and singularities), differential geometry, and group theory. And in this sense there is an unfortunate lack of problems involving foliations, dynamical systems and differential analysis. Yet, these areas frequently involve "local" problems, which I have generally avoided here, e.g. there are no problems directly about wild imbeddings or decompositions.

I will try to keep the list up to date if all of you will endeavor to send me preprints of (partial) solutions and/or interesting new problems.

0. Notation. B^n is the unit ball in R^n, and S^{n-1} is its boundary. RP^n and CP^n are real and complex projective n-space; in particular CP^2 has intersection form $\langle 1 \rangle$, and, changing orientation, $-CP^2$ has form $\langle -1 \rangle$. $T^n = S^1 \times \cdots^n \times S^1$. $L(p, q)$ is a lens space.

F_g (or sometimes F or F^2) always denotes a closed surface of genus g. $\Sigma(p, q, r)$ is the Brieskorn manifold, or link of the isolated singularity, i.e., $\Sigma(p, q, r) = S^5 \cap \{(x, y, z) \in C^3 \mid x^p + y^q + z^r = 0\}$. $\Delta(t)$ is the Alexander polynomial of the knot in question.

Definitions. Most definitions are given in the problems, but here are a few which are widely used and perhaps not sufficiently widely known.

Two knots $f_i\colon S^n \to S^{n+2}$, $i = 0, 1$, are *concordant* (sometimes cobordant) if the f_i extend to an imbedding $F\colon S^n \times I \to S^{n+2} \times I$. A knot is *slice* if it is concordant to the unknot. A knot $K = f(S^1) \subsetneq S^3$ is *ribbon* if f extends to an immersion $f\colon B^2 \to S^3$ whose singularities are always of the form:

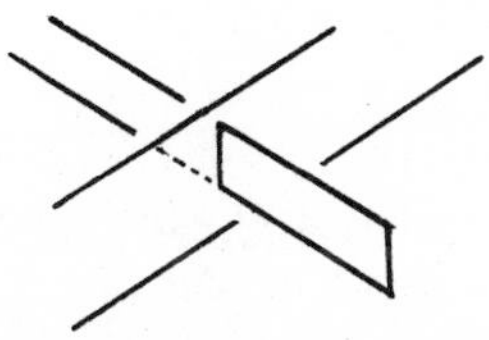

The 4-*ball genus* of K is the minimal genus of a smooth surface F in B^4 with $\partial F = K$.

An unoriented knot K in S^3 is *amphicheiral* if K is isotopic to its reflection across an S^2, which is equivalent to the existence of an orientation reversing homeomorphism $h: S^3 \to S^3$ such that $h(K) = K$. If K is oriented, we can require either that h preserves or reverses the orientation of K; K satisfies the latter case only if K equals its concordance inverse.

Let $f: S^1 \times B^2 \to S^3$ be a trivialization of the normal disk bundle of $K = f(S^1 \times 0)$, for which $f(S^1 \times (1, 0))$ lies on the Seifert surface of K (equivalently, $f(S^1 \times (1, 0))$ represents $0 \in H_1(S^3 - K; Z)$). This is called the 0-framing, and framing n is obtained from $n \in \pi_1(SO(2))$ (in this case $f(S^1 \times (1, 0))$ should wind around K n-times as in a right-handed screw). If a 2-handle is added to B^4 along K with framing n, then the boundary is the result of *n-surgery* on S^3 along K. *Dehn* surgery is more general; one is allowed to remove $f(S^1 \times B^2)$ from S^3 and sew it back in by any homeomorphism of the boundary.

DEFINITION. A 3-manifold M is *sufficiently large* if there exists a 2-sided surface $F \neq S^2$ in M with $F \cap \partial M = \partial F$ such that the inclusion of F into M induces a monomorphism on π_1. (See [Hempel, Ann. of Math. Studies, no. 86, Princeton Univ. Press, Princeton, N.J., 1976].)

DEFINITION. A 3-manifold M^3 is *atoroidal* if every imbedding $(S^1 \times I, \partial) \to (M, \partial)$ is inessential (not injective on π_1) or is homotopic rel ∂ to a map into ∂M, and if every imbedding $T^2 \to M$ is inessential or homotopic to a map into the boundary.

DEFINITION. A closed, 3-manifold M is *hyperbolic* if it has a Riemannian metric with constant, negative sectional curvature; this implies the universal cover of M is hyperbolic 3-space, and $\pi_1(M)$ is a subgroup of $PSL(2, C)$ (the group of isometries of hyperbolic 3-space). If $\partial M \neq \emptyset$, let $\partial_1 M$ consist of the torus components, and $\partial_2 M$ all other components of ∂M; then M is *hyperbolic* if $M - \partial_1 M$ has a complete Riemannian metric with finite volume, and constant, negative sectional curvature, and $\partial_2 M$ totally geodesic; the universal cover of $M - \partial_1 M$ is then hyperbolic 3-space minus disjoint copies of half-space, one for each component of $\partial_2 M$.

1. Knot theory. Earlier lists of problems about knot theory are given by Fox in [*Topology of* 3-*manifolds*, Prentice-Hall, 1962], by Neuwirth in his book, [*Knot groups*, Ann. of Math. Studies, no. 56, Princeton Univ. Press, Princeton, N.J., 1965] (see the update in [V.P.I. Top. Conf., Lecture Notes in Math., vol. 375, Springer-Verlag, Berlin, pp. 209–230, 1973]) and by Birman in her book, [*Braids, links and mapping class groups*, Ann. of Math. Studies, no. 82, Princeton Univ. Press, Princeton, N.J., 1974].

Problem 1.1 (*Lickorish*). *Conjecture*: Given a knot K, any band connected sum with an unknot is still a knot. This follows from the *Conjecture*: genus(K) + genus$(L) \leqq$ genus$(K \#_b L)$.

REMARKS. The knot group H of $K \#_b$ (unknot) has a quotient which is obtained from the group G of K by adding one new generator and one new relation in which the exponent sum of the new generator is 1. It follows from ([Gerstenhaber and Rothaus, Proc. Nat. Acad. Sci. U.S.A. **48** (1962), 1531–1533], also see Problem 5.7) that if G has a nonabelian quotient Q which imbeds in a compact connected Lie group, then H has a quotient which contains Q, and hence is nonabelian. The first conjecture is therefore true for knots K whose group G has this property, in particular those with $\Delta(t) \neq 1$ and those with G residually finite (Problem 3.33).

Problem 1.2. (A) (T. Matumoto) Suppose the band connected sum of a trivial link (of two components) is the trivial knot. Is the band isotopic to the trivial band?

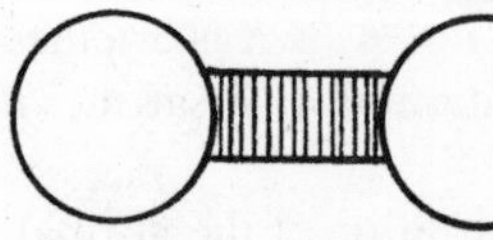

(B) (H. Morton) Suppose we attach a half-twisted band to the unknot and get the unknot. Is the band isotopic to the trivial band?

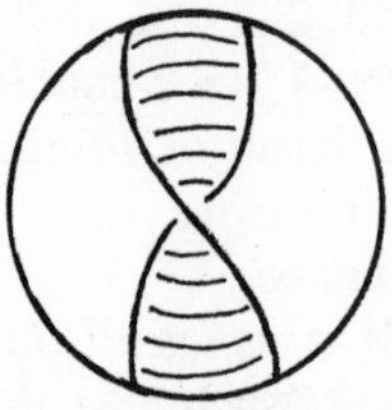

REMARK. Note that these problems are relevant to the question of whether an imbedded T^2 (or RP^2) in R^4 with 4 (or 3) critical points is unknotted (see Problem 4.30).

Problem 1.3 (*Gordon*). Is there a knot K such that if a crossing is changed as below, one gets the unknot in one case and the unlink in the other?

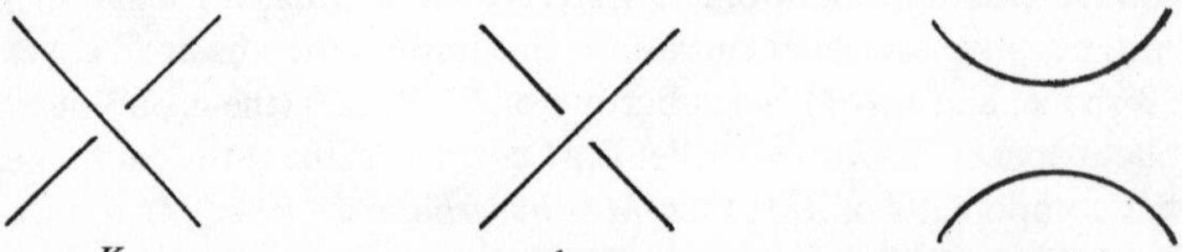

REMARKS. Such a knot K bounds a smooth imbedded 2-ball D in B^4 whose complement is a homotopy circle since $(B^4, D) \times I$ is unknotted. There are examples of the latter, e.g., doubles of slice knots [Gordon and Sumners, Math. Ann. **217** (1975), 47–52]. The answer is no if the answer to Problem 1.2(A) is yes.

Problem 1.4 (*J. Milnor*). Is the unknotting number of the (p, q)-torus knot, $(p, q) = 1$, equal to $(p - 1)(q - 1)/2$?

REMARKS. The unknotting number is the minimum number of crossings which must be changed to get the unknot. For the algebraic geometric background for this problem, see [Milnor, *Singular points of complex hypersurfaces*, Princeton Univ. Press, Princeton, N.J., 1968, pp. 81–95].

The number $(p - 1)(q - 1)/2$ is half the degree of the Alexander polynomial of the (p, q)-torus knot. However, it is not even true in general that the unknotting number is $\geq \frac{1}{2}\deg(\Delta(t))$, for the knot 8_{20} ([Alexander and Briggs, Ann. of Math.

(2) **28** (1927), 562–586], also [Rolfsen, *Knots and links*, Publish or Perish, Boston 1976]) has unknotting number 1 and $\Delta(t) = 1 - 2t + 3t^2 - 2t^3 + t^4$.

Problem 1.5 (*Van Buskirk*). *Conjecture*: K is amphicheiral iff K is invariant under reflection through the origin.

REMARKS. True for knots with $\leqq 10$ crossings. Note that a knot K is invariant under reflection through a plane (hence amphicheiral) iff $K = J \# - J$ for some knot J. For such knots, the conjecture is true.

Problem 1.6 (*Montesinos*). *Conjecture*: Each invertible knot is strongly invertible.

REMARKS. A knot K is invertible if there exists an orientation preserving homeomorphism μ of S^3 which takes K to itself but reverses the orientation of K. It is strongly invertible if μ is also an involution; in this case, μ is equivalent to a rotation about an axis [Waldhausen, Topology **8** (1969), 81–91]. (Added in proof. False, W. Whitten.)

Problem 1.7 (*J. Stallings*). Characterize those braids β of n strands whose closed braid $\hat{\beta}$ is the unknot. Specifically, for σ_i the generators of the braid group, let σ_{ij}, $i < j$, be defined by

$$\sigma_{ij} = (\sigma_i\sigma_{i+1} \cdots \sigma_{j-2})\sigma_{j-1}(\sigma_i\sigma_{i+1} \cdots \sigma_{j-2})^{-1}.$$

The braid σ_{ij} crosses the ith and jth strands in front of all those strands in between. *Conjecture*: $\hat{\beta}$ is the unknot iff $\hat{\beta}$ is conjugate to a product of $n - 1$ elements of the form σ_{ij} or σ_{ij}^{-1}.

Problem 1.8 (*J. Stallings*). Suppose β is a word in the generators $\sigma_1, \cdots, \sigma_{n-1}$ and their inverses in the braid group B_n. If the length of β is minimal over all words representing the same element of B_n, call β "minimal". *Conjecture*: If the last letter of a minimal word β is σ_i^ε, then the word $\beta\sigma_i$ is again minimal ($\varepsilon = \pm 1$).

Problem 1.9 (*Fox and Birman*). Let G be the knot group of a nontrivial knot K and let $\mu \in G$ be represented by a meridian. Let $N(\mu^2)$ be the normal closure of μ^2 in G. *Conjecture*: $G/N(\mu^2)$ is never abelian, i.e., is never $Z/2$.

REMARKS. Let H be the kernel of the obvious homomorphism $G \to Z/2$ and note that the normal closure of μ^2 in H is still $N(\mu^2)$. Then $H/N(\mu^2)$ is of index 2 in $G/N(\mu^2)$ so the conjecture is that $H/N(\mu^2)$ is never trivial. But H is the knot group of K in the double branched cover M^3; if $H/N(\mu^2)$ is trivial, then M^3 is a homotopy 3-sphere which must be fake since by Waldhausen there is no involution on S^3 with knotted fixed point set.

Problem 1.10 (*L. Moser*). Is there a geometric characterization of knots whose groups have one relator?

REMARKS. The groups of 2-bridge knots are presented on 2 generators and one relator where the generators are meridians. The groups of torus knots also are presented on 2 generators with one relator but the generators are not meridians.

Problem 1.11 (*Cappell and Shaneson*). Is every knot, whose group is generated by 2 meridians, actually a 2-bridge knot? Same for n meridians and n-bridge knots.

REMARK (BAILEY). If yes, then Fox's "bushel basket" of homotopy 3-spheres [*Topology of 3-manifolds*, Prentice-Hall, Englewood Cliffs, N.J., 1962, pp. 213–216] contains only S^3 (see [G. Burde, Canad. J. Math. **14** (1971), 84–89]).

Problem 1.12 (*J. Simon*). Let $G_K = \pi_1(S^3 - K)$. *Conjecture*: If there is a nontrivial epimorphism $\phi\colon G_L \twoheadrightarrow G_K$, then

(A) $n(G_L) > n(G_K)$ where $n(G)$ is the minimum number of (meridian?) generators. (Added in proof. False, e.g. the group of the torus knot $(3p, 2)$, p odd, maps onto the trefoil knot group (Hartley-Murasugi, Canad. J. Math. (to appear).)

(B) $\text{genus}(L) \geqq \text{genus}(K)$.

REMARK. (B) is known if genus $K = \frac{1}{2}\deg \Delta_K(t)$ or if $\phi(l_L) = (l_K)^n$, l = longitude.

(C) *Conjecture*: Given K, there exists a number N_K such that any chain of epimorphisms of knot groups $G_K \twoheadrightarrow G_{L_1} \twoheadrightarrow G_{L_2} \twoheadrightarrow \cdots \twoheadrightarrow G_{L_n}$ with $n \geqq N_K$ contains an isomorphism.

REMARK. This implies knot groups are Hopfian (Problem 3.33). Knot groups seem like the right place to start, but the conjecture could be made for compact 3-manifold groups.

(D) *Conjecture*: Given K, there exist only finitely many knot groups G for which there is an epimorphism $G_K \twoheadrightarrow G$.

Problem 1.13 (*J. Simon*). *Conjecture*: Let K be a (p, q)-cable about a nontrivial knot K_0 such that $|p| = 1$ or 2, and let L be a knot such that $\pi_1(S^3 - K) \cong \pi_1(S^3 - L)$; then $S^3 - K$ and $S^3 - L$ are homeomorphic.

REMARKS. The conjecture covers the only remaining case where it is not yet known whether the group of a prime knot in S^3 determines the complement up to homeomorphism. If we draw the (p, q)-torus knot on $\partial(S^1 \times B^2)$ (the knot represents $q \in H_1(S^1 \times B^2)$ and we assume $|q| \neq 1, 0$) and tie $S^1 \times B^2$ into a knot K_0 so that $S^1 \times (1, 0)$ is homologically trivial in $S^3 - K_0$, then the resulting knot is the (p, q)-cable about K_0.

The knot is composite or cable iff the complement of K, $C^3(K)$, admits a proper imbedding of an annulus A that is essential in the sense that (i) $\pi_1(A) \to \pi_1(C^3(K))$ is monic, and (ii) A cannot be pushed into $\partial C^3(K)$ by a homotopy fixing ∂A; then K is composite if a boundary curve of A generates $H_1(C^3(K))$ and is cable otherwise [J. Simon, Ann. of Math. (2) **97** (1973), 1–13].

Suppose $\pi_1(C^3(K)) \cong \pi_1(C^3(L))$. If $C^3(K)$ has no essential annulus, then $C^3(K)$ is homeomorphic to $C^3(L)$ [C. D. Feustel, Trans. Amer. Math. Soc. **217** (1976), 1–43, Theorem 10]. If in addition K has Property P (see Problem 1.15), then K and L are equivalent knots. Assume now that $C^3(K)$ has an essential annulus A. If K is composite, then L is also composite and their prime factors are equivalent, e.g., granny and square knots (Feustel and Whitten, *Groups and complements of knots*, Canad. J. Math. (to appear)).

Suppose K is cable. Then K is a torus knot if $\pi_1(C^3(K))$ has a nontrivial center [G. Burde and H. Zieschang, Math. Ann. **167** (1966), 169–175] or if $C^3(K)$ admits no essential imbedding of a torus (C. D. Feustel, loc. cit.); in this case K is equivalent to L. Hence assume K_0 is nontrivial. If $|p| \geqq 3$ or K_0 has Property P, then K is equivalent to L [Feustel and Whitten, loc. cit.]; otherwise L is a $(\pm p, \pm q)$-cable about a knot L_0 such that $C^3(K_0) \cong C^3(L_0)$, and we arrive at the conjecture.

Note that the papers [Feustel and Whitten, loc. cit.] and [J. Simon, Proc. Amer. Math. Soc. **57** (1976), 140–142] show that the conjectures, "all knots have Property P" and "complements of prime knots are determined by their groups", are nearly equivalent.

Problem 1.14 (*J. Simon*). Characterize those knots K in S^3 for which the commutator subgroup G' of $\pi_1(S^3 - K)$ has infinite weight (is not normally generated by a finite number of elements). *Conjecture*: K has infinite weight if K has a companion of winding number zero.

REMARKS. The conjecture is true for the untwisted double of any knot.

Construct a knot K by putting a knot J in a solid torus T and tying T in a knot J'; then J' is called a companion of K and J is a satellite of K. K represents an integer in $H_1(T; Z)$ which is called the winding number of J'.

If G' has finite weight, then the Alexander polynomial of K is monic. If G' has weight $\geqq 2$, then K has Property R (Problem 1.16).

Problem 1.15. Does every nontrivial knot K have Property P; that is, does Dehn surgery on K always give a nonsimply connected manifold?

REMARKS. Knots with Property P include: *torus knots* [Seifert], [J. Hempel, Proc. Amer. Math. Soc. **15** (1964), 154–158], [Bing and Martin, Trans. Amer. Math. Soc. **155** (1971), 217–231], [Gonzalez-Acuña, Bol. Soc. Mat. Mexicana **15** (1970), 58–79]; *twist knots* [Bing and Martin], [Gonzalez-Acuña], [R. Riley, Quart. J. Math. **25** (1974), 273–283]; *composite knots* [D. Noga, Math. Z. **101** (1967), 131–141], [Bing and Martin], [Gonzalez-Acuña]; *doubled knots* [Bing and Martin], [Gonzalez-Acuña]; *weakly splittable knots* [A. Connor, Thesis, Univ. of Georgia, 1969]; most *cable knots* [J. Simon, *Topology of manifolds*, Georgia Conf., 1969], [Gonzalez-Acuña]; some *pretzel knots* [Simon], [Riley]; some *2-bridge knots* [Riley], [E. J. Mayland, *A class of 2-bridge knots with Property* P, preprint]; and other knots [Y. Nakagawa, Publ. Res. Inst. Math. Sci., vol. 10, no. 2, 1975], [J. P. Neuzil, Trans. Amer. Math. Soc. **204** (1975), 385–406], [J. Simon, Trans. Amer. Math. Soc. **160** (1971), 467–473], [R. A. Litherland, *Surgery on knots in solid tori*, Cambridge Univ., preprint].

Problem 1.16. Does every nontrivial knot K have Property R, that is, does surgery on K with framing 0 always give a manifold other than $S^1 \times S^2$, as expected?

REMARKS. If surgery gives $S^1 \times S^2$, then K has trivial Alexander polynomial, is prime, is not a doubled knot [L. Moser, Pacific J. Math. **53** (1974), 519–523], is a slice of an unknotted S^2 in S^4 [R. Kirby and P. Melvin, *Slice knots and Property* R, and Addendum by Freedman and Taylor, Invent. Math.], and does not have a genus one unknotted Seifert surface [H. W. Lambert, *Longitude surgery on genus* 1 *knots*, Proc. Amer. Math. Soc. **63** (1977), 359–362].

Does 0-surgery on K give a manifold which is not even a homotopy $S^1 \times S^2$? This is equivalent, using infinite cyclic covers, to the question: If G is a knot group and $l \in [G, G]$ is the longitude, is $[G, G]$ not normally generated by l? (Note that the normal closure of l in G coincides with the normal closure of l in $[G, G]$, since l commutes with a meridian, which generates $G/[G, G] = Z$.) Also, see Problem 5.7.

A homology 3-sphere contains a knot K whose 0-surgery produces $S^1 \times S^2$ iff the homology 3-sphere is the boundary of a 4-manifold made with a 0-handle, a 1-handle and a 2-handle. For these homology 3-spheres the appropriate conjecture is that 0-surgery on K alone produces $S^1 \times S^2$.

Problem 1.17 (*R. Edwards, after F. Laudenbach and V. Poenaru*). Suppose K is a nontrivial knot in S^3 with longitude l, tubular neighborhood N, and group

$G=\pi_1(S^3-K)$. It is an appealing conjecture, made by Poenaru (and others?), that

(A) (algebraic version): $l \in G$ cannot be a product of conjugates of itself, with zero exponent sum; that is, there do not exist $a_1, \cdots, a_{2n} \in G$ such that $l = a_1^{-1} l a_1 a_2^{-1} l^{-1} a_2 \cdots a_{2n}^{-1} l^{-1} a_{2n}$ (the exponents of l alternate here only for convenience).

(B) (geometric version): There cannot be an imbedding of a sphere-with-$(2n+1)$-holes F^2 in $S^3 - \text{int } N$, with the $2n+1$ boundary components of F imbedded onto parallel copies of l in ∂N so that their algebraic sum in ∂N is l.

REMARKS (i) Clearly (A) $\Rightarrow$ (B). Conversely, (B) $\Rightarrow$ (A), for given that (A) fails, there is a *map* $f\colon (F^2, \partial F^2) \to (S^3 - \text{int } N, \partial N)$, imbedding ∂F^2 onto $2n+1$ copies of l with algebraic sum l in ∂N. Let M_K denote the 3-manifold obtained by doing 0-framed surgery on S^3 along K. Apply the Sphere Theorem to the capped-off map f taking S^2 into M_K, to get an imbedded, nonseparating S^2 in M_K. Such an S^2 must represent a generator of $H_2(M_K) = Z$, and so it provides, after puncturing, the desired geometric surface F^2 of (B).

(ii) If l normally generates $[G, G]$ (cf. previous problem), then (A) is false, as one sees by writing $l \in [[G, G], [G, G]]$ as $l = \Pi_j [c_j, d_j]$, $c_j, d_j \in [G, G]$, and then writing each c_j and d_j as a product of conjugates of l. In other words, if the Poenaru Conjecture holds for a knot K, then K has homotopy Property R.

(iii) This question is related to the previous one, for if M_K is homotopy-equivalent to $S^1 \times S^2$ (M_K as above), then K bounds in S^3 the punctured sphere F^2 as in (B). (Is the converse true? Cf. the final remarks of Problem 5.7.) In the study of 4-manifolds, this question arises from the following question: Suppose W^4 is a Mazur-like contractible 4-manifold constructed by attaching a 2-handle to $S^1 \times B^3$ along a degree 1 curve Γ in $S^1 \times \partial B^3$, and suppose $\partial W^4 = S^3$. Is Γ necessarily unknotted in $S^1 \times S^2$?

(iv) F. Laudenbach has shown that the conjecture is true for $n = 1$.

Problem 1.18 (*J. Martin*). (A) Suppose T_1 and T_2 are solid tori with $T_2 \subset \text{int } T_1$, such that the wrapping number of T_2 in T_1 is nonzero, and the winding number of T_2 in T_1 is zero. Can T_2 be removed and sewn back differently (Dehn surgery) so that theresult is still a solid torus?

REMARKS. Winding number zero means the homomorphism $H_1(T_2) \to H_1(T_1)$ is zero; nonzero wrapping number means that T_2 does not lie in a 3-ball in T_1. An example of such a T_2 would provide a knot without Property P (Problem 1.15).

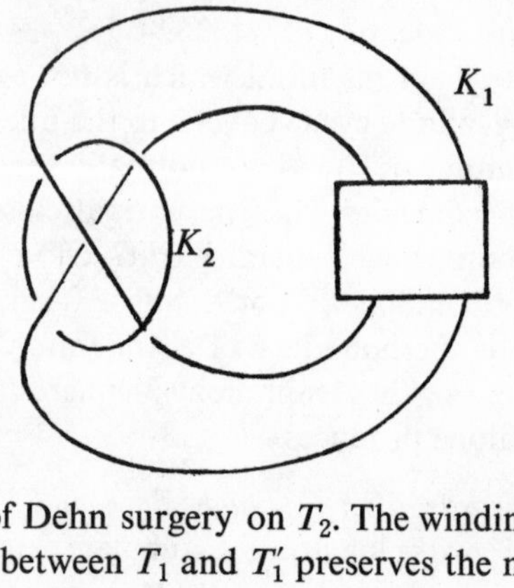

(For let T_1' be the result of Dehn surgery on T_2. The winding number zero implies that the diffeomorphism between T_1 and T_1' preserves the meridian. Imbed T_1 in a knotted fashion in S^3 so that T_2 is knotted in S^3. Then Dehn surgery on T_2 gives S^3.)

(B) Let K_1 and K_2 be unknots in S^3 as drawn above, and assume that K_1 and K_2 are geometrically linked but algebraically unlinked. Can an example be found so that if the overcrossing in K_1 is changed to an undercrossing, then K_1 remains unknotted?

REMARKS. An example for (B) provides one for (A). For $+1$-surgery on K_2 changes the crossing in K_1. If K_1 is unknotted in both cases, then choose $T_1 = S^3 - K_1$ and $T_2 = K_2 \times B^2$, satisfying (A).

(C) Let T be the solid torus and imagine T standardly imbedded in each of S^3 and $S^1 \times S^2$. Is there a simple closed curve J in T such that

(1) J does not bound a disk in T, and

(2) J bounds a disk in each of S^3 and $S^1 \times S^2$?

REMARKS. If so, there is a knot failing Property P or Property R. For perform $+1$-surgery on J in T and get either (i) a solid torus, or (ii) a cube with a knotted hole (= knot complement). If (i), proceed as in (A). If (ii) then $S^1 \times S^2$ is the union of a solid torus and a cube with a knotted hole; that is, it is the result of 0-surgery on the knot making the knotted hole.

Problem 1.19 (*S. Akbulut and R. Kirby*). *Conjecture*: If 0-framed surgery on two knots gives the same 3-manifold, then the knots are concordant.

REMARKS. This is true if one knot is the unknot (Kirby-Melvin, see Problem 1.16). If homotopy 4-spheres are spheres, then it is true if one knot is slice. In general all known concordance invariants of the two knots are the same; this is true even if we assume only that the 0-surgeries give homology cobordant 3-manifolds.

Problem 1.20 (*Giffen and Siebenmann*). A Seifert surface F for a knot K in S^3 is called free if $\pi_1(S^3 - F)$ is free (equivalently, $S^3 - F$ is an open handlebody). The construction in [Fox, *A quick trip...*, Topology of 3-Manifolds, Prentice-Hall, 1962, p. 140] yields such a free Seifert surface.

(A) What is the smallest genus among free Seifert surfaces of K? Call this the free genus. Relate the free genus to other invariants of knots. Note that the free genus seems to be arbitrarily large for genus 1 knots; consider untwisted doubles.

(B) Which knots bound an incompressible free Seifert surface? Fibered knots do. Some knots do not [H. C. Lyon, Proc. Amer. Math. Soc. **35** (1972), 617–620].

(C) Trotter gives examples of nonunique free incompressible Seifert surfaces [Ann. of Math. Studies, no. 84, Princeton Univ. Press, Princeton, N.J., 1975, pp. 51–62]. Are these examples nonunique (up to isotopy?) when pushed into B^4?

Problem 1.21 (*C. McA. Gordon*). Let L be a nontrivial link in S^3 with two unknotted components C_1, C_2, and linking number zero. Take the k-fold branched cover over C_1. *Conjecture*: The k lifts of C_2 form a nontrivial link.

REMARK. If it is trivial, then there are many counterexamples to the Smith conjecture (Problem 3.38) for Z/k actions [Gordon, *Uncountably many stably trivial strings in codimension two*, Quart. J. Math., 1977].

Problem 1.22 (*Montesinos*). Find a set of moves on links of S^3 so that two links have the same 2-fold branched covering space iff it is possible to pass from one link to the other using this set of moves.

Problem 1.23 (*J. Montesinos*). If a knot $K \subset S^3$ is amphicheiral, then the 2-fold covering space branched over K is symmetric (has an orientation reversing diffeomorphism). Is the converse true? (*Conjecture*: No.)

Problem 1.24 (*Fox and Perko*). Does every simple 4-fold branched cover of a knot K have precisely three distinct branch curves?

REMARKS. A simple 4-fold branched cover corresponds to a representation of $\pi_1(S^3 - K)$ onto S_4 with meridians going to elements of order 2. Every orientable, closed 3-manifold is a 3-fold (irregular) branched cover of S^3 over some K, and that representation of $\pi_1(S^3 - K)$ onto $D_3 = S_3$ lifts to a representation onto S_4 which is simple or takes meridians to elements of order 4 [K. Perko, Ann. of Math. Studies, no. 84, Princeton Univ. Press, Princeton, N.J., 1975, pp. 47–50].

Problem 1.25 (*Cappell and Shaneson*). Let M_α be an irregular p-fold dihedral cover of a knot α. Let $\alpha_0, \alpha_1, \cdots, \alpha_r$, $r = (p-1)/2$, be the branching curves in M_α where α_0 has branching index 1. Let $\nu_{i0} = l(\alpha_i, \alpha_0)$. Prove that $\nu_{i0} \equiv 2(4)$ if M_α is a $Z/2$-homology sphere.

REMARKS. For 2-bridge knots, $\nu_{i0} \equiv 2(4)$ [K. Perko, Invent. Math. **34** (1976), 81]. Cappell and Shaneson have shown that $\sum_{i=1}^{r} \nu_{i,0} \equiv (p-1)\ (4)$ for a $Z/2$-homology sphere. Consequently, $\nu_{i,0} \equiv 2\ (4)$ if $p \equiv 3\ (4)$ and if 2 generates $Z_p^*/\{\pm 1\}$. For $p = 3$, the formula reduces to $\nu_{10} \equiv 2\ (4)$ [Cappell and Shaneson, Bull. Amer. Math. Soc. **81** (1975), 559–561].

Problem 1.26 (*K. Murasugi*). Suppose the first homology group of the 2-fold cyclic branched cover of a knot $\alpha \subset S^3$ is Z/p (hence $p = |\Delta_\alpha(-1)|$), and let M_α be the irregular p-fold dihedral cover of α. *Conjecture*: If M_α is a Z-homology sphere, then $\sum_{i=1}^{r} \nu_{i,0} \equiv \sigma(\alpha)$ (mod 8) where $\sigma(\alpha)$ is the signature of α and $\nu_{i,0}$ is defined above.

REMARK. The conjecture holds with *equality* for 2-bridge knots ([Hartley-Murasugi, Canad. J. Math., 1977], also see [K. A. Perko, Invent. Math. **34** (1976), 77–82]).

Problem 1.27 (*D. Goldsmith*). Do there exist distinct prime knots K and K' in S^3 all of whose cyclic branched covers are homeomorphic?

Problem 1.28 (*D. Goldsmith*). Let $M^3 \to^\pi S^3$ be an n-fold cyclic branched cover of S^3 along a knot K. Let A be an unknot in $S^3 - K$. If K is a closed braid about A, then $\pi^{-1}(A)$ is a fibered knot in M^3. Is the converse true?

Problem 1.29 (*Cappell and Shaneson*). Is every closed, oriented 3-manifold the dihedral covering space of a ribbon knot?

Problem 1.30 (*Cappell and Shaneson*). Are the classical PL and TOP knot concordance groups the same?

REMARKS. Clearly $C^{\mathrm{PL}} \to C^{\mathrm{TOP}}$ is onto. This question may be easier than the hauptvermutung for $B^2 \times R^2$.

Problem 1.31 (*Y. Matsumoto*). Let $\mathcal{H}_A = \{$all knots in all homology 3-spheres which bound PL acyclic 4-manifolds, modulo homology cobordism of pairs$\}$. Is the natural map $C^{\mathrm{PL}} \to \mathcal{H}_A$ an isomorphism?

Problem 1.32 (*C. McA. Gordon*). Does the classical knot concordance group contain any nontrivial elements of finite order other than 2?

Problem 1.33. If K is a slice knot, is K a ribbon knot? (See Problem 4.22.)

Problem 1.34 (*A. Casson*). Find an algorithm for determining whether a knot is slice or ribbon.

REMARK. There is an algorithm for the (harder?) problem of determining whether a knot is unknotted [W. Haken, Acta Math. **105** (1961), 245–375].

Problem 1.35 (*Cappell and Shaneson*). The μ-invariant formula of Cappell-Shaneson [Bull. Amer. Math. Soc. **81** (1975), 559–561] detects nonribbon knots. Does it detect nonslice knots as well? Relate this to the Casson-Gordon invariant.

Problem 1.36 (*L. Taylor*). If a knot has Alexander polynomial equal to one, is it a slice knot?

REMARK. Any such knot is algebraically slice, i.e., there is a basis so that the Seifert matrix has the form $\begin{pmatrix}0 & A\\ B & C\end{pmatrix}$ (L. Taylor). These knots have no metacyclic covers, so the Casson-Gordon method does not apply.

Here is a possible generalization to links. A boundary link is a link L whose components bound disjoint imbedded surfaces $F_1, \cdots, F_r$ in S^3. Note that if a knot has a Seifert matrix of the form $\begin{pmatrix}0 & 0\\ B & C\end{pmatrix}$, then its Alexander polynomial is one. Hence, define a *good* boundary link to be one for which there is a summand $A_i \subset H_1(F_i; Z)$ such that $2 \dim A_i = \dim H_1(F_i; Z)$ and the intersection of every element of A_i and every other element of $H_1(F_j; Z)$ is zero for all i, j.

Question. Is a good boundary link slice?

Note that a slice link is not necessarily a boundary link [N. Smythe, Ann. of Math. Studies, no. 60, Princeton Univ. Press, Princeton, N.J., 1966, pp. 69–72].

Problem 1.37 (*A. Casson*). (A) The knot

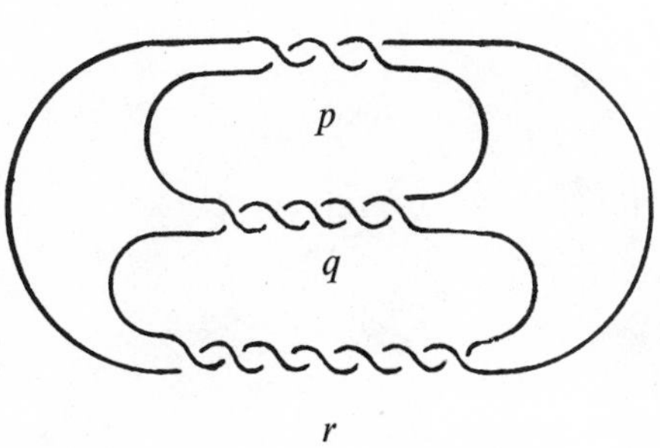

($p = -3, q = 5, r = 7$ in illustration) has Alexander polynomial 1 if p, q, r are odd and $qr + rp + pq = -1$. Is it slice?

(B) The double branched covering of this knot is the Brieskorn homology sphere $\Sigma(|p|, |q|, |r|)$. Does it bound a homology ball?

REMARKS. An affirmative answer to (A) implies that $\Sigma(p, q, r)$ bounds a $Z/2$-homology ball.

If the Brieskorn sphere $\Sigma(|2bc + 1|, |2a(b - d) + 1|, |2d(c - a) + 1|)$ bounds a homology ball for some numbers a, b, c, d with $ad - bc = 1$, then the homology class (a, b) in some homology $S^2 \times S^2$ is representable by an imbedded S^2. For example, if $\Sigma(3, 5, 7)$ bounds, then $(2, 3)$ is representable. (See Problem 1.42.)

Problem 1.38 *Conjecture* (unlikely): The untwisted double of a knot is slice $\Leftrightarrow$ the knot is slice.

Problem 1.39 (*A. Casson*). Drawn below is the Whitehead link and an untwisted double of the Whitehead link.

This construction can be iterated by replacing

 by  or  ;

call the nth iterate W_n. Is any W_n null-concordant?

REMARKS. If so, then there exists a proper homotopy $S^3 \times R$ which is not $S^3 \times R$. If not, there exists an end of a 4-manifold, $Q \cong^P S^2 \times S^2 - \text{pt}$, which is fake (A. Casson). It is also interesting to know if W_n is null-concordant in some contractible 4-manifold.

The manifold Q can be described as follows. Let $X \subset S^1 \times \text{int}B^2$ be the continuum of Whitehead [see his Works, Volume II, various papers at the beginning], defined as $X = \bigcap_{i>0} \alpha^i(S^1 \times B^2)$, where $\alpha: S^1 \times B^2 \to S^1 \times B^2$ is the imbedding shown below (α should have 0-twisting about its core):

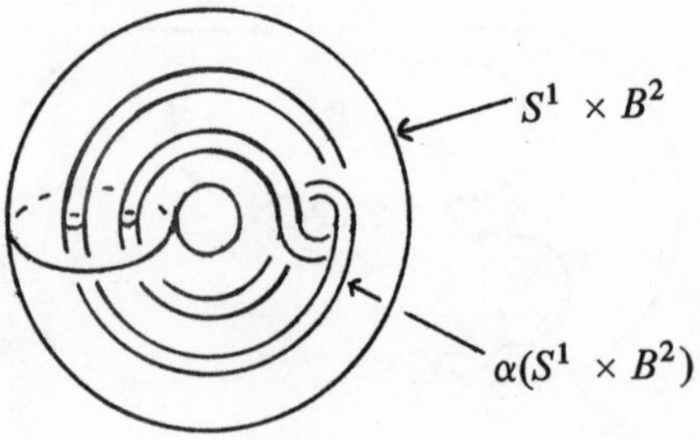

Regarding $S^2 \times S^2$ as the union of a 0-handle B_0^4, two 2-handles $B_1^2 \times D^2$, $B_2^2 \times D^2$ and a 4-handle B_3^4, define a compact subset $C = B_3^4 \cup cX_1 \cup cX_2$, where each X_i is a Whitehead continuum constructed in int $B_i^2 \times \partial D^2 \subset \partial B_3^4$, the *cocore* attaching tube of the ith 2-handle, and where X_i is coned off to the center of the 2-handle $B_i^2 \times D^2$. Then $Q = S^2 \times S^2 - C$.

The open subset $B_1^2 \times \text{int}D^2 - cX_i$ of the open 2-handle $B_i^2 \times \text{int}\, D^2$ is called a *flexible* 2-handle by Casson; this is the simplest such. (The others are constructed by replacing X by a Whitehead-like continuum gotten by ramifying at each stage the original Whitehead embedding α.) Any such flexible 2-handle is proper homotopy equivalent, rel ∂, to the standard open 2-handle $B^2 \times \text{int } D^2$. One fundamental question is whether this can ever be a diffeomorphism. The manifold Q can be regarded as the union of an open 0-handle int B_0^4 and two flexible 2-handles.

The compact subset $C = S^2 \times S^2 - Q$ is cell-like (i.e., homotopic to a point in any arbitrarily small neighborhood) and satisfies the cellularity criterion (i.e., Q is 1-connected at ∞; see [D. R. McMillan, Ann. of Math. (2) **79** (1964), 327–337]). However, C is smoothly cellular ($\equiv$ the intersection of a nested sequence of smooth 4-balls) $\Leftrightarrow$ some W_n is null-concordant.

Problem 1.40. (A) Let K be a torus knot. Is the genus of K equal to the 4-ball genus of K?

REMARKS. The 4-ball genus is the minimal genus of a bounding surface in B^4. The genera are equal for $(2, q)$-torus knots. The genus of the (p, q)-torus knot is $(p-1)(q-1)/2$. Since $n \in H_2(CP^2; Z) = Z$ is represented by an imbedded 2-sphere smooth except for a cone on the $(n-1, n)$-torus knot, the Remark in Problem 4.36 gives a lower bound for the 4-ball genus of the $(n-1, n)$-torus knot.

(B) L. Rudolph has shown that every link in S^3 is isotopic in S^3 to one which bounds a nonsingular complex analytic curve in $B^4 \subset C^2$. What can be said about the minimal genus of such a curve?

Problem 1.41. (A) (Akbulut and Kirby) Let M_K^4 be constructed by adding a 2-handle to B^4 along a knot K with the 0 framing. Define the 0-shake genus of K to be the minimal genus of a smooth, imbedded surface in M_K representing the generator of $H_2(M_K)$. Does the 0-shake genus equal the 4-ball genus of K? Probably not.

REMARK. The r-shake genus of K (obvious definition) can be less than the 4-ball genus, for $r \neq 0$ [S. Akbulut, *On 2-dimensional homology classes of 4-manifolds*, Proc. Cambridge Philos. Soc. **82** (1977), 99–106].

(B) (Giffen) What is the minimal genus of a smooth, imbedded surface in ∂M_K representing the generator of $H_2(\partial M_K)$? *Conjecture*: It equals the genus of K.

REMARKS (L. TAYLOR). The conjecture is true for fibered knots (even in homology spheres); however, the conjecture fails in general for knots in homology spheres. Perhaps it would be easier to show that the Seifert surface of K union the 2-handle is incompressible in ∂M_K. If so, or if the conjecture holds, then every knot has Property R (Problem 1.16).

Problem 1.42 (*Y. Matsumoto*). Does the following link in S^3 bound a smooth punctured sphere in B^4? If so, $(2, 3) \in H_2(S^2 \times S^2; Z)$ is represented by a smooth S^2. Can it be represented by a torus?

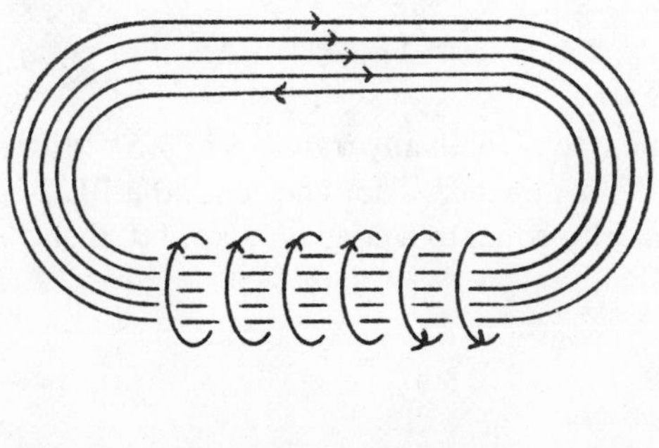

REMARKS. This is the simplest unsolved case one encounters in trying to represent $(2, 3) \in H_2(S^2 \times S^2; Z)$ by a smooth imbedded S^2. This can be done iff such a link as above, with $2 + 2k$ circles in one group and $3 + 2l$ circles in the other group oriented to give (2, 3), bounds a smooth, imbedded punctured S^2 in B^4. The above is the case $k = 2$, $l = 1$. The simpler cases are ruled out by the Murasugi-Tristram inequality

$$|\sigma_p| + |\eta_p| \leqq \mu - 1$$

where σ_p is the p-signature of the link, η_p is the nullity ($\equiv$ dim-rank of the associated form) and μ is the number of components of the link. For $p=2$ we get $|6| + |2+2l| \leqq |5 + 2k + 2l - 1|$, which rules out the cases $k \leqq l$, l arbitrary. For $p = 3$ we get $|5| + |1 + 2k| \leqq |5 + 2k + 2l - 1|$, ruling out k arbitrary, $l = 0$.

Problem 1.43 (*M. Scharlemann*). Are there knots f: $S^1 \to S^3$ such that for any locally flat concordance F: $S^1 \times I \to S^3 \times I$ the map $\pi_1(S^3 - f(S^1)) \to \pi_1(S^3 \times I - F(S^1 \times I))$ is injective? *Conjecture*: This is true for torus knots.

REMARK. This is true for torus knots if F must be a fibered concordance.

Problem 1.44 (*L. Kauffman*). Does link concordance imply link homotopy? (Added in proof, April 1, 1977; Yes, Giffen and (independently) Goldsmith.)

The following definitions are used in the next three problems. Given a knot K in S^3, an "algebraically-one strand" is a way of imbedding an unknotted $T = S^1 \times B^2$ in S^3, containing K, so that K and $p \times \partial B^2$, $p \in S^1$, link algebraically once. Thus K goes algebraically once around T. There is a similar definition for algebraically-l strand.

A (k, l)-twist on K is obtained by taking some algebraically-l strand, i.e., some T, and twisting T k full times in a right-handed direction around $S^1 \times 0$.

Problem 1.45 (*S. Akbulut*). *Conjecture*: There exist a knot K and an algebraically-one strand such that no matter what knot is tied in the strand (in T), the new knot is not slice in any homotopy 4-ball (with $\partial = S^3$).

REMARK. If the conjecture is true, there exists a knot in the boundary of a contractible 4-manifold which does not bound an imbedded PL 2 ball (see Problem 4.21). Specifically, suppose K and T satisfy the conjecture.

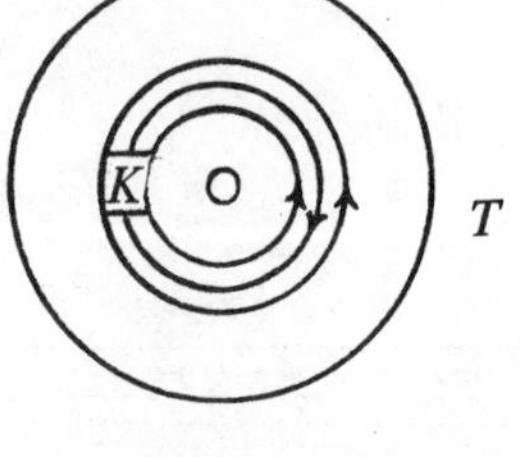

Then if we add a 2-handle to K (with any framing) in $\partial(S^1 \times B^3)$, we get a contractible manifold W and $S^1 \times p$, $p \in \partial B^3$, does not bound a PL 2-ball. For, if there is a PL 2-ball, with singularity equal to a cone on a knot J, then K with J tied in it is slice in a homotopy 4-ball.

Problem 1.46 (*S. Akbulut and R. Kirby*). *Conjecture*: Given a knot K with Arf invariant zero, there is a $(\pm 1, \pm 1)$-twist changing K into:

(A) An algebraically slice knot (Seifert matrix concordant to zero). Very likely true.

(B) A slice or ribbon knot. Perhaps.

(C) The unknot. Surely false.

There is a better chance for the conjecture that K is concordant to a knot K' which, after a $(\pm 1, \pm 1)$-twist, satisfies (A) or (B) or (C).

REMARK. This problem is motivated by: if every homotopy CP^2 is diffeomorphic to CP^2, then ± 1 surgery on K gives a homology 3-sphere bounding an acyclic 4-manifold W with $\pi_1(\partial W) \twoheadrightarrow \pi_1(W)$, iff K is concordant to K' and there is a $(\pm 1, \pm 1)$-twist on K' giving a ribbon knot.

Problem 1.47 (*A. Casson*). Given any Arf invariant zero knot K, is it possible to change K to the unknot by a series of $(1, \pm 1)$-twists on K?

REMARKS (A. CASSON). Yes, if K is ribbon or slice (since we can connect sum with a ribbon knot and get a ribbon knot). Any K can be reduced to a connected sum of granny knots. If the granny knot could be changed to the unknot with 17 $(1, \pm 1)$-twists, then there exists an even, index 16, $\beta_2 = 18$ closed 4-manifold (in $17(-CP^2)$ $\#2\, CP^2$, try to represent $(1, 1, \cdots, 1, 3, 3)$ by a smooth imbedded S^2; in $\#2\, CP^2$, $(3, 3)$ is represented by PL sphere with a singularity equal to a cone on the granny knot, and the $17 - CP^2$'s allow 17 $(1, \pm 1)$-twists to unknot the granny knot).

Problem 1.48 (*J. Levine*). A general question is what groups π are fundamental groups of the complement of some knotted S^2 in S^4?

Recall that a group π has weight 1 if it is normally generated by one element, and deficiency one if it has a presentation $\{x_1, \cdots, x_n, t \colon R_1, \cdots, R_n\}$ with one more generator than relation.

(1) Given π such that $H_1(\pi) = Z$, π has weight one and deficiency one, then π is the group of an S^2 in a homotopy 4-sphere [Kervaire, Bull. Soc. Math. France **93** (1965), 225–271, see p. 233]. Which of these are realizable by knots in S^4? π is realizable if the induced presentation of the trivial group defined by setting $t = 1$ is trivializable by Andrews-Curtis moves.

(2) Let $\Lambda = Z[t, t^{-1}]$ and let the Λ-module A of an $S^2 \to S^4$ be π'/π'' with the induced action of $\pi/\pi' = Z$. Which Λ-modules are realizable? If A is Z-torsion free (implied by deficiency one) the answer is known since there are enough deficiency one π to get all such A's.

Problem 1.49 (*S. Lomonaco*). Does there exist a smooth, prime S^2 in S^4 such that the deficiency of the fundamental group of its complement is < 0?

REMARKS. The deficiency is always $\leqq 1$ because the complement is a homology S^1. E. Artin [Abh. Math. Sem. Univ. Hamburg **4** (1925), 174–177] constructed knots of deficiency one by spinning, and Fox constructed a knot of deficiency 0 in [*A quick trip* $\cdots$, Topology of 3-manifolds, Prentice-Hall, Englewood Cliffs, N.J., 1962]. J. Levine and C. Giffen have recently found many nonprime knots of arbitrarily large negative deficiency, by connect-summing arbitrarily many knots of deficiency zero.

Problem 1.50 (*Gordon*). Can a branched cyclic cover of a (locally flat) knot $S^n \to S^{n+2}$ ever be a $K(\pi, 1)$, for $n \geqq 2$?

Problem 1.51 (*Kawauchi*). Suppose a knotted 2-sphere Σ^2 in S^4 is of Dehn's type (i.e., the homomorphism $\pi_2(\partial N) \to \pi_2(S^4 - \Sigma^2)$ is trivial where N is a tubular neighborhood of Σ^2). Is Σ^2 algebraically unknotted (i.e., is $S^4 - \Sigma^2 \simeq S^1$)? Geometrically unknotted?

REMARKS. The infinite cyclic cover of $S^4 - \Sigma^2$ is acyclic; if in addition $\pi_3(S^4 - \Sigma^2) = 0$ is satisfied, then Σ^2 is algebraically unknotted [A. Kawauchi, *A partial Poincaré duality theorem for topological infinite cyclic coverings and applications to higher dimensional knots*, Osaka J. Math., 1977]. If there exists an algebraically knotted Σ^2 of Dehn's type, it would have infinitely many ends.

2. Surfaces. The following problems use these definitions: F_g is a closed, orientable surface of genus $g \geqq 2$, which bounds a handlebody N_g ($= B^3 \cup g$ 1-handles); $\mathcal{M}_g$ (the mapping class group) is the group of isotopy classes of orientation preserving homeomorphisms of F_g to F_g; $\mathrm{Sp}(2g, Z)$ is the group of $2g \times 2g$ symplectic integer matrices; $\lambda: \mathcal{M}_g \to \mathrm{Sp}(2g, Z)$ is the natural homomorphism mapping each element of $\mathcal{M}_g$ to the induced automorphism on $H_1(F_g; Z)$; κ_g = kernel λ.

Problem 2.1 (*Birman*). (A) Is κ_g finitely generated? (*Conjecture*: No.)

(B) It is known [J. Powell, *Two theorems on the mapping class group of a surface*, Proc. Amer. Math. Soc.] that κ_g is generated by isotopy classes of

(i) Dehn twists about separating curves, and

(ii) if $g \geqq 3$, Dehn twists $t_c t_{c'}^{-1}$ about pairs c, c', of homologous, disjoing, nonseparating curves.

Is the subgroup of κ_g which is generated by maps of type (i) of finite index in κ_g for any $g \geqq 3$? (For $g = 2$, it is the full group.)

REMARK. Chillingworth [Math. Ann. **199** (1972), 152] gives examples of homeomorphisms in κ_g not generated by type (i) maps.

Problem 2.2 (*Birman*). Find explicit representations of κ_g or $\mathcal{M}_g$ which do not factor through the homomorphism λ.

REMARKS. Since $\mathcal{M}_g$ is residually finite [E. Grossman, J. London Math. Soc. **9** (1974), 160–164], such representations surely exist. In particular, κ_g has nontrivial homomorphisms onto $Z/2$, but these representations are not straightforward [J. Birman and R. Craggs, *The μ-invariant of 3-manifolds, and certain structural properties of the group of homeomorphisms of a closed, oriented 2-manifold*, Trans. Amer. Math. Soc. **237** (1978), 283–309].

Problem 2.3 (*D. Johnson*). If $h: F_g \to F_g$ is a homeomorphism with $h_* = \mathrm{id}$ on $H_1(F_g)$, does h extend over some 3-manifold?

REMARK. For $g = 1$, h extends iff trace $h_* = 2$ iff h_* is conjugate to $\left(\begin{smallmatrix} 1 & 0 \\ k & 1 \end{smallmatrix}\right)$ iff h is conjugate to a k-fold Dehn twist around some curve.

Problem 2.4 (*Birman*). Let α be the obvious homomorphism

$$\mathcal{N}_g \xrightarrow{\alpha} \mathrm{Aut}(\pi_1(N_g))$$

where $\mathcal{N}_g$ is the group of isotopy classes of orientation preserving homeomorphisms of N_g. Is kernel(α) finitely generated? (*Conjecture*: No.)

REMARKS. $\mathcal{N}_g$ is known to be finitely generated [Suzuki, *On homeomorphisms of a 3-dimensional handlebody*, Canad. J. Math. **29** (1977), 111–124]. E. Luft has proved that kernel(α) is generated by Dehn twists on properly imbedded 2-balls in N_g.

Problem 2.5 (*Birman*). (A) Suppose that F_g is imbedded in S^3 as a Heegard surface, with $S^3 = U \cup V$ and $U \cap V = F_g$. Let A (resp. B) be the subgroup of $\mathcal{M}_g$ of maps which extend over U (resp. V). Let C be the centralizer of $A \cap B$ in A. Can we express every element $\alpha \in A$ as $\alpha = \gamma\delta$ where $\gamma \in C$ and $\delta \in A \cap B$?

(B) Suppose $\mathcal{P}$, $\mathcal{P}' \subset A \cap B$ have finite order p. Suppose also that $\mathcal{P}' = \alpha\mathcal{P}\alpha^{-1} = \beta\mathcal{P}\beta^{-1}$ for some $\alpha \in A$, $\beta \in B$. *Conjecture*: There exists $\delta \in A \cap B$ such that $\mathcal{P}' = \delta\mathcal{P}\delta^{-1}$.

REMARKS. The conjecture is true for $p = 2$. A positive answer for any $p \geqq 3$ would imply the Smith conjecture (see Problem 3.28) for period p (Birman). Note that an affirmative answer to part (A) would imply an affirmative answer to the conjecture for every p.

Problem 2.6 (*W. Thurston*). Can the group of symmetries of $\pi_1(F) = \pi_1(F_g)$ be represented geometrically as a group of homeomorphisms of F? More precisely, since

$$\pi_0(\text{Homeo } F) \cong \text{Aut}(\pi_1 F)/\text{Inn Aut}(\pi_1 F) = \text{Out Aut}(\pi_1 F),$$

is there a right inverse σ for π_0:

$$\text{Homeo}(F) \underset{\sigma}{\overset{\pi_0}{\rightleftarrows}} \pi_0\,(\text{Homeo}(F)), \qquad \pi_0\sigma = \text{id}.$$

REMARK. If we replace F by its unit tangent bundle TF, this geometric representation is possible, i.e., there is a natural representation $\pi_0(\text{Homeo } F) \to \text{Homeo } TF$ (consider the Cheeger homeomorphism in [Gromov, *Three remarks on geodesic flow*, preprint]).

The problem is also interesting (and well known) for subgroups (particularly finite) of $\pi_0(\text{Homeo } F)$. It is known for solvable, finite groups and for infinite groups with two ends.

Problem 2.7 (*W. Thurston*). Characterize the subgroups of $\pi_0(\text{Diff } F)$ which act as translations of some complex geodesic in Teichmüller space.

Problem 2.8 (*W. Meeks*). For what genus g does every periodic diffeomorphism $h: F_g \to F_g$ have an invariant circle?

REMARKS. Yes for $g \leqq 10$, no for $g = 11$, yes for some $g > 11$, and no for an infinite number of g. For $g = 11$, there is "essentially" one diffeomorphism h, of order 30, with no invariant circle [W. Meeks, *Existence of circles invariant under diffeomorphisms of finite order*].

3. 3-manifolds. There are problems in John Hempel's book [*3-manifolds*, Ann. of Math. Studies, no. 86, Princeton Univ. Press, Princeton, N.J., 1976] in the adjacent list by Waldhausen in these PROCEEDINGS, and in the lists of Fox, Neuwirth and Birman (§1).

Problem 3.1 (*Poincaré*). *Conjecture*: Every homotopy 3-sphere Σ^3 is homeomorphic to S^3.

Here are some presumably easier conjectures:

(A) The suspension of Σ^3 is homeomorphic to S^4.

REMARK. The double suspension is S^5 [Siebenmann, *Topology of manifolds*, Cantrell and Edwards, ed., Markham, 1970, p. 77].

(B) $(\Sigma^3 - \text{point}) \times R$ is diffeomorphic to R^4.

(C) (Poenaru) There exist two contractible open subsets, U_1 and U_2, of Σ^3 such that $\Sigma^3 = U_1 \cup U_2$ and U_1 and U_2 are imbeddable in R^3 (see [Poenaru, Bull. Amer. Math. Soc. **80** (1974), 1203–1204]). (Added in proof. This is equivalent to the Poincaré conjecture [D. R. McMillan, Jr., Bull. Amer. Math. Soc. **76** (1970), 942–964].)

(D) Σ^3 imbeds (smoothly?) in S^4. This implies

(D′) Σ^3 bounds a contractible 4-manifold, and, if the imbedding is smooth,

(D″) Σ^3 does not bound a smooth, almost parallelizable 4-manifold of index 8 mod 16.

(E) If Σ^3 admits an involution (and is irreducible), then Σ^3 is S^3.

(F) The connected binding of some open book decomposition (Problem 3.13) of Σ^3 lies in a 3-ball. (This is equivalent to the Poincaré conjecture.)

Furthermore, the Poincaré conjecture generalizes to $\pi(\Omega)$ simple homotopy equivalent closed, orientable, 3-manifolds are homeomorphic.

REMARKS. This is known for irreducible, sufficiently large manifolds [Waldhausen, Ann. of Math. (2) **87** (1968), 56–88], and for lens spaces [M. M. Cohen, *A course in simple homotopy theory*, Springer-Verlag, Berlin, 1973].

Problem 3.2. Let M^3 be a closed, orientable, irreducible 3-manifold with infinite fundamental group. If M^3 is not sufficiently large, must it have a finite cover which is sufficiently large?

Problem 3.3 (*Jaco*). Let M^3 be a compact 3-manifold. Are there at most a finite number of homotopy equivalent 3-manifolds? What if we restrict to the case where M is orientable and sufficiently large? (See [Jaco and Shalen, Invent. Math. **38** (1976), 55–88].)

Problem 3.4 (*J. Stallings*). Is an irreducible h-cobordism from RP^2 to itself a product?

REMARKS. This is known if its double cover is $S^2 \times I$ [G. R. Livesay, Ann. of Math. (2) **78** (1963), 582–593], and [J. H. Rubenstein, Proc. Amer. Math. Soc. **60** (1976), 317–320]. The open case is also unknown: Is a proper homotopy $RP^2 \times R$ homeomorphic to $RP^2 \times R$ (assume irreducibility of the manifold)? An affirmative answer implies that every topological involution of S^3 with two fixed points is standard.

Problem 3.5. Suppose M^3 is irreducible and $\pi_1 M^3$ is infinite with nontrivial center. (A) (W. Thurston) Is M^3 a Seifert fiber space?

REMARKS. Yes if M is sufficiently large [Waldhausen, Topology **6** (1967), 505–517]. If yes in general, then we get an affirmative answer to Problem 2.6 for finite groups. For let G be a finite subgroup of $\pi_0(\text{Homeo } F^2)$. G acts freely on the unit tangent space TF^2 with quotient space M^3 satisfying the conditions above. If M^3 is a Seifert fiber space, there is a natural, regular, branched covering of its base space with G as the group of deck transformations; the branched covering space is F^2.

(B) (W. Jaco) Is the center of $\pi_1 M^3$ finitely generated?

REMARK. Yes, if $\pi_1 M^3$ contains a (closed) surface group.

Problem 3.6. (A) Is the homology 3-sphere obtained by ± 1-surgery on a knot always prime?

(B) If so, do they all have finite covers which are either sufficiently large or S^3?

REMARK. This is a special case of Problem 3.2 if π_1 is infinite.

(C) Find a prime homology 3-sphere which is not obtained by ± 1-surgery on a knot. Who knows a nonprime example which is not?

REMARK. The Brieskorn homology spheres $\Sigma(2, 3, 6k \pm 1)$ and $\Sigma(p, q, pq \pm 1)$ are so obtained [J. Harer and, for the latter case, Seifert].

(D) (B. Clark) Is there a homology 3-sphere (or any 3-manifold) which can be obtained by n-surgery on an infinite number of distinct knots?

REMARKS. Lickorish gave an example obtained from two knots [Proc. Amer. Math. Soc. **60** (1977), 296–298]. Problems 1.15 and 1.16 ask whether this is possible for S^3 (Property P) or $S^1 \times S^2$ (Property R). It seems likely that the Poincaré homology sphere can be obtained only from $+1$-surgery on the right-handed trefoil knot (or, reversing orientation, from -1-surgery on the left-handed trefoil knot).

Problem 3.7. Let M^3 be a closed, irreducible 3-manifold with infinite fundamental group (hence a $K(\pi, 1)$).

(A) Is the universal cover always R^3?

(B) Does the universal cover always imbed in R^3?

(C) Is the universal cover always simply connected at infinity?

REMARKS. Yes for (A) if M or some finite cover is irreducible and sufficiently large [Waldhausen, Ann. of Math. (2) **87** (1968), 56–88]. Yes for (B) and (C) implies yes for (A).

Problem 3.8 (*Jaco*). A manifold M is said to have a manifold compactification if there exist a compact manifold N and an imbedding $\varphi: M \to N$ with $\varphi(\text{int}M) = \text{int}N$. Let $M(G)$ be the unique covering space determined by the conjugacy class of a subgroup G of $\pi_1(M)$. If M is a compact 3-manifold and G is a finitely generated subgroup, then does $M(G)$ admit a manifold compactification?

REMARKS. It is yes if G is the fundamental group of an incompressible surface [ibid.] or if G is peripheral (i.e., a subgroup of the image of $\pi_1(\partial M)$) [J. Simon, Michigan Math. J. **23** (1976), 245–256].

Problem 3.9. What closed, orientable 3-manifolds admit a round handle decomposition?

REMARKS. A round k-handle is $S^1 \times B^k \times B^{2-k}$ attached along $S^1 \times (\partial B^k) \times B^{2-k}$, $k = 0, 1, 2$. John Morgan has shown that a 3-manifold M with a round handle decomposition must be a Seifert fiber space or $\pi_1(M)$ must contain a subgroup isomorphic to $Z \oplus Z$. Partial converse: Seifert fiber spaces obviously have such decompositions. In dimensions > 3, round handle decompositions exist iff the Euler characteristic is zero [D. Asimov, Ann. of Math. (2) **102** (1975), 41–54].

Problem 3.10 (*Hilden and Montesinos*). Every closed, orientable 3-manifold can be constructed as follows; let F_1 and F_2 be disjoint, compact surfaces (not necessarily orientable) in S^3. Take three copies of $(S^3; F_1, F_2)$, called $S^3_a, S^3_b, S^3_c, F_{1a}, \cdots$, etc. Split S^3_a along F_1, S^3_b along F_1 and F_2, and S^3_c along F_2. Then glue one side of F_1 in S^3_a to the other side in S^3_b, and one side of F_2 in S^3_c to the other side in S^3_b. *Question*: Can the surfaces be chosen to be orientable?

Problem 3.11. Classify imbeddings of orientable surfaces F_g in S^3.

REMARKS. For S^2, the Schoenflies theorem classifies. For T^2, any imbedding bounds an $S^1 \times B^2$ [Alexander, Proc. Nat. Acad. Sci. U.S.A. **10** (1924), 6], and [Fox, Ann. of Math. (2) **49** (1948), 462–470] so the classification "reduces" to knot theory.

Any imbedding of F_g into R^3 is unknotted as soon as some projection to a coordinate axis has only a single local maximum (or minimum) [H. Morton, *A criterion for an embedded surface in* R^3 *to be unknotted*, also Notices Amer. Math. Soc. **24** (1977), A303. Abstract #77T-G56].

Any imbedding is ε-isotopic to a real algebraic variety in R^3 [Seifert, Math. Z. **41** (1936), 1–17]; this is true in general in codimension one.

Call a pair (S^3, F) prime if there is no pairwise connected sum $(S^3, F) \cong (S^3, F_1) \# (S^3, F_2)$. Are prime decompositions unique? Yes for genus 2 [Y. Tsukui, Yokohama Math. J. **18** (1970), 93–104; **23** (1975), 63–75]; also [S. Suzuki, Math. Japon. **20** (1975), 65–83; Hokkaido Math J. **4** (1975), 179–195].

Problem 3.12. Given M^3, with a Riemannian metric, consider all smooth maps $f\colon S^2 \to M^3$ such that $f \not\simeq 0$. There exists one of least area (J. Sacks and K. Uhlenbeck) which is immersed (R. Gulliver). Is it imbedded? (ADDED IN PROOF. Yes, or it double covers an imbedded RP^2 (Meeks and Yau).)

REMARK. This would give a differential geometric proof of the sphere theorem.

Problem 3.13.

THEOREM. *Every closed orientable* M^3 *contains a fibered knot* K, *i.e., there exists a fibration* $f\colon M - K \to S^1$ *and* f *is standard on a deleted tubular neighborhood of* K [*Gonzalez-Acuña, R. Myers, Notices Amer. Math. Soc.* **22** (1975), A651].

This is Winkelnkemper's open book decomposition but with connected binding. Note that K is homologically trivial in M.

Question (*Rolfsen*). What elements of $\pi_1(M)$ are represented by fibered knots? Links? Note that if an element is represented, so is any nonzero power.

Problem 3.14 (*W. Thurston*). (A) *Conjecture*: Every irreducible, closed 3-manifold, with infinite fundamental group which contains no subgroup isomorphic to $Z \oplus Z$, has a hyperbolic structure (see definitions, §0). Assume, if necessary, that some finite cover is sufficiently large.

(Added in proof, March 1, 1977: Thurston has a proof if, in addition, the 3-manifold has an incompressible surface which is not a fiber of a fibration over S^1.)

(B) *Conjecture*: Suppose G acts properly and discontinuously on a contractible 3-manifold with compact quotient. Suppose also that G has no subgroup isomorphic to $Z \oplus Z$. Then G is conjugate to a discrete group of isometries of hyperbolic 3-space.

REMARK. Clearly, (B) $\Rightarrow$ (A). Furthermore, (B) holds iff ((A) and every such G is residually finite) holds. The condition $Z \oplus Z$ is not in G, is necessary.

(C) *Conjecture* (*case when* $\partial M \neq \varnothing$): Suppose (i) M^3 is irreducible, (ii) $\pi_1(M)$ is infinite and every $Z \oplus Z$ in $\pi_1 M$ is peripheral, (iii) $\partial M = \partial_1 M \cup \partial_2 M$ where each component of $\partial_1 M$ is a torus, and each component of $\partial_2 M$ has negative Euler characteristic, (iv) ∂M is incompressible and every annulus A with $\partial A \subset \partial M$ can be deformed, rel ∂, into ∂M. Then $M - \partial_1 M$ has a complete hyperbolic structure, with finite volume, such that $\partial_2 M$ is totally geodesic.

Problem 3.15 (*W. Thurston*). Let M^3 be orientable and let $G \subset \pi_1(M^3)$ be the homomorphic image of $\pi_1(F)$ where genus $F = 2$, e.g., any 2-generator subgroup of $\pi_1(M)$.

(A) Does G have 2-fold symmetry (involution) which comes from 180° rotation about the axis below?

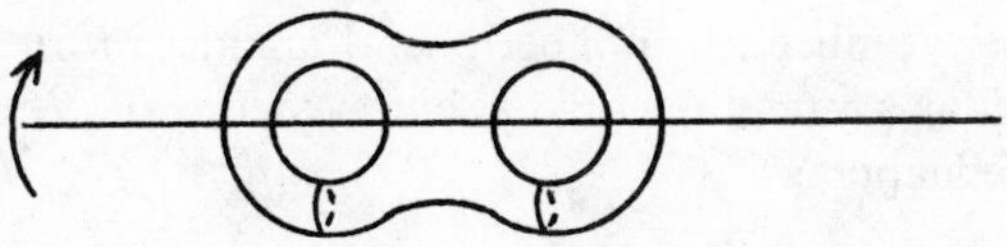

This is true for subgroups of PSL(2, C) (which is the group of isometries of hyperbolic 3-space, the universal cover of any 3-manifold with negative curvature).

(B) Suppose $G \subset \pi_1(M^3)$ has generators a, b with a conjugate to b in $\pi_1(M^3)$. Does G have an additional 2-fold symmetry t where $t(b) = a$? True for 2-bridge knot complements and 3-manifolds with Heegard decompositions of genus 2.

Problem 3.16 (*W. Thurston*). Is there a reasonable real-valued function C on the set of 3-manifolds which measures the complexity of $\pi_1(M^3)$? C should have the following properties:

(a) if M_1 is a k-fold cover of M_2, then $C(M_1) = kC(M_2)$;

(b) $C(M_1 \# M_2) = C(M_1) + C(M_2)$;

perhaps (c): if f: $M_1 \to M_2$ has positive degree, then $C(M_1) \geqq C(M_2)$;

perhaps (d): $C(M) = \text{volume}(M)$ if M is either $(\text{SL}(2, R))^\sim$ or hyperbolic 3-space modulo a discrete group.

Note that (a) and (b) imply that $C(M) = 0$ if M fibers over S^1 with monodromy of finite order, or if M fibers over a torus, or if M is covered by S^3.

REMARKS. Via the axiom of choice, there appears to be an existence theorem for a function C satisfying (a), (b) and (d). First consider the property of a 3-manifold M,

(*) If M_1 and M_2 are homeomorphic finite covering spaces of M, then the degrees of the coverings are equal.

[**Problem 3.16**(A). What 3-manifolds satisfy (*)? In particular, suppose M^3 is not commensurable with $F_g \times S^1$ or a T^2 bundle over S^1; then does M^3 satisfy (*)?]

Next, call M_1 and M_2 commensurable if they have homeomorphic finite covers.

Define C as follows: C must obviously be zero on M if M does not satisfy (*); C depends only on the prime components of M; if M is prime and satisfies (*), then we define C on the commensurability class of M by first defining C on M by (d) or else arbitrarily (axiom of choice here), and then if M' is commensurable with M, let $C(M') = (k/l)C(M)$ where there exists N, a k-fold cover of M and an l-fold cover of M'.

Also, see a forthcoming paper of Milnor and Thurston.

Problem 3.17 (*Jaco*). Suppose M^3 is compact, orientable, and irreducible, and that K is a positive integer. Does there exist at most a finite number (up to homeomorphism of M) of isotopy classes of incompressible surfaces in M having Euler characteristic $\geqq -K$?

REMARKS. Without the restriction on the Euler characteristic, the answer is no. Also, even with the restriction it is necessary to allow equivalence up to homeomorphism of M [Jaco, Canad. J. Math. **22** (1970), 553–568]. Possibly an affirmative answer is contained in [Haken, Acta Math. **105** (1961), 245–375].

Problem 3.18 (*Jaco*). Up to homotopy, does there exist at most a finite number of essential maps of F_g into M^3 where M^3 is atoroidal?

REMARKS. Recall that a map is essential if it induces an injection $\pi_1(F_g) \to \pi_1(M)$ (F_g = closed, orientable surface), and M is atoroidal if M contains no essential, properly imbedded, nonperipheral annuli or tori. The answer is no without the atoroidal assumption [ibid.], but even then it seems one should be able to describe the mappings.

Problem 3.19. Which immersed 2-spheres in R^3 bound immersed 3-balls?

REMARK. This is solved for one less dimension by S. Blank (see [V. Poenaru, Séminaire Bourbaki, Exposé 342, Lecture Notes in Math., Springer-Verlag, Berlin and New York, 1967/68] and [George Francis, Michigan Math. J. **17** (1970), 377–383]).

Problem 3.20. Under what conditions does a closed, orientable 3-manifold M smoothly imbed in S^4?

REMARKS. The torsion of $H_1(M; Z)$ must be of the form $T \oplus T$ [W. Hantzch, Math. Z. **43** (1938), 38–58]. If $H_1(M; Z) \cong Z$, then the quadratic form of M is null-concordant; in particular the signature is zero and the Alexander polynomial $A(t)$ is of the form $f(t)f(t^{-1})$ [Milnor, *Infinite cyclic coverings*, Conference on the Topology of Manifolds, Prindle, Weber and Schmidt, Boston, 1968, pp. 115-133], [A. Kawauchi, *On quadratic forms of* 3-*manifolds* (to appear); *$\tilde{H}$-cobordisms*, Osaka J. Math. **13** (1976), 567–590]. If M is a $Z/2$-homology sphere, then the μ-invariant must be zero (see [Hirzebruch, *Differentiable manifolds and quadratic forms*, Marcel Dekker, New York, 1972]).

Problem 3.21. Let X be an acyclic 2-complex and M_0^3 an abstract regular neighborhood of X; $\partial M_0 = S^2$, so cap off to get a homology 3-sphere M^3. Find an effective way to compute the Rohlin-invariant of M^3 in terms of X and its regular neighborhood.

Problem 3.22 (*Birman and Montesinos*). Do lens spaces admit minimal Heegard splittings of genus > 1? *Conjecture*: Yes.

Problem 3.23 (*Jaco*). A sufficiently large 3-manifold is atoroidal if it contains no essential, nonperipheral annuli or tori. What groups appear as π_1 of an atoroidal manifold?

REMARK. Such manifolds are determined by their fundamental groups (Johannson).

Problem 3.24 (*H. Hilden and J. Montesinos*). Is every homology 3-sphere the double branched covering of a knot in S^3?

REMARKS. It is known that $S^1 \times S^1 \times S^1$ is not a double branched covering (Fox) and that $S^1 \times F_g$ (F_g = surface of genus g) is not a double branched covering (Montesinos), but the arguments depend on a nontrivial first homology group.

Problem 3.25 (*J. Birman*). Let K be a knot in S^3 and $M(K)$ its 2-fold branched covering space. To what extent do topological properties of M determine K? More generally, describe the equivalence class $[K]$ of K under the relation $K_1 \approx K_2$ if $M(K_1)$ is homeomorphic to $M(K_2)$.

REMARKS. (1) If K is a 2-bridge knot, then $M(K)$ determines K.

(2) If $M(K)$ is composite, then K is composite [Kim and Tollefson, *Splitting* PL *involutions on nonprime* 3-*manifolds*, Michigan Math. J.].

(3) The bridge index of $K \leqq$ Heegard genus of M [J. Birman and H. Hilden, *Heegard splittings of branched coverings of* S^3, Trans. Amer. Math. Soc. **213** (1975), 315–352].

(4) There are examples of distinct prime 3-bridge knots which have homeomorphic 2-fold covering spaces [J. Birman, Gonzalez-Acuña and J. Montesinos, *Heegard splittings of prime* 3-*manifolds are not unique*, Michigan Math. J. **23** (1976), 97–104]. In particular, the Brieskorn manifold $\Sigma(2, 3, 11)$ is the 2-fold branched cover of the (3, 11)-torus knot and the knot below (Akbulut).

Problem 3.26 (*J. Birman and J. Montesinos*). Every lens space is a 2-fold cover of S^3 branched over a unique 2-bridge knot or link. Can it be a 2-fold cover of S^3 branched over any other knot or link? *Conjecture*: No.

For the following five problems of A. Durfee, P. Orlik and R. Randell, let $\mathscr{L}_k$ be

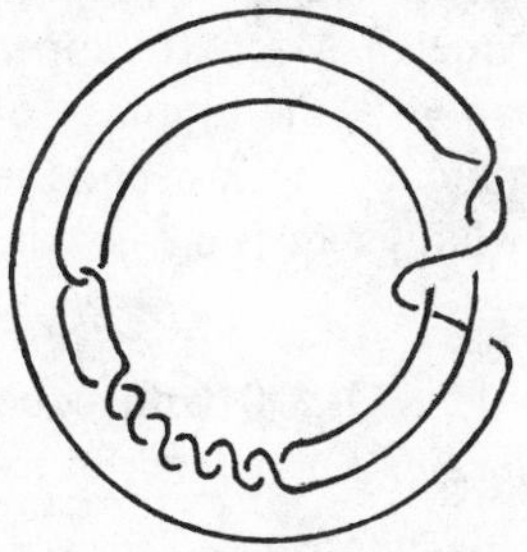

the closed, orientable 3-manifolds which occur as links of isolated singularities of complex, analytic surfaces of complex codimension k in C^{2+k}; in codimension > 1 the singularity should be normal. (Note that any analytic singularity is equivalent after an analytic change of coordinates to an algebraic singularity.) To fix notation when $k = 1$, let f: $C^3 \to C$ be a polynomial with an isolated singularity at the origin, and let $K = f^{-1}(0) \cap \varepsilon S^5$ for small ε. Then the map $f/|f|: S^5 - K \to S^1$ is a bundle map with fiber called F^4.

Problem 3.27. Which closed, orientable 3-manifolds belong to $\mathscr{L}_1$? to $\mathscr{L}_k$, $k > 1$?

REMARKS. It is a classical result that a 3-manifold belongs to $\mathscr{L}_k$ iff it is the boundary of a plumbing on a finite, connected graph, not necessarily simple, with orientable surfaces at the vertices and negative definite intersection form [Hirzebruch, Séminaire Bourbaki, Exposé 250, Lecture Notes in Math., Springer-Verlag, Berlin and New York, 1962/63]. These are a special case of Waldhausen's *Graphenmannigfaltigkeiten* (see [Invent. Math. **3** (1967), 308–333; **4** (1967), 87–117]).

$S^1 \times S^1 \times S^1$ is not a link in any codimension [Sullivan, Topology **14** (1975), 275–278]. Other 3-manifolds, such as the lens space $L(3, 1)$, occur as links only in codimension greater than one [Durfee, Proc. Sympos. Pure Math., vol. 29, American Mathematical Society, Providence, R.I., pp. 441–448].

What can be said about groups which occur as fundamental groups in $\mathscr{L}_1$?

Problem 3.28. Is every 3-manifold K in $\mathscr{L}_1$ irreducible?

REMARKS. Mumford [Inst. Hautes Études Sci. Publ. Math. #9, 1961], also [F. Hirzebruch, Séminaire Bourbaki, Exposé 250, Lecture Notes in Math., Springer-Verlag, Berlin and New York, 1962/63] proved that if $\pi_1(K) = 0$, then $K = S^3$. If V admits an S^1 action, then K is a Seifert manifold and Waldhausen [Invent. Math. **3** (1967), 308–333; **4** (1967), 87–117] proved that K is irreducible. The simplest class to attack next is provided by Wagreich [Topology **11** (1972), 51–72] where $\pi_1(K)$ is solvable and K has a circular plumbing graph.

Problem 3.29. To what extent does the Seifert matrix on F^4 determine the topology of the singularity? That is, does it determine K up to diffeomorphism? F up to diffeomorphism? Up to isotopy?

REMARKS. For $f: C^n \to C$, $n > 3$, the Seifert matrix determines the topology of the singularity completely [Durfee, Topology **13** (1974), 47–59].

Problem 3.30. A polynomial $G(t, z_0, \cdots, z_n)$ is called a μ-homotopy between G_0 and G_1 if $G_t(z_0, \cdots, z_n) = G(t, z_0, \cdots, z_n)$ has an isolated singularity with constant Milnor number μ for all t in a connected open set in C containing 0 and 1. Lê Dũng Tráng and C. P. Ramanujan [Amer. J. Math. (to appear)], also [Lê, Asterisque **7**, **8**, (1973), 183–192] show that for $n \neq 2$, the topology of G_0 and G_1 is identical, i.e., $(\varepsilon S^5, K_{G_0})$ is pairwise diffeomorphic to $(\varepsilon S^5, K_{G_1})$ and the bundles over S^1 are isomorphic. Prove this for $n = 2$ (what is missing is the 4-dimensional h-cobordism theorem).

Problem 3.31. Prove that index $F^4 \leqq 0$ with equality only for a nonsingular point. (See [Durfee, *The signature of smoothings of complex surface singularities*].)

Problem 3.32. What is the Whitehead group Wh(G) of a 3-manifold group $G = \pi_1(M^3)$ for M compact irreducible?

REMARKS. (1) Suppose G is infinite. Then Wh$(G) = 0$ if M^3 is sufficiently large, but is unknown otherwise, e.g., if M^3 has a finite cover which is sufficiently large. In fact, infinite 3-manifold groups belong to the class of finitely presented, torsion free groups, and there is no such group known with nonzero Whitehead group or reduced projective class group $\tilde{K}_0(Z(G))$.

(2) If G is finite, then Wh(G), when known, is cyclic.

Problem 3.33. (A) (W. Thurston) Does every 3-manifold group G have a faithful representation in GL$(4, R)$?

REMARKS. If so, G is residually finite (part (B)). Since PSL$(2, C) \subsetneq$ GL$(4, R)$, this is true for hyperbolic 3-manifolds (see §0).

(B) Is G residually finite?

REMARKS. See [Hempel, *3-manifolds*, Ann. of Math. Studies, no. 86, Princeton Univ. Press, Princeton, N.J., 1976, pp. 176–184] for an extensive discussion. (Residually finite means that for each $g \neq 1$, there exists a representation λ of G to a finite group for which $\lambda(g) \neq 1$.) (Added in proof, March 1, 1977: Thurston has probably shown that $G = \pi_1(M^3)$ is residually finite if M or a finite cover is sufficiently large.)

(C) Is G Hopfian?

REMARKS. Residually finite implies Hopfian (for finitely generated groups). Hopfian means that every epimorphism $G \to G$ is monic.

(D) Is the Frattini subgroup of G trivial?

REMARKS. Yes, if G is a knot group; if the Frattini subgroup is nonzero for an orientable, compact, irreducible, sufficiently large 3-manifold, then the 3-manifold must be a Seifert fiber space (in which case the Frattini subgroup is cyclic) (R. B. J. T. Allenby, J. Boler, B. Evans, L. Moser, and C. Y. Tang). The Frattini subgroup F is the intersection of the maximal subgroups; equivalently, $g \in F$ if every set of generators of G containing g does not need g.

Problem 3.34 (*S. Smale*). *Conjecture*: $\mathrm{Diff}^+(S^3)$ is homotopy equivalent to $SO(4)$.

REMARK $\pi_0(\mathrm{Diff}^+(S^3)/SO(4)) = 0$ [Cerf, Lecture Notes in Math., vol. 53, Springer-Verlag, 1968]. (Added in proof, March 1, 1977: A. Hatcher has announced a proof of the conjecture.)

Problem 3.35 (*A. Hatcher*). Compute $\pi_0(\mathrm{Diff}(L^3))$, the space of diffeomorphisms of a lens space.

REMARK. $\pi_0(\mathrm{Diff}(RP^3)) = Z/2$.

Problem 3.36 (*W. C. Hsiang*). *Conjecture*: $\mathrm{rank}(\pi_1 \mathrm{Diff}(L(p, q)) \otimes Q) \geqq \frac{1}{2}(p-1)$.

REMARKS. For lens spaces of dim $\geqq 5$, the corresponding statement is true [W. C. Hsiang and B. Jahren, *On the rational homotopy groups of the diffeomorphism groups of lens spaces*, preprint]. These elements are detected by Atiyah-Singer invariants.

Note that $\mathrm{rank}(\pi_1 \mathrm{Iso}(L(p, q)) \otimes Q) \leqq 2$ (W. C. Hsiang and B. Jahren). If the conjecture is true, it means that the diffeomorphism group of lens spaces is generally very different from the group of isometries (for any Riemannian metric) and the H-space of homotopy equivalences. Compare the case of 3-manifolds whose universal cover is hyperbolic 3-space (Problem 3.14).

Problem 3.37 (*C. B. Thomas*). Classify free actions of finite groups on S^3.

(A) *Existence*. If a finite group Γ acts freely on S^3, then Γ is, up to direct product with a cyclic group of coprime order, one of the following types:

(a) Z/rZ,

(b) an extension of a cyclic group by one of order 2^k,

(c) a generalized binary tetrahedral group T_v^* of order $8 \cdot 3^v$,

(d) the binary octohedral group O^* of order 48, or the binary icosahedral group of order 120,

(e) a certain split extension of $Z/(2n+1)Z$ by the binary dihedral group of order 8 (see [J. Milnor, Amer. J. Math. **79** (1957), 623–630], and [R. Lee, Topology **12** (1973), 183–199]).

All these groups except for type (e) admit faithful representations in $SO(4)$, i.e., have free, linear actions on S^3 [J. Wolf, *Spaces of constant curvature*, 3rd. ed, Publish or Perish, pp. 224–227]. No nonlinear actions are known.

What about groups of type (e)? If $SO(4) \simeq \mathrm{Diff}^+S^3$ (see Problem 3.34), then these groups do not act freely on S^3 (C. B. Thomas, these PROCEEDINGS).

(B) *Uniqueness.* The absence of nonlinear examples suggests:

(i) If Γ acts freely on S^3, is S^3/Γ homotopy equivalent to the quotient of a linear action?

REMARKS. Yes for Z/rZ, the binary dihedral groups D^*_{8k} (in (b) above), and T^*_v, $v \geqq 2$. Yes for the remaining groups if a sequence of obstructions in $\pi_i(\mathrm{Diff}^+S^3/SO(4))$ vanish; so again there is a reduction to the Smale conjecture (Problem 3.34) [R. Lee and C. B. Thomas, Bull. Amer. Math. Soc. **79** (1973), 211–215], and [Thomas, Math. Ann. (1977)].

We can ask if S^3/Γ is simple homotopy equivalent or even homeomorphic to the quotient of a linear action, but this looks hard for arbitrary Γ. It may be easier for the following special cases:

(ii) Compute the Reidemeister torsion of an arbitrary free Z/rZ-action; in particular, is every cyclic quotient simple homotopy equivalent to a lens space $L(r, q)$?

REMARKS. Yes, trivially, for $r = 2, 3, 4, 6$. At the Poincaré complex level, one can realize geometrically Reidemeister torsions distinct from those of the $L(r, q)$.

(iii) Let $Z/2^kZ$ act freely on S^3. Is the quotient homeomorphic to a lens space?

REMARKS. Yes for $k = 1, 2, 3$ [G. R. Livesay, Ann. of Math. (2) **72** (1960), 603–611], [P. M. Rice, Duke Math. J. **36** (1968), 749–751], [G. X. Ritter, Trans. Amer. Math. Soc. **181** (1975), 195–212].

(iv) If D^*_8 acts freely on S^3, is the quotient homeomorphic to the unique linear quotient?

REMARKS. By (i) and the vanishing of $\mathrm{Wh}(D^*_8)$ [M. Keating, Mathematika **20** (1973), 59–62], the quotient is simple homotopy equivalent to the linear quotient, and its double cover is homeomorphic to $L(4, \pm 1)$ (P. M. Rice, ibid.).

Problem 3.38 (*C. Giffen*). *The Smith Conjecture* (K, r): If K is a smooth, nontrivial knot in S^3, then K is not the fixed point set of a homeomorphism $h: S^3 \to S^3$ of least period r $(r > 1)$.

REMARKS. The conjecture first appeared in [P. A. Smith, Ann. of Math. (2) **40** (1939), 690–711]. Note that K must be smooth, for there are wild knots which are fixed point sets [Montgomery and Zippin, Proc. Amer. Math. Soc. **5** (1954), 460–465] and [Bing, Ann. of Math. (2) **80** (1964), 78–93]; however, they bound wild disks. The conjecture is known for (K, r) if r is even [Waldhausen, Topology **8** (1969), 81–91], or if K is a torus knot [Giffen, Thesis, Princeton Univ., Princeton, N.J., 1964], and [Fox, Michigan Math. J. **14** (1967), 331–334], or if K is a 2-bridge knot (and others) (Cappell and Shaneson), or if K is a cable knot, a cable braid, or a double of a knot (R. Myers), or other special cases of (K, r).

(A) *Covering Conjecture* (K, r): If K is a smooth, nontrivial knot in a homotopy sphere Σ^3, then the r-fold cyclic branched covering $\Sigma_r(K)$ over K is not simply connected if $r > 1$.

REMARKS. This is equivalent to the conjecture that $\pi_1(\Sigma^3 - K)/\langle \mu^r \rangle \not\cong Z/rZ$ for $\mu =$ meridian of K. Also the Covering Conjecture (K, r) for all K, r implies the Smith Conjecture (K, r) for all K, r and the converse holds if the Poincaré conjecture is true. The Covering Conjecture is known for doubles of any nontrivial knot in Σ^3 [Giffen, Illinois J. Math. **11** (1967), 644–646], see also [Gordon, *Uncountably many stable strings in codimension* 2, Quart. J. Math., 1977] and cases covered in [Kinoshita, Osaka Math. J. **10** (1958), 43–52] and [Fox, Osaka Math. J. **10** (1958), 31–35].

(B) Let $F(S^3, K) = \{h: S^3 \to S^3 \mid h = \text{id on } K\}$ with *C-O*-topology. The path component $F(S^3, K)_0$ of the identity contains no elements of finite order if K is nontrivial [Giffen, Bull. Amer. Math. Soc. **73** (1967), 913–914]. Thus a counterexample to the Smith Conjecture implies $\pi_0(F(S^3, K)) (\cong F(S^3, K)/F(S^3, K)_0)$ has an element of order $r > 1$. Let $\Gamma(S^3, K) \equiv \text{Aut}(\pi_1(S^3 - K), \mu, \lambda)/I_\mu$ be the group of automorphisms of $\pi_1(S^3 - K)$ which fix a meridian-longitude pair μ, λ, divided by the normal subgroup, I_μ, generated by the inner automorphism conjugation-by-μ. Then $\Gamma(S^3, K) \cong \pi_0 F(S^3, K)$ (Giffen).

Algebraic Conjecture (K, r): $\Gamma(S^3, K)$ contains no element of least order r, $r > 1$.

REMARKS. Clearly this conjecture implies the Smith Conjecture (K, r), and the reverse is true for fibered knots; also there is some information about $\pi_0(F(S^3, K))$ (Giffen).

Problem 3.39. Let PL(M^3) be the group of PL homeomorphisms of a compact 3-manifold. (A) (Tollefson) Does PL(M) have only finitely many conjugacy classes of finite cyclic subgroups of given order? (If yes, for S^3, then the Smith Conjecture holds.) What about finite subgroups?

(B) (Giffen) Suppose M admits no S^1 action. Is it possible for PL(M) to contain an infinite torsion subgroup?

(C) (Thurston) Is there a bound to the order of finite subgroups of PL(M)? Note that this is true for surfaces F_g since a finite subgroup represents faithfully in $GL(2g - 2, Z/3) = \text{Aut}(H_1(F_g; Z/3))$.

Problem 3.40 (*Nielsen*). Let h be a homeomorphism of a 3-manifold M such that the nth iterate h^n is homotopic to the identity. When does there exist a map g homotopic to h such that g^n = identity?

REMARKS. For $n = 2$ and M a closed, orientable 3-manifold fibered over S^1 with fiber F, there exist examples of such h which are not homotopic to any involution (for F = torus, [F. Raymond and L. Scott, *Failure of Nielsen's Theorem in higher dimensions*, Ark. Mat. (to appear)] and for genus $F > 1$, J. L. Tollefson).

For n a prime and M an orientable, closed 3-manifold fibered over S^1 with genus $F > 1$, there always exists a periodic g (homotopic to h) whenever either $H_1(M; Q) = Q$ or $u > 2$ and M is a Seifert fiber space [Tollefson, Trans. Amer. Math. Soc. **223** (1976), 223–234, and later erratum]. (Added in proof, April 1, 1977): Consider the natural map

$$\text{Diff}(M^3) \xrightarrow{\pi_0} \pi_0(\text{Diff}(M^3)).$$

Thurston has shown that there exists an inverse ρ such that $\pi_0 \rho = \text{id}$ if M^3 is closed, irreducible, sufficiently large and $\pi_1(M^3)$ is infinite and contains no $Z \oplus Z$, i.e., if M^3 is hyperbolic (see Problem 3.14); furthermore, $\pi_0(\text{Diff } M^3)$ is finite under the same assumptions.

Problem 3.41 (*Montesinos*). Is there a 3-manifold with an infinite number of nonequivalent involutions with S^3 as orbit space?

REMARK. Given N, there exists a manifold with more than N such involutions (Montesinos).

Problem 3.42 (*J. L. Tollefson*). Is every periodic homeomorphism of a 3-manifold homotopic to a periodic PL homeomorphism of the same period?

REMARKS. Yes for homeomorphisms which are locally nice so that the quotient is locally triangulable, for then we can triangulate the 3-manifold so that the original homeomorphism is PL. Tameness of the fixed point set often implies this niceness [E. Moise, Trans. Amer. Math. Soc. (to appear)].

Problem 3.43 (*Casson*). Does every homology 3-sphere H with an orientation reversing diffeomorphism have Rohlin invariant zero?

REMARKS. If not, then $H \# H \cong H \# (-H)$, which bounds an acyclic manifold, giving an element of order two in θ^3_H = homology bordism classes of homology 3-spheres; this triangulates higher dimensional manifolds (see Problem 4.4). Brieskorn homology 3-spheres $\Sigma(p, q, r)$ do not even admit orientation reversing homotopy equivalences. Probably the homology 3-spheres arising as irregular branched covers of amphicheiral knots have Rohlin invariant zero; cyclic covers do have invariant zero.

Problem 3.44 (*W. Meeks*). Let M^3 be a closed, orientable 3-manifold with universal cover R^3; then $G = \pi_1(M)$ acts on R^3 with quotient M^3. Let Γ_1 and Γ_2 be graphs in R^3 which are invariant under G, with $\pi_1(R^3 - \Gamma_i)$ free, $i = 1, 2$. Let N_1 and N_2 be equivariant regular neighborhoods of Γ_1 and Γ_2. *Conjecture*: ∂N_1 is isotopic to ∂N_2.

4. 4-manifolds. Perhaps the main problem is to find some pathology which is not predicted by analogy with higher dimensional (surgery) theory (as the Cappell-Shaneson fake RP^4 is), nor possibly explained by 3-dimensional pathology (e.g., a fake $S^3 \times R$ may be a fake 3-sphere cross R; similarly with Casson's strange ends). Such pathology might take the form of nondiffeomorphic but homotopy equivalent, closed, simply connected, 4-manifolds (Problem 4.11), or a Rohlin-invariant-zero homology 3-sphere which does not bound a contractible (or even acyclic) 4-manifold (Problem 4.2), or a homology 3-sphere M such that $M \# M$ does not bound an acyclic 4-manifold (Problem 4.4), or nonexistence of a closed, TOP, almost parallelizable 4-manifold of index 8 (Problem 4.7).

Problem 4.1. *Existence.* What integral, symmetric, unimodular, bilinear forms are the intersection forms of simply connected, closed 4-manifolds?

REMARKS. See [Milnor-Husemoller, *Symmetric bilinear forms*, Springer-Verlag, New York, 1973] for the algebraic background and further references. Odd, indefinite forms are represented by connected sums of copies of CP^2 and $-CP^2$, but little is known otherwise. In particular, is $E_8 \oplus \langle 1 \rangle$ (the odd, definite form of index 9), or $E_8 \oplus E_8 \oplus n\left(\begin{smallmatrix}0&1\\1&0\end{smallmatrix}\right)$, $n \leqq 2$, or Γ_{16} (the other index 16, even, definite form) represented by a manifold? By Rohlin's Theorem [Freedman and Kirby, these PROCEEDINGS] any closed, *smooth*, almost parallelizable (implied by simply connected and even) M^4 satisfies index $M^4 = 0 \bmod 16$; this rules out E_8 for example.

$$E_8 \oplus E_8 \oplus 3\begin{pmatrix}0&1\\1&0\end{pmatrix} \cong \Gamma_{16} \oplus 3\begin{pmatrix}0&1\\1&0\end{pmatrix}$$

is represented by the complex Kummer surface.

Any such form is represented by a simply connected, smooth, M^4 with ∂M^4 equal to a homology 3-sphere; hence, the next problems.

Problem 4.2. Which homology 3-spheres (with Rohlin invariant 0) bound contractible (or acyclic) 4-manifolds?

REMARKS. Essentially nothing is known, except for some families of homology 3-spheres which do bound contractible 4-manifolds, e.g (Casson and Harer), all Brieskorn spheres $\Sigma(p, q, r)$ where (p, q, r) equals (2, 3, 13), or (2, 3, 25), or $(p, ps + 1, ps + 2)$ for p odd, or $(p, ps - 1, ps - 2)$ for p odd, or $(p, ps - 1, ps + 1)$ for p even and s odd (and always $p > 0$, $s > 0$); in fact, these bound a 4-manifold with one 0-handle, one 1-handle, and one 2-handle, except (2, 3, 25) which uses two 1-handles and two 2-handles.

Akbulut's candidate for a homology 3-sphere which does not bound an acyclic 4-manifold is $\Sigma(2, 3, 11)$ which can be obtained by $+1$-surgery on the knot below.

$+1.$

The homology 3-sphere $\Sigma(2, 7, 13)$, which can also be obtained by $+1$-surgery on the (2, 7)-torus knot, lies in the Kummer surface, bounding a manifold with form Γ_{16} on one side, and the even form $3\left(\begin{smallmatrix}0&1\\1&0\end{smallmatrix}\right)$ on the other. Does it bound an even 4-manifold of rank $\leqq 4$?

Problem 4.3 (*S. Kaplan*). Does every homology 3-sphere bound an even, definite 4-manifold?

REMARK. Many of the interesting examples are links of isolated singularities in complex surfaces, and hence bound negative definite forms [see Hirzebruch, Séminaire Bourbaki, Exposé 250, Lecture Notes in Math., Springer-Verlag, Berlin and New York, 1962/63].

Problem 4.4. Find a homology 3-sphere H of Rohlin invariant one such that $H \# H$ bounds a PL acyclic 4-manifold.

REMARKS. If such an H exists, manifolds of dimension $\geqq 6$ are triangulable ([D. Galewski and R. Stern, Bull. Amer. Math. Soc. **82** (1976), 916–918] or [T. Matumoto, Thesis, Orsay, 1976], and R. Edwards' triple suspension theorem). In our current state of ignorance the following conjecture is conceivable: H bounds an acyclic 4-manifold iff $H \# H$ does (H not necessarily of Rohlin invariant one).

Problem 4.5 (*A. Casson*). Which rational homology 3-spheres bound rationally acyclic 4-manifolds?

REMARKS. If M^3 bounds a rationally acyclic W^4, then the linking form on $H_1(M; Z)$ must be null concordant (implying $|H_1(M; Z)|$ = square). If $H_1(W; Z)$ is cyclic, then the Casson-Gordon invariants must be ± 1 or 0. The question is already interesting for lens spaces. $L(m^2, q)$, m odd, bounds a W^4 if $q = km \pm 1$ with k, m coprime, or if $q = (m \pm 1)d$ with $d|2m \mp 1$, or if $q = (m \pm 1)d$ or $(2m \mp 1)(m \pm 1)/d$ with $d|m \pm 1$ and d odd [Casson and Gordon, *Concordance of classical knots*, preprint].

Problem 4.6 (*M. Freedman*). (A) Let $f: (M, \partial M) \to (X, \partial X)$ be a degree 1 normal map from a smooth (or TOP) 4-manifold to a Poincaré space. Suppose $f|\partial$ is a $Z[\pi_1(X)]$-equivalence. A surgery obstruction

$$\sigma(f) \in \Gamma_4(Z[\pi_1(X)] \to Z[\pi_1(X)]) \cong L_4(\pi_1(X))$$

is defined. If $\sigma(f) = 0$, is f normally bordant to a homotopy equivalence rel ∂?

(B) Can (A) be reduced to an equivalent question about link concordance?

REMARKS. An affirmative answer to (A) would yield "all" sought-after closed, smooth (or TOP) 4-manifolds (compare Problem 4.1).

There are Γ-group surgery problems with vanishing obstructions $\sigma(f) \in \Gamma_4(Z[Z] \to Z[e])$, which are not normally bordant rel ∂ to homotopy equivalences (this follows from [Casson and Gordon, *Cobordism of classical knots*, Orsay preprint], and [Cappell and Shaneson, Ann. of Math. (2) **99** (1974), 277–348]). There are also problems with no solution and zero obstruction in $\Gamma_4(Z[\{a, b: aba = bab\}] \to^{\gamma} Z[Z])$ where $\gamma(a) = \gamma(b) = 1$. Can a similar failure occur for a Wall group problem as in (A)?

Casson has shown (in forthcoming work on "Casson handles") that if certain sequences of links contain a slice link (see Problem 1.39) then simply connected surgery will work in dimension 4. Since slicing these links is itself a nonsimply connected surgery problem, there is hope that a nonsimply connected generalization of Casson's work would yield a "universal surgery problem" (i.e., a problem set up to slice a certain link) whose solution would be equivalent to the solution of all 4-dimensional Wall group problems with zero obstructions.

Problem 4.7. Is there a TOP, closed, almost parallelizable 4-manifold of index 8?

REMARKS. Yes, if one can prove TOP transversality in the missing case when a 4-dimensional preimage is expected (see [M. Scharlemann, Invent. Math. **33** (1976), 1–14]) (does Sullivan's proof that TOP manifolds are Lipschitz (dim $\neq 4, 5$) help here?), or if a Rohlin invariant one homology 3-sphere bounds an acyclic 4-manifold, perhaps by being TOP imbedded in S^4.

Problem 4.8. Does there exist a manifold proper homotopy equivalent (or even homeomorphic) to $S^3 \times R$ but not diffeomorphic?

REMARK. Casson has shown that either such a manifold exists or another manifold, $Q^4 \simeq pS^2 \times S^2 - \text{pt}$, exists having a fake end (see Problem 1.39).

Problem 4.9 (*M. Cohen*). Does there exist a 4-dimensional h-cobordism with nontrivial Whitehead torsion?

Problem 4.10 (*W. Thurston*). If a closed, orientable 4-manifold M^4 is a $K(\pi, 1)$, must the Euler characteristic be $\geqq 0$?

REMARK. If M^4 has nonpositive curvature then it is a $K(\pi, 1)$ and $\chi(M) \geqq 0$ [S. Chern, Abh. Math. Sem. Univ. Hamburg **20** (1955), 117–126]. This argument fails in higher dimensions. The Hopf conjecture is that sectional curvature $\leqq 0$ implies that $(-1)^k\chi(M^{2k}) \geqq 0$.

Is it possible that $(-1)^k\chi(M^{2k}) \geqq 0$ for all $M = K(\pi, 1)$?

Problem 4.11. *Uniqueness.* If two closed, simply connected 4-manifolds are homotopy equivalent, are they homeomorphic? If so, is the homotopy equivalence homotopic to the homeomorphism?

REMARKS. They are h-cobordant; thus they are diffeomorphic after connected sum with enough $S^2 \times S^2$'s [C. T. C. Wall, J. London Math. Soc. **39** (1964), 141–149]. Without the simply connected assumption, simply homotopy equivalence does not even imply a smooth, normal bordism (Problems 4.13–4.15; also [Cappell and Shaneson, Comment. Math. Helv. **45** (1971), 500–528]).

If the manifolds are not closed, but have homeomorphic boundaries and the homeomorphism extends to a homotopy equivalence, then we can ask if the homeomorphism extends to a homeomorphism of the 4-manifolds (unlikely; see Problem 4.16).

Kodaira has described [Ann. of Math. (2) **78** (1963), 563–626] some possible counterexamples, logarithmic transforms of elliptic surfaces. Here is a description for topologists (A. Kas): Let $S \to^{\pi} CP^1 = S^2$ be an elliptic surface with an analytic projection to CP^1 such that $\pi^{-1}(\text{point}) = \text{torus } T^2$ except for a finite number of points. For example, we construct the Kummer surface by first taking the quotient of $T^4 = S^1 \times S^1 \times S^1 \times S^1$ by the involution which reflects each circle. The involution has 16 fixed points, so the quotient is a manifold except for 16 singularities equal to the cone on RP^3. Replace the cone by the cotangent disk bundle of S^2, whose boundary is RP^3 with the right orientation. This is the Kummer surface, and π is constructed from the projection of T^4 on any $S^1 \times S^1$. A similar construction using $S^2 \times T^2$, and 180° rotation on S^2, gives the "half-Kummer" surface, which is known to be diffeomorphic to $CP^2 \# 9(-CP^2)$.

The logarithmic transform is essentially S^1 cross the construction in Seifert fiber space theory in which a nonsingular S^1 fiber is replaced by a singular fiber of multiplicity m.

Let D be a 2-ball in CP^1 so that $\pi^{-1}(D) = S^1 \times S^1 \times D$. The logarithmic transform $L_a(m)$ of S is $L_a(m) = S - \pi^{-1}(\text{int } D) \cup_h S^1 \times S^1 \times D$ where a is the center of D and $h\colon S^1 \times S^1 \times \partial D \to \partial(S - \pi^{-1}(\text{int } D)) = S^1 \times S^1 \times \partial D$ is given by

$$h(\theta_1, \theta_2, \varphi) = \begin{pmatrix} 1 & 0 & 0 \\ 0 & 0 & -1 \\ 0 & 1 & m \end{pmatrix} \begin{pmatrix} \theta_1 \\ \theta_2 \\ \varphi \end{pmatrix}, \qquad \theta_i \text{ and } \varphi \in [0, 2\pi].$$

The map $\pi\colon S - \pi^{-1}(\text{int } D) \to CP^1 - \text{int } D$ extends to $L_a(m)$ by defining π on $S^1 \times S^1 \times D$ by $\pi(\theta_1, \theta_2, r\varphi) = r(\theta_2 + m\varphi)$ where $r \in [0, 1]$.

The logarithmic transform of the half-Kummer surface is still a rational surface and hence is still diffeomorphic to the half-Kummer surface. However, two logarithmic transforms, $L_a(m)$ followed by $L_b(n)$, $(m, n) = 1$, is homotopy equivalent but not known to be diffeomorphic to the half-Kummer surface. Similarly one logarithmic transform of the Kummer surface is homotopy equivalent, but not known to be diffeomorphic.

Another case is the nonsingular quintic in CP^3, which is known to be homotopy equivalent to $\# 9CP^2 \# 44(-CP^2)$ but not necessarily homeomorphic.

It is not hard to construct examples of homotopy equivalent, simply connected, closed 4-manifolds, and presumably they are diffeomorphic. The point to these examples is that it is of independent interest to construct the expected diffeomorphism.

Problem 4.12 (*Mandelbaum and Moishezon*). Suppose M_1^4 and M_2^4 are closed, simply connected, and homotopy equivalent. Is $M_1 \# \pm CP^2$ homeomorphic to $M_2 \# \pm CP^2$?

REMARKS. The choice of CP^2 or $-CP^2$ may be crucial. The right choice gives an odd, indefinite intersection form so that we have a homotopy equivalence with a connected sum of copies of $\pm CP^2$. Is there a homeomorphism? Mandelbaum and Moishezon [Topology **15** (1976), 23–40, and later preprints] construct a diffeomorphism, for the case $+CP^2$, for a large class of complex surfaces.

Problem 4.13 (*Cappell and Shaneson*). There are homotopy RP^4's which are not diffeomorphic to RP^4 [Cappell and Shaneson, Ann. of Math. (2) **104** (1976), 61–72]. Which of these homotopy RP^4's are homeomorphic (diffeomorphic), and which are homeomorphic to RP^4?

REMARK. Some of the homotopy RP^4's have double covers which are diffeomorphic to S^4 (Akbulut and Kirby). Find an elegant way to describe these exotic involutions on S^4.

Problem 4.14 (*Cappell and Shaneson*). (A) There is a homotopy equivalence $h: S^2 \times RP^2 \to S^2 \times RP^2$ constructed by

$$S^2 \times RP^2 \to (S^2 \times RP^2) \vee S^4 \xrightarrow{\mathrm{id} \vee \alpha} S^2 \times RP^2$$

where α generates $\pi_4(S^2)$. h is not homotopic to a PL homeomorphism because it has a nontrivial normal invariant (the induced normal map $h^{-1}(RP^2) \to RP^2$ has a nontrivial Kervaire invariant). Is h homotopic to a homeomorphism?

(B) Construct a homotopy $S^2 \times RP^2$ by removing a normal B^3-bundle of RP^1 and sewing in a suitable $(T^3 - B^3)$-bundle over S^1. Is the manifold or the homotopy equivalence exotic?

REMARKS. Note that the existence of the (PL) exotic homotopy equivalence in part (A) implies that the manifold above is s-cobordant to $S^2 \times RP^2$.

Problem 4.15 (*Y. Matsumoto*). Is Scharlemann's "fake" $S^1 \times S^3 \# S^2 \times S^2$ an exotic manifold or an exotic self-homotopy equivalence of $S^1 \times S^3 \# S^2 \times S^2$ [M. Scharlemann, Duke Math. J. **43** (1976), 33–40]?

REMARK. After connected sum with $S^2 \times S^2$, Scharlemann's manifold is diffeomorphic to $S^1 \times S^3 \# 2(S^2 \times S^2)$ [Fintushel and Pao, *Identification of certain 4-manifolds with group actions* Proc. Amer. Math. Soc. (to appear)].

Problem 4.16 (*S. Akbulut and R. Kirby*). Does every diffeomorphism of the boundary of a contractible 4-manifold X^4 extend over X^4?

REMARKS. If not, there is a counterexample to the relative h-cobordism theorem in dimension 5. Here is a candidate for a diffeomorphism which does not extend: In the symmetric link below, we can add 2-handles (to B^4) to both circles with framing 0. The boundary of this 4-manifold has an obvious involution obtained by switching circles. Let the contractible manifold X^4 be obtained by surgering one of the two obvious 2-spheres; X^4 is a well-known Mazur manifold.

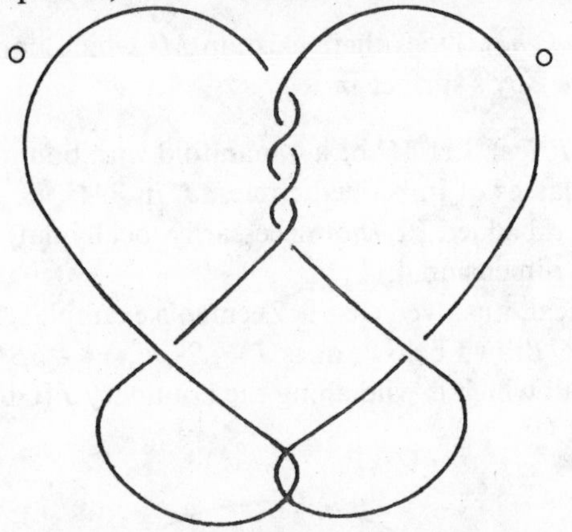

Problem 4.17 (*W. Thurston*). Can a homology 4-sphere ever be a $K(\pi, 1)$? Who knows an example of a rational homology 4-sphere which is a $K(\pi, 1)$?

REMARKS. Many homology 3-spheres are $K(\pi, 1)$'s, e.g., any Brieskorn $\Sigma(p, q, r)$ with p, q, r pairwise coprime and infinite fundamental group.

Problem 4.18. Does every simply connected, closed 4-manifold have a handlebody decomposition without 1-handles? Without 1- and 3-handles?

REMARK. Because there are nontrivial groups G which cannot be trivialized by adding the same number of generators and relations [Gerstenhaber and Rothaus, Proc. Nat. Acad. Sci. U.S.A. **48** (1962), 1531–1533], there are contractible 4-manifolds V^4, with $\pi_1(\partial V^4) = G$, that require 1-handles (Casson). On the other hand, nonsingular complex hypersurfaces in CP^3 need no 1-handles [L. Rudolph, Topology **14** (1976), 301–303], or even 3-handles (Harer, Kas and Kirby, or Akbulut). This is true (Mandlebaum) for complete intersections (the intersection of n hypersurfaces in CP^{n+2} which are in general position). If 1-handles are unnecessary, then there is a geometric proof [Mandlebaum and Moishezon, *Numerical invariants of links*] of some of Rohlin's inequalities [Functional Anal. Appl. **5** (1971), 39–48] using Tristram's work [Proc. Cambridge Philos. Soc. **66** (1969), 251–264].

Problem 4.19. (A) Does any integral, unimodular, symmetric, bilinear form contain a characteristic element α (i.e., $\alpha \cdot x = x \cdot x$ (mod 2) for all x) such that $\alpha \cdot \alpha$ = index.

REMARK. Yes for indefinite forms and definite forms of rank $\leqq 16$.

(B) Does every orientable, closed, smooth M^4 smoothly imbed in R^7?

REMARKS. (i) M^4 smoothly imbeds in $R^7 \Leftrightarrow$ there exists $\alpha \in H_2(M; Z)$ such that $\alpha \cdot x = x \cdot x$ (mod 2) for all $x \in H_2(M; Z)$, and $\alpha \cdot \alpha = \text{index}(M^4)$ [J. Boechat and A. Haefliger, *Essays on topology and related topics*, Memoires dédiés à Georges de Rham, Springer-Verlag, Berlin, 1970]. The simply connected case is already interesting, where characteristic elements (integral duals to the second Stiefel-Whitney class) are known to exist, with $\alpha \cdot \alpha \equiv \text{index}(M^4)$ (mod 8).

(ii) M^4 PL imbeds in R^7 [M. Hirsch, Proc. Cambridge Philos. Soc. **61** (1965), 657–658].

(iii) M^4 PL imbeds in $R^6 \Leftrightarrow M^4$ is a spin manifold (in which case the imbedding can be chosen to have a single nonlocally flat point). M^4 smoothly imbeds in $R^6 \Leftrightarrow M^4$ is spin and index $M^4 = 0$ (Cappell and Shaneson).

Problem 4.20 (*R. Fenn*). Does there exist an M^4 which immerses in R^6 with just one triple point (like Boy's surface in R^3)?

Problem 4.21 (*T. Price*). Let M^4 be a 4-manifold with boundary. Characterize the following isotopy classes of imbedded circles, J, in ∂M^4.

(A) J bounds an imbedded PL (not necessarily locally flat) 2-ball in M^4. (This is "Dehn's lemma" in dimension 4.)

REMARK. A classical, unsolved case is Zeeman's example: Add a handle to $S^1 \times B^3$ along the curve C drawn below; does $J = S^1 \times *$, $* \in \partial B^3$, bound a PL 2-ball? J does bound a 2-ball which is wild along the boundary J [Giffen, *F-isotopy implies I-equivalence*].

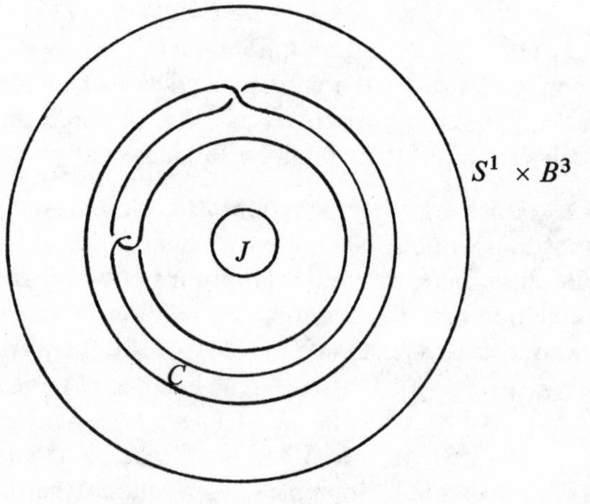

(B) J bounds a smooth immersed 2-ball with null-homologous image.

(C) J bounds an imbedded, orientable, smooth surface F^2 in M^4 with $H_1(F; Z) \to H_1(M; Z)$ being the zero map.

REMARKS. (A) $\Rightarrow$ (B) $\Rightarrow$ (C). There are other possible definitions of null-homologous singularities, e.g., define the "self-intersection number" of an immersion $f\colon B^2 \to M^4$ with double points $p_1, \cdots, p_k$ to be $\sum_{i=1}^k \varepsilon_i(g_i + g_i^{-1}) \in Z[H_1(M)]$ where $\varepsilon_i = \pm 1$ is the sign of p_i, and g_i is represented by the image of an arc in B^2 joining the two points in $f^{-1}(p_i)$; then (D) J bounds a smooth immersed 2-ball with self-intersection number zero. Note that (B) $\Rightarrow$ (D) $\Rightarrow$ (C). One can replace homology by homotopy throughout.

Problem 4.22 (*Gordon*). Let Σ^3 be a homology 3-sphere which bounds an acyclic 4-manifold V^4 such that $\pi_1(\Sigma^3) \to \pi_1(V^4)$ is surjective. Let K be a knot in Σ^3. Define K to be homotopically ribbon in V if there is a smoothly imbedded B^2 in V, $\partial B^2 = K$, such that $\pi_1(\Sigma^3 - K) \to \pi_1(V - B^2)$ is surjective.

(A) Does "K slice in V" imply "K homotopically ribbon in V"?

(B) Does (A) hold at least for contractible V?

REMARK. The classical "slice implies ribbon" conjecture (Problem 1.33) splits into two parts, "slice implies homotopically ribbon (in B^4)" and "homotopically ribbon implies ribbon".

Problem 4.23. Let f: $S^2 \to CP^2$ be a smooth imbedding which represents the generator of $H_2(CP^2; Z)$. *Conjecture*: $(CP^2, f(S^2))$ is pairwise diffeomorphic to (CP^2, CP^1). Perhaps, $f(S^2)$ is even isotopic to CP^1.

REMARK. The conjecture may be easier than the 4-dimensional Poincaré conjecture which implies it.

Problem 4.24 (*H. Gluck*). Let $K^2 \subsetneq S^4$ be a knotted 2-sphere. If we cut out a tubular neighborhood of K^2 and sew it back in with a "twist" (i.e., the nontrivial element of $\pi_1(SO(3)) = Z/2$ gives the "twist" $S^1 \times S^2 \to S^1 \times S^2$), then the resulting manifold is a homotopy 4-sphere. Must it be S^4?

REMARK (P. MELVIN). This is equivalent to asking whether $(CP^2, CP^1) \# (S^4, K^2)$ is pairwise diffeomorphic to (CP^2, CP^1) (see the previous Problem 4.23).

Problem 4.25 (*Y. Matsumoto*). Does there exist a smooth, compact W^4, homotopy equivalent to S^2, which is spineless, i.e., contains no PL imbedded S^2 representing the generator of $H_2(W)$?

REMARK. There is such an example for T^2 instead of S^2 [Y. Matsumoto, Bull. Amer. Math. Soc. **81** (1975), 467–470].

Problem 4.26 (*L. Taylor*). Construct a fake Hopf bundle by realizing $\Gamma = 3\gamma_0 + \gamma_1 + \cdots + \gamma_8$ in $CP^2 \# 8(-CP^2)$ by a PL imbedded sphere and taking a regular neighborhood (γ_i is the generator of $H_2(\pm CP^2)$; the "Hopf bundle" is $B^4 \cup$ 2-handle attached to the trefoil knot with $+1$ framing). Twice the core of this "Hopf bundle" can be represented by a smoothly imbedded double torus. Can it be represented by a torus? A sphere?

REMARKS. $r\Gamma$ cannot be represented by a smoothly imbedded sphere if $r = 1$ [Kervaire-Milnor, Proc. Nat. Acad. Sci. U.S.A. **47** (1961), 1651–1657] or if $r \geqq 3$ [Tristram, Proc. Cambridge Philos. Soc. **66** (1969), 251–264]; [W. C. Hsiang and R. H. Szczarba, Proc. Sympos Pure. Math., vol. 22, Amer. Math. Soc., Providence, R.I., 1970, pp. 97–103]. The double branched cover of this "Hopf bundle" along the imbedded surface can be used to construct a spin manifold of index 16 and betti number 22 (double torus), 20 (torus), or 18 (sphere).

Problem 4.27. (A) (S. Weintraub) Does every simply connected, closed 4-manifold have a basis for H_2 consisting of PL imbedded 2-spheres? Smooth?

REMARK. Yes, in the PL case, if there is a 2-dimensional spine, or if there are no 1-handles (see Problem 4.18).

(B) (R. Fenn) Which even elements of $\pi_2(RP^2 \times T^2) \cong Z$ can be represented by PL imbedded spheres?

REMARK. Odd elements cannot be represented [R. Fenn, Proc. Cambridge Philos. Soc. **74** (1973), 251–256].

Problem 4.28 (*Y. Matsumoto*). Construct M^4 by adding two 2-handles with 0-framing to B^4 along the link below.

(A) Can the generators of $H_2(M^4)$ be represented by a smoothly imbedded wedge of two 2-spheres?

(B) Can one of the generators be represented by a smoothly imbedded S^2?

(C) In $M^4 \# kS^2 \times S^2$, k arbitrary, can a "half-basis" of $H_2(M^4)$ be represented by $k + 1$ smoothly imbedded, disjoint 2-spheres?

REMARKS. (A) ⇒ (B) ⇒ (C). The generators of $H_\alpha(M^4)$ are represented by two Casson flexible handles, i.e., by an open subset proper homotopy equivalent to $S^2 \times S^2$ – point (see Problem 1.39). A no for (A) implies the Casson handles are exotic.

(A) implies that ∂M^4 bounds a contractible 4-manifold, and (B) implies that ∂M^4 bounds an acyclic 4-manifold (in both cases by surgering an S^2). (C) holds iff ∂M^4 bounds an acyclic smooth 4-manifold.

Problem 4.29 (*M. Kato*). If F is a smooth, orientable, closed surface in S^4, is $H_2(\pi_1(S^4 - F); Z) = 0$?

REMARK. If so, then the class of fundamental groups of surface complements in S^4 is exactly Kervaire's class of finitely presented groups of weight one with $H_i(G; Z) = H_i(Z, Z)$, $i = 1, 2$, which occur as knot groups in higher dimensions. That Kervaire's groups are realized by surfaces follows from arguments of T. Yajima [Osaka J. Math. **6** (1969), 435–446; Proc. Japan Acad. **46** (1970), 997–1000].

Problem 4.30. Given an imbedding of T^2 in R^4 with 4 critical points (with respect to projection to a coordinate axis), is it standard (up to isotopy)?

REMARK. Observe the example

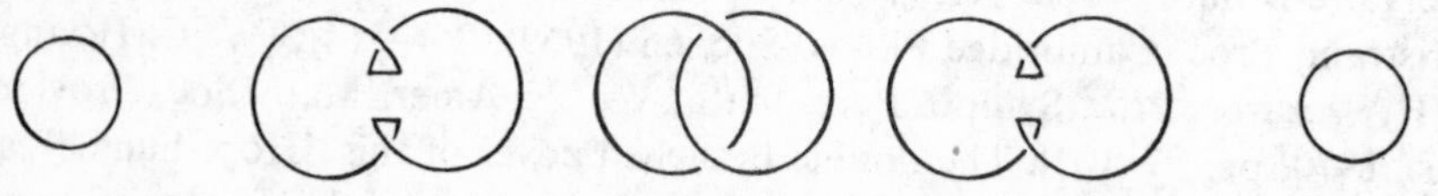

which *is* standard.

Problem 4.31. *Conjecture*: Any locally flat surface in a 4-manifold has a normal bundle.

REMARKS. This is true for codimension two imbeddings in other dimensions [Kirby and Siebenmann, Geometric Topology (Utah Conference, 1974), Lecture Notes in Math., vol. 438, Springer-Verlag, 310–324].

Problem 4.32 (*Schoenflies*). *Conjecture*: If S^3 is PL imbedded in S^4, then its closed complements are PL 4-balls.

REMARK. Note that they are TOP 4-balls since the S^3 is (PL) locally flat.

Problem 4.33 (*R. Fenn*). If S^3 is locally flatly imbedded in $S^2 \times S^2$, does it bound a TOP 4-ball? One may wish to assume a smooth imbedding.

REMARK. Suppose the Casson continuum [A. J. Casson (to appear)] is cellular in $S^2 \times S^2$; then its end is $S^3 \times R$. For the converse we need an affirmative answer to this problem.

Problem 4.34. What is $\pi_0(\text{Diff}(S^4))$? Indeed, is $\text{Diff}(S^4) \simeq O(5)$?

REMARK. The involution of S^4 giving the Cappell-Shaneson fake RP^4 may not be isotopic to the antipodal map (see Problem 4.13).

Problem 4.35 (*A. Hatcher*). On the torus T^n, $n \geqq 5$, there are many homeomorphisms concordant but not isotopic to the identity [A. Hatcher, these PROCEEDINGS]. Are there such examples on T^4?

Problem 4.36. (A) *Conjecture*: The minimal genus of a smooth imbedded surface in CP^2 representing $n \in H_2(CP^2; Z) = Z$ is $(n-1)(n-2)/2$.

REMARKS. Any nonsingular algebraic curve in CP^2 of degree n has genus $(n-1)(n-2)/2$. The minimal genus is at least $n^2/4 - 1$ if n is even, and at least $(n^2(h^2-1)/4h) - 1$ if $n \equiv 0 \bmod h$ and h is a power of an odd prime [Rohlin, Functional Anal. Appl. **5** (1971), 39–48], and [Hsiang and Szczarba, Proc. Sympos. Pure Math., vol. 22, Amer. Math. Soc., Providence, R.I., 1971, pp. 97–103].

(B) Let S be a simply connected complex, algebraic surface and let $\alpha \in H_2(S; Z)$. Suppose α is represented by an algebraic curve of genus g (given by the formula $2g - 2 = -c_1 \cdot \alpha + \alpha \cdot \alpha$ where c_1 = first Chern class). *Conjecture*: g is the minimal genus of any smoothly imbedded surface representing α.

REMARKS. This conjecture is risky. A counterexample could come from a simply connected surface S which is diffeomorphic to $M^4 \# CP^2$ with odd indefinite intersection form on M, for then all primitive, ordinary classes in $S \# -CP^2$ are represented by 2-spheres [Wall, J. London Math. Soc. **39** (1964), 131–140]. Apparently, no such surface is known (see next problem).

Problem 4.37. A simply connected, complex analytic surface S may contain exceptional curves (complex CP^1's with self-intersection -1) in which case S is diffeomorphic to $M^4 \# r(-CP^2)$. Can S be decomposed as a connected sum in any other way?

REMARK (YAU). If $\beta_2(S) = 1$, then $S = CP^2$; otherwise S has an indefinite intersection form and S decomposes homotopically.

Problem 4.38 (*A. Kas*). Let F^2 be a closed orientable 2-manifold. Let $\gamma_1, \cdots, \gamma_n$ be imbedded circles in F^2. Let $\tau_i: F^2 \to F^2$ $(i = 1, \cdots, n)$ be a left-handed Dehn (Lickorish) twist about the circle γ_i. Assume that $\tau_n \circ \cdots \circ \tau_2\tau_1: F^2 \to F^2$ is isotopic to the identity. *Conjecture*: Let $E \subset H_1(F^2, Q)$ be the subspace spanned by the γ_i $(i = 1, \cdots, n)$. Then the alternating intersection form is nondegenerate on E.

REMARKS. This conjecture gives a topological proof of the hard Lefschetz theorem for complex algebraic surfaces. Let S be an algebraic surface and let $F \subset S$ be a generic hyperplane section. The hard Lefschetz theorem states that intersection with F defines an isomorphism from $H_3(S, Q)$ to $H_1(S, Q)$ [Zariski, *Algebraic surfaces*, 2nd ed., Mumford's appendix to Chapter VI, pp. 151–152].

Problem 4.39 (*T. Matumoto*). Find new examples of complex surfaces with positive index.

REMARKS. The only known simply connected example is CP^2. There are nonsimply connected examples due to Hirzebruch [*Topological methods in algebraic geometry*, 2nd ed., Appendix 1, p. 164], also [A. Borel, Topology **2** (1963), 14–122] and [Kodaira, J. Analyse Math. **19** (1967), 207–215].

Notice that if S is a complex surface with $c_1 \equiv 0\ (2)$, then there exists a holomorphic line bundle $\frac{1}{2}K$ such that

$$\text{index}(S) = -8(2 \dim H^0(S, \mathcal{O}(\tfrac{1}{2}K)) - \dim H^1(S; \mathcal{O}(\tfrac{1}{2}K))).$$

It follows that $H^1(S, \mathcal{O}(\frac{1}{2}K)) = 0$ implies that index $S < 0$.

Problem 4.40 (A) (*Moishezon*). Let C be an algebraic curve in CP^2 which has only ordinary quadratic singularities. *Conjecture*: $\pi_1(CP^2 - C)$ is commutative.

REMARKS. A point p is an ordinary quadratic singularity (or ordinary double point) if there are local coordinates near $p = (0, 0)$ such that C is given locally by $x^2 + y^2 = 0$. It would follow that if C is irreducible of degree n, then $\pi_1(CP^2 - C) = Z/n$. The conjecture is known for small n and large n.

Using a theorem of Severi, now considered unproven, Zariski published a proof of the conjecture [Zariski, Amer. J. Math. **51** (1929); also *Algebraic surfaces*, p. 210]. Thus the conjecture can be proved by proving a topological analogue of Severi's theorem. Let F_g be a compact, Riemann surface of genus g. Call a smooth immersion $g: F \to CP^2$ semialgebraic if there exists a point $q \in CP^2 - g(F)$ such that the projection (along lines through q) $\pi: g(F) \to CP^1$ satisfies, (1) $\pi g: F \to CP^1$ is holomorphic of, say, degree n, (2) in local coordinates, $\pi g(z) = z$ or z^2, (3) g and πg are in general position, and (4) if d is the number of double points of $g(F)$, then $g = \frac{1}{2}(n-1)(n-2) - d$.

(B). Let $g_1, g_2: F \to CP^2$ be two semialgebraic immersions of degree n. Find an ambient isotopy of CP^2 carrying $g_1(F)$ to $g_2(F)$. (True for $n \leqq 3$.)

5. Miscellany.

Problem 5.1 (*M. Cohen*). Let $\mathcal{P}$ and $\mathcal{P}'$ be finite presentations (with the same deficiency) of a given group π. Let $K_{\mathcal{P}}$ and $K_{\mathcal{P}'}$ be the 2-dim CW-complexes associated to these presentations. Consider the assertions:

(A) $K_{\mathcal{P}} \simeq K_{\mathcal{P}'}$ (homotopy equivalence),

(B) $K_{\mathcal{P}} \wedge\!\!\!\searrow K_{\mathcal{P}'}$ (simple homotopy equivalence),

(C) $K_{\mathcal{P}} \underset{3}{\wedge\!\!\!\searrow} K_{\mathcal{P}'}$ (simple homotopy equivalence by moves of dimension $\leqq 3$),

(D) $\mathcal{P}$ can be changed to $\mathcal{P}'$ by extended Andrews-Curtis moves (i.e., we can change the presentation $\mathcal{P} = \{x_1, \cdots, x_n : R_1, \cdots, R_m\}$ in these ways

(i) $R_i \to R_i^{-1}$,

(ii) $R_i \to R_i R_j$, $i \neq j$,

(iii) $R_i \to w R_i w^{-1}$, w any word,

(iv) add generator x_{n+1} and relation $w x_{n+1}$.)

Note. Redundant relations *cannot* be added. Then (D) ⇒ (C) ⇒ (B) ⇒ (A) and (C) ⇒ (D) [P. Wright, Trans. Amer. Math. Soc. **208** (1975), 161–169]. (A) fails for the trefoil group [Dunwoody, Bull. London Math. Soc. **4** (1972), 151–155, and recent preprint] and for many finite abelian groups [W. Metzler, J. Reine Angew. Math. **285** (1976), 7–23].

Question. What other relations hold?

Problem 5.2 (*Lickorish*). Let K be a contractible finite 2-complex.

(A) *Conjecture* (*Zeeman*): $K \times I$ collapses to a point.

(B) *Conjecture*: K 3-deforms to a point, i.e., there exists a 3-complex L such that $K \swarrow L \searrow$ pt.

(C) *Conjecture*: The unique 5-dim regular neighborhood of K^2 in R^5 is B^5.

(D_0) *Conjecture*: Any presentation of the trivial group can be changed to the trivial presentation by Andrews-Curtis moves.

REMARKS. Conjecture (A) implies the Poincaré conjecture. Conjecture (C) is equivalent to knowing whether the boundary is S^4. (A) $\Rightarrow$ (B) $\Rightarrow$ (C) and (B) $\Leftrightarrow$ (D_0). The analogue of Conjecture (D_0) is false for nontrivial groups (see (D) of Problem 5.1). Possible counterexamples are $\{a, b\colon a^{-1}b^2a = b^3, b^{-1}a^2b = a^3\}$ and $\{a, b, c\colon [a, b]b = [b, c]c = [c, a]a = 1\}$. It is not known whether the regular neighborhoods in R^5 of the corresponding 2-complexes are B^5.

Problem 5.3 (*R. Fenn*). Is there an acyclic 2-complex which does not imbed in R^4?

REMARK. All n-complexes with H^n cyclic imbed in R^{2n}, $n \geqq 3$.

Problem 5.4 (*J. H. C. Whitehead*). Is every subcomplex of an aspherical 2-complex aspherical? Assume finiteness if you wish.

REMARKS. Aspherical means $\pi_k = 0$ for $k > 1$. For partial results, see [W. H. Cockcroft, Proc. London Math. Soc. **4** (1954), 375–384], and [J. F. Adams, J. London Math. Soc. **30** (1955), 482–488].

Problem 5.5. (A) (*Lickorish*) *Conjecture*: Any linear subdivision of an n-simplex collapses simplicially.

REMARK. True for $n \leqq 3$ [Chillingworth, Proc. Cambridge Philos. Soc. **63** (1967), 354–357].

(B) (Goodrick) *Conjecture*: Any linear subdivision of a star-like n-cell in R^n collapses simplicially.

REMARK. True for $n < 3$. Not true for a triangulation of a topological n-cell [Goodrick, Proc. Cambridge Philos. Soc. **64** (1968), 31–36].

Problem 5.6. What more can be said about nonsingular real algebraic varieties in RP^2, RP^3 or RP^4?

REMARKS. A number of pretty results about algebraic invariants of such varieties have been established (by complexifying) in a series of papers by Gudkov, Arnold, Rohlin, Kharlamov and Zvonilov in *Functional analysis and its applications* during the 1970's; there is also Gudkov's survey [Uspehi Mat. Nauk **29** (1974), 3–79 = Russian Math. Surveys **29** (1974), 1–79].

Problem 5.7 (*M. Freedman*). Let G be a nontrivial group, $G * Z$ the free product, and $G * Z/r$ the free product with one relation r. *Conjecture* (Kervaire): $G * Z/r$ is nontrivial.

REMARKS. Any counterexample must satisfy: (i) G is perfect, (ii) the degree of t (= generator of Z) in the relation r is ± 1 (proof: abelianize).

The conjecture is true if G has a normal subgroup of finite index [Gerstenhaber and Rothaus, Proc. Nat. Acad. Sci. U.S.A. **48** (1962), 1531–1533]; perhaps knot groups always have normal subgroups of finite index (Problem 3.33(B)).

It seems rare that both natural maps $A \to A * B/r$ and $B \to A * B/r$ have nontrivial kernels. This happens if $A = Z/2$, $B = Z/3$ and $r = ab$. Are there any other examples of a different flavor? Specifically if A is torsion-free and B nontrivial, is either A or B into $A * B/r$ always an injection?

Here is a relation to knot theory. Let M_K^3 be the result of surgery on a knot K with 0-framing. There is a natural map $f_K\colon M_K \to S^1 \times S^2$ and the closer K is to being trivial, the closer f_K is to a homotopy equivalence, e.g., f_K always induces an isomorphism on integral homology, and induces a $Z[Z]$-homology isomorphism exactly when the Alexander polynomial is trivial. $H_2(\pi_1(S^3 - K)/l) = 0$ iff M_K is diffeomorphic to $S_1 \times S^2 \# H^3$ where H^3 is a homology 3-sphere (l represents the longitude of K). Since $\pi_1(M_K) = \pi_1(S^3 - K)/l$ is normally generated by a meridian, the conjecture would imply that H^3 above could be replaced by a homotopy 3-sphere. (See Problems 1.16 and 1.17.)

Problem 5.8. Does there exist a graph G such that for any imbedding $f\colon G \to R^3$, $f(G)$ contains a nontrivial knot?

REMARK. It suffices to consider $G = C_n$ = complete graph on n-vertices. (Added in proof. Yes, for $n = 7$, John Conway.)

UNIVERSITY OF CALIFORNIA, BERKELEY

Proceedings of Symposia in Pure Mathematics
Volume 32, 1978

SOME PROBLEMS ON 3-MANIFOLDS

FRIEDHELM WALDHAUSEN

Except for the first section, the problems discussed are all from the general area of Heegaard diagrams and Heegaard splittings.

1. Nonsufficiently large 3-manifolds. Let M be a closed orientable 3-manifold which is irreducible (every PL 2-sphere in M bounds a 3-ball in M) and has infinite fundamental group; such an M is known to be an Eilenberg-Mac Lane space. M is called *sufficiently large* if a large number of theorems applies to it; equivalently, if there is an embedding of a closed 2-manifold whose fundamental group is nontrivial and injects. Unfortunately this is not always the case.

The first such examples are Seifert fibre spaces whose decomposition surface is the 2-sphere, with exactly three exceptional fibres, and with the added condition that H_1 be finite (and π_1 infinite) **[18]**; there are infinitely many such. Though it is quite inconceivable that these should be the only nonsufficiently large 3-manifolds, no new ones have been exhibited so far.

One way to search for new examples, emphasized by R. P. Osborne in particular, is to try surgery on a knot. Indeed, given a knot, in general, one may expect almost any surgery on this knot to produce a manifold which is irreducible and has infinite fundamental group. If on the other hand the surgery produces a sufficiently large 3-manifold, there must exist, in the knot space, an incompressible surface of a particular kind; namely an incompressible surface which is either closed, or has its boundary curves in the isotopy class of curves (on the boundary of the knot space) used in the surgery. One would expect this to hold in fewer cases. The construction of the known examples of nonsufficiently large 3-manifolds, can be interpreted to fit this program (at least some of them can be obtained by surgery on torus knots). In general, the main problem involved in this program is a classification of incompressible surfaces in a knot space. (In the case of a torus knot, the

AMS (*MOS*) *subject classifications* (1970). Primary 55A99; Secondary 55A40.

knot space is a Seifert fibre space, so this classification is comparatively easy **[17]**.)

Concerning properties of nonsufficiently large 3-manifolds in general, one may try to extrapolate from the known examples. Specifically, there is the so-called Fenchel conjecture, the theorem that any Fuchsian group has a torsionfree subgroup of finite index, e.g., **[26]**. An immediate consequence of this is that any Seifert fibre space has a finite covering space which is a fibre bundle. In particular then any of the known nonsufficiently large 3-manifolds has a finite covering space which is sufficiently large. One may ask if this continues to hold for the unknown examples. The question naturally splits into two subquestions, of unequal likelihood:

If M is as before, must it be true that $\pi_1 M$ contains a nontrivial subgroup $\pi_1 F$ where F is a closed 2-manifold?

By results of Jaco **[8]** and Scott **[12]** the answer must be affirmative if $\pi_1 M$ contains any subgroup whatsoever which is finitely generated, of infinite index, and not free. It is hard to believe that this could fail.

If $\pi_1 M$ contains $\pi_1 F$ as before, must there exist a finite covering $M' \to M$ so that $\pi_1 F \cap \pi_1 M' \to \pi_1 M'$ is induced by an embedding $F' \to M'$?

The question can be asked in just this form for sufficiently large 3-manifolds, and it is striking to notice how little is known about it. The real problem seems to be if $\pi_1 M$ contains a sufficient number of subgroups of finite index, or for that matter, any such subgroup at all. No one has devised a method yet how to get $\pi_1 M$ to act on finite sets (except in special cases, e.g., fibrations over S^1 **[9]**). The relevance of the latter problem is emphasized by a group theoretic result of Scott **[13]** which in the case at hand implies that the answer is affirmative if $\pi_1 F$ is the intersection of a set of subgroups of finite index. Indeed this latter fact can easily be seen directly. Namely if $\pi_1 F \to \pi_1 M$ is induced by a map $f: F \to M$, the hypothesis implies that there is a finite covering space $M' \to M$ and a lifting $f': F \to M'$ so that, if $U(f'(F))$ denotes a regular neighborhood, the map $\pi_1 F \to \pi_1 U(f'(F))$ is surjective and hence bijective. Making the boundary of the regular neighborhood incompressible, one then finds the required F' as a component.

Given that M has a finite covering space which is sufficiently large, it is natural to try carrying over to it some of the results available for sufficiently large 3-manifolds, e.g., that homotopy equivalences can be deformed to homeomorphisms. This has indeed been done in a few of the known cases, by what may be called equivariant surgery in the covering space **[2]**. Though the manifolds considered were extremely special, the effort required was considerable.

2. Heegaard diagrams and Heegaard splittings. Let M be a connected closed orientable 3-manifold, it will be natural to assume that M is in fact oriented. We think of M as smooth. Let f be a nice Morse function on M, that is, f is a smooth real-valued function, the critical points are nondegenerate, and at each critical point the value equals the index. The critical points of index 0 or 3 are of no interest whatsoever, so we assume they are minimal in number (that is, there is just one of either kind).

Define $F = f^{-1}(1\frac{1}{2})$; it is a closed oriented 2-manifold. Let $V = f^{-1}[1\frac{1}{2}, \infty)$ and $W = f^{-1}(-\infty, 1\frac{1}{2}]$. Then V and W are 'handlebodies'.

Let $v \subset V$ denote the system of properly embedded 2-disks given by the cores of the 2-handles determined by f; it is referred to as a *system of meridian disks*. V may also be considered as a regular neighborhood of $F \cup v$, plus a single 3-ball attached. Let similarly $w \subset W$ denote the system of properly embedded 2-disks given by the cocores of the 1-handles (equivalently, by the cores of the 2-handles of the dual function $f' = 3 - f$).

Up to trivial alteration (i.e., deformation of nice Morse functions) f is determined by the quadruple $(M, F; v, w)$. This quadruple is referred to as a *Heegaard diagram* of M. It is, in turn, determined up to isomorphism by the oriented 2-manifold F and the ordered pair of systems of curves ∂v and ∂w in F.

With easy modifications, the above carries over to manifolds with boundary ∂M and functions f such that $f(\partial M) \subset \{-1, 4\}$. The case of more general functions f is not without interest (e.g., bridge presentations of a knot give rise to such functions on the knot space) but it will not be considered here. The quadruple $(M, F; v, w)$ may again be referred to as a Heegaard diagram, not of M this time but of the triple $(M, f^{-1}(-1), f^{-1}(4))$.

By a *Heegaard splitting* of M (resp., of $(M, f^{-1}(-1), f^{-1}(4))$ if $\partial M \neq \emptyset$) we shall mean any oriented F arising in the way described, the notation (M, F) will be used. The *genus* of the Heegaard splitting (M, F) is by definition the number $g(F)$, the genus of F. If $(M, F; v, w)$ is a Heegaard diagram, (M, F) will be called the underlying Heegaard splitting.

Heegaard diagrams have less 'random structure' than functions, or handle decompositions, or triangulations. Thus among the effective ways to present 3-manifolds (*all* of them, not just a special class) they appear to be the most efficient.[1] Still if one wishes to put Heegaard diagrams to any use one must face the fact that there are far too many of them. In particular, from any given Heegaard diagram one may construct others, by the process of 'handle sliding'. By definition, this process does not alter the underlying Heegaard splitting, and it is generated by (i) isotopy of v, resp. w, (ii) sliding one component of v, resp. w, over another. It is a pleasant fact, not too hard to prove, that conversely any two Heegaard diagrams with the same underlying Heegaard splitting can be transformed one into another by handle sliding. Thus a Heegaard splitting may be identified with an equivalence class of Heegaard diagrams, the equivalence relation being generated by handle sliding.

Concerning Heegaard splittings, it is an interesting fact, pointed out by Stallings [**15**], that, up to isomorphism, these may be characterized algebraically. For simplicity it will be assumed that M is a closed manifold. Let (M, F) be a Heegaard splitting and let V, W be the pair of handlebodies (ordered by the orientations of M and F) with $V \cup W = M$, $V \cap W = F$. The inclusions of F induce a pair of surjective maps $\pi_1 F \to \pi_1 V$, $\pi_1 F \to \pi_1 W$, well defined up to conjugation. The assertion is that

$$(M, F) \mapsto (\pi_1 F \to \pi_1 V, \pi_1 F \to \pi_1 W)$$

[1]There are at least two more ways to effectively present closed orientable 3-manifolds:

(i) By surgery on a framed link in S^3. Here the equivalence relation is known, that is, if two surgeries give the same 3-manifold, one knows how to transform the two framed links one into another (Craggs, Kirby). Analysis of the equivalence relation is practically untouched however, and it does not seem easier than the classification problems discussed below.

(ii) As branched coverings of S^3, or even 3-fold branched coverings, branched over a knot (Hilden, Hirsch, Montesinos). Here even the equivalence relation is unknown.

is a bijection of isomorphism classes. Details to this were provided by Jaco [7]; as these details are excessively complicated, here is a simpler way of verifying the assertion.

LEMMA (CF. [7]). *Let X be a wedge of n 1-spheres, and $\pi_1 F \to \pi_1 X$ any map. Then there are a handlebody V', an isomorphism $F \to \partial V'$, and a map $V' \to X$ so that $F \to V' \to X$ induces $\pi_1 F \to \pi_1 X$, up to conjugation.*

In fact, let $\pi_1 F \to \pi_1 X$ be induced, up to conjugation, by a map $f_0 \colon F \to X$, say. In the jth 1-sphere in X, let p_j be a point different from the basepoint. Identify F to $F \times 0$ in $F \times [0,1]$ and extend f_0 to f so that $f_1 = f | F \times 1$ is in general position with respect to $\bigcup p_j$. Form Y from $F \times I$ by attaching a 2-handle at each component of $f_1^{-1}(\bigcup p_j)$. Then f may be extended to $g \colon Y \to X$ so that (i) the core of each 2-handle is mapped into $\bigcup p_j$, (ii) for any component G_i of ∂Y other than $F \times 0$, $g(G_i) \subset X - \bigcup p_j$. Since $X - \bigcup p_j$ is contractible one may now form V' from Y by attaching to each G_i a handlebody V_i, in any way whatsoever, and extend g by mapping V_i into $X - \bigcup p_j$.

Thus the surjectivity part of the above assertion has been established. To see the injectivity, let $F = \partial V$ and $F = \partial V'$, and let $\pi_1 V \to \pi_1 V'$ be an isomorphism so that $\pi_1 F \to \pi_1 V \to \pi_1 V'$ and $\pi_1 F \to \pi_1 V'$ are the same, up to conjugation. Then the loop theorem (or better, its elementary version available for handlebodies **[24]**) and the Alexander trick show that the identity on F extends to an isomorphism $V \to V'$ which itself is unique up to isotopy.

Stallings also showed **[15]**, given (M, F), M is a homotopy 3-sphere if and only if the map $\pi_1 F \to \pi_1 V \times \pi_1 W$ is surjective. In view of the fact (which was not available yet when **[15]** was written) that for any genus there is only one isomorphism class of Heegaard splittings of S^3, cf. below, one has as a corollary that the Poincaré conjecture is equivalent to the group theoretic conjecture that for any g there is only one isomorphism class of surjections $\pi_1 F \to \Phi_1 \times \Phi_2$ where Φ_1 and Φ_2 denote free groups of rank $g = g(F)$. Interesting though this fact is, philosophically, it has not been possible so far to use it in any way.

3. Classification problems for Heegaard splittings. From any Heegaard splitting one may obtain a new one by 'standard handle addition'. Since by definition (M, F) is just a particular kind of manifold pair, the result of a standard handle addition may simply be described as the connected sum of (M, F) with a (or 'the') genus one Heegaard splitting of S^3. Again it is a pleasant fact, the theorem of Reidemeister **[11]** and Singer **[14]**, that any two Heegaard splittings of M are 'stably equivalent' i.e., equivalent under the equivalence relation generated by isomorphism (in fact, isotopy) and standard handle addition.

One approach to the classification of 3-manifolds is thus to start from Heegaard diagrams, which may be classified 'upon inspection'. By imposing on these the equivalence relation of handle sliding, one obtains Heegaard splittings; and by further imposing stable equivalence one obtains the (isomorphism classes of) manifolds themselves. One may thus try to classify Heegaard splittings first, and then proceed from this. The former will be discussed in the next section; the latter leads to various interesting problems, a sample of which is given below.

It is convenient to call a Heegaard splitting *minimal* if it cannot be obtained, by a standard handle addition, from a Heegaard splitting of lower genus. Here is a list of some known, respectively unknown, facts.

The 3-sphere has, up to isomorphism, precisely one Heegaard splitting of any genus $g \geqq 0$ [**19**]. The only minimal one is that of genus 0.

It is not known if a minimal Heegaard splitting of a lens space must have genus 1. In fact this is unknown even for projective 3-space. although the argument of [**20**] seems close to establishing it.

Some lens spaces have two isomorphism classes of Heegaard splittings of genus 1, obtainable from each other by re-orienting F. By taking connected sums one should expect to mess up the nonuniqueness so that it is no more due to just orientation phenomena. Renate Engmann [**3**] has shown that such 'essential' nonuniqueness does in fact occur.

Indeed, nonuniqueness does not depend on connected sum phenomena either: There is a prime manifold (in fact, there are infinitely many such) with two minimal Heegaard splittings (of genus 2) that are not isomorphic as unoriented manifold pairs [**1**].

Here are some problems.

Show that M has only finitely many isomorphism classes (or even isotopy classes) of minimal Heegaard splittings.

Given any of these, give a procedure to obtain the others.

Is it true that any two minimal Heegaard splittings of M have the same genus? An example of P. Schupp shows that the corresponding question for group presentations has the answer 'no'.

Let (M, F) and (M, F') be two Heegaard splittings of M, both of genus g. Do they become isomorphic when the genus is raised, by standard handle additions, to $2g$, say? Is there only one isomorphism class of Heegaard splittings of genus $2g$?

4. Decision problems for Heegaard splittings. In reality these are problems about Heegaard diagrams. For example, given a Heegaard diagram $(M, F; v, w)$, how can one find out if the underlying Heegaard splitting (M, F) is minimal? Is there a way to alter $(M, F; v, w)$ to a canonical Heegaard diagram, or one of a finite set of such, from which the answer may be read off by inspection? Similarly, given $(M, F; v, w)$ and $(M', F'; v', w')$ one may want to know if their underlying Heegaard splittings are isomorphic, and in particular, say, if (M, F) is isomorphic to a Heegaard splitting of S^3.

There is a notion of complexity for a Heegaard diagram. It is best to discuss first the analogous notion of complexity for curves on the boundary of a handlebody.

Let V be a handlebody, and v a system of meridian disks in V. Let the components of v be numbered $v_1, \cdots, v_n$ and let a normal direction to each v_i be chosen. If V has a basepoint, off v, any based loop k in V gives rise to a word $v(k)$ in the alphabet $\{v_1, v_1^{-1}, v_2, \cdots, v_n^{-1}\}$ to record its encounters with v. In this way v determines a basis of $\pi_1 V$, and the element of $\pi_1 V$ represented by k is given, in this basis, by the reduced word $\bar{v}(k)$ associated to $v(k)$. We define the geometric, resp. algebraic, length of k to be the number of letters in the word $v(k)$, resp. $\bar{v}(k)$; it is denoted $l(k, v)$, resp. $\bar{l}(k, v)$. If k is not a single loop but a finite set of such, we define its geometric (resp., algebraic) length by adding those of the individual loops. The minimal geometric length $l(k)$ is defined to be the minimum of the numbers $l(k, v)$ as v varies; similarly the minimal algebraic length $\bar{l}(k)$ is defined. If V has no basepoint, these considerations still apply when elements of $\pi_1 V$ are replaced by conjugacy classes of such elements.

Suppose v is changed to another system of meridian disks, v', in the following special way. Namely a single component v_j of v may be replaced by a disk v'_j *in the complement of* v; furthermore the components of v' may be re-indexed, and some of the normal directions altered. If V has a basepoint, the result of the replacement $v \mapsto v'$ may be interpreted in two ways. Firstly we may say that the basis of $\pi_1 V$ is replaced by another one, taken from a particular finite list. The other interpretation is that $\pi_1 V$ has been subjected to a particular kind of automorphism, called a *T-transformation* by Whitehead **[23]**. The second interpretation still makes sense when V is unbased and when elements of $\pi_1 V$ are replaced by conjugacy classes. The substitution $v \mapsto v'$ will be referred to as a geometric T-transformation.

THEOREM (WHITEHEAD **[23]**). (1) *Let k be a finite collection of (based, resp., unbased) loops in V. Suppose the algebraic length $\bar{l}(k, v)$ can be made smaller by some automorphism of $\pi_1 V$. Then it can be made smaller by a T-transformation.*

(2) *Suppose k and k' are such that their algebraic lengths are minimal, i.e., $\bar{l}(k, v) = \bar{l}(k)$ and $\bar{l}(k', v) = \bar{l}(k')$. Suppose there is an automorphism which takes the set of elements of $\pi_1 V$ (resp., conjugacy classes) represented by k into that represented by k'. Then this automorphism may be written as a sequence of T-transformations none of which increases the algebraic length.*

(Incidentally, Whitehead's theorem can be extended to cover the case of a finite set of finitely generated subgroups, cf. **[23**, p. 97]; the theorem is just the case of a set of cyclic subgroups. The proof uses 3-dimensional topology and is an extension of Whitehead's method: Instead of mapping curves into $\#_n(S^1 \times S^2)$, as Whitehead does, one maps $\#_k(S^1 \times S^2)$, or a finite number of such. In the context of Heegaard diagrams this extension is of no interest however.)

Suppose now that k is a system of mutually disjoint simple closed curves in the boundary ∂V. In this case one is interested in having the geometric length of k as small as possible. This can indeed be achieved.

THEOREM (ZIESCHANG **[25]**). *Suppose that $l(k, v)$ is strictly bigger than the minimal algebraic length $\bar{l}(k)$. Then $l(k, v)$ can be made smaller by a geometric T-transformation. In particular the minimal geometric length equals the minimal algebraic length.*

Similarly the second part of Whitehead's theorem has a (weak) geometric analogue.

Zieschang's proof is a delicate analysis of the situation. It seems to be of some importance that there is an alternative, somewhat crude, proof which is based on a trick of Whitehead **[22]**. In the present situation the trick amounts to temporarily admitting 'meridian surfaces' other than disks. The argument will be described below, after another notion has been introduced.

Suppose there is an embedded 2-disk D in V with the properties (i) $D \cap \partial V$ is a single arc c, and either $c \cap k = \emptyset$, or $c \subset k$, (ii) $D \cap v$ is a single arc, equal to $\mathrm{Cl}(\partial D - c)$. Let v_i be the component of v that contains $D \cap v$; let $U(v_i \cup D)$ be a regular neighborhood. Then $\mathrm{Cl}(\partial U(v_i \cup D) - \partial V)$ consists of three disks, one of which is parallel to v_i. We are interested in the other two. For precisely one of these, call it v_i', it is true that $v' = (v - v_i) \cup v_i'$ is again a system of meridian disks. The substitution $v \mapsto v'$ will be called a *geometric T-transformation* that is *special with respect to k*. Whether or not there exists a special T-transformation to decrease the geometric length of k can be found out by searching for the arc c (a 'wave' in the terminology of [**16**]).

As to the theorem, suppose first that $l(k, v) > \bar{l}(k, v)$. Then a special T-transformation can be found (with $c \subset k$) to decrease $l(k, v)$. So suppose $l(k, v)$ equals $\bar{l}(k, v)$, but the latter is not minimal. By Whitehead's theorem there exists a geometric T-transfmoration $v \mapsto w$ to decrease the algebraic length. If $l(k, w) > \bar{l}(k, w)$ one could try now to perform a special T-transformation to decrease $l(k, w)$; while this is certainly possible, it might happen however that the algebraic length increases again, so nothing is in fact gained. Here is where Whitehead's trick comes in. Namely instead of using the arc c for a special T-transformation, one uses it to 'add a handle' to w, i.e., one replaces the component w_j of w containing ∂c, by the annulus component of $\mathrm{Cl}(\partial U(w_j \cup c) - \partial V)$. Then $l(k, w)$ goes down by 2, but $\bar{l}(k, w)$ is unaltered because the set of reduced words from which it is computed is unaltered. Furthermore the new w is still disjoint to v because of the initial hypothesis $l(k, v) = \bar{l}(k, v)$. Proceeding in this way, w will eventually have been replaced by w' with $l(k, w') = \bar{l}(k, w') = \bar{l}(k, w)$, but the component w_j' of w' is some complicated 2-manifold rather than a disk. On the other hand, w' is still a system of 2-sided 2-manifolds, not separating V, and n in number. So we can construct a map from $\pi_1 V$ onto a free group of rank n, and the kernel of this map contains $\pi_1 w_j'$. But any surjective endomorphism of a finitely generated free group is an isomorphism. So $\mathrm{Im}(\pi_1 w_j')$ is the trivial subgroup of $\pi_1 V$. So the loop theorem applies, and w_j' can be dismantled to a system of disks. Keeping a suitable one of these, the theorem is proved.

REMARK. There does not seem any reason to suppose that in general the geometric length of k can be made minimal by special T-transformations only. There is one special case however where this can be done. This special case is when no component of k is contractible in V, and the minimal algebraic length of k equals the number of components. This was pointed out by Whitehead in the final paragraph of [**22**]; it depends on a certain technical result of that paper.

Let now $(M, F; v, w)$ be a Heegaard diagram, and for simplicity assume M is closed. One may replace these data by the equivalent data $(F; \partial v, \partial w)$. Supposing that ∂v and ∂w are in general position, one defines the *complexity* $c(F; \partial v, \partial w)$ to be the number of intersection points of ∂v and ∂w. Given F, and given any upper bound, there is only a finite number, up to isomorphism, of Heegaard diagrams whose complexity is below this upper bound.

By definition, the complexity $c(F; \partial v, \partial w)$ coincides with the geometric length $l(\partial w, v)$ considered before. It may happen that $l(\partial w, v)$ can be made smaller by a geometric T-transformation applied to v. Whether or not this happens can be found out by inspection of $(F; \partial v, \partial w)$; the test is especially simple if one looks for special T-transformations (the search for a 'wave' in the terminology of [**16**]). We say $(F; \partial v, \partial w)$ is *minimal* if neither $l(\partial w, v)$ nor $l(\partial v, w)$ can be made smaller by a geometric T-transformation; and we say $(F; \partial v, \partial w)$ is a *weak minimum* if neither can be made smaller by a special T-transformation.

With this terminology we can now give sharper versions of the problems stated in the beginning of this section. These are:

Show that any Heegaard splitting is underlying to only finitely many minimal Heegaard diagrams. Given any of these, describe a procedure to obtain the others.

Let (M, F) be the underlying Heegaard splitting of $(M, F; v, w)$. Suppose $(F; \partial v, \partial w)$ is minimal (or even a weak minimum only). Suppose (M, F) is not minimal. Show that $(M, F; v, w)$ has a cancelling pair of handles.

Let $(F; \partial v, \partial w)$ be a Heegaard diagram of S^3 which is minimal (or even a weak minimum only). Show that the complexity of $(F; \partial v, \partial w)$ equals the genus of F. (Note this amounts to the assertion that the method to produce a minimal Heegaard diagram from a given one is in fact an algorithm to recognize S^3.)

Concerning the status of these problems, nothing is known about the first two, and very little is known about the third one: Whitehead has shown the assertion is true in the very special case where one assumes that one of v and w is 'standard' already, this is the remark above; the argument has been reproduced in **[16]**. It is interesting to note that a computer check has been run on the third problem **[16]**. One million Heegaard diagrams of S^3 were examined; no exotic weak minimum was found.[2]

5. The Heegaard genus. It is convenient here to consider 3-manifolds with boundary (possibly empty) but only Heegaard splittings of the triple $(M, \partial M, \emptyset)$, in the notation of §2. With this qualification, the *Heegaard genus* $g(M)$ of M is defined to be the smallest integer g so that M has some Heegaard splitting of genus g.

For example, of the manifolds M without boundary spheres, $g(M) = 0$ characterizes S^3, and $g(M) = 1$ characterizes lens spaces, $S^1 \times S^2$, and $S^1 \times D^2$.

So far only one nontrivial fact is known about the Heegaard genus, a beautiful argument of Haken **[4]**, that $g(M)$ is additive for connected sum.

Let $r(M)$ denote the minimum number of generators for $\pi_1 M$. One has the inequality $g(M) \geqq r(M)$. For want of better knowledge one may ask the question

Is it true that $g(M) = r(M)$?

It is amusing to contemplate this question on the background of the unresolved status of the Poincaré conjecture (the case $r(M) = 0$ of the question). Put in a fancy way, the content of the Poincaré conjecture is that all the difficulties inherent in attacking it are due to the internal structure of the 3-sphere. Therefore if one assumes the Poincaré conjecture is wrong, it seems reasonable to expect the above question to become easier if one restricts attention to submanifolds $M \subset S^3$, e.g., knot spaces. If on the other hand one assumes the Poincaré conjecture is true, there does not seem to be any reason why the question should be easier to decide for submanifolds $M \subset S^3$.

6. A space of Heegaard splittings. From current problems of 3-dimensional topology there is no justification to consider such a notion, except maybe a vague feeling that it could be useful in work on the Smale conjecture on $\mathrm{Diff}(S^3)$. The following definition was concocted by analogy with a notion that is useful in

[2]The reviewer of reference **[16]** reports that an exotic weak minimum has been found [Math. Rev. **53**, Abstr. 9219 (1977)]. It thus appears that the consideration of weak minima is of little interest.

studying higher concordances (the 'expansion space' of [**21**]). To understand the definition one should note that there are really two ways in which one wants to alter Heegaard splittings: by isotopy and by handle addition. As these are heterogeneous notions, they should be kept apart. Thus the space envisaged should be a bisimplicial set (or a simplicial category, for that matter). From this one may then obtain a simplicial set, and hence a homotopy type, in any of various ways, all ultimately equivalent.

The Heegaard splittings of a given (say, closed orientable) 3-manifold M are the objects of a category $h(M)$ in an obvious way: a morphism in $h(M)$ is a standard handle addition, or composition of such, performed on one Heegaard splitting to yield another. Still the definition of morphism needs interpretation. Firstly, the notion of handle additon is supposed to be very rigid. One way to have this rigidity is always to refer to the standard D^3 in which the standard punctured torus is embedded in the standard way, and then for a handle addition use a specific embedding of D^3 in M. Secondly, the order of handle additions must be discussed: It may happen that one handle is attached on top of another. In this case the two can be attached only in that particular order. On the other hand one can envisage two handle additions being performed simultaneously, far away from each other. In this case we insist that either one of the two could be attached first, and it does not matter which one; thus we have a commutative square in $h(M)$.

More generally, for any nonnegative integer k one can define a category $h(M)_k$: an object is a k-parameter family of Heegaard splittings of M (with parameter domain the k-simplex Δ^k) and a morphism is a k-parameter family of standard handle additions, or a composition of such. The same remarks as above apply: a k-parameter family of handle additions involves a k-parameter family of embeddings of D^3 (with its additional structure), and two k-parameter families of handle additions are considered to be in a particular order only if such order is forced, in the sense above, at one point at least of the parameter domain.

DEFINITION. $h(M)_\cdot$ is the simplicial category which in degree k is $h(M)_k$.

The known results on the classification of Heegaard splittings can be rephrased to say that such or such a space is connected, or not connected, as the case may be. For example the Reidemeister-Singer theorem says that $h(M)_\cdot$ is connected for any M, and the classification of Heegaard splittings of S^3 is equivalent to the statement that $h^g(S^3)_\cdot$ is connected for any g, where $h^g(M)_\cdot$ denotes the simplicial subcategory of $h(M)_\cdot$ given by the Heegaard splittings of genus at most g.

The problem is of course if one can say anything about the higher homotopy groups. For example, is $h(M)_\cdot$ contractible?

To conclude, one way of forming the 'nerve' and then forcing the extension condition gives the following Kan simplicial set representing the homotopy type of $h(M)_{\cdot\cdot}$. An n-simplex consists of:

(i) a simplicial subdivision of the geometric n-simplex,

(ii) for each Δ^k in this subdivision, a continuous family of Heegaard splittings over the interior of Δ^k,

(iii) for each face $d_i\Delta^k$, as one approaches this face from the interior of Δ^k, the data of a $(k - 1)$-parameter family of standard handle cancellations (or composition of such) and finally, at the last moment, the actual cancellation,

(iv) for each face of a face, a compatibility condition.

REFERENCES

1. J. S. Birman, F. González-Acuña and J. M. Montesinos, *Heegaard splittings of prime* 3-*manifolds are not unique* (preprint).

2. H. Boehme, *Fast genügend grosse irreduzible* 3-*dimensionale Mannigfaltigkeiten*, Invent. Math. **17** (1972), 303–316.

3. R. Engmann, *Nicht-homöomorphe Heegaard-Zerlegungen vom Geschlecht* 2 *der zusammenhängenden Summe zweier Linsenräume*, Abh. Math. Sem. Univ. Hamburg **35** (1970), 33–38. MR **44** #1033.

4. W. Haken, *Some results on surfaces in* 3-*manifolds*, Studies in Modern Topology, No. 5, pp. 39–98, Math. Assoc. Amer. (distributed by Prentice-Hall, Englewood Cliffs, N. J.), 1968.

5. P. Heegaard, *Forstudier til en topologisk teori för de algebraiske Sammenhäeng* (Dissertation, Univ. of Copenhagen, 1898; published by det Nordiske Forlag Ernst Bojesen, Copenhagen, 1898).

6. ———, *Sur l'Analysis situs*, Bull. Soc. Math. France 44 (1916), 161–242. (translation of [5]).

7. W. Jaco, *Heegaard splittings and splitting homomorphisms*, Trans. Amer. Math. Soc. **144** (1970), 365–379. MR **40** #6555.

8. ———, *Finitely presented subgroups of three-manifold groups*, Invent. Math. **13** (1971), 335–346. MR **45** #9325.

9. L. P. Neuwirth, *Knot groups*, Ann. of Math. Studies, no. 56, Princeton Univ. Press, Princeton, N. J., 1965.

10. C. D. Papakyriakopoulos, *Some problems on* 3-*dimensional manifolds*, Bull. Amer. Math. Soc. **64** (1958), 317–335.

11. K. Reidemeister, *Zur dreidimensionalen Topologie*, Abh. Math. Sem. Univ. Hamburg **9** (1933), 189–194.

12. G. P. Scott, *Finitely generated* 3-*manifold groups are finitely presented*, J. London Math. Soc. (2) **6** (1973), 437–440.

13. ———, *Ends of pairs of groups*, Univ. Liverpool, Oct. 1976.

14. J. Singer, *Three-dimensional manifolds and their Heegaard-diagrams*, Trans. Amer. Math. Soc. **35** (1933), 88–111.

15. J. R. Stallings, *How not to prove the Poincaré conjecture*, Ann. of Math. Studies, no. 60, Princeton Univ. Press, Princeton, N. J., 1966, pp. 83–88.

16. I. A. Volodin, V. E. Kuznetsov, A. T. Fomenko, *The problem of discriminating algorithmically the standard three-dimensional sphere*, Russian Math. Surveys **29** (1974), 72–168.

17. F. Waldhausen, *Eine Klasse von* 3-*dimensionalen Mannigfaltigkeiten*. I, II, Invent. Math. **3** (1967), 308–333; ibid. **4** (1967), 87–117. MR **38** #3880.

18. ———, *Gruppen mit Zentrum und* 3-*dimensionale Mannigfaltigkeiten*, Topology **6** (1967), 505–517. MR **38** #5223.

19. ———, *Heegaard-Zerlegungen der* 3-*Sphäre*, Topology **7** (1968), 195–203. MR **37** #3576.

20. ———, *Über Involutionen der* 3-*Sphäre*, Topology **8** (1969), 81–91. MR **38** #5209.

21. ———, *Algebraic K-theory of topological spaces*. I, these PROCEEDINGS, part 1, pp. 35–60.

22. J. H. C. Whitehead, *On certain sets of elements in a free group*, Proc. London Math. Soc. (2) **41** (1936), 48–56; Collected Works Vol. II, Pergamon Press, New York, 1962, pp. 69–77.

23. ———, *On equivalent sets of elements in a free group*, Ann. of Math. **37** (1936), 782–800; Collected Works Vol. II, Pergamon Press, New York, 1962, pp. 79–97.

24. H. Zieschang, *Über einfache Kurven auf Vollbrezeln*, Abh. Math. Sem. Univ. Hamburg **25** (1962), 231–250. MR **26** #6957.

25. ———, *On simple systems of paths on complete pretzels*, Amer. Math. Soc. Transl. (2) **92** (1970), pp. 127–137.

26. H. Zieschang, E. Vogt and H. D. Coldewey, *Flächen und ebene diskontinuierliche Gruppen*, Lecture Notes in Math., Vol. 122, Springer-Verlag, Berlin and New York, 1970.

UNIVERSITÄT BIELEFELD